Mechanics of Materials

Formulations and Solutions
with Python

Series in Computational Methods

Series Editor: Gui-Rong Liu *(University of Cincinnati, USA)*

Published

Vol. 3   *Mechanics of Materials: Formulations and Solutions with Python*
         by G R Liu

Vol. 1   *Numbers and Functions: Theory, Formulation, and Python Codes*
         by G R Liu

Series in Computational Methods
Volume 3

Mechanics of Materials

Formulations and Solutions with Python

G. R. Liu
University of Cincinnati, USA

NEW JERSEY • LONDON • SINGAPORE • GENEVA • BEIJING • SHANGHAI • TAIPEI • CHENNAI

Published by

World Scientific Publishing Co. Pte. Ltd.
5 Toh Tuck Link, Singapore 596224
USA office: 27 Warren Street, Suite 401-402, Hackensack, NJ 07601
UK office: 57 Shelton Street, Covent Garden, London WC2H 9HE

Library of Congress Cataloging-in-Publication Data
Names: Liu, G. R. (Gui-Rong), author.
Title: Mechanics of materials : formulations and solutions with Python /
    G. R. Liu, University of Cincinnati, USA.
Description: Singapore ; Hackensack, NJ ; London : World Scientific, 2025. |
    Series: Series in computational methods ; vol. 3 | Includes bibliographical references and index.
Identifiers: LCCN 2024028348 | ISBN 9789811294525 (hardcover) |
    ISBN 9789811294532 (ebook for institutions) | ISBN 9789811294549 (ebook for individuals)
Subjects: LCSH: Strength of materials--Computer simulation. | Python (Computer program language)
Classification: LCC TA405 .L554 2025 | DDC 620.1/120113--dc23/eng/20241101
LC record available at https://lccn.loc.gov/2024028348

British Library Cataloguing-in-Publication Data
A catalogue record for this book is available from the British Library.

Copyright © 2025 by World Scientific Publishing Co. Pte. Ltd.

All rights reserved. This book, or parts thereof, may not be reproduced in any form or by any means, electronic or mechanical, including photocopying, recording or any information storage and retrieval system now known or to be invented, without written permission from the publisher.

For photocopying of material in this volume, please pay a copying fee through the Copyright Clearance Center, Inc., 222 Rosewood Drive, Danvers, MA 01923, USA. In this case permission to photocopy is not required from the publisher.

For any available supplementary material, please visit
https://www.worldscientific.com/worldscibooks/10.1142/13878#t=suppl

Desk Editors: Nambirajan Karuppiah/Steven Patt

Typeset by Stallion Press
Email: enquiries@stallionpress.com

About the Author

Gui-Rong Liu received his Ph.D. from Tohoku University, Japan, in 1991. He was a Postdoctoral Fellow at Northwestern University, USA, from 1991 to 1993. He was a Professor at the National University of Singapore until 2010. He is currently a Professor at the Department of Aerospace Engineering and Engineering Mechanics, University of Cincinnati, USA. He was the Founder of the Association for Computational Mechanics (Singapore) (SACM) and served as the President of SACM until 2010. He served as the President of the Asia-Pacific Association for Computational Mechanics (APACM) (2010–2013) and an Executive Council Member of the International Association for Computational Mechanics (IACM) (2005–2010; 2020–2026). He authored a large number of journal papers and books, including two bestsellers, *Mesh Free Method: Moving Beyond the Finite Element Method* and *Smoothed Particle Hydrodynamics: A Meshfree Particle Methods*. He is the Editor-in-Chief of the *International Journal of Computational Methods* and served as an Associate Editor for *IPSE* and *MANO*. He is the recipient of numerous awards, including the Singapore Defence Technology Prize, NUS Outstanding University Researcher Award, NUS Best Teacher Award, APACM Computational Mechanics Award, JSME Computational Mechanics Award, ASME Ted Belytschko Applied Mechanics Award, Zienkiewicz Medal from APACM, AJCM Computational Mechanics Award, Humboldt Research Award, and SACM Medal from the Association of Computational Mechanics (Singapore). He has been listed as one among the world's top 1% most influential scientists (Highly Cited Researchers) by Thomson Reuters for a number of years.

Contents

About the Author   v

1. Introduction   1
   1.1 Computational methods 1
   1.2 Why start and contribute to this book series 2
   1.3 Mechanics of materials: Essential for safe and effective use of materials 2
   1.4 Who may read this book? 3
   1.5 Codes used in this book 4
   1.6 Use of external modules or dependences 5
   1.7 Use of help() 7
   References 8

2. Preliminaries: Number, Vector, Tensor, and Functions   9
   2.1 Numbers 10
   2.2 Vectors: Collection of numbers 12
       2.2.1 Vectors defined in a computer 12
       2.2.2 Basic properties of vectors 13
       2.2.3 Attributes of a vector in a computer 14
       2.2.4 The standard basis vectors 15
       2.2.5 Basic arithmetic operations for vectors 16
       2.2.6 Other arithmetic operations for vectors 17
   2.3 Dot product of vectors, producing a scalar 18
       2.3.1 Definition: A scalar produced using two vectors 18
       2.3.2 Direction cosine of two vectors 19
       2.3.3 Inner product associated norm 20
       2.3.4 Orthogonality and orthogonal projection 20
       2.3.5 Derivation of the law of cosines 21

	2.3.6	Power, energy, and work-done	22
	2.3.7	Python examples	22
		2.3.7.1 Dot product of an arbitrary pair of vectors	22
		2.3.7.2 Dot product of a pair of parallel vectors	23
		2.3.7.3 Dot product of two orthogonal vectors	23
		2.3.7.4 Direction cosine of two vectors	23
	2.3.8	Types of norms of a vector	24
	2.3.9	Direction cosine of an arbitrary vector	24
2.4	Outer product of two vectors, producing a matrix	25
	2.4.1	Definition of outer product	25
	2.4.2	Rank-one matrix	26
	2.4.3	Python example	27
2.5	Cross product of vectors, producing a vector	27
	2.5.1	Definition	28
	2.5.2	Formulation	28
	2.5.3	Volume of a parallelepiped	29
	2.5.4	Differences between the dot product and cross product	30
	2.5.5	Derivation of formulas	30
	2.5.6	Examples of computing cross products	33
2.6	Examples of vector operations and visualization	34
	2.6.1	Creation of vectors using the standard basis vectors, and scaling vectors	35
	2.6.2	Addition of vectors	36
	2.6.3	Element-wise multiplication and dot product of two vectors	37
	2.6.4	Orthogonal vectors	39
2.7	Coordinate transformation	41
2.8	Vectors in mechanics	42
	2.8.1	Notation for coordinate systems	42
	2.8.2	Description of vectors	42
	2.8.3	Equivalence in representation, first-order tensor	43
	2.8.4	Tensor concept in general	44
	2.8.5	Definition of tensor in mechanics of materials	44
2.9	Transformation rules	45
	2.9.1	Definition of coordinate transformation	45
	2.9.2	Representation of a tensor in a coordinate	46
	2.9.3	Essential rule to obtain values on a coordinate axis	46
2.10	Transformation matrix	47
	2.10.1	Two-dimensional (2D) cases	47
	2.10.2	General definition of transformation matrix for 3D cases	49

	2.10.3	Transformation via yaw, pitch, and roll in 3D	50
	2.10.4	Property of transformation matrix, a unitary matrix	52
	2.10.5	Numpy code for transformation matrices	54
	2.10.6	Examples: Unitary property of transformation matrix	55
	2.10.7	Examples: Basic property preservation by transformation matrix	56
	2.10.8	Example: Transformation matrix on arbitrary matrices	57
2.11	Rotation matrix	58
	2.11.1	Derivation for 2D cases	58
	2.11.2	Relationship between transformation and rotation matrix	60
2.12	Function, an object varying over space	60
2.13	Gradient of a scalar function, a vector	61
	2.13.1	Definition	61
	2.13.2	Python demonstration	62
2.14	Gradient of a vector of function, a matrix	64
	2.14.1	Definition	64
	2.14.2	Python examples	65
	2.14.3	Code function for gradient of vector functions	66
2.15	Divergence of a vector function, a scalar	67
	2.15.1	Definition	67
	2.15.2	Python code and examples	68
2.16	Curl of a vector function, a vector	69
	2.16.1	Definition	69
	2.16.2	Python code and examples	70
2.17	Remarks	71
References	72

3. Forces and Stresses   73

3.1	Introduction	73
3.2	Example: A bar subjected to a point force	74
3.3	Example: A bar subjected to a distributed force	76
	3.3.1 Derivation of equilibrium equation	76
	3.3.2 Stress solution to 1D equilibrium equation	77
3.4	General measure of force vectors in 3D space	78
	3.4.1 Graphic representation of force vectors	78
	3.4.2 Components of a force, orthogonal projection	78
	3.4.3 Equivalence of force vector and its components	78
	3.4.4 Force vector in transformed coordinates	79
3.5	General measure of stresses in 3D solids	80
	3.5.1 General definition of stress	80

	3.5.2	Shear stresses on a surface	81
	3.5.3	Stress representation in 3D solids	83
	3.5.4	Shear stress equivalence, symmetric tensor	84
	3.5.5	Dimensionality and stresses	85
3.6	Tractions on an arbitrary surface		86
	3.6.1	Formulation	87
	3.6.2	Stress vector projection	87
	3.6.3	Area projection	88
	3.6.4	Traction on a boundary surface	89
	3.6.5	Stress vector projection, again	90
		3.6.5.1 Normal stress on a surface	90
		3.6.5.2 Shear stress on a surface	91
		3.6.5.3 The magnitude of the shear stress on a surface	91
	3.6.6	Example: Stresses on a given surface	91
3.7	Coordinate transformation of second order tensor (stress)		95
	3.7.1	Formulation by simple inspection	96
	3.7.2	Inverse transformation	97
	3.7.3	Python example: Stress tensor under coordinate transformation	97
		3.7.3.1 Use of matrix formulation	98
		3.7.3.2 Use of tensor operation	99
	3.7.4	Transformation of stress tensor in 2D	99
		3.7.4.1 Example: Uniaxial stress after a coordinate transformation	100
		3.7.4.2 Example: A general stress state under coordinate transformation	103
3.8	Principal stresses		105
	3.8.1	Eigenvalue problem	106
	3.8.2	Characteristic equation, polynomial	107
	3.8.3	Python code for finding principal stresses and their directions	108
	3.8.4	Example: Computing the principal stresses and principal directions	108
3.9	Stress invariants		110
3.10	Special stresses		110
	3.10.1	Octahedral stress	110
	3.10.2	The von Mises stress	111
	3.10.3	The strain energy stress	112
	3.10.4	Python code for computing various stresses	112
	3.10.5	Example: Computation of various stresses	114
3.11	Mohr circles, geometric representation		115
	3.11.1	Formulation for 2D stress states, double angle formula	116

	3.11.2 Python code for coordinate transformation for 2D stresses, double angle	117
	3.11.3 Principal stresses for 2D stress states	117
	3.11.4 Example: Principal stresses of 2D stress states	118
	3.11.5 Mohr circle for 2D and stress states	119
	3.11.6 Sign convention for 2D Mohr circles	120
	3.11.7 Python code for generating 2D Mohr circles	121
	3.11.8 Example: Mohr circle of a uniaxial stress state	122
	3.11.9 Example: Mohr circle of a general 2D stress state	125
	3.11.10 Mohr circle for 3D stress states	126
	3.11.11 Python code for generating 3D Mohr circles	128
	3.11.12 Example: Mohr circle of a three-dimensional (3D) stress state	129
3.12	Determination of the principal axis angle	131
	3.12.1 Transformation matrices for Yaw, Pitch, and Roll	131
	3.12.2 Stress transformation via Yaw, Pitch, and Roll	132
	3.12.3 Diagonalization of a stress tensor in 3D	133
	3.12.4 Python code to find Yaw, Pitch, and Roll angles to diagonalize stress tensors	134
	3.12.5 Example: Diagonalize a stress tensor via Yaw, Pitch, and Roll	135
3.13	Remarks	136
References		137

4. Displacements and Strains — 139

4.1	Introduction	139
4.2	Displacement vector, representations	140
4.3	Displacement vector in transformed coordinates	141
	4.3.1 Forward transformation	141
	4.3.2 Inverse transformation	142
4.4	Vector of displacement functions	142
	4.4.1 Displacement functions, a vector function	142
	4.4.2 Code using tensor operation for transformation	144
4.5	Strains derived via kinematic relations	145
	4.5.1 An ideal uniform bar under tension	145
	4.5.2 Normal strains	147
	4.5.3 Shear strains	147
4.6	Strains, derivatives of displacement functions	148
	4.6.1 Displacement gradient	148
	4.6.2 Small strains, derived via displacement gradient	150
	4.6.3 Example: Computation of small strains	151
	4.6.4 Rotation-displacement relations	152

- 4.6.5 Large strains, derived via displacement gradient 153
- 4.6.6 Example: Computation of large strains 155
- 4.6.7 Large strains derived via conventional deformation gradient 155
  - 4.6.7.1 Deformation gradient, definition 157
  - 4.6.7.2 Deformation gradient, in displacements 157
  - 4.6.7.3 Cauchy–Green deformation tensor 158
  - 4.6.7.4 Green–Lagrange strain tensor 158
- 4.7 Strain–displacement relation, a general derivation 159
  - 4.7.1 Normal strain along a fiber, magnification factor 160
    - 4.7.1.1 Evaluation of the length of a fiber in a deformed solid 160
    - 4.7.1.2 Engineering strain of a given fiber 161
    - 4.7.1.3 Magnification factor of a fiber, large normal strain 161
    - 4.7.1.4 Magnification factor and the strain components 162
    - 4.7.1.5 Direction cosine of a fiber after deformation 163
  - 4.7.2 Shear strains caused by large deformation 164
  - 4.7.3 Example: Symbolic computation of strains 165
  - 4.7.4 Example: Computation of displacements, normal and shear strains 167
- 4.8 Coordinate transformation of strain tensor 171
- 4.9 Principal strains 173
- 4.10 Strain invariants 174
- 4.11 Example: Strain components after coordinate rotation 174
- 4.12 Example: Computation of various strains 175
- 4.13 Compatibility equations 177
  - 4.13.1 2D problems 177
  - 4.13.2 3D problems 178
- 4.14 Remarks 178
- References 179

5. Elasticity Tensor and Constitutive Equations 181

- 5.1 Simplest case: 1D tests 181
- 5.2 Comparison between Young's modulus and spring constants .. 183
- 5.3 General formulation 184
- 5.4 Coordinate transformation of fourth-order tensor 184
- 5.5 Codes for fourth order tensor transformation 185
- 5.6 Reduction of elastic constants 186
  - 5.6.1 Reduction due to stress and strain symmetry, $81 \to 36$. 186

		5.6.2	Reduction due to the smoothness of the strain energy density	187

- 5.6.2 Reduction due to the smoothness of the strain energy density 187
  - 5.6.2.1 Strain energy density in solids 187
  - 5.6.2.2 Symmetry from the smoothness of the strain energy density 188
  - 5.6.2.3 Reduction due to strain energy density symmetry, $36 \to 21$ 189
- 5.6.3 Tensor vs. matrix, a discussion 190
- 5.7 The Voigt notation 190
- 5.8 Codes for conversion between a 2D matrix to fourth-order tensor 191
- 5.9 Monoclinic materials, $21 \to 13$ 193
- 5.10 Orthotropic materials, $13 \to 9$ 193
- 5.11 Transversely isotropic materials, $9 \to 5$ 193
- 5.12 Cubic materials, $5 \to 3$ 194
- 5.13 Isotropic materials, $3 \to 2$ 194
  - 5.13.1 For 3D problems 194
  - 5.13.2 For stress and strain invariants 197
  - 5.13.3 For 2D plane stress problems 197
  - 5.13.4 For 2D plane strain problems 198
- 5.14 Mean and deviatoric stress and strain tensors 199
- 5.15 Compliance matrix 200
  - 5.15.1 Compliance matrix for orthotropic materials 200
  - 5.15.2 Python code for elasticity matrix for orthotropic materials (3D problems) 201
  - 5.15.3 Plane stress problems 203
  - 5.15.4 Python code for elasticity matrix for orthotropic materials (2D plane stress) 203
  - 5.15.5 Plane strain problems 204
  - 5.15.6 Python code for elasticity matrix for orthotropic materials (2D plane strain) 206
  - 5.15.7 Compliance matrix for isotropic materials 207
  - 5.15.8 Derivation of the C matrix for isotropic materials 207
  - 5.15.9 Compliance equations for 2D problems 208
  - 5.15.10 Example: Proof of relation between G, E and ν 209
- 5.16 A comprehensive example: Displacement, strain and stress analysis 212
  - 5.16.1 Setting of the problem 212
  - 5.16.2 Solution to the problem 213
    - 5.16.2.1 Analysis of the problem 213
    - 5.16.2.2 Find the displacement functions 214
    - 5.16.2.3 Compute the gradient of the displacement ... 215

		5.16.2.4	Find the strain tensor functions 215

- 5.16.2.4 Find the strain tensor functions 215
- 5.16.2.5 Compute the normal strains 216
- 5.16.2.6 Compute the shear strains 216
- 5.16.2.7 Compute the principal strains and strain invariants 217
- 5.16.2.8 Displacements in the new coordinates system 218
- 5.16.2.9 Strains in the new coordinates system 218
- 5.16.2.10 Compute the stresses 218
- 5.16.2.11 Compute the principal stresses, and stress invariants 219
- 5.16.2.12 Rotate the coordinates by 30° about z-axis, and find the stresses at point B in the new coordinates system 220

5.17 Thermal expansion effects 220
    5.17.1 Thermal strains 220
    5.17.2 Strain-stress-temperature relation 221
    5.17.3 Stress-strain-temperature relation 221

5.18 Examples for elasticity tensor transformation 221
    5.18.1 Examples on arbitrary symmetric 6 by 6 matrix 221
    5.18.2 Examples on elasticity matrices for steel 224
    5.18.3 Examples on elasticity matrices for steel for 2D plane stress 226
    5.18.4 Examples on elasticity matrices for steel for 2D plane strain 227
    5.18.5 Examples on elasticity matrices for carbon/epoxy composite 228

5.19 Remarks 230
References 231

6. Equilibrium Equations   233

6.1 Introduction 233
6.2 Equilibrium equation in terms of stresses 234
    6.2.1 3D free-body diagram 234
    6.2.2 The standard sign convention 235
    6.2.3 Equilibrium equation in x-direction 236
    6.2.4 Equilibrium in y- and z-directions 237
    6.2.5 Equilibrium equation in indicial notation 237
    6.2.6 Equilibrium equation in 2D and 1D 238
    6.2.7 Equilibrium equation via divergence of stress tensor function 238
6.3 Equilibrium equations in displacements 240
    6.3.1 General formulation in 3D 240

	6.3.2	Formulations of 2D problems		242
		6.3.2.1	Plane strain problems	242
		6.3.2.2	Plane stress problems	243
	6.3.3	Formulations of 1D bar problems		244
		6.3.3.1	1D yz-constrained problem	244
		6.3.3.2	1D y-constrained problem	245
		6.3.3.3	1D bar problem	245
6.4	Boundary conditions			246
	6.4.1	Displacement boundary conditions		246
	6.4.2	Stress boundary conditions		246
6.5	On techniques for solving solid mechanics problems			247
6.6	Remarks			247
References				248

7. Solution to Axially Loaded Bars   249

7.1	Governing equation: Second-order differential equation		249
7.2	Solution procedure		250
	7.2.1	Example: A bar subjected to a force at its right end	250
	7.2.2	Python code for deriving solutions for bars	252
7.3	Example: Constant body force and concentrated force at the right end		253
	7.3.1	Method 1: Step-by-step approach integration	253
	7.3.2	Method 2: Use the Sympy differential equation solver	256
	7.3.3	Python code for plotting the solutions	256
7.4	Example: Linear body force and concentrated force at the right end		260
7.5	Example: A displacement imposed at the right end		261
7.6	Bars subjected to discontinuous body forces		264
	7.6.1	Distributed forces	264
	7.6.2	Point force	265
7.7	Statically indeterminate bars: Reaction force		266
7.8	Thermal stress		267
	7.8.1	Thermal displacement	267
	7.8.2	Thermal forces	268
	7.8.3	Computing thermal stresses	269
		7.8.3.1 Approach 1: Use reaction and external forces	269
		7.8.3.2 Approach 2: Use the combined displacement	270

7.9 Example: Stress in an overconstrained bar loaded mechanically and thermally 270
7.10 Example: Thermal stress in an overconstrained bar 273
7.11 Example: A composite bar subjected to both mechanical and thermal loads 275
    7.11.1 Solution to (1) 276
    7.11.2 Solution to (2) 278
    7.11.3 Solution to (3) 280
    7.11.4 Solution to (4) 281
7.12 Remarks 282
Reference 282

8. Solution to Bended Beams   283

8.1 Governing equation, fourth-order differential equation 284
    8.1.1 Thin beam theory 284
    8.1.2 Normal stress and the resulted moment 285
    8.1.3 On second moment of area of beam cross-section 288
    8.1.4 Equilibrium equation for beams 289
8.2 Boundary conditions for beams 291
    8.2.1 Possible BCs for beams 291
    8.2.2 Admissible BCs for beams 292
8.3 Example: A cantilever beam subjected to a force at its right-end 292
8.4 A Python beam solver 294
    8.4.1 Python code of beam solver 295
    8.4.2 Example: Clamped-clamped beams 297
    8.4.3 Use of the Sympy differential equation solver 299
8.5 Python code for finding the maximum and minimum values 300
8.6 Python code for plotting the solution marked with extreme values 302
8.7 Solutions at specific locations 303
8.8 Example: Simple-simple supported beams 304
8.9 Example: Simple-clamped beams 310
8.10 Example: Free-clamp supported beams 312
8.11 Example: Clamped-clamped beam subjected to a point force at the middle span 317
8.12 Plot piecewise solutions 319
8.13 Example: Simply supported beam with two spans 321
    8.13.1 For span-1 322
    8.13.2 For span-2 323
    8.13.3 Continuity and equilibrium conditions 324
    8.13.4 Final solutions 325

8.14	Plot piecewise solutions		326
8.15	Example: A symmetric frame subjected to a uniformly distributed load		328
	8.15.1	For member-1	330
	8.15.2	For member-2	331
	8.15.3	Continuity and equilibrium conditions at point B	332
	8.15.4	Final solutions	333
8.16	Plot piecewise solutions		335
8.17	Thermal stress		337
	8.17.1	Setting of the problem	337
	8.17.2	Governing equation for deflection caused by the temperature gradient	338
	8.17.3	Solution of deflection resulting from temperature gradient	339
	8.17.4	Solutions of indeterminate beams in temperature gradient	340
	8.17.5	Example: Simple-clamped beams subjected to temperature gradient	340
		8.17.5.1 Find the general displacement solution for an s-c beam	341
		8.17.5.2 Add in the displacement caused by temperature gradient	341
		8.17.5.3 Find the combined displacement using BCs	341
		8.17.5.4 Plot the distribution of solutions	342
	8.17.6	Example: Beams simple-simple supported subjected to temperature gradient	344
		8.17.6.1 Find the general displacement solution for an s-s beam	344
		8.17.6.2 Add in the displacement caused by temperature gradient	344
		8.17.6.3 Find the combined displacement	344
		8.17.6.4 Plot the distribution of solutions	345
8.18	Remarks		347
Reference			347

9.	Torsion of Bars		349
9.1	Bars with circular cross-section		351
	9.1.1	Problem description	351
	9.1.2	Assumptions	351
	9.1.3	Displacements on the cross-section	352
	9.1.4	Shear strains and stresses on the cross-section	353

	9.1.4.1	Divergence of the stress states in a circular bar	355
	9.1.4.2	Curl of the stress vector field in a circular bar	355
9.1.5	Distribution of shear stress on the cross-section		356
9.1.6	Boundary condition on the lateral surface of the bar		357
9.1.7	Equilibrium on the cross-section, twist-torque relation		358
9.1.8	On polar second moment of area		359
9.1.9	Twist-angle and torque relations		360
9.1.10	Continuous formulation		361
9.1.11	Bars with hollow cylindrical cross-section		361
9.1.12	Bars with thin-wall cross-sections		362
9.1.13	Example: A shaft with hollow circular cross-section subject to a torque		363
9.1.14	Example: A shaft with two segments subject to two torques		365

9.2 Bars with enclosed thin-wall cross-sections ... 368
- 9.2.1 Concept of shear flow ... 369
- 9.2.2 Torque and shear-flow relation ... 370
- 9.2.3 Twist-angle and torque relation ... 371
- 9.2.4 Example: Shear stress in a thin wall bar with box cross-section ... 372
- 9.2.5 A comparison of exact and thin-wall formulas ... 375

9.3 Bars with non-circular cross-sections ... 375
- 9.3.1 Setting of the problem ... 375
- 9.3.2 Assumed displacements ... 376
- 9.3.3 Shear strains in non-circular bars ... 377
- 9.3.4 Equilibrium equations at any point in torsion bar and stress function ... 378
- 9.3.5 Boundary conditions for the stress function ... 379
- 9.3.6 Shear stress on the cross-section by the boundary ... 381
- 9.3.7 Equilibrium conditions for the stress function ... 382
- 9.3.8 Linear elastic solution ... 383
- 9.3.9 Techniques for constructing the stress function ... 384
- 9.3.10 Example: Bars with elliptical cross-section ... 384
- 9.3.11 Comparison of torsional stiffness between elliptic and circular bars ... 386
- 9.3.12 Example: Bars with equilateral triangle cross-section ... 386

9.4 Prandtl's elastic membrane analogy for torsion ... 387
- 9.4.1 General formulation ... 387
- 9.4.2 A bar with narrow rectangular cross-section ... 389

		9.4.3	Bars with cross-sections consisting of narrow rectangles	392
9.5			Bars with rectangular cross-section	393
		9.5.1	Strategy for solution	393
		9.5.2	Method of separation of variables	394
		9.5.3	Final solution of the stress function	396
		9.5.4	Shear stress solution	396
		9.5.5	Torque twist-angle solution and equivalent polar 2nd moment of area	396
		9.5.6	The maximum shear stress	397
		9.5.7	Python code for computing k_1	397
			9.5.7.1 An example	398
			9.5.7.2 Cut-off terms in the solution series	398
			9.5.7.3 Values as function of b/h	399
		9.5.8	Comparison of equivalent J and polar moment of area J_z	400
		9.5.9	Comparison of torsional stiffness between rectangular and square bars	401
		9.5.10	Comparison of torsional stiffness between circular and square bars	403
		9.5.11	Code to compute the stress function on rectangular cross-section	403
		9.5.12	Plot of the stress function on rectangular cross-section	404
		9.5.13	Code to compute the shear stresses on rectangular cross-section	405
		9.5.14	Plot of the shear stresses on rectangular cross-section	405
		9.5.15	Example: An I-beam	407
9.6			Hollow thin-wall torsion members, use of membrane analogy	408
		9.6.1	Solution strategy	408
		9.6.2	Formula for twist angle	410
		9.6.3	Torque stress relation	411
		9.6.4	Equivalent polar 2nd moment of area	411
		9.6.5	Example: Bars with thin-wall circular cross-sections	411
9.7			Remarks	414
References				414
10.	Property of Areas			415
10.1	Area of areas			415
	10.1.1	Integral formulas		415
	10.1.2	Area of a composite area		417
10.2	First moment of area			418
	10.2.1	Definition		418

	10.2.2	First moment of a composite area	419
	10.2.3	Example: The first moment of area of a rectangle	419
	10.2.4	Centroid of an area	421
	10.2.5	Centroid for composite areas	422
	10.2.6	Example: Centroid of an I-beam	422
	10.2.7	Example: Compute the first moment of an area, using centroid	424
	10.2.8	First moment of area, coordinate rotation	425
10.3	Second moment of area	426
	10.3.1	Definition	426
	10.3.2	Second moment of area for regular areas	426
	10.3.3	Second moment of area, coordinate translation	428
	10.3.4	Second moments of composite areas	429
	10.3.5	Example: Second moments of an I-beam	429
	10.3.6	Example: Second moments of an S-beam	431
	10.3.7	Second moment of area, coordinate rotation	433
	10.3.8	Principal second moments of area	434
	10.3.9	Example: Second moments of area after coordinate rotation	434
	10.3.10	Example: The principal second moments of area	435
	10.3.11	Use of eigenvalue solver	435
10.4	Polar moment of area	438
	10.4.1	Perpendicular axis theorem	438
	10.4.2	Proof of the perpendicular axis theorem	438
	10.4.3	Other useful cases	438
	10.4.4	Equivalent polar second moment of area	439
10.5	Remarks	439
Reference	440

11. Normal Stress by Moments — 441

11.1	Loading plane	441
11.2	Setting of the problem	443
11.3	Symmetrical bending	443
	11.3.1 Superposition of stresses	443
	11.3.2 Neutral axis of the combined normal stress	445
	11.3.3 Example: A cantilever beam subjected to a force and a moment	445
	11.3.4 Use gr.solver1D4() to find the moment	447
	11.3.5 Example: A cantilever T-beam subjected to a force and a moment	449
11.4	Beams under non-symmetrical bending	452

	11.4.1 General formula for computing normal stress distribution	452
	11.4.2 Neutral axis: general formula	454
	11.4.3 Example: A cantilever L-beam subjected to a force and a moment	455
	11.4.3.1 Solution to 1 and 2	456
	11.4.3.2 Solution to 3: Determine the maximum stress of the beam, using the formulas for non-symmetric bending	456
	11.4.3.3 Solution to 4: Find the principal axis of the cross-section	457
11.5	Effects of error in loading plane angle	459
	11.5.1 Example: A tall beam with rectangular cross-section	459
	11.5.1.1 Python code for the solution	460
	11.5.1.2 An important finding	461
	11.5.2 Underlying reasons for error magnification	461
11.6	Remarks	463
References		463

12. Shear Stress by Shear Force   465

12.1	Setting of the problem	466
12.2	General formulation	467
	12.2.1 Free-body diagram	467
	12.2.2 Equilibrium conditions	468
	12.2.3 Formula for shear stress	468
	12.2.4 Shear-flow resulted from shear force	469
12.3	Shear stress on rectangular cross-section	469
	12.3.1 Parabolic distribution	469
	12.3.2 Alternative approach	471
12.4	Shear stress on thin-wall cross-sections	472
	12.4.1 Setting of the problem	472
	12.4.2 Assumptions	473
	12.4.3 Free-body diagram and formulation	473
	12.4.4 Shear-flow on thin wall cross-sections	474
12.5	Example: Shear stress on the cross-section of a C-beam	474
	12.5.1 Shear-flow and stress at any point in the flange	475
	12.5.2 Shear stress at any point s in the web	476
12.6	Example: Shear stress on the cross-section of an L-beam	477
12.7	Shear stress on an I-beam cross-section	479
	12.7.1 Problem analysis	479
	12.7.2 Calculation procedure	480

- 12.7.2.1 Shear-flow and stress at any point in the flange 480
- 12.7.2.2 Shear-flow and stress at point D in the flange 480
- 12.7.2.3 Shear stress at point D in the web 481
- 12.7.2.4 Shear stress at any point s in the web 481
- 12.7.2.5 Shear stress at D' on the top of the bottom flange 482
- 12.7.3 Example: Python code for computing the shear-flow and stress in an I-beam 483
  - 12.7.3.1 Data preparation 483
  - 12.7.3.2 Compute shear-flow and stress at D in the top flange 484
  - 12.7.3.3 Compute the shear-flow and stress at D on the top of the web 484
  - 12.7.3.4 Compute shear-flow and stress at any s in web 484
  - 12.7.3.5 Compute shear stress at D' at the top of the bottom flange 485
- 12.7.4 Plot of shear-flow in the web 485
- 12.7.5 Compute forces in channels 486
- 12.8 Shear-center of a cross-section 489
  - 12.8.1 Setting of the problem 489
  - 12.8.2 Strategy to find the shear-center 490
  - 12.8.3 Procedure to find the shear-center 491
  - 12.8.4 Example: Find the formula to compute e for a C-beam 491
  - 12.8.5 Example: Python code for computing the shear-center location for a box-beam 492
    - 12.8.5.1 Dividing the box to two open cross-sections .. 493
    - 12.8.5.2 Splitting the shear force 494
    - 12.8.5.3 General purpose code for rectangular box sections 495
    - 12.8.5.4 Data preparation 498
    - 12.8.5.5 Distribution profile of the shear-flow 498
  - 12.8.6 Example: The shear-center location for a C-beam 499
- 12.9 Remarks 500
- References 500

13. Energy Methods   501

- 13.1 Energy principles and fundamental concepts 501
- 13.2 Principle of virtual work by virtual displacement 502

13.2.1 Example: A loaded linear elastic spring 503
13.2.2 Evaluation of potential energy 503
13.2.3 Imposition of infinitely small virtual displacement 505
13.2.4 Compatibility condition 505
13.2.5 Potential energy principle for 1-DOF systems 506
13.2.6 Results of potential energy principle: Equilibrium
equation 506
13.2.7 On compatibility equation 507
13.2.8 Clapeyron's theorem 507
13.2.9 Formulas of potential energy for 1-DOF systems 507
13.2.10 Formulas of potential energy for various members 509
13.2.11 Strain energies U_N for bars with varying cross-section . . 509
13.2.12 Strain energies for beams with varying cross-section . . . 510
       13.2.12.1 From moment, U_M 510
       13.2.12.2 From shear force, U_V 510
13.2.13 Strain energies U_T for bars under torsion 511
13.2.14 Summary of formulas for U for various structural
members 512
13.3 Principle of virtual work by virtual force 512
13.3.1 Application of an infinitely small virtual force 513
13.3.2 Evaluation of complementary energy 513
13.3.3 Equilibrium condition 515
13.3.4 Complementary energy principle for 1-DOF systems . . . 515
13.3.5 Compatibility equation 516
13.3.6 On equilibrium equation 516
13.3.7 General formulas for displacements of 1-DOF
members 516
13.3.8 Example: Cantilever beam subjected to a concentrated
force at its tip 517
13.3.9 Comparison of PEP and CEP, linear elastic systems . . . 521
13.3.10 Comparison of PEP and CEP, nonlinear systems 522
13.4 Strain energy formulation for 3D solids 522
13.4.1 Mathematical derivation 522
       13.4.1.1 Work done by external forces 523
       13.4.1.2 Strain energy expression 524
       13.4.1.3 Compatibility conditions 525
13.4.2 Pictorial derivation 525
13.4.3 Strain energy density for 1D problem 525
       13.4.3.1 General formulas 526
       13.4.3.2 Formulas for linear elastic materials 528
13.4.4 Various formulas for strain energy density for
3D solids 529

- 13.5 Potential energy principle for structures 530
  - 13.5.1 Virtual work via virtual displacement for structures ... 530
  - 13.5.2 Formulas of potential energy principle for structures ... 530
  - 13.5.3 Example: A two-bar truss subjected to two loads 531
    - 13.5.3.1 Setting of the problem 531
    - 13.5.3.2 Solutions to the problem 532
- 13.6 Complementary energy formulation for 3D solids 537
  - 13.6.1 General formulations 537
  - 13.6.2 Equilibrium conditions 537
  - 13.6.3 Complementary energy 1D problems 538
- 13.7 Complementary energy principle for structures 539
  - 13.7.1 Virtual work via virtual forces for structures 539
  - 13.7.2 Formulas of complementary energy principle for structures 539
  - 13.7.3 Example: Linear spring–mass system 540
  - 13.7.4 Example: Nonlinear spring–mass system 543
  - 13.7.5 Example: A two-bar truss subjected to two loads 544
    - 13.7.5.1 Setting of the problem 544
    - 13.7.5.2 Solution to the problem 544
- 13.8 Displacements of statically determinate structures 546
  - 13.8.1 Principles for linear elastic structure members 546
  - 13.8.2 Principles for structures with multiple multiple-type members 547
  - 13.8.3 Examples: Cantilever beam subjected to two concentrated forces 548
  - 13.8.4 Examples: Cantilever beam subjected to distributed force 551
  - 13.8.5 Fictitious unit load method 553
  - 13.8.6 Example: Cantilever beam subjected to distributed force 554
  - 13.8.7 Example: A planar truss structure with three bars 555
- 13.9 Statically indeterminate structures 557
  - 13.9.1 Issues in dealing with indeterminate structures 557
  - 13.9.2 Example: Cantilever beam with a simple support at the other end 558
- 13.10 Remarks 559
- References 559

14. Failure Criteria and Failure Analysis — 561

- 14.1 Basic concept 562
- 14.2 A case for this study: Yield analysis of a steel shaft 562
  - 14.2.1 Setting of the problem 562

	14.2.2	Formulas for computing the stresses	562
	14.2.3	Python code to compute the stresses	564
14.3	Maximum principal stress criterion		565
	14.3.1	Yield surface	566
	14.3.2	Application to the case study problem	567
14.4	Maximum principal strain criterion		567
	14.4.1	Yield Surface	568
	14.4.2	Application to the case study problem	568
14.5	Strain-energy density criterion		569
	14.5.1	Strain-energy density	570
	14.5.2	Application to the case study problem	570
14.6	Yielding criteria for ductile materials		571
	14.6.1	Maximum shear-stress (Tresca) criterion	571
	14.6.2	Tresca yield surface	571
	14.6.3	Application to the case study problem	572
	14.6.4	Distortional energy density criteria	573
		14.6.4.1 von Mises criterion	573
		14.6.4.2 Application to the case study problem	574
		14.6.4.3 Octahedral shear-stress criterion	574
		14.6.4.4 Application to the case study problem	575
		14.6.4.5 Difference between the von Mises and Tresca criteria	575
	14.6.5	π-plane	576
	14.6.6	Deviatoric planes	576
14.7	Yield criterion for cohesive materials		577
	14.7.1	Mohr–Coulomb yield function	577
	14.7.2	Determination of material parameters	577
	14.7.3	Mohr–Coulomb yield surface	578
	14.7.4	Drucker–Prager yield criterion	578
14.8	Comparison of different criteria		579
14.9	Remarks		581
References			582

Index   585

Chapter 1

Introduction

1.1 Computational methods

This book is part of a series that delves into the exciting world of computational methods. Our goal is to equip you with a solid foundation in this field, combining theoretical principles with practical applications.

The series encompasses general methods and techniques used throughout science, technology, engineering, and math (STEM) education, as well as research in various scientific and engineering disciplines. Think of it as your one-stop encyclopedia for computational methods, covering the theory, formulation, and even the code you'll need to get started. We've designed the series to be accessible. Readers who have completed elementary and middle school (roughly nine years of education) can begin with the foundational volumes on mathematics. Once you progress through the mechanics volumes, you'll be equipped to tackle research and design projects that leverage computational methods.

Taking advantage of the rapid advancements in computer technology, we'll heavily integrate code examples throughout the series. This allows you to see concepts, theories, and formulations come to life through practical examples with clear visualizations. Initially, we'll primarily use Python for the fundamental volumes, potentially introducing other languages for more advanced topics. By effectively utilizing code, you can dedicate more time to understanding the core principles. Let the computer handle the complex calculations and tedious derivations, allowing you to focus on the bigger picture.

This approach empowers you to explore the fascinating world of computational methods with a strong theoretical grounding and practical skills.

1.2 Why start and contribute to this book series

The editor-in-chief of this book series, Dr. G.R. Liu, has been working in areas related to computational methods for over 40 years. He developed his first FEM code for nonlinear problems in 1980 and has since published more than 600 journal papers and 12 monographs in this area. After all these years of studying, using, and developing computational methods, he started to think about a means to help other interested individuals learn computational methods in a more effective, systematic, and smoother way. He has concluded that developing this book series is the best way to achieve this objective.

1.3 Mechanics of materials: Essential for safe and effective use of materials

Materials are used to build devices and structures for various purposes in our everyday life. This book introduces the subject of **mechanics of materials**, which is the most basic and essential for safe and effective use of materials in the fields of sciences and engineering. When a device (or structure) is loaded, the members in it experience force, stress, and strain, and they undergo deformation. If the device is not designed properly, it may either fail to carry the desired load or become too heavy with material not fully utilized. The subject that studies the **relationships** among the **forces, stresses, strains, and deformation** of the **basic structural members of materials** is called mechanics of materials.

This book covers the **fundamental principles** of mechanics of materials, focusing on the mechanical behavior of structural members under various types of loads, including **axial loading, bending, shearing, and torsion**. The members can have various shapes and can be constrained in different ways. Concepts of **energy** and **failure criteria** will also be studied.

This book introduces the **tensor** concept at the beginning because it is the most fundamental concept for all subjects in mechanics. Traditionally, the tensor concept is introduced at a more advanced course, typically in elasticity because of its difficulty level. Based on the author's experience, however, it is far better to teach this concept in the introductory course on mechanics. Students may not be able to fully comprehend all the details, but with the help of Python codes, the basic concepts can be reasonably well set. Given time, these concepts will be fully established, leading to a rigorous and solid foundation for future studies/courses. The principles of mechanics of materials are explained in detail, formulated, and demonstrated with **numerous examples** and illustrations. **Python** is intensively used for deriving formulations, finding solutions symbolically or numerically, and plotting the results graphically.

Proper connection between theory and **real-world engineering problems** will be made through these examples.

The materials of this book can be digested through different forms of the **learning process**, including classroom teaching, online courses, and even self-study. Since Python codes are provided, readers can easily see how the theory is formulated and how the solutions are obtained in terms of formulas, numerical values, and graphs. Readers may also deepen their understanding via playing with codes and even further develop **their own codes** for solving other related problems.

The book is written in **Jupyter Notebook** format, so that the description of theory, formulation, and coding can all be done in a unified document. This provides an environment for easy reading, exercise, practice, and further exploration.

This book is written in reference to Refs. [1–3], which were the textbooks when the author was a university student, Refs. [4,5], which were frequently used as reference books during his PhD and research work, and Refs. [6–9], which were used as textbooks and reference books when teaching at universities. Some example problems given in these textbooks are used in this book.

1.4 Who may read this book?

This book is written for beginners interested in learning computational methods for solving problems in nature, engineering, and science. Readers include high school students, university students, graduate students, researchers, and professionals in any discipline. Engineers and practitioners may also find the book useful in establishing systematic basic concepts in computational methods. For beginners, it may be a good idea to first read or skim over Volume 1 of this book series: *Numbers and Functions — Theory, Application, and Python Codes* and *Volume 2: Calculus: Formulations and Solutions with Python.*

This volume contains a substantial amount of advanced concepts in the mechanics of materials, which are marked with * or **. This provides better completeness of the topic. When this book is used as a textbook for an entry-level course on mechanics of materials, these materials may be treated as optional. In the past 40 years of teaching, the author often found that there would always be quite a number of students having more time and capable of learning more materials than the curriculum defines. Therefore, providing additional starred materials in the book can be beneficial to many students. When this book is used as a textbook for an advanced course on mechanics of materials, these starred materials may be directly used. More advanced materials will be covered in the next planned volume of this book series: *Solid Mechanics — Cartesian Coordinates*.

1.5 Codes used in this book

Readers who purchased the book may contact the author directly at liugr100@gmail.com (mailto:liugr100@gmail.com) to request a soft copy of the book in Jupyter Notebook format with codes (which may be updated) for free for academic use after registration. If the book is used as a textbook, instructors may contact the author directly for sharing some of the homework assignments and examination papers with codes.

The conditions for use of the book and codes developed by the author, in both hard and soft copies, are as follows:

1. Users are entirely at their **own risk** when using any part of the codes and techniques. The codes are written primarily for proof of concept and not necessarily for efficiency and robustness. Many of the codes are **not** thoroughly tested by third parties.
2. The book and codes are only for your **own use**. You are not allowed to further distribute them without permission from the author.
3. There will not be any user support.
4. Proper reference and **acknowledgment** must be given for the use of the book, codes, ideas, and techniques.

These codes are often run with various **external packages/modules**. Therefore, care is needed when using these codes because the behavior of the codes often depends on the versions of Python and all these packages/modules. When the code does not run as expected, **version mismatch** could be one of the problems. When this book was written, the versions of Python and some of the packages/modules were as follows:

- Python 3.9.16 :: Anaconda, Inc.,
- Jupyter Notebook 6.1.5.

When issues are encountered while running a code, readers may need to check the versions of the packages/modules used. If Anaconda Navigator is used, the versions of all these packages/modules installed in the Python environment are listed when the Python environment is highlighted. You can also check the version of a package in a code cell of the Jupyter Notebook. For example, to check the version of the current environment of Python, one may use the following:

```
1  !python -V                              #! is used to execute an external command
```

Python 3.9.16

```
1  !jupyter notebook --version
```

6.1.5

If the version is indeed an issue, one would need to either **modify the code** to fit the version or install the **correct version** on your system. It is very useful to query the web using the error message, and solutions or leads can often be found. **Online AI tools**, such as ChatGPT, Gemini, and Copilot can also be quite helpful. This is the approach that the author uses most of the time facing issues in running a code.

This book will not discuss how to use Python. There is plenty of literature openly available online. Interested readers may also refer to Chapters 2 and 3 in Ref. [10] for a **concise description** of using Python for scientific computations. Since Python is quite easy to learn, one may simply use the codes and examples given in this book and jump-start learning Python in the process of studying the technical subjects of this book.

1.6 Use of external modules or dependences

To use Python and codes provided in this volume, we import the necessary modules and functions. The following are the most essential ones:

```
1  import sys                                           # import "sys" module
2
3  sys.path.append('../grbin')                          # Relative directory.
4                        # Or absolute folder like 'F:\\xxx\\...\\code'
5
6  #Author's own grcodes module is placed in folder grbin
7  import grcodes as gr
```

To view the codes in the imported module, one may just use the following code by uncommenting it (which may produce a long output):

```
1  import inspect
2  #source_code = inspect.getsource(gr)    # to view everything in gr module
3  source_code = inspect.getsource(gr.cheby_T) # to view any function in gr
4  print(source_code)
```

```
def cheby_T(n, x):
    '''Generate the first kind Chebyshev polynomials of degree (n-1) '''
    if    n == 0: return sp.S.One
    elif n == 1: return x
    else:        return (2*x*cheby_T(n-1,x)-cheby_T(n-2,x)).expand()
```

Alternatively, one may use any **text editor** to view and change the codes. To use any code function, say printx() in the imported module gr, for example, use gr.printx. Following is an example:

```
1  x = 8
2  gr.printx('x')  # when gr. is used, the code function is from the grcodes
```

x = 8

To avoid frequent import of lengthy external modules, we put all the frequently used modules in the "commonImports.py" file, as follows:

```
1   from __future__ import print_function
2   import numpy as np                        # for numerical computation
3   import sympy as sp                        # sympy module for computation
4   import numpy.linalg as lg                 # numpy linear algebra module
5   import scipy.linalg as sg                 # scipy linear algebra module
6   import scipy.integrate as si
7   from scipy.stats import ortho_group
8   import importlib
9   import itertools
10  import inspect
11  import csv
12  import pandas as pd
13
14  from grcodes import drawArrow, plotfig, printM, printx # frequently used
15
16  import math as ma
17  from sympy import sin, cos, symbols, lambdify, init_printing
18  from sympy import pi, Matrix, sqrt, oo, integrate, diff, Derivative
19  from sympy import MatrixSymbol, simplify, nsimplify, Function
20  from sympy import factor, expand, nsimplify, Matrix, ordered, hessian
21  from sympy.plotting import plot as splt
22  init_printing(use_unicode=True)   # for latex-like quality print format
23
24  from matplotlib.ticker import MultipleLocator
25  import matplotlib.pyplot as plt           # for plotting figures
26  import matplotlib as mpl
```

In the beginning of each Jupyter Notebook (Chapter), we simply import everything in commonImports. This reduces the need to frequently import modules in the coding process:

Introduction

```
1  #Often used external modules are in commonImports placed in folder grbin
2  #Place cursor in this cell, and press Ctrl+Enter to import dependences.
3
4  import sys                              # for accessing the computer system
5  sys.path.append('../grbin/')   # Change to the directory in your system
6
7  from commonImports import *            # Import dependences from '../grbin/'
8  import grcodes as gr                       # Import the module of the author
9  importlib.reload(gr)                     # When grcodes is modified, reload it
10
11 from continuum_mechanics import vector
12 from continuum_mechanics.solids import sym_grad, strain_stress
13 init_printing(use_unicode=True)          # For Latex-like quality printing
14
15 # Digits in print-outs
16 np.set_printoptions(precision=4,suppress=True,
17                     formatter={'float_kind': '{:.4e}'.format})
```

In general, importing the same module multiple times does no harm. In fact, Python ignores all the later importations if the module is already in the cache. Due to this, if one makes changes to the imported module, the module needs to be reloaded for the modification to take effect. This is done using importlib.reload(). For example:

```
1  # grcodes was imported as gr, reload it when grcodes is modified
2  importlib.reload(gr);
```

1.7 Use of help()

To find more details on what a module or an object in the module does, use help() after importing. For example:

```
1  help(gr.solver1D4)
```

Help on function solver1D4 in module grcodes:

solver1D4(E, I, by, l, v0, θ0, v1, θ1, V0, M0, V1, M1, key='c-c')
    Solves the Beam Equation for integrable distributed body force:
        u,x4=-by(x)/EI, with various boundary conditions (BCs):
        s-s, c-c, s-c, c-s, f-c, c-f.
    Input: EI: bending stiffness factor; by, body force; l, the length
           of the beam; v0, θ0, V0, M0, deflection, rotation,
           (internal) shear force, (internal) moment at x=0;
           v1, θ1, V1, M1, are those at x=1.
    Return: u, v_x, v_x2, v_x3, v_x4 up to 4th derivatives of v

Note that in the code cells given in the book, necessary **comments** (starting with "#") are used to provide additional explanations. These comments are placed on the right-hand side of the cells to avoid disrupting the reading flow too much while still being available for any help when needed.

References

[1] J.M. Gere and S.P. Timoshenko, *Mechanics of Materials*, Van Nostrand Reinhold Company, New York, 1972.

[2] S. Timoshenko and J.N. Goodier, *Theory of Elasticity*. McGraw Hill, New York, 1970. http://books.google.com/books?id=yFISAAAAIAAJ&dq=theory+of+elasticity&ei=ICiKSsr3G4jwkQS bxMyPCg.

[3] Z.L. Xu, *Elasticity*, Vols. 1 & 2. People's Publisher, China, 1979.

[4] J.D. Achenbach, *Wave Propagation in Elastic Solids*, 1984.

[5] Y.C. Fung, *Foundations of Solid Mechanics*, 1968.

[6] A.P. Boresi, K. Chong and J.D. Lee, *Elasticity in Engineering Mechanics*. John Wiley & Sons, New York, 2010.

[7] F.P. Beer, *Statics and Mechanics of Materials*, 2011.

[8] A.P. Boresi, R.J. Schmidt and O.M. Sidebottom, *Advanced Mechanics of Materials*. John Wiley & Sons, New York, 1985.

[9] T.H.G. Megson, *Structural and Stress Analysis*, 1996 (https://books.google.com/books?id=qZykmEz4nzkC).

[10] G.R. Liu, *Machine Learning with Python: Theory and Applications*. World Scientific, New York, 2023.

Chapter 2

Preliminaries: Number, Vector, Tensor, and Functions

```
 1  # Place cursor in this cell, and press Ctrl+Enter to import dependences.
 2  import sys                        # for accessing the computer system
 3  sys.path.append('../grbin/')   # Change to the directory in your system
 4
 5  from commonImports import *      # Import dependences from '../grbin/'
 6  import grcodes as gr              # Import the module of the author
 7  importlib.reload(gr)              # When grcodes is modified, reload it
 8
 9  from continuum_mechanics import vector
10  init_printing(use_unicode=True)       # For latex-like quality printing
11  np.set_printoptions(precision=4,suppress=True)  # Digits in print-outs
```

Mechanics of materials deals with and quantifies a family of objects, including forces, displacements, stresses, and strains. These objects are constructed as **tensors** that can be studied systematically with predictable behaviors.

This chapter provides some preliminary mathematical topics related to objects in mechanics of materials, as well as their mathematical treatment. This includes numbers, vectors, the tensor concept, coordinate representation, coordinate transformation, and various differential operations on functions. Readers familiar with these topics may skip this chapter or quickly skim over it. Readers having difficulty following the discussions in this chapter may simply skim over it for some ideas, considerations, and terminologies. He/she may read in more detail the relevant parts of this chapter when studying the later chapters.

We start with the most elementary object: numbers, which are 0th-order tensors and used to form higher-order tensors.

2.1 Numbers

In dealing with mechanics problems in a scientific manner, we must ensure that variables and parameters are **quantified precisely**. For this, we use numbers and operate on them in a number of ways. Therefore, understanding the behavior of numbers subject to various operations is fundamentally important to ensure that what we obtain is **consistent, predictable, and reliable**.

Generally speaking, a number is a single element in a **field** (https://en.wikipedia.org/wiki/Field_(mathematics)) defined with the **standard arithmetic operations** (addition "+", subtraction "−", multiplication "×" (often omitted in expression), and defined-division "÷" (alternatively "/" or, sometimes, ":"), all satisfy necessary conditions, including commutativity, associativity, and distributivity for addition). In all our operations in mechanics, we require that all these conditions hold for all numbers, which may stand alone or as a part of other objects like vectors, matrices, or in general tensors.

A number can be a real number or a complex number. Real numbers form the set $\mathbb{R}$ that is a collection of numbers of different types including zero, all integers, all rational numbers, and all irrational numbers, such as $\sqrt{2}, \pi, e$, and $\frac{1+\sqrt{5}}{2} \approx 1.618$ (the golden ratio). All these numbers can be positive or negative, filling the entire one-dimensional (1D) coordinate space $\mathbb{R}$. With addition and scalar multiplication, it is a real vector space. Figure 2.1 shows a schematic 1D line space where real numbers live. Space $\mathbb{R}$ is closed under standard arithmetic operations, collectively noted as $\ast$ (addition, subtraction, multiplication, or defined-division). This means that any of these operations on any numbers in $\mathbb{R}$ produces a new number d:

$$d = a \ast b \quad \forall a \quad \text{and} \quad b \in \mathbb{R} \tag{2.1}$$

which will still be in $\mathbb{R}$. This enables the most fundamental property: **closure**, implying that no numbers disappear under any of these operations. Thus, a number becomes a building block for more complicated mathematical objects, such as tensors.

Clearly, $\mathbb{R}$ is an ordered field because all the real numbers can be ranked based on their values. The square of a real number is always positive.

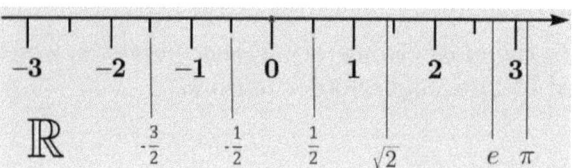

Figure 2.1. A 1D coordinate space where real numbers live. Image adapted from Wikimedia Commons, the free media repository, public domain, by Phrood commonswiki.

In theory, $\mathbb{R}$ is boundless (bounded by $\pm\infty$) and seamless. In a computer, however, it is bounded and with gaps due to bit limitations in computers. For example, in Numpy, this information for np.float64 can be found using:

```
1  print("Largest float in NumPy = ", np.finfo(np.float64).max)
2  print("Smallest float in NumPy =", np.finfo(np.float64).min)
3  print("Number of significant decimal digits = ",
4        np.finfo(np.float64).precision)
```

```
Largest float in NumPy =  1.7976931348623157e+308
Smallest float in NumPy = -1.7976931348623157e+308
Number of significant decimal digits =  15
```

Real numbers include integers by default. Due to its exactness in digital representation in computers (within its bits limit [1]), an integer is also defined as the integer type, which is often used for indexing, ordering, and ranking.

All complex numbers form the set $\mathbb{C}$. It is, in a way, an extension of the real set $\mathbb{R}$, formed by introducing the imaginary number: $i = \sqrt{-1}$. Any member in $\mathbb{C}$ is made up of real numbers a and b in the form of $c = a+ib$. Set $\mathbb{C}$ is closed under the above-defined arithmetic operations, **transcendental operations**, such as exponents with generally a base in $\mathbb{C}$ and a power in $\mathbb{C}$ and their inverse operation: various logarithms. Set $\mathbb{C}$ is known as **algebraically closed**, implying that any polynomial of degree $n > 0$ defined in $\mathbb{C}$ has always n roots in $\mathbb{C}$. Simply put: no root will disappear, ensuring solution existence, when algebraic operations are used.

More detailed discussion on the closure properties of numbers and their representation in a computer can be found in Ref. [1]. The takeaway is as follows:

1. When these defined arithmetic operations are used, we should be able to find a solution if the problem is well posed. In other words, a numerical **solution exists** under the defined operations if the solution exists physically.
2. When algebraic operations are involved, we may need to switch to complex numbers, and the solution could be multiple. In such a case, we may need to choose a physically meaningful one(s).
3. The most commonly encountered failure in numerical computations is undefined division: **division by zero**.
4. The second most commonly encountered is **under- or overflow** because of the physical limits of finite digits in a computer.

Thus, when failing to obtain a solution, one may check these related issues. All problems posed in mechanics of materials prohibit zero division. When zero division or under- and overflow occurs, one may need to check the formulations and perform dimensional analysis of equations, data scaling, or normalization [1].

2.2 Vectors: Collection of numbers

2.2.1 *Vectors defined in a computer*

Generally speaking, a vector is a collection of numbers (real or complex) in properly arranged structures so that its behavior can be managed, analyzed, and computed in a consistent manner for desired purposes. Each of these numbers is often called an **element** or **component**. Depending on the field of applications, precise definition may vary. In the framework of mechanics in general, it is built from numbers vertically or horizontally. It becomes a building block for more complicated mathematical objects, such as matrices and tensors.

As a convention, we use **boldface** to denote a vector or a matrix, which is a collection of vectors.

From the theory of standard linear algebra, a vector can have two forms: line-up numbers called a **row-vector** or a stack of numbers called a **column-vector**. Row- and column-vectors behave differently when used in mathematical derivations and operations, in which dimensionality match is required. A vector in Python is simply a **1D array** by default. The number of components in a vector is called length and can be queried using the Python function len(). Its dimension is 1, which can be confirmed using ndim(). Consider an array **v** with n numbers, or $len(v) = n$. It has a shape of $(n,)$, as shown in the following example.

```
1  v = np.array([5, 2, 8])        # create an 1D array with length 3
2
3  print(f'Array:{v}, its dimension:{v.ndim}, length:{len(v)} and',
4        f'shape:{v.shape}')
```

Array:[5 2 8], its dimension:1, length:3 and shape:(3,)

It is an **ndarray** class in numpy. Readers may use help() to find out the details about it, including the attributes and methods it has.

```
1  # help(v)   # reader may use this to find out the details
```

Note that a numpy array behaves differently from the row- or column-vector in standard linear algebra, where the **transpose** of a row-vector becomes a column-vector and vice versa. However, transposing a numpy 1D array has no effect, as shown in the following example.

Preliminaries: Number, Vector, Tensor, and Functions

```
1  print(f'Array v.T:{v.T}, its dimension:{v.T.ndim}, length:{len(v.T)}',
2        f'and shape:{v.T.shape}')
```

Array v.T:[5 2 8], its dimension:1, length:3 and shape:(3,)

A row-vector needs, when it must be used, to be specifically defined as a matrix with only one row with a shape of $(1, n)$. It is a two-dimensional (2D) matrix but has only one row.

```
1  v_row = v[None,:] # adds in a new dimension without elements
2
3  print(f'Array v_row:\n{v_row}, its dimension:{v_row.ndim},',
4        f' length:{len(v_row)}, and shape:{v_row.shape}')
```

Array v_row:
[[5 2 8]], its dimension:2, length:1, and shape:(1, 3)

Similarly, if a column-vector is needed, one must specifically define a matrix with only one column with a shape of $(n, 1)$.

```
1  v_row = v[:,None] # adds in a new dimension without elements
2
3  print(f'Array v_row:\n{v_row}, its dimension:{v_row.ndim}',
4        f', length:{len(v_row)}, and shape:{v_row.shape}')
```

Array v_row:
[[5]
 [2]
 [8]], its dimension:2, length:3, and shape:(3, 1)

With this kind of specific definition, the rules of operations follow largely those defined in standard linear algebra. Care is needed when using the Numpy 1D arrays because of some different behavior. A more detailed discussion can be found in Section 2.9.13 in Ref. [2].

2.2.2 *Basic properties of vectors*

Vectors are widely used in science and engineering, including mechanics of materials. The two most basic features of a vector are its **magnitude** and **direction**. The former is represented graphically by the length and the latter by an arrow. When we treat a vector as an object in isolation, these two features are sufficient to define the vector.

A vector in an n-D coordinate space $\mathbb{R}^n$ can have a general form of

$$\mathbf{v} = \begin{bmatrix} v_1 \\ v_2 \\ \vdots \\ v_n \end{bmatrix} \tag{2.2}$$

where $v_i, i = 1, 2, \ldots, n$, is the ith element of the vector on the ith axis of the generalized **Cartesian coordinate** system. The zero vector has zero in all its components.

Direction of a vector: The direction of a vector is defined by all its elements in reference to the coordinate system. Each v_i is the value on the ith axis of the Cartesian coordinate system where the vector is defined. It lives essentially in a 1D coordinate space. This means that a vector in $\mathbb{R}^n$ is defined by n numbers each in $\mathbb{R}^1$ with $\text{ndim}(\mathbf{v}) = 1$ and $\text{len}(\mathbf{v}) = n$.

Magnitude of a vector: The magnitude or physical length of a vector must be a non-negative real number. It is denoted as $\|\mathbf{v}\|_2$ and is computed using

$$\|\mathbf{v}\|_2 \equiv \sqrt{v_1^2 + v_2^2 + \cdots + v_n^2} = \sqrt{\sum_{i=1}^{n} v_i^2} \tag{2.3}$$

where $\|\mathbf{v}\|_2$ is also known as the L2 norm and is a physical measure of the length of the vector. It is zero if and only if all the components are zero. In general, we can use different types of norms, but L2 is the default. Thus, we often drop the subscript 2 and denote the magnitude of a vector using $\|\mathbf{v}\|$.

Unit vectors: The unit vector of a vector is denoted as $\bar{\mathbf{v}}$. It has a magnitude of 1, but with the same direction as $\mathbf{v}$. It is also called a normalized vector. It is obtained using

$$\bar{\mathbf{v}} = \frac{\mathbf{v}}{\|\mathbf{v}\|} \tag{2.4}$$

Let us visualize vectors in 2D graphs using Python.

2.2.3 Attributes of a vector in a computer

```
1  # Define a vector in Python, in 3D x, y, z coordinates.
2  np.random.seed(8)    # use a seed to make the random number repeatable
3  v = np.array([4,2,5])                        # w.r.t origin (0, 0, 0)
4
5  print('The data type of vector v in Python:', type(v))
6  print('The size (number of elements) of v =', np.size(v))
7  print('The shape (structure of data) of v:', np.shape(v))
8  print('The vector v is:', v, ', and its 3rd element=',v[2])
9  print('The direction of v is defined by all its elements.')
10
11 length_v = np.sqrt(np.sum(v*v))
12 v_bar = v/length_v
13
14 print(f"The length or magnitude of v = {length_v:.4f}")
15 print(f"The unit vector of v is: {v_bar}")
16 print(f"Length of unit vector = {np.sqrt(np.sum(v_bar*v_bar))}")
```

```
The data type of vector v in Python: <class 'numpy.ndarray'>
The size (number of elements) of v = 3
The shape (structure of data) of v: (3,)
The vector v is: [4 2 5] , and its 3rd element= 5
The direction of v is defined by all its elements.
The length or magnitude of v = 6.7082
The unit vector of v is: [0.5963 0.2981 0.7454]
Length of unit vector = 1.0
```

Note that the choice of the orientation of the coordinate system will produce **differences in the presentation** of the vectors. This relates to the coordinate transformation of vectors, which will be discussed in detail later in this chapter.

Vectors may change with the location in mechanics problems. There is an area of study called **function** and functional analysis dealing with vectors that vary over space. This study is widely applied in the modeling and simulation of solid mechanics and fluid dynamics problems [3–5]. The locational variation is represented in terms of functions [1]. In this book, we will simply use the concept of functions.

2.2.4 *The standard basis vectors*

Consider a space $\mathbb{R}^n$. We can define a set of n unit vectors that can used as basis vectors. This special set of unit vectors is known as the **standard basis vectors** (SBVs), and it is of importance in vector analysis. It can be expressed as follows:

$$\mathbf{i}_1 = \begin{bmatrix} 1 \\ 0 \\ \vdots \\ 0 \end{bmatrix}; \quad \mathbf{i}_2 = \begin{bmatrix} 0 \\ 1 \\ \vdots \\ 0 \end{bmatrix}; \quad \ldots; \quad \mathbf{i}_n = \begin{bmatrix} 0 \\ 0 \\ \vdots \\ 1 \end{bmatrix} \quad (2.5)$$

Clearly, each one of these SBVs has length 1 by the definition of Eq. (2.3). These SBVs form a set of vectors for the coordinate system, and they are always **linearly independent**. This means that any one of these basis vectors in the set cannot be written as a linear combination of the remaining vectors in the same set. This can be easily proven by simple observation. We see clearly that in any of the basis vectors, there is always only one entry of unit 1, and all the other entries are zero. Take any basis vector, say the ith. Its ith entry is 1. All the remaining basis vectors have only zero at their ith entry. Therefore, it is not possible to find a linear combination of the remaining basis vectors to produce a vector that has a nonzero value at the ith entry. Thus, the ith SBV is linearly independent of the remaining SBVs. **This means that any vector in the space can be written as a linear combination of the SBVs.**

2.2.5 Basic arithmetic operations for vectors

Consider a space $\mathbb{R}^n$ that contains all possible vectors with length n and each of these n elements are in $\mathbb{R}$. We now define the basic properties and operations for these vectors.

1. **The zero vector:** The zero vector **0** is a vector with **all** its elements as 0. A nonzero vector shall have at least one nonzero element.
2. **Scaling:** For a given arbitrary vector **v** in $\mathbb{R}^n$, a new vector **w** can be created through scaling the vector **v** by multiplying an arbitrary scaling factor $\alpha \in \mathbb{R}$:

$$\mathbf{w} = \alpha\mathbf{v} = \begin{bmatrix} \alpha v_1 \\ \alpha v_2 \\ \vdots \\ \alpha v_n \end{bmatrix} \quad \forall \alpha \in \mathbb{R} \quad (2.6)$$

where $v_i, i = 1, 2, \ldots, n$, is the ith element of the vector. The scaling is applied to each of the elements of **v**. Thus, the ith element of the resultant new vector **w** becomes $\alpha v_i (i = 1, 2, \ldots, n)$, each of which is a number in $\mathbb{R}$. Even if $\alpha = 0$, the resulting **w** is still in $\mathbb{R}^n$ because $\mathbb{R}^n$ has a zero vector by definition. Therefore, the new vector **w** is always a member of $\mathbb{R}^n$. Since $\alpha v_i = v_i \alpha$ based on the commutative property of standard arithmetic operation of multiplication for numbers, we thus have $\alpha \mathbf{v} = \mathbf{v}\alpha$. This implies that the scalar and the vector are also commutative. This allows us to move the scalar numbers back and forth when needed.

3. **Addition:** Given two arbitrary vectors **v** and **u** both in $\mathbb{R}^n$, a new vector **w** is created by adding them up:

$$\mathbf{w} = \mathbf{v} + \mathbf{u} = \mathbf{u} + \mathbf{v} = \begin{bmatrix} v_1 + u_1 \\ v_2 + u_2 \\ \vdots \\ v_n + u_n \end{bmatrix} \quad \forall \mathbf{v} \text{ and } \mathbf{u} \in \mathbb{R}^n \quad (2.7)$$

The addition is done to each of the corresponding elements of these two vectors **v** and **u**. Thus, the resultant new vector **w** is also in $\mathbb{R}^n$ because each of its elements, $(v_i + u_i)$, is a number in $\mathbb{R}$. Due to the commutative property of standard arithmetic operation of addition for scalar numbers, vectors in addition operation are also commutative.

4. **Linear combination:** With the definition of the zero vector and the above two basic operations defined in items 2 and 3, it is clear that the space $\mathbb{R}^n$ formed by the vectors is closed under addition and scaling. Any linear combination of any two vectors $\mathbf{u} \in \mathbb{R}^n$ and $\mathbf{v} \in \mathbb{R}^n$ with an arbitrary scaling factor $\alpha \in \mathbb{R}$ and an arbitrary scaling factor $\beta \in \mathbb{R}$, $\mathbf{w} = \alpha\mathbf{u} + \beta\mathbf{v}$, is in $\mathbb{R}^n$. Therefore, the $\mathbb{R}^n$ of vectors is called a linear

vector space. Its members are all vectors with n elements, which are all in $\mathbb{R}$. By default, all vectors dealt with in this book have all the above basic operations defined and are vector spaces.

Note that a vector space does not mean a space formed just by vectors. The space of real numbers $\mathbb{R}$ is also a vector space because it is closed under addition and scaling based on the defined standard arithmetic operations. The essential point is that any linear combination of members in the space shall remain in the same space. Matrices can also form vector spaces.

Now, any arbitrary vector $\mathbf{v}$ in $\mathbb{R}^n$ can be expressed as multiples of the SBVs:

$$\mathbf{v} = \begin{bmatrix} v_1 \\ v_2 \\ \vdots \\ v \end{bmatrix} = v_1 \begin{bmatrix} 1 \\ 0 \\ \vdots \\ 0 \end{bmatrix} + v_2 \begin{bmatrix} 0 \\ 1 \\ \vdots \\ 0 \end{bmatrix} + \cdots + v_n \begin{bmatrix} 0 \\ 0 \\ \vdots \\ 1 \end{bmatrix} \quad (2.8)$$

$$= v_1 \mathbf{i}_1 + v_2 \mathbf{i}_2 + \cdots + v_n \mathbf{i}_n$$

where $v_i (i = 1, 2, \ldots, n)$ are scalar numbers that are the n independent elements of $\mathbf{v}$, and $\mathbf{v}$ is a linear combination of all the SBVs.

Note that we omitted the standard $\times$ symbol for multiplication of real numbers to avoid confusion because in vector operations, $\times$ has an entirely different meaning, known as the **cross product**, which is discussed later in the section on cross product of vectors.

2.2.6 *Other arithmetic operations for vectors*

For vectors, we often perform also the following element-wise operations.

1. Subtraction: Similarly, a new vector $\mathbf{w}$ can be created by subtracting a vector $\mathbf{u}$ from another vector $\mathbf{v}$ all in $\mathbb{R}^n$:

$$\mathbf{w} = \mathbf{v} - \mathbf{u} = -\mathbf{u} + \mathbf{v}$$
$$= (v_1 - u_1)\mathbf{i}_1 + (v_2 - u_2)\mathbf{i}_2 + \cdots + (v_n - u_n)\mathbf{i}_n \; \forall \mathbf{v} \text{ and } \mathbf{u} \in \mathbb{R}^n \quad (2.9)$$

The subtraction is only applied to each of the corresponding elements of $\mathbf{v}$ and $\mathbf{u}$. Thus, the resultant new vector $\mathbf{w}$ is also in $\mathbb{R}^n$ because each of its elements, $(v_i - u_i)$, is in $\mathbb{R}$. Subtraction operations are also commutative for the same reason. Note that subtraction is simply negated addition and can be induced by a scaling operation with $\alpha = -1$ and the defined addition.

2. **Element-wise multiplication (also known as the Hadamard product or Schur product):** Given two vectors **v** and **u** both in $\mathbb{R}^n$, we perform

$$\mathbf{w} = \mathbf{v} * \mathbf{u} = \mathbf{u} * \mathbf{v}$$
$$= (v_1 u_1)\mathbf{i}_1 + (v_2 u_2)\mathbf{i}_2 + \cdots + (v_n u_n)\mathbf{i}_n \quad \forall \mathbf{v} \text{ and } \mathbf{u} \in \mathbb{R}^n \quad (2.10)$$

As the name suggests, the multiplication is performed, respectively, on each of the corresponding elements of these two vectors **v** and **u**. Thus, the resultant new vector **w** is also in $\mathbb{R}^n$ because each of its elements, $(v_i u_i)$, is in $\mathbb{R}$. We use the symbol $*$ to denote element-wise multiplication, following the Python convention. Symbols $\odot$ and $\circ$ are also used in the literature for this operation.

3. **Element-wise division:** Given two vectors **v** and **u** both in $\mathbb{R}^n$, we perform

$$\mathbf{w} = \mathbf{v}/\mathbf{u} = (v_1/u_1)\mathbf{i}_1 + (v_2/u_2)\mathbf{i}_2 + \cdots + (v_n/u_n)\mathbf{i}_n$$
$$\forall \mathbf{v} \text{ and } \mathbf{u} \in \mathbb{R}^n \text{ with } u_i \neq 0 \quad (2.11)$$

As the name suggests, the division is performed, respectively, on each of the corresponding elements of these two vectors **v** and **u**. Thus, the resultant new vector **w** is also in $\mathbb{R}^n$ because each of its elements, (v_i/u_i) when $u_i \neq 0$, is in $\mathbb{R}$. We use the symbol / to denote element-wise division, following Python's convention. Note that element-wise division can often fail because all the elements in **u** have to be nonzero for the defined operation.

2.3 Dot product of vectors, producing a scalar

Dot product (or inner product) is essential for all linear algebraic operations and will be used frequently throughout the entire book. It is the most important operation to produce tensors for mechanics of materials. Let us introduce it in detail.

2.3.1 Definition: A scalar produced using two vectors

Consider two nonzero vectors **u** and **v** both in $\mathbb{R}^n$, as shown in Fig. 2.2.

To find the angle θ between them, we define an operation using these two vectors, called the dot product or inner product:

$$\boxed{\mathbf{u} \cdot \mathbf{v} = \|\mathbf{u}\| \, \|\mathbf{v}\| \cos \theta} \quad (2.12)$$

where $\mathbf{u} \cdot \mathbf{v}$ is computed using multiplications of their corresponding elements:

$$\boxed{\mathbf{u} \cdot \mathbf{v} = (u_1 v_1) + (u_2 v_2) + \cdots + (u_n v_n) \quad \forall \mathbf{u} \text{ and } \mathbf{v} \in \mathbb{R}^n} \quad (2.13)$$

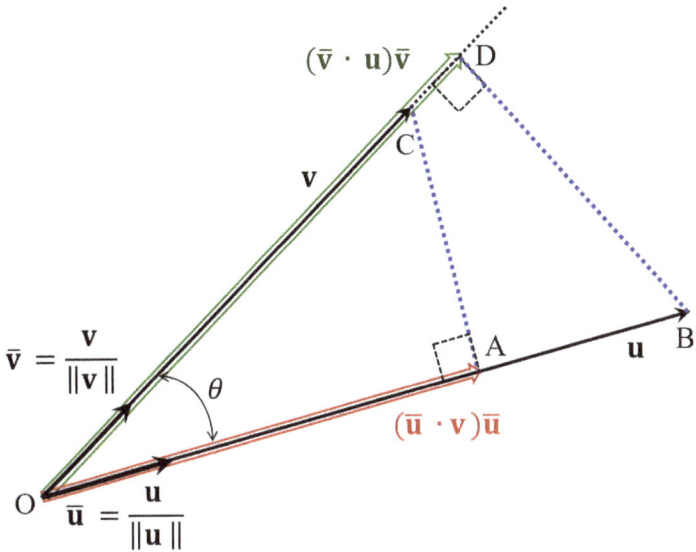

Figure 2.2. Two vectors **u** and **v** and the angle θ between them.

It is the sum of elements of the vector of element-wise multiplication defined above. The result is a scalar number in $\mathbb{R}$. For this reason, dot product is sometimes called the scalar product.

It is obvious that the dot product is commutative: $\mathbf{u} \cdot \mathbf{v} = \mathbf{v} \cdot \mathbf{u}$. This is also called symmetry. If the vectors are complex, it are conjugated symmetry: $\mathbf{u} \cdot \mathbf{v} = \overline{\mathbf{v} \cdot \mathbf{u}}$. Widely used notations for dot product are given as follows:

$$\begin{aligned}
\mathbf{u} \cdot \mathbf{v} &= \mathbf{v} \cdot \mathbf{u} && \text{dot product notation} \\
&= \langle \mathbf{u}, \mathbf{v} \rangle = \langle \mathbf{v}, \mathbf{u} \rangle && \text{inner product notation} \\
&= \mathbf{v}^\top \mathbf{u} = \mathbf{u}^\top \mathbf{v} && \text{matrix product notation} \\
&= (v_1 u_1) + (v_2 u_2) + \cdots + (v_n u_n) && \text{for computation}
\end{aligned} \quad (2.14)$$

All these notations are widely used in the literature and will be used in this book interchangeably.

2.3.2 Direction cosine of two vectors

Since we already know how to compute the norm of a vector using Eq. (2.3) and it will not be zero for any nonzero vector, the angle θ can now be computed using Eqs. (2.12) and (2.13), which gives

$$\boxed{\cos \theta = \underbrace{\cos(\mathbf{u}, \mathbf{v})}_{\text{direction cosine}} = \frac{\mathbf{u} \cdot \mathbf{v}}{\|\mathbf{u}\| \, \|\mathbf{v}\|} = \bar{\mathbf{u}} \cdot \bar{\mathbf{v}}} \quad (2.15)$$

This gives an important concept of **direction cosine** of two vectors, which is used very often in mechanics problems. The angle θ is found using

$$\theta = \arccos\left(\frac{\mathbf{u} \cdot \mathbf{v}}{\|\mathbf{u}\| \, \|\mathbf{v}\|}\right) = \arccos(\bar{\mathbf{u}} \cdot \bar{\mathbf{v}}) \qquad (2.16)$$

Care is needed when computing θ using the foregoing equation because θ can be positive or negative due to the multivalued nature of the cosine function.

To make a better sense of the dot product defined, let us look at a few special cases.

2.3.3 Inner product associated norm

First, assume $\mathbf{u}$ and $\mathbf{v}$ are identical. In this case, $\cos\theta = 1$ because the in-between angle is zero, and Eq. (2.12) becomes

$$\mathbf{v} \cdot \mathbf{v} = \|\mathbf{v}\| \, \|\mathbf{v}\| = \|\mathbf{v}\|^2$$
$$\Longrightarrow \|\mathbf{v}\| = \sqrt{\mathbf{v} \cdot \mathbf{v}} = \sqrt{v_1 v_1 + v_2 v_2 + \cdots + v_n v_n} \qquad (2.17)$$

which is exactly Eq. (2.3). The inner product has an associated norm, which is the L2 norm. Thus, the inner product is often used to compute the L2 norm of a vector.

Suppose $\mathbf{u}$ and $\mathbf{v}$ are in the same direction but have different magnitudes. In this case, we still have $\cos\theta = \cos 0 = 1$ because the in-between angle is zero, and Eq. (2.12) gives

$$\mathbf{u} \cdot \mathbf{v} = \|\mathbf{u}\| \, \|\mathbf{v}\| \quad \text{when } \mathbf{u} \parallel \mathbf{v} \qquad (2.18)$$

These two vectors are parallel. In this case, the inner product of these two vectors is the product of their norms.

When $\mathbf{u}$ and $\mathbf{v}$ are in the opposite direction, $\cos\theta = \cos 180° = -1$, they are also parallel, but

$$\mathbf{u} \cdot \mathbf{v} = -\|\mathbf{u}\| \, \|\mathbf{v}\| \qquad (2.19)$$

2.3.4 Orthogonality and orthogonal projection

Assume $\mathbf{u}$ is perpendicular (orthogonal) to $\mathbf{v}$. In this case, $\cos\theta = \cos\pm 90° = 0$ because their in-between angle is either $\pi/2$ or $-\pi/2$, and Eq. (2.12) leads to

$$\mathbf{u} \cdot \mathbf{v} = \|\mathbf{u}\| \, \|\mathbf{v}\| \times 0 = 0 \qquad (2.20)$$

Vectors that satisfy this inner product property are mutual orthogonal vectors. The inner product is often used as tool to check the orthogonality among vectors. Equation (2.20) gives also a clear definition for **orthogonality** of two vectors.

It is obvious that the inner product of the set of SBVs satisfy

$$\langle \mathbf{i}_i, \mathbf{i}_j \rangle = \mathbf{i}_i \cdot \mathbf{i}_j = \mathbf{i}_i^\top \mathbf{i}_j = \mathbf{i}_j^\top \mathbf{i}_i = \boxed{\delta_{ij} = \begin{cases} 1 & i = j \\ 0 & i \neq j \end{cases}} \quad (2.21)$$

It is the so-called **Kronecker delta function** defined in the box.

Let $\mathbf{u}$ be a unit vector. In this case, $\|\mathbf{u}\| = 1$, we have

$$\mathbf{u} \cdot \mathbf{v} = \underbrace{\|\mathbf{u}\|}_{1} \|\mathbf{v}\| \cos\theta = \|\mathbf{v}\| \cos\theta \quad (2.22)$$

This is the orthogonal projection of $\mathbf{v}$ on $\bar{\mathbf{u}}$! It is the red arrow shown in Fig. 2.2. The magnitude (length) of the projected vector is

$$|\overline{OA}| = \mathbf{v} \cdot \bar{\mathbf{u}} = \|\mathbf{v}\| \cos\theta \quad (2.23)$$

This can only happen when $\angle OAC$ is a right angle, and hence it is an orthogonal projection.

Similarly, one can project $\mathbf{u}$ on $\mathbf{v}$, and the projected vector is the green vector shown in Fig. 2.2. This time, the magnitude of the projected vector is

$$|\overline{OD}| = \mathbf{u} \cdot \bar{\mathbf{v}} = \|\mathbf{u}\| \cos\theta \quad (2.24)$$

This implies that $\angle ODB$ is a right angle, and $\mathbf{u}$ is orthogonally projected on $\mathbf{v}$.

Dot products will be frequently used for **orthogonal projections** in this book.

2.3.5 *Derivation of the law of cosines*

You may have studied the law of cosines for arbitrary triangles in high school. Using the dot product, we can easily derive this law.

Consider two given vectors $\mathbf{u}$ and $\mathbf{v}$ with an angle θ in between. Let $\mathbf{w}$ be the difference vector: $\mathbf{w} = \mathbf{u} - \mathbf{v}$. Thus, $\mathbf{u}$, $\mathbf{v}$, and $\mathbf{w}$ form a triangle. If we want to calculate the length of $\mathbf{w}$, we can use its dot product and the standard arithmetic operations:

$$\begin{aligned} \mathbf{w} \cdot \mathbf{w} &= (\mathbf{u} - \mathbf{v}) \cdot (\mathbf{u} - \mathbf{v}) = \mathbf{u} \cdot \mathbf{u} - \mathbf{u} \cdot \mathbf{v} - \mathbf{v} \cdot \mathbf{u} + \mathbf{v} \cdot \mathbf{v} \\ &= \|\mathbf{u}\|^2 - \mathbf{u} \cdot \mathbf{v} - \mathbf{u} \cdot \mathbf{v} + \|\mathbf{v}\|^2 = \|\mathbf{u}\|^2 - 2\mathbf{u} \cdot \mathbf{v} + \|\mathbf{v}\|^2 \end{aligned} \quad (2.25)$$

which gives the law of cosines:

$$\|\mathbf{w}\|^2 = \|\mathbf{u}\|^2 + \|\mathbf{v}\|^2 - 2\|\mathbf{u}\|\|\mathbf{v}\| \cos \theta \quad (2.26)$$

2.3.6 *Power, energy, and work-done*

In science and engineering, an inner product is often the outcome of a pair of complementary vectors. For example:

- The **power** of an object is the inner product of the force vector and the velocity vector that the object possesses.
- The **angular momentum** is the inner product of moment of inertia and the angular velocity.
- The **traction** on the boundary of a solid is the inner product of the stress vector and the outward normal vector of the boundary.
- The **strain energy** (density) in a stressed solid is the inner product of the stress and strain tensors.
- The **work-done** on an object is the inner product of the force vector applied on it and the displacement vector of the object [3].

2.3.7 *Python examples*

2.3.7.1 *Dot product of an arbitrary pair of vectors*

```
1  # dot product of two arbitrary vectors
2  u = np.arange(1,6)                          # numpy 1D array
3  print(f"Vector u: {u}; u·u = {np.dot(u,u)}, norm(u) = {lg.norm(u)**2}")
4
5  v = np.ones(len(u)) * 2                     # another numpy 1D array of same length
6  print(f"Vector v: {v}")
7
8  udv, vdu = np.dot(u, v), np.dot(v, u)
9  print(f"Symmetry: u·v = {udv}; v·u = {vdu}")
10 print(f"Alternative calculation: u·v = {u.dot(v)}; v·u = {u.dot(v)}")
```

```
Vector u: [1 2 3 4 5]; u·u = 55, norm(u) = 55.0
Vector v: [2. 2. 2. 2. 2.]
Symmetry: u·v = 30.0; v·u = 30.0
Alternative calculation: u·v = 30.0; v·u = 30.0
```

2.3.7.2 Dot product of a pair of parallel vectors

```
1  # dot product of two parallel vectors
2  u = np.arange(1,6)                          # numpy 1D array
3  v = u * 2                                   # another but parallel array
4  print(f"Vector u: {u}; Vector v parallel u: {v}")
5
6  udv = np.dot(u, v)
7  print(f"u·v = {udv}; norm(u)*norm(v) = {lg.norm(u)*lg.norm(v)}")
```

Vector u: [1 2 3 4 5]; Vector v parallel u: [2 4 6 8 10]
u·v = 110; norm(u)*norm(v) = 110.0

2.3.7.3 Dot product of two orthogonal vectors

```
1  # dot product of two orthogonal vectors
2  u = np.arange(1,3)                          # numpy 1D array
3  v = np.array([-u[1], u[0]])                 # make v that is orthogonal to u
4
5  # create an orthogonal vector in higher dimension:
6  #v = np.random.randn(len(u))    # create a random vector of same length
7  #v-= v.dot(u)*u/lg.norm(u)**2                # make it orthogonal
8
9  print(f"Vector u: {u}; \nIts orthogonal vector: {v}")
10 print(f"u·v = {np.dot(u, v)}")              # check orthogonality
```

Vector u: [1 2];
Its orthogonal vector: [-2 1]
u·v = 0

2.3.7.4 Direction cosine of two vectors

```
1  # Direction cosine of two orthogonal vectors
2  np.random.seed(8)
3  n = 6
4  u = np.random.randn(6)       # create a random vector in numpy 1D array
5  v = np.random.randn(6)                      # create another
6
7  u_norm = lg.norm(u)
8  v_norm = lg.norm(v)
9
10 cos θ = v.dot(u)/(u_norm*v_norm)            # The direction cosine
11 θ = np.arccos(cosθ)                         # the angle in between
12
13 print(f"The direction cosine of u and v = {cos θ}")
14 print(f"The angle between u and v = {θ}rad, {np.rad2deg(θ)}°")
```

The direction cosine of u and v = 0.43319936686066546
The angle between u and v = 1.1227568333736386rad, 64.3292279717825°

2.3.8 Types of norms of a vector

Without detailed definition and discussion, readers may simply take a look at the means to compute other types of norms of a vector. Readers may encounter these norms in the literature.

```
1  # compute other norms of a vector
2  u = np.arange(1,6)            # numpy 1D array
3  p = 3                         # for Lp-norms, one may change it
4
5  print(f"Vector u: {u}; p={p}")
6  print(f"Infinity norms: ‖u‖∞={lg.norm(u,np.inf)}; "
7        f" ‖u‖-∞={lg.norm(u,-np.inf)}")
8
9  print(f"L2 norm: ‖u‖2={lg.norm(u)}; ‖u‖2={lg.norm(u,ord=2)}")
10 print(f"L1 norm: ‖u‖1={lg.norm(u,ord=1)}")
11 print(f"Lp norm: ‖u‖p={lg.norm(u, ord=p)}")
```

Vector u: [1 2 3 4 5]; p=3
Infinity norms: ‖u‖∞=5.0; ‖u‖-∞=1.0
L2 norm: ‖u‖2=7.416198487095663; ‖u‖2=7.416198487095663
L1 norm: ‖u‖1=15.0
Lp norm: ‖u‖p=6.082201995573399

2.3.9 Direction cosine of an arbitrary vector

Consider an arbitrary vector **v** defined in a Cartesian coordinate. It is from point A with coordinates (x_A, y_A, z_A) to point B with coordinates (x_B, y_B, z_B) in three-dimensional (3D) coordinate space, as shown in Fig. 2.3.

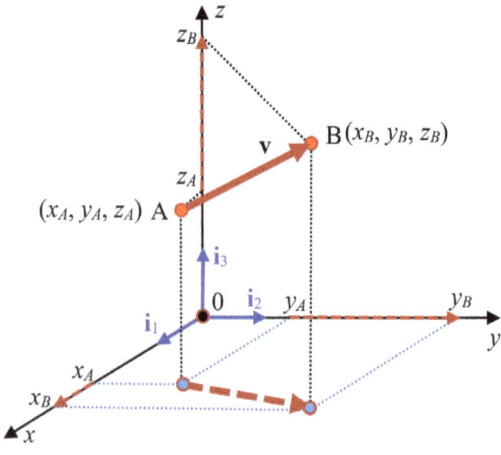

Figure 2.3. An arbitrary vector **v** in 3D Cartesian coordinate space.

Using Eq. (2.15), the direction cosine of vector **v** and the x-axis can be computed using

$$n_1 = \cos(\mathbf{v}, x) = \frac{\mathbf{v}}{\|\mathbf{v}\|} \cdot \mathbf{i}_1 = \frac{1}{\|\mathbf{v}\|} \underbrace{\mathbf{v} \cdot \mathbf{i}_1}_{x_B - x_A} = \frac{x_B - x_A}{\|\mathbf{v}\|} \qquad (2.27)$$

Here, we used the fact that $\mathbf{v} \cdot \mathbf{i}_1$ is the orthogonal projection of **v** on $\mathbf{i}_1$, which gives $x_B - x_A$.

The same applies to the direction cosines of **v** with the y-axis and **v** with the z-axis. We thus obtain in summary:

$$\boxed{\begin{aligned} n_1 &= \cos(\mathbf{v}, x) = \frac{x_B - x_A}{\|\mathbf{v}\|} \\ n_2 &= \cos(\mathbf{v}, y) = \frac{y_B - y_A}{\|\mathbf{v}\|} \\ n_3 &= \cos(\mathbf{v}, z) = \frac{z_B - z_A}{\|\mathbf{v}\|} \end{aligned}} \qquad (2.28)$$

The norm of **v** is computed using

$$\|\mathbf{v}\| = \sqrt{(x_B - x_A)^2 + (y_B - y_A)^2 + (z_B - z_A)^2} \qquad (2.29)$$

For 2D problems, say in the x–y plane, we need to use the first two equations given above. These formulas will be frequently used in this book.

2.4 Outer product of two vectors, producing a matrix

2.4.1 Definition of outer product

An inner product of two vectors produces a scalar, as we have seen in the previous section. In contrast, the outer product of two vectors produces a matrix. It has applications in linear algebra, particularity in matrix operations. In this book, we will use it to generate the so-called displacement gradient matrix that leads to strain–displacement relations.

Let two vectors $\mathbf{u} \in \mathbb{R}^m$ and $\mathbf{v} \in \mathbb{R}^n$ be

$$\begin{aligned} \mathbf{u} &= \begin{bmatrix} u_1 & u_2 & \cdots & u_m \end{bmatrix}^\top \\ \mathbf{v} &= \begin{bmatrix} v_1 & v_2 & \cdots & v_n \end{bmatrix}^\top \end{aligned} \qquad (2.30)$$

The outer product $\mathbf{u} \otimes \mathbf{v}$ becomes a matrix in $\mathbb{R}^{m \times n}$:

$$
\underset{m \times n}{\mathbf{A}} = \underset{m \times 1}{\mathbf{u}} \otimes \underset{n \times 1}{\mathbf{v}}
$$

$$
= \begin{bmatrix} u_1 \\ u_2 \\ \vdots \\ u_m \end{bmatrix} \begin{bmatrix} v_1 & v_2 & \cdots & v_n \end{bmatrix} = \begin{bmatrix} u_1 v_1 & u_1 v_2 & \cdots & u_1 v_n \\ u_2 v_1 & u_2 v_2 & \cdots & u_2 v_n \\ \vdots & \vdots & \ddots & \vdots \\ u_m v_1 & u_m v_2 & \cdots & u_m v_n \end{bmatrix} \quad (2.31)
$$

2.4.2 Rank-one matrix

The ijth element in $\mathbf{A}$ is $u_i v_j$. The outer product is same as the matrix multiplication of $\mathbf{uv}^\top$. Thus, the shapes of $\mathbf{u}$ and $\mathbf{v}$ are always compatible for this operation. This type of matrix is known as a **rank-one matrix** because it has essentially one independent row or column.

Now, if we use the SBVs to write vectors $\mathbf{u}$ and $\mathbf{v}$, as in Eq. (2.8), we shall have the following:

$$
\begin{aligned}
\mathbf{u} &= u_1\, \mathbf{i}_1 + u_2\, \mathbf{i}_2 + \cdots + u_m\, \mathbf{i}_m \\
\mathbf{v} &= v_1\, \mathbf{i}_1 + v_2\, \mathbf{i}_2 + \cdots + v_n\, \mathbf{i}_n
\end{aligned}
\quad (2.32)
$$

The outer product $\mathbf{u} \otimes \mathbf{v}$ can be re-written as follows:

$$
\underset{m \times n}{\mathbf{A}} = \underset{m \times 1}{\mathbf{u}} \otimes \underset{n \times 1}{\mathbf{v}} = u_1 v_1 \underbrace{\mathbf{i}_1 \otimes \mathbf{i}_1}_{\mathbf{I}_{11}} + u_1 v_2 \underbrace{\mathbf{i}_1 \otimes \mathbf{i}_2}_{\mathbf{I}_{12}}
$$
$$
+ u_2 v_1 \underbrace{\mathbf{i}_2 \otimes \mathbf{i}_1}_{\mathbf{I}_{21}} + \cdots + u_m v_n \underbrace{\mathbf{i}_m \otimes \mathbf{i}_n}_{\mathbf{I}_{mn}}
\quad (2.33)
$$

where $\mathbf{I}_{ij}$ is called standard basis matrix (SBM), given by the outer products of the SBVs:

$$
\underset{m \times n}{\mathbf{I}_{ij}} = \underset{m \times n}{\mathbf{i}_i \otimes \mathbf{i}_j}; \quad i = 1, 2, \ldots, m;\quad j = 1, 2, \ldots, n \quad (2.34)
$$

Any matrix in $\mathbb{R}^{m \times n}$ can be write as a linear combination of $\mathbf{I}_{ij}$.

2.4.3 Python example

Let us first look at an example.

```
1  u = np.arange(1,4)                                    # numpy 1D array
2  v = np.ones(5) * 2                                    # numpy 1D array
3  print(f"Vector u: {u}; Vector v: {v}")
4
5  A, B = np.outer(u, v), np.outer(v, u)
6  print(f"v ⊗ u: \n{B}; its rank = {lg.matrix_rank(B)}")
7  print(f"u ⊗ v: \n{A}; its rank = {lg.matrix_rank(A)}")
8
9  uc = u.reshape(u.shape[0],1)                          # change u to a column-vector
10 vc = v.reshape(v.shape[0],1)                          # change v to a column-vector
11 print(f"uc@vc: \n{uc@vc.T}")                          # numpy matrix multiplication
```

```
Vector u: [1 2 3]; Vector v: [2. 2. 2. 2. 2.]
v ⊗ u:
[[2. 4. 6.]
 [2. 4. 6.]
 [2. 4. 6.]
 [2. 4. 6.]
 [2. 4. 6.]]; its rank = 1
u ⊗ v:
[[2. 2. 2. 2. 2.]
 [4. 4. 4. 4. 4.]
 [6. 6. 6. 6. 6.]]; its rank = 1
uc@vc:
[[2. 2. 2. 2. 2.]
 [4. 4. 4. 4. 4.]
 [6. 6. 6. 6. 6.]]
```

2.5 Cross product of vectors, producing a vector

A dot product produces a scalar, and an outer product produces a matrix. Is there a vector operation that produces a vector? Yes. A cross product does this. It is also called the vector product. In this book, it is used to produce the curl of stress-vector functions leading to the torque on a torsional bar.

An excellent wiki page (https://en.wikipedia.org/wiki/Cross_product) is available on this topic. This section briefly outlines the major points with demonstrations in Python codes. Our notation follows largely that in the wiki page so that readers can make easy connections.

2.5.1 Definition

Given two linearly independent vectors **a** and **b** both in $\mathbb{R}^3$, the cross product is denoted as **a** × **b** and often reads "a crosses b". The outcome is a vector **c** also in $\mathbb{R}^3$ perpendicular (or orthogonal) to both **a** and **b**. Therefore, it is normal to the plane of $\mathbb{R}^2$ spanned by **a** and **b** in $\mathbb{R}^3$. Together with **a** and **b**, it spans a space of $\mathbb{R}^3$, as shown in Fig. 2.4.

The direction of the cross product has only two choices: either positive or negative. This is why the cross product is called a pseudo vector. The positive direction of the cross product is defined by the so-called right-hand rule, as shown on the right in Fig. 2.4. When **b** crosses **a**, the resultant vector is in the negative (downwards) direction. The right-hand rule is followed throughout this book.

2.5.2 Formulation

When these two vectors are not linearly independent, meaning they are in-line or collinear, their cross product is a zero vector. This is also true when either of these two vectors is a zero vector.

The magnitude of the cross product equals the area of the parallelogram formed by **a** and **b**, as shown in Fig. 2.5.

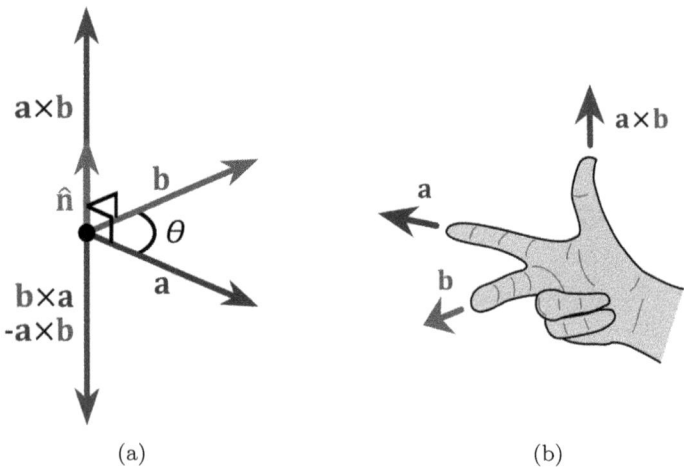

(a) (b)

Figure 2.4. (a) Cross product of two vectors. (b) The right-hand rule determines the positive direction of the cross product of two vectors.
Source: https://en.wikipedia.org/wiki/Cross_product#/media/File:Right hand rule cross_product.svg, made by Acdx, under the CC BY-SA 3.0 license.

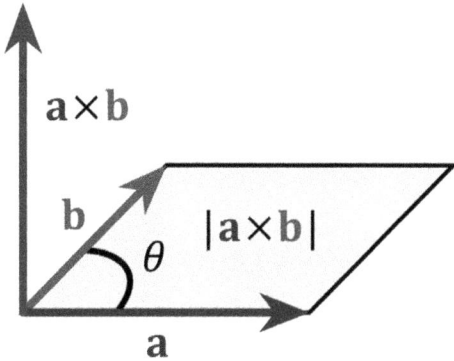

Figure 2.5. Cross product of two vectors forms a parallelogram. The amplitude of the cross product is the area of the parallelogram.
Source: (https://commons.wikimedia.org/wiki/File:Cross product parallelogram.svg).

The formula for the cross product definition is given by

$$\mathbf{c} = \mathbf{a} \times \mathbf{b} = \underbrace{\|\mathbf{a}\| \, \|\mathbf{b}\| \sin(\theta)}_{\|\mathbf{c}\|} \, \mathbf{n} \tag{2.35}$$

where $0° \leq \theta \leq 180°$ is the angle measured from $\mathbf{a}$ to $\mathbf{b}$, and $\mathbf{n}$ is the unit vector perpendicular to the plane spanned by $\mathbf{a}$ and $\mathbf{b}$ following the right-hand rule. As seen in Eq. (2.35), the positivity of the direction of $\mathbf{c}$ is controlled by the sine function, which is an odd function. Therefore, if $\mathbf{b}$ crosses $\mathbf{a}$, θ will be negative, and the direction will reverse. In other words, cross product is anti-commutative.

2.5.3 *Volume of a parallelepiped*

The volume V of a parallelepiped formed by three vectors $\mathbf{a}$, $\mathbf{b}$, and $\mathbf{c}$ as its edges can be computed using a combination of a cross product and a dot product, called **scalar triplet product**, as shown in Fig. 2.6.

The triplet product can be done in any of the three orders, and they all gives a scalar:

$$\mathbf{a} \cdot (\mathbf{b} \times \mathbf{c}) = \mathbf{b} \cdot (\mathbf{c} \times \mathbf{a}) = \mathbf{c} \cdot (\mathbf{a} \times \mathbf{b}) \tag{2.36}$$

Note that the scalar triplet product can be negative, so the volume of the parallelepiped should be the absolute value of the scalar:

$$V = |\mathbf{a} \cdot (\mathbf{b} \times \mathbf{c})| \tag{2.37}$$

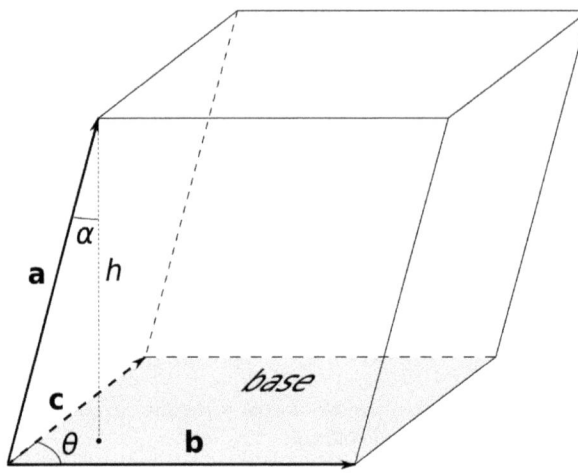

Figure 2.6. A parallelepiped defined using three vectors. Image from Wikimedia Commons in public domain, by Jitse Niesen.

2.5.4 Differences between the dot product and cross product

1. The magnitude of the cross product is computed using the sine of the angle between two vectors. Thus, the cross product can be a measure of perpendicularity of two vectors. In contrast, the dot product is a measure of parallelism, as mentioned earlier.
2. Given two vectors with unit length (unit vectors), if they are perpendicular, the cross product has a magnitude of 1, and if the two are parallel, the magnitude is zero. Conversely, the dot product of two perpendicular unit vectors is zero and is 1 if they are parallel.
3. The magnitude of the cross product of two unit vectors gives the sine of the angle between two vectors θ, which is always positive by definition (θ is 0°~180° and measured counterclockwise). In contrast, the dot product of two unit vectors gives the cosine of θ, and hence θ can be positive or negative.

2.5.5 Derivation of formulas

Since the cross product involves direction change, the computation becomes more complicated. For easy understanding, we first examine the cross products of the SBVs. Let us consider 3D spaces, and denote these three SBVs

Preliminaries: Number, Vector, Tensor, and Functions

as $\mathbf{i}$, $\mathbf{j}$, and $\mathbf{k}$. Based on the right-hand rule, we have

$$\mathbf{i} \times \mathbf{j} = \mathbf{k}, \quad \mathbf{j} \times \mathbf{i} = -\mathbf{k}$$
$$\mathbf{j} \times \mathbf{k} = \mathbf{i}, \quad \mathbf{k} \times \mathbf{j} = -\mathbf{i} \tag{2.38}$$
$$\mathbf{k} \times \mathbf{i} = \mathbf{j}, \quad \mathbf{i} \times \mathbf{k} = -\mathbf{j}$$

Note the negative sign resulting from anti-commutativity. In addition, because of collinearity, we have

$$\mathbf{i} \times \mathbf{i} = \mathbf{j} \times \mathbf{j} = \mathbf{k} \times \mathbf{k} = \mathbf{0} \tag{2.39}$$

Using these SBVs, the cross product of any two vectors $\mathbf{a}$ and $\mathbf{b}$ can be, respectively, written as

$$\mathbf{a} = a_1 \mathbf{i} + a_2 \mathbf{j} + a_3 \mathbf{k}$$
$$\mathbf{b} = b_1 \mathbf{i} + b_2 \mathbf{j} + b_3 \mathbf{k} \tag{2.40}$$

where a_i, a_j, and a_k are the elements of $\mathbf{a}$ and b_i, b_j, and b_k are the elements of $\mathbf{b}$, respectively, on its corresponding SBVs.

Now, following the standard arithmetic operations and the definition of cross products, we have

$$\begin{aligned}\mathbf{c} = \mathbf{a} \times \mathbf{b} &= (a_1 \mathbf{i} + a_2 \mathbf{j} + a_3 \mathbf{k}) \times (b_1 \mathbf{i} + b_2 \mathbf{j} + b_3 \mathbf{k}) \\ &= a_1 b_1 (\mathbf{i} \times \mathbf{i}) + a_1 b_2 (\mathbf{i} \times \mathbf{j}) + a_1 b_3 (\mathbf{i} \times \mathbf{k}) \\ &\quad + a_2 b_1 (\mathbf{j} \times \mathbf{i}) + a_2 b_2 (\mathbf{j} \times \mathbf{j}) + a_2 b_3 (\mathbf{j} \times \mathbf{k}) \\ &\quad + a_3 b_1 (\mathbf{k} \times \mathbf{i}) + a_3 b_2 (\mathbf{k} \times \mathbf{j}) + a_3 b_3 (\mathbf{k} \times \mathbf{k}) \end{aligned} \tag{2.41}$$

Using Eqs. (2.38) and (2.39), the foregoing equation becomes

$$\begin{aligned}\mathbf{c} = \mathbf{a} \times \mathbf{b} &= a_1 b_1 \mathbf{0} + a_1 b_2 \mathbf{k} - a_1 b_3 \mathbf{j} \\ &\quad - a_2 b_1 \mathbf{k} + a_2 b_2 \mathbf{0} + a_2 b_3 \mathbf{i} \\ &\quad + a_3 b_1 \mathbf{j} - a_3 b_2 \mathbf{i} + a_3 b_3 \mathbf{0} \\ &= \underbrace{(a_2 b_3 - a_3 b_2)}_{c_1} \mathbf{i} + \underbrace{(a_3 b_1 - a_1 b_3)}_{c_2} \mathbf{j} + \underbrace{(a_1 b_2 - a_2 b_1)}_{c_3} \mathbf{k} \\ &= c_1 \mathbf{i} + c_2 \mathbf{j} + c_3 \mathbf{k}\end{aligned} \tag{2.42}$$

We observe a nice duality formula for computing the elements of vector $\mathbf{c}$ if we re-denote $c_1 = c_{23}$, $c_2 = c_{31}$, $c_3 = c_{12}$:

$$c_{ij} = a_i b_j - a_j b_i \tag{2.43}$$

When $i, j = 1, 2, 3$, the above duality formula gives exactly three distinct values, which we denote as c_1, c_2, c_3 to express vector **c**.

Alternatively, **c** can be expressed in the determinant of a matrix as follows:

$$\underbrace{\mathbf{a} \times \mathbf{b}}_{\mathbf{c}} = \begin{vmatrix} \mathbf{i} & \mathbf{j} & \mathbf{k} \\ a_1 & a_2 & a_3 \\ b_1 & b_2 & b_3 \end{vmatrix}$$

$$= \underbrace{\begin{vmatrix} a_2 & a_3 \\ b_2 & b_3 \end{vmatrix}}_{c_1} \mathbf{i} - \underbrace{\begin{vmatrix} a_1 & a_3 \\ b_1 & b_3 \end{vmatrix}}_{c_2} \mathbf{j} + \underbrace{\begin{vmatrix} a_1 & a_2 \\ b_1 & b_2 \end{vmatrix}}_{c_3} \mathbf{k}$$

(2.44)

Moreover, we can use the following matrix-vector formulation in computation.

$$\underbrace{\mathbf{a} \times \mathbf{b}}_{\mathbf{c}} = \begin{bmatrix} 0 & -a_3 & a_2 \\ a_3 & 0 & -a_1 \\ -a_2 & a_1 & 0 \end{bmatrix} \begin{bmatrix} b_1 \\ b_2 \\ b_3 \end{bmatrix}$$

(2.45)

where the matrix is anti-symmetric because the cross product is anti-commutative.

The cross product has applications in various areas, including computational geometry, evaluation of angular momentum and torque, and calculation of Lorentz forces, calculation of torques by forces, just to name a few. It is often used to determine the direction normal to any pair of vectors, which is needed in many operations, such as coordinate transformation in the finite element methods [3] for stress analysis in structures. Interested readers may refer to the wiki page (https://en.wikipedia.org/wiki/Cross_product) and the references therein for more detailed formulations, properties, history, and applications.

Note that we did not discuss about cross products in higher dimensions. This is because it would get into the concept of **exterior algebra** or **tensor algebra**. From Eq. (2.43), we see that our formula works just right for 3D when the dimension gets higher, 4D, for example, i and j shall vary from 1 to 4. In such cases, the duality formula would produce a lot more terms than 4, implying it becomes a tensor for higher dimensions. Discussion on exterior or tensor algebra is beyond the scope of this book. We thus stop here.

2.5.6 Examples of computing cross products

We now write some simple codes to examine cross products.

```
1  a = [2., 1, 3]
2  b = [5., 3, 2]
3
4  c1 =  a[1]*b[2]-a[2]*b[1]
5  c2 = -(a[0]*b[2]-a[2]*b[0])
6  c3 =  a[0]*b[1]-a[1]*b[0]
7  print(f"elements of vector c = a x b: c1={c1}, c2={c2}, c3={c3}")
8
9  c_ab = np.cross(a, b)
10 c_ba = np.cross(b, a)
11 print(f"c = a x b: {c_ab}\nc = b x a: {c_ba}")
12 print(f"c dot a: {c_ab.dot(a):.4f}, c dot b: {c_ab.dot(b):.4f}")
```

```
elements of vector c = a x b: c1=-7, c2=11.0, c3=1.0
c = a x b: [-7. 11.   1.]
c = b x a: [ 7. -11.  -1.]
c dot a: 0.0000, c dot b: 0.0000
```

We observe all the major properties: 1. **a** × **b** is negative **b** × **a**; 2. **c** is normal to both **a** and **b** because these dot products vanish.

```
1  # matrix-vector formulation:
2  A = np.array([[    0, -a[2],  a[1]],
3                 [ a[2],     0, -a[0]],
4                 [-a[1],  a[0],     0]])
5
6  print(f"elements of vector c = a x b = {A@b}")
```

```
elements of vector c = a x b = [-7. 11.   1.]
```

```
1  θ = np.arccos(np.dot(a/lg.norm(a), b/lg.norm(b)))  # θ can be negative
2  Area = abs(lg.norm(a)*lg.norm(b)*np.sin(θ))         # thus, take abs()
3
4  print(f"Area of the parallelogram via definition: {Area:.4f}")
5  print(f"Area via lg.norm: {lg.norm(c_ab):.4f}")
```

```
Area of the parallelogram via definition: 13.0767
Area via lg.norm: 13.0767
```

We confirm another property that the norm of **c** is the area of the parallelogram. We can do the same using randomly generated vectors.

```
1   n = 3
2   np.random.seed(8)
3   a = np.random.randn(n)           # randomly generated vector
4   b = np.random.randn(n)
5
6   c_ab, c_ba = np.cross(a, b), np.cross(b, a)
7   print(f"c = a x b: {c_ab}\nc = b x a: {c_ba}")
8   print(f"c dot a: {c_ab.dot(a):.4f}, c dot b: {c_ab.dot(b):.4f}")
9
10  theta = np.arccos(np.dot(a/lg.norm(a), b/lg.norm(b)))
11  Area = abs(lg.norm(a)*lg.norm(b)*np.sin(theta))
12  print(f"Area of the parallelogram via definition: {Area:.4f}")
13  print(f"Area via lg.norm: {lg.norm(c_ab):.4f}")
```

```
c = a x b: [-1.8414  2.4794  1.3034]
c = b x a: [ 1.8414 -2.4794 -1.3034]
c dot a: 0.0000, c dot b: -0.0000
Area of the parallelogram via definition: 3.3522
Area via lg.norm: 3.3522
```

For these randomly generated vectors, we observe again the same properties.

2.6 Examples of vector operations and visualization

Now, we present examples to demonstrate vector operations using Python code. In order to have the results easily visualizable, we study vectors in 2D space.

For vectors in 2D space $\mathbb{R}^2$, we shall have two basis vectors which can be represented in graphs. In this case, the standard basis vectors, or SBVs, become

$$\mathbf{i}_1 = \begin{bmatrix} 1 \\ 0 \end{bmatrix}; \quad \mathbf{i}_2 = \begin{bmatrix} 0 \\ 1 \end{bmatrix} \tag{2.46}$$

Any vector in $\mathbb{R}^2$ can now be expressed using these basis vectors, as shown in the following example:

$$\begin{bmatrix} v_1 \\ v_2 \end{bmatrix} = v_1 \begin{bmatrix} 1 \\ 0 \end{bmatrix} + v_2 \begin{bmatrix} 0 \\ 1 \end{bmatrix} = v_1 \mathbf{i}_1 + v_2 \mathbf{i}_2 \tag{2.47}$$

Since v_1 and v_2 are scalar numbers in $\mathbb{R}$, vector $\begin{bmatrix} v_1 \\ v_2 \end{bmatrix}$ lives in $\mathbb{R}^2$. This can be demonstrated using the following codes.

2.6.1 *Creation of vectors using the standard basis vectors, and scaling vectors*

```
1  o1 = np.array([0,0])
2  i_1 = np.array([1.,0])                            # basis vector i1
3  i_2 = np.array([0.,1])                            # basis vector i2
4
5  v = np.array([3., 4])                             # any given vector
6  print('elements:', v[0], ' and ', v[1])           # vector elements
7  v1 = v[0]*i_1 + v[1]*i_2
8  print('v=', v, '; v_formed by two SBVs = v1', v1)
9  print('Norm of i_1=', lg.norm(i_1), ' Norm of i_2=',
10        lg.norm(i_2), ' Norm of v=', lg.norm(v))
11
12 print('L1-norm of v1=', lg.norm(v1,1), ' L2-norm of v1=',
13       lg.norm(v1), ' Inf-norm of v1=', lg.norm(v1,np.inf))
14
15 print('Normalized i_1=', i_1/lg.norm(i_1), ' Normalized i_2=',
16       i_2/lg.norm(i_2),' Normalized v=', v/lg.norm(v))
17
18 v2 = np.array([-2, 3])
19 v4 = -v2
20 v3 = v1/lg.norm(v1)
21 v5, v6= 0.5 *v1, -0.5 *v2
22 print('v2=',v2,' -v2=',v4,' .5v1=',v5,' -.5v2=',v6)
23
24 fig, axis = plt.subplots(1,2,figsize=(6.5,3))#3.8,1.8))#
25 Lb1 = drawArrow(axis,o1,i_1,color='r',iax=0,label='Basis')
26 Lb2 = drawArrow(axis,o1,i_2,color='r',iax=0,label='None')
27 Lb3 = drawArrow(axis,o1,v1,color='b',iax=0,label='v1=$2i_1+5i_2$')
28 Lb4 = drawArrow(axis,o1,v3,color='k',iax=0,label='Normalized v1')
29 Lb5 = drawArrow(axis,o1,v2,color='c',iax=0,label='v2')
30 Lb6 = drawArrow(axis,o1,v4,color='m',iax=0,label='$-$v2')
31 axis[0].legend(handles=[Lb1,Lb3,Lb4,Lb5,Lb6], loc='lower left')
32 plotfig(axis,iax=0, title='SBVs and scaled vectors',xylim=5)
33
34 Lb1 = drawArrow(axis,o1,i_1,color='r',iax=1,label='Basis')
35 Lb2 = drawArrow(axis,o1,i_2,color='r',iax=1,label='None')
36 Lb3 = drawArrow(axis,o1,v1,color='b',iax=1,label='v1=$2i_1+5i_2$')
37 Lb4 = drawArrow(axis,o1,v2,color='c',iax=1,label='v2')
38 Lb5 = drawArrow(axis,o1,v5,color='k',iax=1,label='0.5*v1')
39 Lb6 = drawArrow(axis,o1,v6,color='m',iax=1,label='$-$0.5*v2')
40 axis[1].legend(handles=[Lb1,Lb3,Lb4,Lb5,Lb6], loc='lower left')
41 plotfig(axis,iax=1, title='SBVs and scaled vectors',xylim=5)
42 plt.savefig('images/ScaledVectors.png', dpi=500)
43 plt.show()
```

```
elements: 3.0 and  4.0
v= [3. 4.] ; v_formed by two SBVs = v1 [3. 4.]
Norm of i_1= 1.0  Norm of i_2= 1.0  Norm of v= 5.0
L1-norm of v1= 7.0  L2-norm of v1= 5.0  Inf-norm of v1= 4.0
Normalized i_1= [1. 0.]  Normalized i_2= [0. 1.]  Normalized v= [0.6 0.8]
v2= [-2  3]  -v2= [ 2 -3]  .5v1= [1.5 2. ]  -.5v2= [ 1.  -1.5]
```

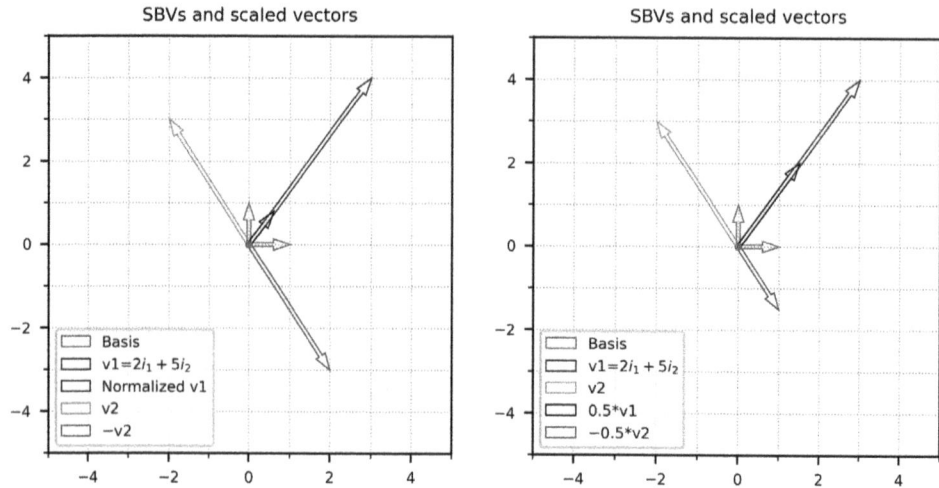

Figure 2.7. Vectors generated in various means.

2.6.2 Addition of vectors

```
1  v1 = np.array([3,-1])
2  v2 = np.array([1, 4])
3  v3 = v1 + v2
4  v4 = v1 - v2
5  print('v1=',v1,' v2=',v2,' v1+v2=',v3,' v1-v2=',v4)
6
7  fig, axis = plt.subplots(1,2,figsize=(6.5,3))#(3.8,1.8))#(10.5,5))
8
9  Lb1 = drawArrow(axis,o1,v1,color='r',iax=0,label='v1')
10 Lb2 = drawArrow(axis,o1,v2,color='b',iax=0,label='v2')
11 Lb3 = drawArrow(axis,o1,v3,color='c',iax=0,label='v3=v1+v2')
12 axis[0].legend(handles=[Lb1,Lb2,Lb3], loc='upper left')
13 plotfig(axis,iax=0, title='Addition of two vectors',xylim = 5)
14
15 Lb1 = drawArrow(axis,o1,v1,color='r',iax=1,label='v1')
16 Lb2 = drawArrow(axis,o1,v2,color='b',iax=1,label='v2')
17 Lb3 = drawArrow(axis,o1,v4,color='c',iax=1,label='v4=v1-v2')
18 axis[1].legend(handles=[Lb1,Lb2,Lb3], loc='upper left')
19 plotfig(axis,iax=1, title='Subtraction of two vectors',xylim=5)
20 plt.savefig('images/AdditionOfVectors.png', dpi=500)
21 plt.show()
```

v1= [3 -1]   v2= [1 4]   v1+v2= [4 3]   v1-v2= [2 -5]

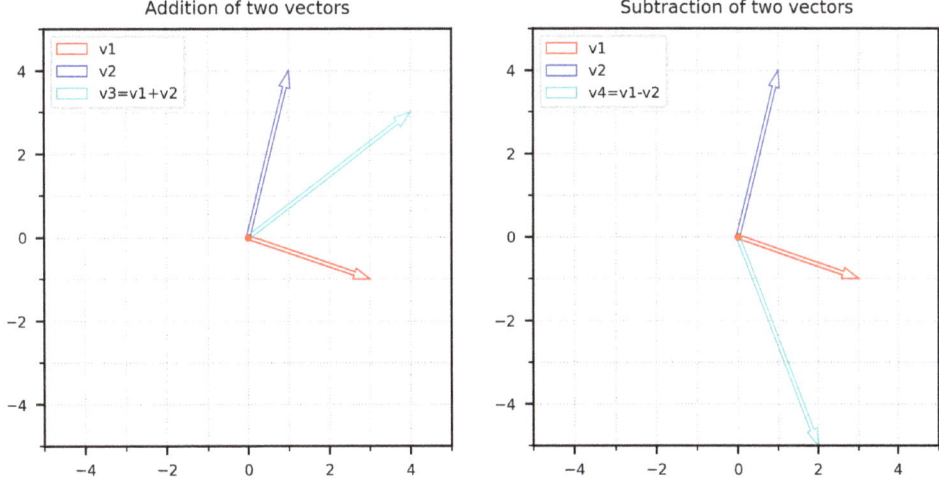

Figure 2.8. Vector additions.

2.6.3 Element-wise multiplication and dot product of two vectors

```
1  # Define two pairs of vectors of the same shape
2  v1 = np.array([3.,-1])
3  v2 = np.array([1., 4])
4  v3 = v1*v2
5
6  print('v1=',v1,' v2=',v2,' v1+v2=',v3)
7  print('numpy dot product of two vectors =',np.dot(v1,v2))
8  print('sum of products of their corresponding elements=',
9        v1[0]*v2[0]+v1[1]*v2[1])
10 print('sum of elements of vector element-wise multiplication =',
11       np.sum(v1*v2))
12 print('')
13
14 v1 = np.array([3.,-1, 4, 8, 2])
15 v2 = np.array([1., 4,-3, 5, 3])
16 v3 = v1*v2
17 v4 = v1/v2
18
19 print('v1=',v1,' v2=',v2,'\n v1*v2=',v3,'\n v1/v2=',v4)
20 print('dot product of two vectors =',np.dot(v1,v2))
21 print('sum of elements of vector element-wise multiplication =',
22       np.sum(v1*v2))
23 print(f"sum of elements of vector element-wise division =",
24       f"{np.sum(v1/v2):.4f}")
```

```
v1= [ 3. -1.]   v2= [1. 4.]   v1+v2= [ 3. -4.]
numpy dot product of two vectors = -1.0
sum of products of their corresponding elements= -1.0
sum of elements of vector element-wise multiplication = -1.0

v1= [ 3. -1.  4.  8.  2.]  v2= [ 1.  4. -3.  5.  3.]
v1*v2= [  3.  -4. -12.  40.   6.]
v1/v2= [ 3.         -0.25       -1.3333  1.6         0.6667]
dot product of two vectors = 33.0
sum of elements of vector element-wise multiplication = 33.0
sum of elements of vector element-wise division = 3.6833
```

It is confirmed that the dot product is the same as the sum of the elements of the vector created via element-wise multiplication of these two vectors, which is defined in Eq. (2.14).

We write the following codes to plot the vectors produced by the element-wise multiplications of two vectors and that of element-wise divisions of two vectors.

```
1  # Define a pair of vectors in 2D space
2  v1 = np.array([3,-1])
3  v2 = np.array([1, 4])
4  v3 = v1*v2
5  v4 = v1/v2
6  print('v1=',v1,' v2=',v2,' v1*v2=',v3,' v1/v2=',v4)
7
8  fig, axis = plt.subplots(1,2,figsize=(6.5,3))#(3.8,1.8))#(10.5,5))
9
10 Lb1 = drawArrow(axis,o1,v1,color='r',iax=0,label='v1')
11 Lb2 = drawArrow(axis,o1,v2,color='b',iax=0,label='v2')
12 Lb3 = drawArrow(axis,o1,v3,color='c',iax=0,label='v3=v1*v2')
13 axis[0].legend(handles=[Lb1,Lb2,Lb3], loc='upper left')
14 plotfig(axis,iax=0, title='element-wise multiplication of\
15 two vectors',xylim = 5)
16
17 Lb1 = drawArrow(axis,o1,v1,color='r',iax=1,label='v1')
18 Lb2 = drawArrow(axis,o1,v2,color='b',iax=1,label='v2')
19 Lb3 = drawArrow(axis,o1,v4,color='c',iax=1,label='v4=v1/v2')
20 axis[1].legend(handles=[Lb1,Lb2,Lb3], loc='upper left')
21 plotfig(axis,iax=1, title='Element-wise division of\
22 two vectors',xylim = 5)
23 plt.savefig('images/Mul_DivOfVectors.png', dpi=500)
24 plt.show()
```

v1= [3 -1]   v2= [1 4]   v1*v2= [3 -4]   v1/v2= [3.   -0.25]

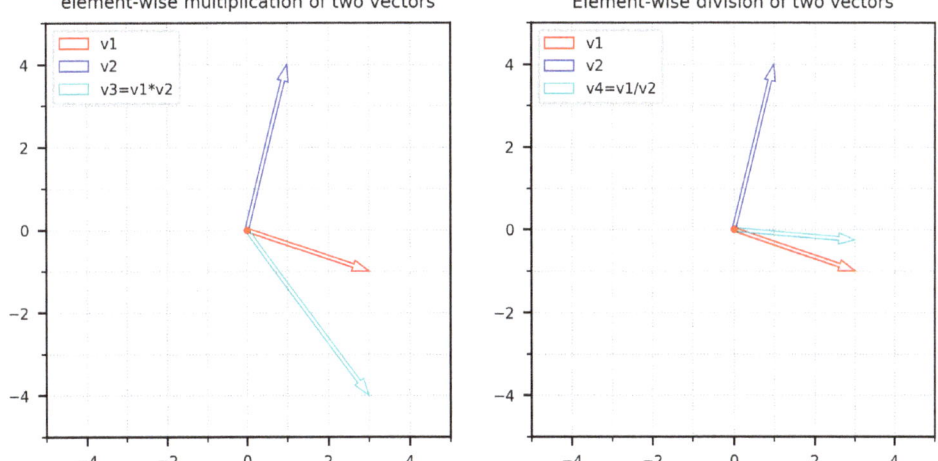

Figure 2.9. Vector multiplication and division.

2.6.4 Orthogonal vectors

As mentioned earlier, if two vectors $\mathbf{v}, \mathbf{u}$ in $\mathbb{R}^n$ are said to be orthogonal, it means that the dot product of them is zero:

$$\mathbf{v} \cdot \mathbf{u} = (v_1 u_1) + (v_2 u_2) + \cdots + (v_n u_n) = 0 \qquad (2.48)$$

The obvious example of orthogonal vectors are the set of the SBVs. Let us write a simple code to demonstrate this.

```
1  n = 4
2  I = np.eye(n)           # Create an identity matrix.
3
4  i_1 = np.hsplit(I,n)[0].reshape(n,)    #[1 0 0 0]
5  i_2 = np.hsplit(I,n)[1].reshape(n,)    #[0 1 0 0]
6  i_3 = np.hsplit(I,n)[2].reshape(n,)    #[0 0 1 0]
7  i_4 = np.hsplit(I,n)[3].reshape(n,)    #[0 0 0 1]
8
9  print(i_1, i_2, i_3, i_4)
10 print('Dot products of pairs of SBVs:\n',np.dot(i_1,i_1),
11        np.dot(i_1,i_2), np.dot(i_1,i_3), np.dot(i_2,i_4))
```

```
[1. 0. 0. 0.] [0. 1. 0. 0.] [0. 0. 1. 0.] [0. 0. 0. 1.]
Dot products of pairs of SBVs:
1.0 0.0 0.0 0.0
```

It is confirmed that the SBVs are **orthogonal basis vectors**.

Orthogonal vectors are not necessarily as "nice looking" as the SBVs. Let us generate more orthogonal vectors randomly using the following snippet.

```
1  # The orth function has already imported before from scipy.linalg
2  A = np.random.rand(5, 5)            # generate a random matrix
3  OA = sg.orth(A)                     # compute orthogonal vectors using A
4
5  print(f"Any column vector is orthogonal to the rest:\n{OA}")
6  print('Dot products of column vectors:\n',np.dot(OA[:,0],OA[:,1]),
7        np.dot(OA[:,0],OA[:,2]),np.dot(OA[:,1],OA[:,2]))
```

```
Any column vector is orthogonal to the rest:
[[-0.5411 -0.0601 -0.1721 -0.5599 -0.6004]
 [-0.3929  0.1741 -0.6865 -0.0543  0.584 ]
 [-0.2489 -0.9268 -0.0437  0.2656  0.082 ]
 [-0.5318  0.3236  0.1047  0.7295 -0.2633]
 [-0.4561  0.0498  0.6973 -0.2844  0.4715]]
Dot products of column vectors:
 1.1102230246251565e-16 -2.220446049250313e-16 -1.8735013540549517e-16
```

As shown, these five vectors look not so nice, but they are orthogonal to each other. They can be used as orthogonal basis vectors and span a 5D vector space.

For vectors in 2D vector space, we can view clearly the orthogonality of vectors.

```
1  # Define two pairs of orthogonal vectors in 2D space
2  v1, v2 = np.array([3,-1]), np.array([1,3])
3  v3, v4 = np.array([2, 3]), np.array([-3,2])
4  print('v1=',v1,' v2=',v2,' v1*v2=',v3,' v1/v2=',v4)
5  print('v1 dot v2=',np.dot(v1,v2),'   v3 dot v4=',np.dot(v1,v2))
6
7  fig, axis = plt.subplots(1,2,figsize=(6.5,3))
8
9  Lb1 = drawArrow(axis,o1,v1,color='r',iax=0,label='v1')
10 Lb2 = drawArrow(axis,o1,v2,color='b',iax=0,label='v2')
11 axis[0].legend(handles=[Lb1,Lb2], loc='upper left')
12 plotfig(axis,iax=0, title='Two orthogonal vectors',xylim = 5)
13
14 Lb1 = drawArrow(axis,o1,v3,color='r',iax=1,label='v3')
15 Lb2 = drawArrow(axis,o1,v4,color='b',iax=1,label='v4')
16 axis[1].legend(handles=[Lb1,Lb2], loc='upper left')
17 plotfig(axis,iax=1, title='Two orthogonal vectors', xylim = 5)
18 plt.savefig('images/OrthogonalVectors.png', dpi=500)
19 plt.show()
```

```
v1= [ 3 -1]  v2= [1 3]  v1*v2= [2 3]  v1/v2= [-3  2]
v1 dot v2= 0    v3 dot v4= 0
```

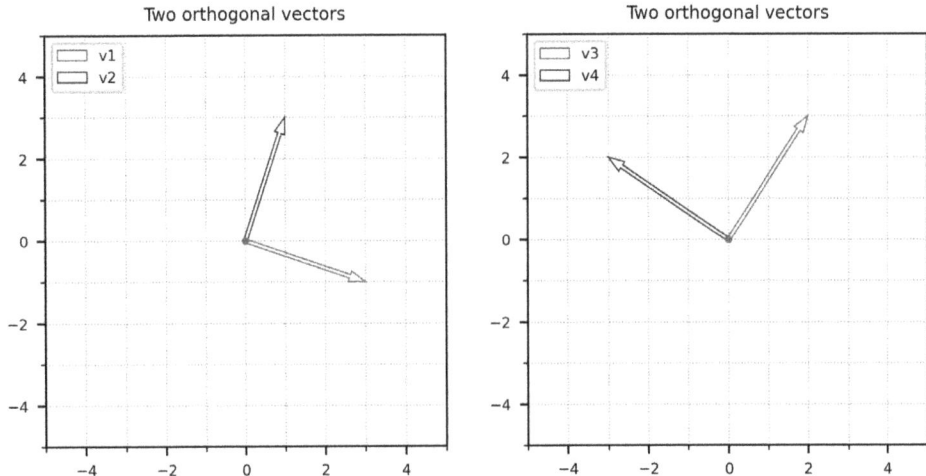

Figure 2.10. Orthogonal vectors in 2D.

2.7 Coordinate transformation

In solving a mechanics problem, we deal with objects including numbers, vectors, and stress tensors. Such an object is usually represented in reference to an orthogonal coordinate system, such as Cartesian, polar, cylindrical, or spherical. We often perform coordinate transformation, which can result in changes in the representation of the object. Coordinate transformation often refers to **rotation** or a **change** from one type of (often orthogonal) coordinate system to another type of (often orthogonal) coordinate system, for example changes from the Cartesian to polar coordinate system. Coordinate transformation of any object occurs usually at a point.

Therefore, when we say coordinate transformation, it does NOT include translation. The change of an object due to the location change is dealt with using the concept of **function** [1].

A number is a tensor of 0th order. The rule of coordinate transformation for a number is the simplest:

> A number does not change when the coordinate transforms. It does not require an index with respect to the coordinates.

This is simply because of the transformation definition: When a number is represented in a coordinate, we have only a single 1D axis, say x, in $\mathbb{R}$, as shown in Fig. 2.1. The number has a single value right on the axis of a given 1D coordinate. In a 1D space, the axis can only be translated along its axis direction and cannot have rotation (otherwise, the space becomes 2D or 3D).

The single value can still be viewed as obtained by an orthogonal projection, but the angle of the projection is zero. Any rotation (with zero angle) of the coordinate has no effect at all to any number. In other words, the transformation matrix $\mathbf{T}$ (discussed in detail later) is simply 1.

Due to the transformation invariant property of a number, it is called a **scalar**.

2.8 Vectors in mechanics

In mechanics, vectors are often encountered important objects made of scalar numbers, each of which represents a component of the vector. Typical ones are the force vector and the displacement vector.

2.8.1 *Notation for coordinate systems*

To quantify a vector in mechanics, we use an orthogonal coordinate system. A vector can then be equivalently described in a number of scalars in an array.

Consider a 3D space $\mathbb{R}^3$, in which we define a Cartesian coordinate system $\mathbf{x}$ that is orthogonal: x-, y-, and z-axis are the three mutually orthogonal axes of $\mathbf{x}$. Each of these axes has a unit vector $\mathbf{i}, \mathbf{j}$, and $\mathbf{k}$ with unit length, respectively.

In indicial rotation, these axes are denoted as x_1, x_2, and x_3, or x_i ($i = 1, 2, 3$), and its unit vectors are denoted as $\mathbf{i}_1$, $\mathbf{i}_2$, and $\mathbf{i}_3$ or $\mathbf{i}_i$ ($i = 1, 2, 3$). The unit vectors are also denoted as $\mathbf{e}_1$, $\mathbf{e}_2$, and $\mathbf{e}_3$ or $\mathbf{e}_i$ ($i = 1, 2, 3$). Indicial notation is more concise and often more convenient in derivations. Hence, we often switch to it in this book when necessary. Readers will be getting used to it after doing this for some time.

2.8.2 *Description of vectors*

Consider generally a vector in mechanics denoted as $\mathbf{v}$ in $\mathbb{R}^3$, as shown in Fig. 2.11. The vector can be a force vector, displacement vector, velocity vector, or acceleration vector.

We present $\mathbf{v}$ uniquely under the Cartesian coordinate system. We first decompose it into two vectors: a vertical component $v_3\mathbf{i}_3$ and an in-plane vector $\mathbf{v}_s$ following the orthogonal projection rule. $\mathbf{v}_s$ is then further decomposed into two vectors: $v_1\mathbf{i}_1$ and $v_2\mathbf{i}_2$ following also the orthogonal projection

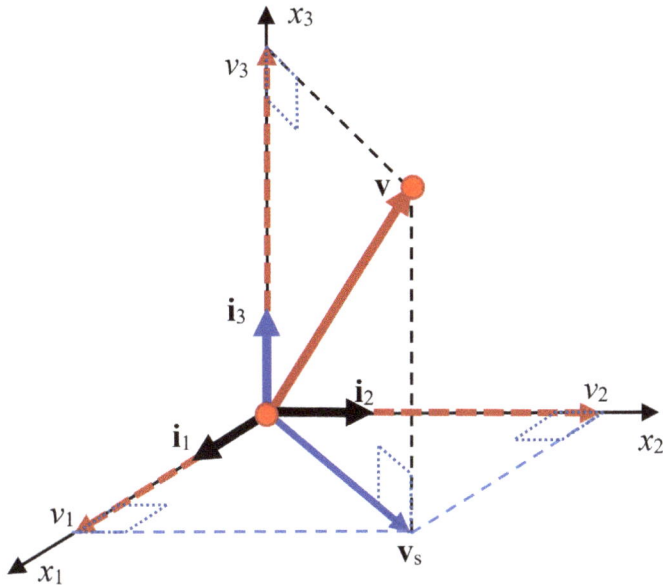

Figure 2.11. Any vector **v** defined in an orthogonal coordinate system $\mathbb{R}^3$ can be decomposed into three components: $v_1\mathbf{i}_1$, $v_2\mathbf{i}_2$, and $v_3\mathbf{i}_3$ following the orthogonal projection rule.

rule. This process results in three components of **v**: v_1, v_2, and v_3, respectively, on the x_1-, x_2-, and x_3-axis. The vector can now be written uniquely as

$$\mathbf{v} = \begin{bmatrix} v_1 \\ v_2 \\ v_3 \end{bmatrix} = \begin{bmatrix} v_1 & v_2 & v_3 \end{bmatrix}^\top \tag{2.49}$$

where $^\top$ stands for transpose. Any change to any of these three components results in another different vector.

2.8.3 Equivalence in representation, first-order tensor

From Fig. 2.11, we found the following:

1. It is clear that this representation of a vector is **unique** under a fixed coordinate, implying that any change, in length or in direction, in **v** will result in unique changes in v_1, v_2, and v_3, and vice versa, as long as the **orthogonal projection** rule is followed. This mutual unique representation ensures the **equivalent** representation, implying **v** and $\begin{bmatrix} v_1 & v_2 & v_3 \end{bmatrix}^\top$ are identical. Such an equivalence is ensured by the orthogonal projection rule in an orthogonal coordinate system.

2. When the coordinate **x** is rotated (transformed) to **x'**, vector **v** is now represented as

$$\mathbf{v} = \begin{bmatrix} v_1' \\ v_2' \\ v_3' \end{bmatrix} \qquad (2.50)$$

The values of v_1', v_2', and v_3' are different from v_1, v_2, and v_3. However, these two basic attributes: length and direction of the vector **v** remain unchanged. The vector under our transformation rule (orthogonal projection on orthogonal coordinates) is a tensor. Since we need only one index to uniquely locate its components, it is called a **first-order tensor**. Since it is in 3D, it is a 3D tensor.

2.8.4 *Tensor concept in general*

Tensor is a term we often hear about, but it confuses many, including the author in the past for a long time. If you do not have a clear idea about it, you are not alone. To put this concept in a short sentence, it states as follows.

A tensor is an object defined in a system, where the change in the values describing the object obeys a certain rule, when a transformation occurs in the system.

This is a very general definition. It implies that there are different types of tensors in different fields or systems with different transformation rules.

A more straightforward question may be: Is a square matrix a tensor? The answer can be either no or yes. If the matrix contains merely stacks of numbers, it is not. If there is a certain rule attached to it, so that the numbers in it change following a system-wide rule when a transformation occurs, it becomes a tensor. A second-order tensor can be in a square matrix form, but a square matrix is not automatically a tensor.

2.8.5 *Definition of tensor in mechanics of materials*

In mechanics of materials (or solid mechanics, continuum mechanics, or elasticity theory), all the objects we deal with are tensors of different orders. Scalars are of zeroth order, forces and displacements are vectors and are first order, stress and strain are second, and elasticity constants are fourth. These objects are defined in an orthogonal coordinate that can transform (rotate or type change). The rule for operating on these objects and to obtain the values to describe them is the **orthogonal projection** in mathematic operation and the **principle of equilibrium** in mechanics law.

The order of the tensor is identified by the number of the indexes the object carries as a subscript in indicial notation. Zeroth-order tensors (scalar numbers) carry no index, first-order (vector) carries 1, second-order (stress) carries 2, and nth n. These indexes refer to the coordinates. Thus, when the coordinate transforms, the components of the object change accordingly.

The order of the tensor is also the times of operations needed to get the coordinate-transformed values, and hence the mathematical formula of each of them has a different form, but these mathematical rules are all obtained following the **orthogonal projection** rule.

The first high-level question is may need to be answered is why we would need such a concept of tensor. This is because with it, the mechanics family can have responsible objects, and all the objects follow the same rule to behave, and thus all these objects can be evaluated consistently and correctly when a transformation takes place. This enables us to do whatever necessary transformations needed to get what we want.

The second high-level question is why we transform the coordinate. This is because we often want to find the values of the tensor objects at one particular direction, or find the critical (maximum, minimum, etc.) values and their corresponding directions.

In summary, in mechanics of materials, we deal with tensors: scalars, vectors (force vector, displacement vector, etc.), stresses, strains, elasticity, just to name most important ones. All these variable objects need to satisfy the equilibrium conditions. They are often described in reference to a coordinate that is orthogonal (at least locally), so that we can quantify and operate on them in a meaningful manner in numerical values or symbolic expressions. When the coordinate transforms, all the values of these variables may change. Therefore, there must be a rule that governs such changes, so that the entire system behaves in a manner for consistent and correct quantification. The objects may have different forms of mathematical representations, including scalar (single number), vector (a stack of numbers), matrix (2D stack of numbers), an elasticity (four-dimensional stack of numbers), but they all are tensors because they all follow the same rule governing the transformation. It is the **orthogonal projection rule** in mathematical operations and the **principle of equilibrium** in mechanics law.

2.9 Transformation rules

2.9.1 *Definition of coordinate transformation*

As mentioned earlier, when we say coordinate transformation, we always refer to **coordinate rotation** or **type change** of the coordinate system. This is because when we study object changes, we are interested in the change at the same location in solids. Therefore, the transformation can only be a rotation or type change because the translational position is fixed. We also require that when the coordinate rotates, the entire coordinate system **rotates rigidly** together.

For changes of objects with the location change, we shall use the usual **function** concept in mathematics, which is a separate issue from coordinate rotation. In this case, the object becomes a variable over space, and we require all the variables being sufficiently smooth functions of the coordinates for differentiations. Functions of variables are another important aspect to be studied, independently of coordinate transformation. This is because many governing law of mechanics are the result of the function behavior of the variables.

To put it more clearly: the *tensor concept deals with the coordinate transformation*, and the *function concept deals with the variation of the object along with the coordinate*.

2.9.2 *Representation of a tensor in a coordinate*

In mechanics of materials, we use the following assumptions for the representation of a tensor in a coordinate system.

1. **Orthogonal coordinate:** We use orthogonal coordinate systems, implying all the axes in the coordinate system are, at least locally, mutually perpendicular.
2. **Right-hand rule:** We require also these axes to follow the conventional right-hand rule: x-cross-y gives z. This is for consistence and convenience purposes.

In this book, we will use the Cartesian coordinate as the default, which is also followed in many Python modules. Other coordinate systems may be used in a similar manner. Conversion between coordinate systems can be done based on the so-called curvilinear coordinates that are at least locally orthogonal.

2.9.3 *Essential rule to obtain values on a coordinate axis*

Orthogonal projection rule: When a tensor is represented in coordinates, we follow the orthogonal projection rule to compute the component values of the tensor, in addition to the equilibrium law of mechanics. This means that the component value of the tensor on an axis is the value obtained always by orthogonally projecting the tensor onto the axis. When the coordinate

2.10 Transformation matrix

2.10.1 *Two-dimensional (2D) cases*

When the coordinate rotates to a new one, the representation of the tensor in the new coordinate changes. To quantify such changes, we must figure out the relationship between these two, the original and the rotated, coordinate systems and in precise mathematical formulas.

For easy visualization, consider first a 2D case. Let the original **x** with two axes be denoted in conventional manner as x and y. The rotated new coordinate **x'** with axes x' and y', as drawn in Fig. 2.12. The prime symbol stands for objects in the new coordinate.

From Fig. 2.12, we find the following relationship via simple geometric examination following the orthogonal projection rule:

$$x' = x\cos\theta + y\sin\theta = x\underbrace{\cos(x',x)}_{\cos\theta} + y\underbrace{\cos(x',y)}_{\cos(90°-\theta)} \quad (2.51)$$

$$y' = -x\sin\theta + y\cos\theta = x\underbrace{\cos(y',x)}_{\cos(90°+\theta)} + y\underbrace{\cos(y',y)}_{\cos\theta}$$

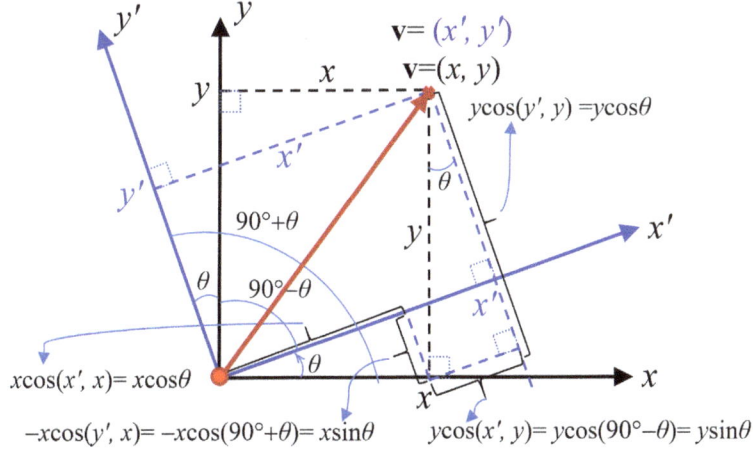

Figure 2.12. A point with coordinate values of x and y in coordinate **x**, and with x', y' in coordinate **x'** that is obtained by rotating coordinate **x** by θ. All the operations here are based on the orthogonal projection rule.

Equations (2.51) can be written in a matrix form as

$$\underbrace{\begin{bmatrix} x' \\ y' \end{bmatrix}}_{\mathbf{x'}} = \underbrace{\begin{bmatrix} \cos(x',x) & \cos(x',y) \\ \cos(y',x) & \cos(y',y) \end{bmatrix}}_{\mathbf{T}} \underbrace{\begin{bmatrix} x \\ y \end{bmatrix}}_{\mathbf{x}} \qquad (2.52)$$

These cosines are called **direction cosines**. Equation (2.52) can be written simply as follows:

$$\mathbf{x'} = \mathbf{T}\mathbf{x} \qquad (2.53)$$

where $\mathbf{x}$ is a vector of the original and $\mathbf{x'}$ is a vector of the rotated coordinate. Matrix $\mathbf{T}$ is called **transformation matrix**. It is clear that matrix $\mathbf{T}$ contains only the angular relationship of these axes of the two original and new coordinates. For 2D cases, it can be written as follows, which is seen more often in the literature:

$$\mathbf{T} = \begin{bmatrix} \cos\theta & \sin\theta \\ -\sin\theta & \cos\theta \end{bmatrix} \qquad (2.54)$$

It is important to note that although the values of the components in $\mathbf{x}$ and $\mathbf{x'}$ are different, they both represent the same point in 2D space (a position vector), as shown in Fig. 2.12.

We emphasize that all these expressions obtained following the **orthogonal projection rule** and using orthogonal coordinates, as shown in Fig. 2.12. Due to this, $\mathbf{T}$ has the following property:

$$\mathbf{T}\mathbf{T}^\top = \begin{bmatrix} \cos\theta & \sin\theta \\ -\sin\theta & \cos\theta \end{bmatrix} \begin{bmatrix} \cos\theta & -\sin\theta \\ \sin\theta & \cos\theta \end{bmatrix}$$

$$= \begin{bmatrix} \cos^2\theta + \sin^2\theta & -\cos\theta\sin\theta + \sin\theta\cos\theta \\ -\sin\theta\cos\theta + \cos\theta\sin\theta & \sin^2\theta + \cos^2\theta \end{bmatrix}$$

$$= \begin{bmatrix} 1 & 0 \\ 0 & 1 \end{bmatrix} = \mathbf{I} \qquad (2.55)$$

This property of $\mathbf{T}$ is called **unitary** or **orthogonality**. It is invertible, and its inverse transformation can be simply done using the transpose of $\mathbf{T}$:

$$\mathbf{T}^{-1} = \mathbf{T}^\top \qquad (2.56)$$

where superscript stands for transpose. It is also easy to confirm that $\mathbf{T}^\top \mathbf{T} = \mathbf{I}$.

Due to this property of **T**, the inverse transformation of Eq. (2.53) becomes

$$\mathbf{x} = \mathbf{T}^\top \mathbf{x}' \tag{2.57}$$

2.10.2 General definition of transformation matrix for 3D cases

With the explicit examination of 2D cases of rotation matrix for coordinate transformation, we are now ready to have a general definition. For convenience of description and formulation, we use indicial notation.

Consider a Cartesian coordinate **x** with axes x_1, x_2, and x_3. It rotates to **x**' with axes x'_1, x'_2, and x'_3, and the values of the components of an object (obtained also via the orthogonal projection rule) will change. We shall discuss the general mathematic rule of transformation to compute the components of the object in the new coordinate **x**'. To this end, we define the rotation matrix for coordinate transformation that bridges the original and new coordinates.

Naturally, the rotation matrix for coordinate transformation, **T**, shall contain the angular relationship of these axes of the two coordinates. This must be true regardless of dimensionality. Based on the orthogonal projection rule, **T** for 3D cases should be formed by the direction cosines of the angle between the axes of the original coordinate x_1, x_2, and x_3 and the rotated coordinates x'_1, x'_2, and x'_3.

$$\mathbf{T} = \begin{bmatrix} \underbrace{\cos(x'_1, x_1)}_{a_{11}} & \underbrace{\cos(x'_1, x_2)}_{a_{12}} & \underbrace{\cos(x'_1, x_3)}_{a_{13}} \\ \underbrace{\cos(x'_2, x_1)}_{a_{21}} & \underbrace{\cos(x'_2, x_2)}_{a_{22}} & \underbrace{\cos(x'_2, x_3)}_{a_{23}} \\ \underbrace{\cos(x'_3, x_1)}_{a_{31}} & \underbrace{\cos(x'_3, x_2)}_{a_{32}} & \underbrace{\cos(x'_3, x_3)}_{a_{33}} \end{bmatrix} \tag{2.58}$$

It contains all the direction cosines of these axes of these two coordinate systems. This definition is a natural extension from the 2D cases discussed in the previous section. Here, we introduced indicial notation $a_{ij}, i, j = 1, 2, 3$, to denote the entries of **T**. The right-top 2×2 sub-matrix is the same **T** given in Eq. (2.52), which can be re-written as

$$\mathbf{T} = \begin{bmatrix} a_{11} = \cos\theta & a_{12} = \sin\theta \\ a_{21} = -\sin\theta & a_{22} = \cos\theta \end{bmatrix} \tag{2.59}$$

2.10.3 Transformation via yaw, pitch, and roll in 3D*

Consider coordinate transformation in 3D by yaw, pitch, and roll defined as follows to create the new coordinates x', y' and z' based on x, y and z:

- **Yaw:** rotate the Cartesian coordinate by θ about x-axis, which gives $\mathbf{T}_x$.
- **Pitch:** rotate the Cartesian coordinate by β about y-axis, which gives $\mathbf{T}_y$.
- **Roll:** rotate the Cartesian coordinate by γ about z-axis, which gives $\mathbf{T}_z$.

By extending Eq. (2.59), these transformation matrices are written as follows:

$$\mathbf{T}_z(\theta) = \begin{bmatrix} \cos\theta & \sin\theta & 0 \\ -\sin\theta & \cos\theta & 0 \\ 0 & 0 & 1 \end{bmatrix} \tag{2.60}$$

$$\mathbf{T}_y(\beta) = \begin{bmatrix} \cos\beta & 0 & -\sin\beta \\ 0 & 1 & 0 \\ \sin\beta & 0 & \cos\beta \end{bmatrix} \tag{2.61}$$

$$\mathbf{T}_x(\gamma) = \begin{bmatrix} 1 & 0 & 0 \\ 0 & \cos\gamma & \sin\gamma \\ 0 & -\sin\gamma & \cos\gamma \end{bmatrix} \tag{2.62}$$

Note that when rotating about the y-axis by the right-hand rule, angle β should be negative.

When yaw, pitch, and roll are performed together, the total transformation matrix $\mathbf{T}$ can be obtained from these three, using matrix multiplication:

$$\mathbf{T} = \mathbf{T}_z(\theta)\,\mathbf{T}_y(\beta)\,\mathbf{T}_x(\gamma)$$

$$= \underbrace{\begin{bmatrix} \cos\theta & \sin\theta & 0 \\ -\sin\theta & \cos\theta & 0 \\ 0 & 0 & 1 \end{bmatrix}}_{\mathbf{T}_z(\theta)\text{ for yaw}} \underbrace{\begin{bmatrix} \cos\beta & 0 & -\sin\beta \\ 0 & 1 & 0 \\ \sin\beta & 0 & \cos\beta \end{bmatrix}}_{\mathbf{T}_y(\beta)\text{ for pitch}} \underbrace{\begin{bmatrix} 1 & 0 & 0 \\ 0 & \cos\gamma & \sin\gamma \\ 0 & -\sin\gamma & \cos\gamma \end{bmatrix}}_{\mathbf{T}_x(\gamma)\text{ for roll}}$$

$$= \begin{bmatrix} \cos\beta\cos\theta & \sin\beta\sin\gamma\cos\theta + \sin\theta\cos\gamma & -\sin\beta\cos\gamma\cos\theta + \sin\gamma\sin\theta \\ -\sin\theta\cos\beta & -\sin\beta\sin\gamma\sin\theta + \cos\gamma\cos\theta & \sin\beta\sin\theta\cos\gamma + \sin\gamma\cos\theta \\ \sin\beta & -\sin\gamma\cos\beta & \cos\beta\cos\gamma \end{bmatrix} \tag{2.63}$$

We write the following code for future use to create this type of transformation matrices in Sympy for symbolic computations.

Preliminaries: Number, Vector, Tensor, and Functions

```
1  def transf_YPRs(θ, about = 'z'):
2      '''Create a transformation matrix for coordinate transformation\
3      Input:  θ, rotation angle, in Sympy \
4              about, the axis of the rotation is about \
5      Return: Sympy matrix of transformation matrix of shape (3,3)
6      '''
7      c, s = sp.cos(θ), sp.sin(θ)
8
9      if about == 'z':
10         # Yaw: rotates about z by θ
11         T = sp.Matrix([[ c, s, 0],
12                        [-s, c, 0],
13                        [ 0, 0, 1]])
14     elif about == 'y':
15         # Pitch: rotates about y by θ
16         T = sp.Matrix([[ c, 0,-s],
17                        [ 0, 1, 0],
18                        [ s, 0, c]])
19     elif about == 'x':
20         # roll: rotates about x by θ
21         T = sp.Matrix([[ 1, 0, 0],
22                        [ 0, c, s],
23                        [ 0,-s, c]])
24     else: # randomly generated unitary matrix as transformation matrix:
25         print("Rotation axis must be given: z, y, or x")
26
27     return T, about
```

Following are examples on how to use this code.

```
1  θ, β, γ = symbols("θ, β, γ")   # define variables for angle of rotation
2  Tz = transf_YPRs(θ, about = 'z')[0]
3  Ty = transf_YPRs(β, about = 'y')[0]
4  Tx = transf_YPRs(γ, about = 'x')[0]
5  print(f'Ty=, {Ty}, \nTx=, {Tx}')
6  Tz
```

Ty=, Matrix([[cos(β), 0, -sin(β)], [0, 1, 0], [sin(β), 0, cos(β)]]),
Tx=, Matrix([[1, 0, 0], [0, cos(γ), sin(γ)], [0, -sin(γ), cos(γ)]])

$$\begin{bmatrix} \cos(\theta) & \sin(\theta) & 0 \\ -\sin(\theta) & \cos(\theta) & 0 \\ 0 & 0 & 1 \end{bmatrix}$$

```
1  T = Tz@Ty@Tx
2  T
```

$$\begin{bmatrix} \cos(\beta)\cos(\theta) & \sin(\beta)\sin(\gamma)\cos(\theta)+\sin(\theta)\cos(\gamma) & -\sin(\beta)\cos(\gamma)\cos(\gamma)+\sin(\ )\sin(\ ) \\ -\sin(\theta)\cos(\beta) & -\sin(\beta)\sin(\gamma)\sin(\theta)+\cos(\gamma)\cos(\theta) & \sin(\beta)\sin(\theta)\cos(\gamma)+\sin(\ )\cos(\ ) \\ \sin(\beta) & -\sin(\gamma)\cos(\beta) & \cos(\beta)\cos(\gamma) \end{bmatrix}$$

```
1  TTT = sp.simplify(T.T@T)    # Test on orthogonality (see next section)
2  print(TTT)
```

Matrix([[1, 0, 0], [0, 1, 0], [0, 0, 1]])

2.10.4 *Property of transformation matrix, a unitary matrix*

Since a_{ij} are obtained by the orthogonal projection rule, **T** is a tensor. It has a matrix form and is in general asymmetric. Since the axes of the coordinate systems are kept mutually orthogonal, we shall have the following orthogonal condition:

$$a_{ij}a_{kj} = \delta_{ik} \tag{2.64}$$

where δ_{ij} is the Kronecker delta defined as

$$\delta_{ij} = \begin{cases} 1, & \text{if } i = j \\ 0, & \text{if } i \neq j \end{cases} \tag{2.65}$$

In Eq. (2.64), we see that index j is repeated which implies a summation:

$$a_{ij}a_{kj} = a_{i1}a_{k1} + a_{i2}a_{k2} + a_{i3}a_{k3} \tag{2.66}$$

It is known as the Einstein notation of summation: any repeated index in a term in an expression implying a summation over the range of the repeated index. As the result of such a summation, the index will disappear and is said to be *contracted*. Therefore, any letter can be used for a repeated index, as long as it differs from other indexes in the term and the range is the same. For example, $a_{ij}a_{kj}$ is the same as $a_{im}a_{km}$, and they all result in the same second-order tensor with indexes ik. For these reasons, a repeated index is called a **dummy index**.

Indexes i and k in Eq. (2.64) are all single in a term in the equation and cannot be contracted. They are called **free-indexes**. Clearly, one can use any other symbol for these free-indexes, as long as it does not interfere with other indexes and the range is the same. For example, $a_{ij}a_{kj} = \delta_{ik}$ is the same as $a_{rj}a_{sj} = \delta_{rs}$.

Since the coordinate rotates rigidly, and coordinate $\mathbf{x}'$ is also orthogonal, we have

$$a_{ij}a_{ik} = \delta_{jk} \tag{2.67}$$

Due to Eq. (2.64), the transformation matrix $\mathbf{T}$ shall satisfy the following orthogonal condition in matrix form, as seen earlier:

$$\mathbf{T}\mathbf{T}^\top = \mathbf{I} \quad \text{or} \quad \mathbf{T}^\top\mathbf{T} = \mathbf{I} \tag{2.68}$$

where $\mathbf{I}$ is a 3×3 identify matrix in $\mathbb{R}^3$. Here, we follow the standard matrix multiplication rule. Eq. (2.68) means that $\mathbf{T}$ is an **unitary matrix**. We have seen this property explicitly for 2D case in the previous section. The unitary property $\mathbf{T}$ implies that:

- The determinant of $\mathbf{T}$ (with the right-hand rule enforced) is 1.
- When it operates on a vector, it will not change the vector length.
- When it operates on a square matrix, it will not change the determinant of the matrix. Therefore, it explains also that why the determinant of a matrix is invariant under coordinate transformation. This can be proven easily: because $\det(\mathbf{TM}) = \det(\mathbf{T})\det(\mathbf{M})$ and $\mathbf{T}$ has a determinant of 1. The same holds if $\mathbf{T}$ operates from the right.

When $\mathbf{T}$ acts on a tensor in $\mathbf{x}$, the new components of the same tensor in the rotated coordinate $\mathbf{x}'$ can be obtained. The length of $\mathbf{x}$ and $\mathbf{x}'$ is the same. This can easily be proven using Eq. (2.17) as follows.

Consider an arbitrary vector $\mathbf{v}$ defined in $\mathbf{x}$. A transformation matrix $\mathbf{T}$ is used to transfer the coordinate $\mathbf{x}$ to $\mathbf{x}'$. The length of the vector $\mathbf{v}'$ in $\mathbf{x}'$ is evaluated as

$$\|\mathbf{v}'\|^2 = \mathbf{v}'^\top \mathbf{v}' = (\mathbf{T}\mathbf{v})^\top \mathbf{T}\mathbf{v} = \mathbf{v}^\top \underbrace{\mathbf{T}^\top \mathbf{T}}_{\mathbf{I}} \mathbf{v} = \mathbf{v}^\top \mathbf{v} = \|\mathbf{v}\|^2 \tag{2.69}$$

The length of the vector is unchanged, which implies that the vector is a first order tensor and its length will not change after the coordinates transform. Note that the component values in $\mathbf{v}'$ will change, but such a change is to ensure the vector direction remained unchanged, when measured in reference to the new coordinates. This **vector property preservation** is ensured by the orthogonal projection rule, which results in a unitary $\mathbf{T}$.

It may be worth to note that this general definition of tensors for mechanics of materials applies, in theory, to arbitrary dimension. In n-dimensions, the ranges of i and j become n, and the dimension for $\mathbf{T}$ will be $n \times n$. The property of $\mathbf{T}$ will still hold, as long as the transformation rules defined in Section 2.9 are followed.

In indicial notation, Eq. (2.53) can be written as

$$x'_i = a_{ij} x_j \qquad (2.70)$$

And Eq. (2.57) can be written as

$$x_i = a_{ji} x'_j \qquad (2.71)$$

The transpose of a matrix is equivalent to **swapping** the indexes of a second-order tensor.

2.10.5 *Numpy code for transformation matrices*

We write a Python function that can be used to generate some typical transformation matrices or a random transformation matrix. With such a matrix, one can rotate the coordinate, resulting in description change of objects, such as vector, defined in the coordinate. This function will be used later for multiple times to perform coordinate transformation.

```
def transferM(theta, about = 'z'):
    '''Create a transformation matrix for coordinate transformation\
    Input theta: rotation angle in degree \
          about: the axis of the rotation is about \
    Return: numpy array of transformation matrix of shape (3,3)'''
    from scipy.stats import ortho_group

    n = 3   # 3-dimensional problem
    c, s = np.cos(np.deg2rad(theta)), np.sin(np.deg2rad(theta))

    if about == 'z':
        # rotates about z by theta
        T = np.array([[ c, s, 0.],
                      [-s, c, 0.],
                      [0.,0., 1.]])
    elif about == 'y':
        # rotates about y by theta
        T = np.array([[ c, 0.,-s],
                      [ 0, 1.,0.],
                      [ s, 0., c]])
    elif about == 'x':
        # rotates about x by theta
        T = np.array([[ 1.,0., 0.],
                      [ 0., c, s],
                      [ 0.,-s, c]])
    else: # randomly generated unitary matrix as transformation matrix:
        T = ortho_group.rvs(dim=n)        # Generate a random matrix
        T[2,:] = np.cross(T[0,:], T[1,:]) # Enforce the right-hand rule

    return T, about
```

2.10.6 Examples: Unitary property of transformation matrix

Let us examine in detail **T** matrices created using transferM(), and check whether the orthogonality and the right-hand rule are observed.

```
1  from scipy.stats import ortho_group
2
3  np.random.seed(8)         # set a seed, random number can be reproduced
4  about, theta = 'random', 30.         # rotation angle and axis
5  Tr, _ = transferM(theta, about = about)    # Or gr.transferM()
6
7  print(f'Transformation tensor {theta:3.2f}°, w.r.t. {about}:\n{Tr}')
8  print(f"Determinant of Tr= {lg.det(Tr):.6f}\n")
9  print(f"Is Tr orthogonal? \n{Tr.dot(Tr.T)}")
10 print(f"Is Tr.T orthogonal? \n{(Tr.T).dot(Tr)}\n")
11
12 print(f"Check right-hand rule:")
13 print(Tr[0,:],np.cross(Tr[1,:], Tr[2,:])) #Check right-hand rule: 1=2X3
14 print(Tr[1,:],np.cross(Tr[2,:], Tr[0,:])) #Check right-hand rule: 2=3X1
15 print(Tr[2,:],np.cross(Tr[0,:], Tr[1,:])) #Check right-hand rule: 3=1X2
```

```
Transformation tensor 30.00°, w.r.t. random:
[[ 0.0411  0.9184  0.3935]
 [-0.6253 -0.2835  0.7271]
 [ 0.7793 -0.276   0.5626]]
Determinant of Tr= 1.000000

Is Tr orthogonal?
[[ 1.  0. -0.]
 [ 0.  1. -0.]
 [-0. -0.  1.]]
Is Tr.T orthogonal?
[[ 1. -0. -0.]
 [-0.  1.  0.]
 [-0.  0.  1.]]

Check right-hand rule:
[0.0411 0.9184 0.3935] [0.0411 0.9184 0.3935]
[-0.6253 -0.2835  0.7271] [-0.6253 -0.2835  0.7271]
[ 0.7793 -0.276   0.5626] [ 0.7793 -0.276   0.5626]
```

A coordinate transformation matrix is unitary, implying that its inverse is its transpose. Let us confirm.

```
1  print(f"Is Tr unitary? {np.allclose(lg.inv(Tr), Tr.T, atol=1e-04)}")
```

```
Is Tr unitary? True
```

2.10.7 Examples: Basic property preservation by transformation matrix

Let us examine basic property preservation feature of a transformation matrix **T**. Consider an arbitrary vector **v**, the basic property is its length and direction. Computing the length of a vector is easy, and the value is unique. However, when computing the angle of the vector care is needed because cosine function is multi-valued, depending on which quadrant the vector is in. In the following code, we set a vector in the first quadrant with an randomly generated angle θ_x. When rotating the coordinate, we use a rotation angle that satisfies $\theta_{\text{rotation}} \leq \theta_x$. With this setting the computations of angles become unique. If such a setting is changed, care is needed in computing the values of angles. With this said, the following code is self-explanatory and easy to follow.

```
1  θ_x = np.random.rand()*90.                    # a random angle in [0, 90.]
2
3  # create a vector in x-y coordinate using θ_x:
4  v = np.array([np.cos(np.deg2rad(θ_x)), np.sin(np.deg2rad(θ_x)), 0.])
5  v = np.random.rand()*v                        # make it with arbitrary length
6  print(f'Original vector v = {v} \n its length = {lg.norm(v)}')
7  print(f'Direction cosine of v = {v/lg.norm(v)}')
8
9  # angle w.r.t the base coordinate x where v is defined originally:
10 θ_x = np.arccos(v[0]/lg.norm(v))*180/np.pi    # unique under the setting
11 print(f'θ of v w.r.t its base coordinate x = {θ_x:.4f}o')
12
13 # rotate the base coordinate x-y about the z-axis to form x'-y' coords:
14 # rotation would be in [0, θ_x]:
15 about = 'z'; θ_rotation = np.random.rand()*θ_x
16
17 Tz, _ = transferM(θ_rotation, about = about)
18 vp = Tz@v
19 print(f'Transformed v = {vp}, \n its length = {lg.norm(vp)}')
20 print(f'Direction cosine of Tv = {vp/lg.norm(vp)}')
21
22 # angle w.r.t the rotated coordinate x':
23 θ_xp=np.arccos(vp[0]/lg.norm(vp))*180/np.pi   # unique under the setting
24 print(f"θ of v' w.r.t rotated coordinate x' = {θ_xp:.4f}o")
25
26 # angle w.r.t the base coordinate x where v is defined originally:
27 θ_x_= θ_xp+θ_rotation
28 print(f"θ of v' w.r.t its base coordinate x = {θ_x_:.4f}o")
29 print(f"Is the angle  unchanged? {abs(θ_x_-θ_x)<1.e-6}")
30 print(f"Is the length unchanged? {abs(lg.norm(v)-lg.norm(Tz@v))<1.e-6}")
```

```
Original vector v = [0.2397 0.3523 0.    ]
 its length = 0.4260917704761922
Direction cosine of v = [0.5625 0.8268 0.    ]
θ of v w.r.t its base coordinate x = 55.7714∘
Transformed v = [0.3281 0.2719 0.    ],
 its length = 0.4260917704761921
Direction cosine of Tv = [0.77   0.6381 0.    ]
θ of v' w.r.t rotated coordinate x' = 39.6493∘
θ of v' w.r.t its base coordinate x = 55.7714∘
Is the angle  unchanged? True
Is the length unchanged? True
```

It is clear that both the basic properties of the vector, original length and direction, remain unchanged under **T**. Since the random angle is set random, readers may execute the code for multiple times, and think about the results obtained.

2.10.8 Example: Transformation matrix on arbitrary matrices

Transformation matrix is unitary, and a unitary matrix does not change the determinant of a matrix it acts on. We shall test this using a randomly generated matrix, which is a lined-up vectors.

```
1  np.random.seed(9)                    # set a seed for reproducibility
2  n = 3
3  M = np.random.randn(n,n)             # Generate a random matrix
4  print(f'Randomly generated matrix M: \n{M}')
5  print(f'Determinant of    M = {lg.det(M)}')
6  print(f'Determinant of Tr@M = {lg.det(Tr@M)}')
```

```
Randomly generated matrix M:
[[ 0.0011 -0.2895 -1.1161]
 [-0.0129 -0.3784 -0.4811]
 [-1.5173 -0.4909 -0.2407]]
Determinant of    M = 0.4230331712701571
Determinant of Tr@M = 0.4230331712701568
```

It is seen that the determinant of the matrix is unchanged, as proven earlier.

Let us check the effect of **T** on the eigenvalues of a matrix. We first look at a symmetric matrix.

```
1  Ms = (M + M.T)/2              # random symmetric matrix
2  print(f'Symmetric random square matrix Ms: \n{Ms}')
3  print(f'Eigenvalues of      Ms = {lg.eig(M)[0]}')
4  print(f'Eigenvalues of Tr@Ms = {lg.eig(Tr@M)[0]}')
```

```
Symmetric random square matrix Ms:
[[ 0.0011 -0.1512 -1.3167]
 [-0.1512 -0.3784 -0.486 ]
 [-1.3167 -0.486  -0.2407]]
Eigenvalues of      Ms = [ 1.2042 -1.603  -0.2191]
Eigenvalues of Tr@Ms = [ 0.5069 -1.5005 -0.5562]
```

It is shown that the eigenvalues of Tr@M are different from those of M. This is because M is a second-order tensor, we need to let **T** act on both sides in a "symmetric" manner. With such a two-side application of **T**, the eigenvalues of the resultant new matrix will be unchanged because **T** is unitary. This is shown in the following code, using an asymmetric matrix.

```
1  M = np.random.randn(n,n)              # random asymmetric square matrix
2  print(f'Random asymmetric square matrix M: \n{M}')
3  print(f'Eigenvalues of      M = {lg.eig(M)[0]}')
4  print(f'After forward & inverse transformation: \n{Tr@M@Tr.T}')
5  print(f'Eigenvalues of Tr@M@Tr.T = {lg.eig(Tr@M@Tr.T)[0]}')
```

```
Random asymmetric square matrix M:
[[-0.6479  0.6359  1.7401]
 [ 0.2967  0.7075  1.8228]
 [ 0.4308  1.5427 -0.9007]]
Eigenvalues of      M = [ 2.0292 -0.7299 -2.1405]
After forward & inverse transformation:
[[ 1.7428  0.3881  0.7523]
 [-0.3824 -2.188  -0.9215]
 [ 1.195   0.136  -0.396 ]]
Eigenvalues of Tr@M@Tr.T = [ 2.0292 -0.7299 -2.1405]
```

With a two-side application of **T** on a matrix, the eigenvalues of the resultant new matrix are unchanged, even if the matrix is a random asymmetric matrix.

This two-side action of **T** will be performed to stress and strain tensors in later Chapters.

2.11 Rotation matrix

2.11.1 *Derivation for 2D cases*

Readers may be familiar with the **rotation matrix**, often denoted as **R**. It is used to rotate a vector in a coordinate system counterclockwise, which results in new components of the vector in the same coordinate system. Assume the

Preliminaries: Number, Vector, Tensor, and Functions

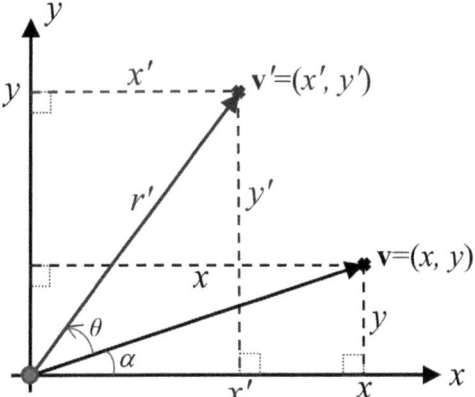

Figure 2.13. A vector with coordinate values of x and y in a coordinate system. It is rotated counterclockwise by θ, which results in new vector with coordinate values x', y' in the same coordinate system. All the operations here are based on the orthogonal projection rule.

vector, before rotation, is denoted as $\mathbf{v} = (x, y)$ in the coordinate system. After it is rotated, it becomes $\mathbf{v}' = (x', y')$, as shown in Fig. 2.13. The prime symbol stands for the new coordinate values.

Upon simple analysis based on Fig. 2.13, we have

$$x' = r'\cos(\alpha + \theta) = \underbrace{r'\cos\alpha}_{x}\cos\theta - \underbrace{r'\sin\alpha}_{y}\sin\theta$$

$$= x\cos\theta - y\sin\theta$$

$$y' = r'\sin(\alpha + \theta) = \underbrace{r'\sin\alpha}_{y}\cos\theta + \underbrace{r'\cos\alpha}_{x}\sin\theta \qquad (2.72)$$

$$= x\sin\theta + y\cos\theta$$

where α is the angle of the vector with respect to the x-axis, and θ is the rotation angle to the vector. The foregoing equation can be written in matrix form:

$$\mathbf{v}' = \underbrace{\begin{bmatrix} x' \\ y' \end{bmatrix}}_{\mathbf{v}'} = \underbrace{\begin{bmatrix} \cos\theta & -\sin\theta \\ \sin\theta & \cos\theta \end{bmatrix}}_{\mathbf{R}} \underbrace{\begin{bmatrix} x \\ y \end{bmatrix}}_{\mathbf{v}} = \mathbf{R}\mathbf{v} \qquad (2.73)$$

where $\mathbf{R}$ is the **rotation matrix** given as

$$\mathbf{R} = \begin{bmatrix} \cos\theta & -\sin\theta \\ \sin\theta & \cos\theta \end{bmatrix} \qquad (2.74)$$

2.11.2 *Relationship between transformation and rotation matrix*

Comparing Eqs. 2.54 and 2.74, it is clear that

$$\mathbf{R} = \mathbf{T}^\top \tag{2.75}$$

This is simply because $\mathbf{T}$ rotates the coordinate with the object (vector) remaining still, but $\mathbf{R}$ rotates the object (vector) within the same coordinate. Thus, the effects of their actions are mutual reverse. Thus, Eq. (2.75) is valid for any higher dimensions.

Due to Eq. (2.75), the rotation matrix has all the properties of the transformation matrix.

One may note that deriving Eq. (2.74) is in fact easier than deriving Eq. (2.54) for 2D cases. However, for higher dimensions, the transformation matrix $\mathbf{T}$ can be much easier to obtain because it can be systematically obtained using the direction cosine concept, as shown in Section 2.3.2. A quite comprehensive article on rotation matrices can be found at the wikepage on Rotation_matrix (https://en.wikipedia.org/wiki/Rotation_matrix).

2.12 Function, an object varying over space

Note that when we talk about coordinate transformations we stand at a point in solid and examine the change of the components of an object in mechanics of materials.

We now discuss about functions that deal with the variation of an object with respect to changes of coordinate values. In a problem of mechanics materials, a variable, such as a component of displacement or stress or strain tensor, can change from location to location. This means that it can vary with the coordinates in a coordinate system. The variation of a variable with respect to coordinates is called **function** of coordinates. It can be defined as an operator $f()$ that takes inputs of coordinates in an n-dimensional space $\mathbb{R}^n$ and produces a new number in $\mathbb{R}$ corresponding to the coordinates. It is a mapping from $\mathbb{R}^n \to \mathbb{R}$. In the 3D Cartesian coordinate system, for example, the function is often expressed as $f(\mathbf{x})$ or $f(x, y, z)$, where $x, y,$ and z are the coordinates as the independent variables of the function f [1].

In higher dimensions, there are multiple components. Each of which can be a function of coordinates. For a displacement vector in 3D solids, for example, it has 3 scalar component functions. It thus forms a vector function of 3 scalar functions. For a stress tensor in 3D solids, we have 9 (6 independent) scalar functions.

Note that the function for an object is a concept for accounting for the variation of the object along the coordinates. It is a separate concept of the coordinate transformation. Therefore, a tensor at a location is still a tensor at any other location in the coordinate system, as long as the same rule is followed for coordinate transformation at locations. However, operations to a tensor object with respect to the coordinates can produce a new tensor object of different order. We will discuss about this in great detail because it happens all the time in mechanics formulation. In our discussion, we assume all the scalar functions involved are sufficiently continuous and smooth and hence differentiable to the required order, allowing us to take limits when necessary.

2.13 Gradient of a scalar function, a vector

A function is a device that takes in an input (or argument) and produces an output, as defined in Ref. [1]. When the argument is a coordinate, the function determines how an variable function changes with the coordinate. For example, a function of a displacement component gives the variation of the component against the coordinate. In general, the argument can be a scalar, a vector, or even a matrix. The function itself can also be a scalar, a vector or a matrix.

Gradient of a scalar function with scalar number argument converts the scalar number (0th first order tensor) to a vector (first order tensor).

Consider a scalar function defined in a domain in the coordinate system. A typical such a function is one of the components of a displacement vector or stress tensor.

2.13.1 *Definition*

The gradient of an arbitrary scalar **differentiable** function $f(\mathbf{x})$ of the independent variables (that is a vector) becomes a vector of three functions defined as

$$\nabla f(\mathbf{x}) := \underbrace{\begin{bmatrix} \frac{\partial}{\partial x} \\ \frac{\partial}{\partial y} \\ \frac{\partial}{\partial z} \end{bmatrix}}_{\nabla} f(\mathbf{x}) = \begin{bmatrix} \frac{\partial f(\mathbf{x})}{\partial x} \\ \frac{\partial f(\mathbf{x})}{\partial y} \\ \frac{\partial f(\mathbf{x})}{\partial z} \end{bmatrix} \qquad (2.76)$$

where ∇ (pronounced "nabla") stands for **gradient operator**, and has a vector form. The components in the resultant vector $\nabla f(\mathbf{x})$ are, respectively, the partial derivatives of the scalar function with respect to these 3 coordinates. The shape of $\nabla f(\mathbf{x})$ is the same as that of $\mathbf{x}$. We see now, f is a scalar number, a 0th order tensor, but its gradient becomes a vector of first order tensor! It can be subjected to the same rule of coordinate transformation for a vector.

2.13.2 Python demonstration

Let us use the following Python codes to demonstrate this. We write first a function to compute the gradient of a given scalar function.

```
def grad_f(f, X):
    '''Compute the gradient of a given scalar function f.
    input: f, sp.Function.
           X, coordinates, array-like of symbols.
    return: grad_f, sp.Matrix, a matrix of 2×1 or 3×1 or len(X)×1.
    ## Example:
    x, y, z = sp.symbols("x, y, z")
    X = [x, y, z]
    f = 8*x**2 + 5*x*y -2*y + 3*y**2 + 2*z**2 - 5*z
    f_g = grad_f(f, X)
    f_g
    '''
    gradf = sp.Matrix([sp.diff(f,xi) for xi in X])
    return gradf
```

```
# Define an arbitrary scalar function of coordinates
x, y, z = symbols("x, y, z")
X = Matrix([x, y, z])

f = Function('f')(x, y, z)
#f = 8*x*sp.sin(y)+5*x**2-2*y*sp.cos(x)+3*y**2-5*z    # alternative
f = 8*x**2 + 5*x*y -2*y + 3*y**2 + 2*z**2 - 5*z
f                                                    # print out
```

$$8x^2 + 5xy + 3y^2 - 2y + 2z^2 - 5z$$

This function f takes any value of x, y, and z in $\mathbb{R}^3$, and produces a single value in $\mathbb{R}$. For example,

```
fv = f.subs({x:1/2, y:1, z: 1/4}).evalf()    # f value for a inputs
fv
```

4.375

The gradient of the scalar function becomes a vector of 3 functions:

```
1  f_g = grad_f(f, X)     # This produces a column vector of 3 functions
2  f_g
```

$$\begin{bmatrix} 16x + 5y \\ 5x + 6y - 2 \\ 4z - 5 \end{bmatrix}$$

```
1  # Or use:
2  from continuum_mechanics import vector
3  f_g = vector.grad(f)   # This produces a column vector of 3 functions
4  f_g                    # only work for 3D
```

$$\begin{bmatrix} 16x + 5y \\ 5x + 6y - 2 \\ 4z - 5 \end{bmatrix}$$

It is a vector because of the definition given in Eq. (2.76). Each of which of 3 functions in the vector is the derivative of $f(\mathbf{x})$ with respect to, respectively, x, y, and z. The same can be obtained using sp.diff() method.

```
1  # Or simply use:
2  f.diff(X)              # This produces a column vector of 3 functions
```

$$\begin{bmatrix} 16x + 5y \\ 5x + 6y - 2 \\ 4z - 5 \end{bmatrix}$$

```
1  print(f_g.shape,X.shape)   # The shapes of f_g & the coordinate vector
```

(3, 1) (3, 1)

It is seen that the shape of the gradient of the scalar function is the same as that of the coordinate vector.

The vector of functions can be simplified by substituting some of the coordinates with fixed values.

```
1  f_gs = f_g.subs({x:1/2})
2  f_gs
```

$$\begin{bmatrix} 5y + 8.0 \\ 6y + 0.5 \\ 4z - 5 \end{bmatrix}$$

It is seen now f_gs is still a vector of the same shape, but the components are no longer functions of x. This is because of the use of subs() function, which simplifies the functions at a given x coordinate. We can simplify it further.

```
1  f_gs = f_gs.subs({y:2})                # fix y
2  f_gs
```

$$\begin{bmatrix} 18.0 \\ 12.5 \\ 4z - 5 \end{bmatrix}$$

We can also get the numerical values of this gradient vector by fixing z:

```
1  f_gv = f_gs.subs({z:1/4}).evalf()      # get the vector values
2  printM(f_gv, 'Values of function gradient',n_dgt=5)   # printed pretty
```

Values of function gradient

$$\begin{bmatrix} 18.0 \\ 12.5 \\ -4.0 \end{bmatrix}$$

Remark: The gradient of a scalar function defined in the coordinate becomes a vector. It has the same shape as the coordinate vector.

2.14 Gradient of a vector of function, a matrix

2.14.1 *Definition*

Gradient of a vector of functions converts a vector (first order tensor) to a matrix (second-order tensor).

Consider a differentiable vector field in 3D Cartesian coordinates defined by a vector of three scalar functions:

$$\mathbf{f}(\mathbf{x}) = f_x(\mathbf{x})\mathbf{i} + f_y(\mathbf{x})\mathbf{j} + f_z(\mathbf{x})\mathbf{k} = \begin{bmatrix} f_x(\mathbf{x}) \\ f_y(\mathbf{x}) \\ f_z(\mathbf{x}) \end{bmatrix} \qquad (2.77)$$

where $\mathbf{i}, \mathbf{j}$ and $\mathbf{k}$ are, respectively, the unit vector on x, y and z axes, $f_i(\mathbf{x})$ ($i = x, y, z$) are the components of the vector function $\mathbf{f}(\mathbf{x})$. The vector function creates a vector field in the domain where these functions are defined.

Each of the component function is a function of the coordinates. The gradient of a vector of functions is defined using the **outer product** of ∇ and $\mathbf{f}(\mathbf{x})$:

$$\nabla \mathbf{f}(\mathbf{x}) := \underbrace{\begin{bmatrix} \frac{\partial}{\partial x} \\ \frac{\partial}{\partial y} \\ \frac{\partial}{\partial z} \end{bmatrix}}_{\nabla} \otimes \underbrace{\begin{bmatrix} f_x(\mathbf{x}) \\ f_y(\mathbf{x}) \\ f_z(\mathbf{x}) \end{bmatrix}}_{\mathbf{f}(\mathbf{x})}$$

$$= \begin{bmatrix} \frac{\partial}{\partial x} \\ \frac{\partial}{\partial y} \\ \frac{\partial}{\partial z} \end{bmatrix} \begin{bmatrix} f_x(\mathbf{x}) & f_y(\mathbf{x}) & f_z(\mathbf{x}) \end{bmatrix} \quad (2.78)$$

$$= \begin{bmatrix} \frac{\partial}{\partial x} f_x(\mathbf{x}) & \frac{\partial}{\partial x} f_y(\mathbf{x}) & \frac{\partial}{\partial x} f_z(\mathbf{x}) \\ \frac{\partial}{\partial y} f_x(\mathbf{x}) & \frac{\partial}{\partial y} f_y(\mathbf{x}) & \frac{\partial}{\partial y} f_z(\mathbf{x}) \\ \frac{\partial}{\partial z} f_x(\mathbf{x}) & \frac{\partial}{\partial z} f_y(\mathbf{x}) & \frac{\partial}{\partial z} f_z(\mathbf{x}) \end{bmatrix}$$

2.14.2 *Python examples*

Let us evaluate the gradient of a vector function, assume each of the component function is differentiable, such as the vector functions given in the following:

```
1  vf = Matrix([y*sp.sin(x), z*sp.sin(y), x*sp.cos(z)])
2  vf
```

$$\begin{bmatrix} y \sin(x) \\ z \sin(y) \\ x \cos(z) \end{bmatrix}$$

Using vector.grad_vec() from the continuum_mechanics module, the gradient of the vector function can be found as follows:

```
1  vf_g_ = vector.grad_vec(vf)   # produces 3 column vectors of 3 functions
2  vf_g_
```

$$\begin{bmatrix} y \cos(x) & 0 & \cos(z) \\ \sin(x) & z \cos(y) & 0 \\ 0 & \sin(y) & -x \sin(z) \end{bmatrix}$$

It is a 3 by 3 matrix. This is because we have three component functions, and the gradient of each of the function is a vector by the definition of Eq. (2.76). These three vectors can form a matrix.

2.14.3 Code function for gradient of vector functions

Note that vector.grad_vec() works only for 3D vector functions. One may use gr.grad_vf() which is given in the following code function. It works for both 2D and 3D vector functions.

```
 1  def grad_vf(vf, X):
 2      '''Compute the gradient of a given vector function, vf
 3      input: vf, sp.Matrix, a column vector of 2 or 3.
 4             X, coordinates, array-like of symbols
 5      return: grad_f, sp.Matrix, a matrix of 2×1 or 3×1 or len(X)×1.
 6      ## Example:
 7      x, y, z = sp.symbols("x, y, z")
 8      vf = sp.Matrix([y*sp.sin(x), z*sp.sin(y), x*sp.cos(z)])
 9      vf_g = grad_vf(vf, X)
10      '''
11      grad_f = sp.Matrix([sp.diff(vf,xi).T for xi in X])
12      return grad_f
```

```
 1  # 3D vector function example:
 2  vf_g = grad_vf(vf, X)                          # or gr.grad_vf(vf)
 3  vf_g
```

$$\begin{bmatrix} y\cos(x) & 0 & \cos(z) \\ \sin(x) & z\cos(y) & 0 \\ 0 & \sin(y) & -x\sin(z) \end{bmatrix}$$

```
 1  # 2D vector function example:
 2  vf2d = Matrix([y*sp.sin(x), z*sp.sin(y)])
 3  vf2d_g = gr.grad_vf(vf2d, X[:2])               # or gr.grad_vf(vf2d, X[:2]
 4  vf2d_g
```

$$\begin{bmatrix} y\cos(x) & 0 \\ \sin(x) & z\cos(y) \end{bmatrix}$$

When the independent variables are set with fixed values, we shall obtain a matrix with fixed values.

```
1  vf_gs = vf_g.subs({x:1/2,y:2,z:1/4}).evalf()
2  vf_gs
3  printM(vf_gs, 'Gradient of vector at a fixed location',n_dgt=5)
```

Gradient of vector at a fixed location

$$\begin{bmatrix} 1.7552 & 0 & 0.96891 \\ 0.47943 & -0.10404 & 0 \\ 0 & 0.9093 & -0.1237 \end{bmatrix}$$

It is still a matrix. The entries become fixed numbers, as expected.

Gradient of vector of functions will be used in deriving the strains in solid using displacement vectors.

2.15 Divergence of a vector function, a scalar

2.15.1 Definition

The divergence of a vector of functions produces a scalar function. It converts a first order tensor object to a 0th order one.

Consider a differentiable vector field in 3D Cartesian coordinates defined by a vector of three scalar functions, given in Eq. (2.77). The divergence of a vector of functions is defined as the **dot product** of ∇ with $\mathbf{f}$:

$$\begin{aligned} \text{div}\, \mathbf{f} &:= \nabla \cdot \mathbf{f} \\ &= \begin{bmatrix} \frac{\partial}{\partial x} & \frac{\partial}{\partial y} & \frac{\partial}{\partial z} \end{bmatrix} \begin{bmatrix} f_x \\ f_y \\ f_z \end{bmatrix} \\ &= \frac{\partial f_x}{\partial x} + \frac{\partial f_y}{\partial y} + \frac{\partial f_z}{\partial z} \\ &= \frac{\partial f_i}{\partial x_i} = f_{i,i} \end{aligned} \qquad (2.79)$$

This is a scalar function because of the contraction of index i. The dot in Eq. (2.79) stands for the dot product of vectors: a del operator vector and a given vector of functions. This is a measure of the vector diverges away from a point. The divergence carries different physical meaning depending on application. For example,

- If the vector of functions is concentration of a species, it is the spatial rate of loss of mass at that point. The compensation to that loss should be the birth rate.

- If the vector is temperature, it is the spatial loss of heat at that point. The compensation to that loss should be the heat source.
- If the vector has stress components (all in the same direction) at a point, it is the loss of stresses in that direction. Such a loss must be compensated externally by an applied body force in that direction. This leads to equilibrium at the point together with the externally applied total body force (force density) there.

2.15.2 *Python code and examples*

Readers may use gr.div_vf() to compute the divergence of a vector function, which is given in the following.

```
1  def div_vf(vf, X):
2      '''Compute the divergence of a vector function vf.
3         input: vf, vector function in, sp.Matrix
4                X, vector of in dependent variables, sp.Matrix
5         return: div_f
6      '''
7      div_f = sp.Add(*[sp.diff(vf[i], X[i]) for i in range(len(X))])
8      return div_f                         # divergence of a vector function
```

Let us demonstrate this using the vector of three simplest functions. Consider first a vector with three stress components in a direction that are all constants (independent of the coordinates x, y, and z):

```
1  x, y, z = symbols('x, y, z')              # define coordinates
2  X = Matrix([x, y, z])
3
4  a, b, c = symbols('a, b, c')              # define constants
5  vc = Matrix([a, b, c])                    # constant vector function
6  vc
```

$$\begin{bmatrix} a \\ b \\ c \end{bmatrix}$$

```
1  # compute the divergence of a vector function
2  div_f = div_vf(vc, X)                     # Or use: vector.div(vc)
3  div_f
```

0

The divergence of the constant stress vector is zero. This is because a field of constant stays still and no loss will occur. This means also that the body force in that direction must be zero. We will see an example in the next chapter.

Consider now a vector with three stress components in a direction that are linear functions of x, y, and z:

```
1  a1, b1, c1 = symbols('a1, b1, c1')              # define constants
2  a2, b2, c2 = symbols('a2, b2, c2')              # define constants
3  a3, b3, c3 = symbols('a3, b3, c3')              # define constants
4
5  vl=Matrix([a1*x+b1*y+c1*z, a2*x+b2*y+c2*z, a3*x+b3*y+c3*z])   # linear
6  vl
```

$$\begin{bmatrix} a_1 x + b_1 y + c_1 z \\ a_2 x + b_2 y + c_2 z \\ a_3 x + b_3 y + c_3 z \end{bmatrix}$$

```
1  # compute the divergence of a vector function
2  div_f = div_vf(vl, X)                           # Or use: vector.div(vl)
3  div_f
```

$$a_1 + b_2 + c_3$$

Thus, if a vector of the stresses in a direction that are all linear functions, its divergence is a constant. This means that the body force in that direction must be a constant, which is the sum of the slopes in the respective coordinates. Therefore, if a solid is subjected to a constant gravity force in a direction, the stresses in that direction is linear.

Generally speaking, the function of stresses is one order higher than the body force function.

This relationship of stress vector divergence and body force may be used to check stress solutions. We shall do so in some of the examples in the following chapters. Most importantly, it offers a convenient way to derive the equilibrium equations for solids.

2.16 Curl of a vector function, a vector

2.16.1 *Definition*

Consider a vector function **f** defined in Eq. (2.77). Using the gradient operator vector ∇, the curl of a vector function can be defined in a similar way as the cross product of two vectors. Using Eq. (2.44), we can write:

$$\underbrace{\nabla \times \mathbf{f}}_{\text{curl } \mathbf{f}} := \begin{vmatrix} \mathbf{i} & \mathbf{j} & \mathbf{k} \\ \frac{\partial}{\partial x} & \frac{\partial}{\partial y} & \frac{\partial}{\partial z} \\ f_x & f_y & f_z \end{vmatrix}$$

$$= \begin{bmatrix} 0 & -\frac{\partial}{\partial z} & \frac{\partial}{\partial y} \\ \frac{\partial}{\partial z} & 0 & -\frac{\partial}{\partial x} \\ -\frac{\partial}{\partial y} & \frac{\partial}{\partial x} & 0 \end{bmatrix} \begin{bmatrix} f_x \\ f_y \\ f_z \end{bmatrix} \quad (2.80)$$

$$= \begin{bmatrix} \frac{\partial f_z}{\partial y} - \frac{\partial f_y}{\partial z} \\ \frac{\partial f_x}{\partial z} - \frac{\partial f_z}{\partial x} \\ \frac{\partial f_y}{\partial x} - \frac{\partial f_x}{\partial y} \end{bmatrix}$$

As shown, curl **f** is a vector. It is a measure of the components of **rate of rotation** or **rate of circulation** in the coordinate directions at a point in a vector function field. In mechanics of materials, a typical example is a bar subject to pure torsion in the axial direction. The rate of the twist angle in the axial direction of the bar is proportional to the curl of the shear stress vector function on the cross-section of the bar. We will discuss this in Chapter 9. Here, let us take a look at the curl of two simple vector functions created earlier: the constant vector function and the linear vector function, which can offer some intuitive understanding of the curl of a vector function field.

2.16.2 *Python code and examples*

```
def curl_vf(vf, X):
    '''Compute the curl of a given vector function vf w.r.t X.
       For 3D only.
    '''
    curl_f = sp.Matrix([sp.diff(vf[2], X[1]) - sp.diff(vf[1], X[2]),
                        sp.diff(vf[0], X[2]) - sp.diff(vf[2], X[0]),
                        sp.diff(vf[1], X[0]) - sp.diff(vf[0], X[1])])
    return curl_f
```

```
curl_F = curl_vf(vc, X)         # Or use: vector.curl(vc)
curl_F                          # Curl of the vector function
```

$$\begin{bmatrix} 0 \\ 0 \\ 0 \end{bmatrix}$$

As shown, 1. the curl of a vector function is a vector; 2. when the vector function field is a constant, the curl will be zero because there is no rotation or circulation at all at any point in a constant field.

```
1  curl_F = curl_vf(vl, X)           # Or use: vector.curl(vl)
2  curl_F                            # Curl of the vector function
```

$$\begin{bmatrix} b_3 - c_2 \\ -a_3 + c_1 \\ a_2 - b_1 \end{bmatrix}$$

For linear vector functions, the curl will be constant in all three directions because these three vector functions vary linearly in its own way, creating rotations at any point in the field. The condition for linear vector function field has no rotation with respect to x-axis would be $b_3 = c_2$, no rotation with respect to y-axis is $a_3 = c_1$, and no rotation with respect to z-axis is $a_2 = b_1$. The proof is given in the following codes.

```
1  vl_0curl = Matrix([a1*x+a2*y+a3*z, a2*x+b2*y+c2*z, a3*x+c2*y+c3*z])
2  vl_0curl                          # Linear vector function without rotation
```

$$\begin{bmatrix} a_1 x + a_2 y + a_3 z \\ a_2 x + b_2 y + c_2 z \\ a_3 x + c_2 y + c_3 z \end{bmatrix}$$

```
1  curl_vf(vl_0curl, X)              # Or use: vector.curl(vl_0curl)
```

$$\begin{bmatrix} 0 \\ 0 \\ 0 \end{bmatrix}$$

This is a curl-free field. The field described in this set of specially designed linear vector functions does not have rotation or circulation.

2.17 Remarks

In concluding this chapter, we mention a few high level remarks.

1. Objects in solid mechanics are constructed as tensors. The scalar variables are of 0th order, vectors are first order, and matrices are second order,

etc. The rules for the construction and transformations are the **equilibrium** in mechanics law and **orthogonal projection** in mathematical operations. Thus, all these objects can behave at any point in solid in a consistent and predictable manner.
2. We proved and demonstrated via code example that the basic properties of a vector (first order tensor), the length and direction, remains unchanged under coordinate transformation. The component values in $\mathbf{v}'$ will change to ensure the vector direction remaining unchanged, when measured in reference to the new coordinates. This **basic property preservation** is ensured by the unitary property of $\mathbf{T}$. This applies to both force and displacement vectors to be discussed in the next chapter.
3. We will see also the similar **basic property preservation** feature of a stress or strain tensor (second order) under coordinate transformation. Their principal stresses and strains (and hence the Mohr circles) remain unchanged under coordinate transformation using $\mathbf{T}$.
4. Functions are in general devices that take objects and produce new objects. Objects in solid mechanics vary over the coordinates defined for the solid, and these can be expressed in forms of functions with coordinates as the independent variables. Mathematical operations, such as different forms of differentiations, are closely related to mechanics laws.

Note also that **integrations** are frequently used in solid mechanics. Examples include derivation of displacements from strain/stresses, and in evaluation of strain energy, etc. Readers may review the concepts and techniques for integration detailed in Ref. [6]. We will use these in later relevant chapters.

All these will be elaborated in great detail in the following chapters.

References

[1] G.R. Liu, *Numbers and Functions: Theory, Formulation, and Python Codes*. World Scientific, New Jersey, 2024.
[2] G.R. Liu, *Machine Learning with Python: Theory and Applications*. World Scientific, New Jersey, 2023.
[3] G.R. Liu and S.S. Quek, *The Finite Element Method: A Practical Course*. Butterworth-Heinemann, 2013.
[4] G.R. Liu and T.T. Nguyen, *Smoothed Finite Element Methods*. Taylor and Francis Group, New York, 2010.
[5] G.R. Liu, *Mesh Free Methods: Moving Beyond the Finite Element Method*. Taylor and Francis Group, New York, 2010.
[6] G.R. Liu, *Calculus: A Practical Course with Python*, World Scientific, New Jersey, 2025.

Chapter 3

Forces and Stresses

Let us import the necessary external modules or dependencies for later use in this chapter.

```
1  # Place cursor in this cell, and press Ctrl+Enter to import dependences.
2  import sys                        # for accessing the computer system
3  sys.path.append('../grbin/')   # Change to the directory in your system
4
5  from commonImports import *      # Import dependences from '../grbin/'
6  import grcodes as gr              # Import the module of the author
7  importlib.reload(gr)              # When grcodes is modified, reload it
8
9  from continuum_mechanics import vector
10 init_printing(use_unicode=True)    # For latex-like quality printing
11 np.set_printoptions(precision=4,suppress=True)  # Digits in print-outs
```

3.1 Introduction

A force is a typical vector familiar to readers in almost all disciplines, not just in mechanics of materials. When a solid material is subject to the action of a force, stresses may result in the material. This chapter examines the force-stress relations.

Types of forces: There are mainly two types of force in terms of its spatial distribution features: concentrated force (or point force), and distributed force.

A **concentrated force** acts at a point in a solid material or at a point on the surface of a solid material. It carries a force unit, typically Newton denoted as $[N]$.

A **distributed force** acts over area or volume in three-dimensional (3D) solids, or volume of two-dimensional (2D) solids, or a line in 2D solids, or volume of one-dimensional (1D) solid. It is also often called **body force**. If it is over a volume, the unit is typically $[N/m^3]$ with $[m]$ (meter) being the typical length unit. If it is over an area, the unit can be $[N/m^2]$ and over a line/curve $[N/m]$.

An often encountered distributed force is the pressure on the surface of a solid, which can be exerted by air, water, or granular materials. Forces distributed over an volume, include the gravity force, centrifugal force, and electromagnetic force over solids.

Note that truly concentrated or point force does not exist in practical applications in engineering. It is an idealized case of a distributed force over a very small area or volume. For conveniences in analysis, we lump it to a point, and treat it as a concentrated or point force, when our interest is at a distance away from it.

To study how a force is related to stress, let us first look at a simple 1D case.

3.2 Example: A bar subjected to a point force

Consider a solid with simple geometry of a straight bar with uniform cross-section made of isotropic linear elastic material. Its dimension of the cross-section is far smaller than its length L. The bar is fixed at the left-end, and subjected to a force at the right-end in the longitudinal direction denoted by the 1D coordinate x, as shown in Fig. 3.1.

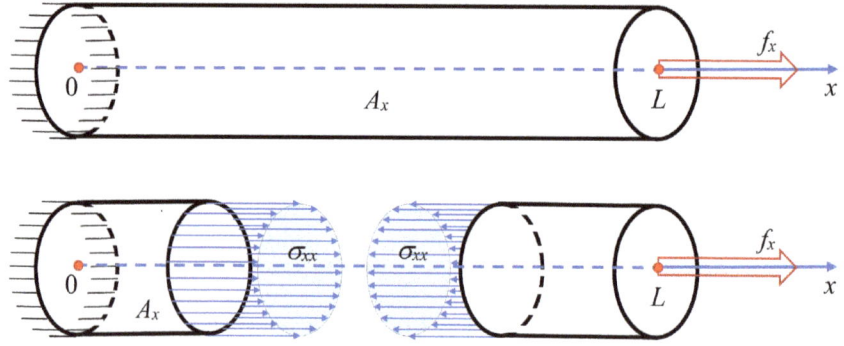

Figure 3.1. A uniform bar with a cross-sectional area of A_x fixed at its left-end and subjected to a force f_x at its right-end. One-dimensional coordinate x is taken along the axial direction of the bar. When a cut is made to expose the cross-section surface, we see the stress σ_{xx}.

Under these conditions, we can assume that stress within the bar is uniform over its cross-section. The stress can be calculated as

$$\sigma_{xx} = \frac{f_x}{A_x} \qquad (3.1)$$

where f_x is the point force in the x-direction. It is applied at a fixed location: $x = L$ and A_x is the area of the cross-section that is **perpendicular** to its axial axis x. Thus, the stress on the cross-section has a unit of typically $[N/m^2]$. The stress distribution along the x coordinate is assumed uniform.

Note here that the stress σ_{xx} carries two subscripts. The first x stands for the surface direction of the cross-section on which the stress is on, and the second x stands for the direction of the stress. We note:

Remark: Stress in solid material depends on both **direction** of the force that causes the stress, and the direction of the **surface** on which the stress is on. It carries two subscripts to fully denote its feature.

With this conventional use of subscripts, the stress is precisely described in a given coordinate system. We call σ_{xx} a **normal stress** because its direction is normal to the cross-section surface.

If our concern is confined to 1D problems, we can drop the two subscripts and denote the stress simply as σ. One may see this notation often in the literature. For us, however, the concern in this chapter is 3D in general, and 1D is just a special case to start our discussion. We shall stick on the two subscripts notation. Also, in general stresses are a second order tensor, and shall always carry two subscripts. We drop the subscripts and use simple notation of σ only when there is no possibility of misleading.

For this 1D case, we have the simple force-stress relationship Eq. (3.1). If the bar is uniform in cross-section, A_x is constant. Since f_x is at a fixed location, it is a constant too (does not change with x). Therefore, σ_{xx} is also constant. Its derivative with respect to x will be zero.

$$\frac{d\sigma_{xx}}{dx} = 0 \qquad (3.2)$$

Note that since our problem is 1D, the shape of the cross-section of the bar is immaterial. For example, if the cross-section is square instead of circular, the foregoing equations shall still hold. All we need is the bar is long compared to its largest cross-sectional dimension, so that the stress can be assumed uniformly distributed over the cross-section.

3.3 Example: A bar subjected to a distributed force

3.3.1 *Derivation of equilibrium equation*

Assume the external force applied to the bar is a body force b_x distributed along the bar axial axis, as shown in Fig. 3.2.

Assume that stress is uniform over its cross-section, but it will not be uniform along its axial direction x because of the distributed body force. The stress will be a function of the coordinate x.

To analyze the stress at a location x in the bar, we assume that the function of stress is sufficiently smooth and hence differentiable at least for once. We can then cut off a free-body with an infinitely small dx, and look at the equilibrium condition for the free-body. The diagram is shown at the bottom of Fig. 3.2. We then put all the forces involved on the free-body diagram.

Sign convention: As a **general rule**, the direction of a force put on a free-body diagram should follow the **standard convention**. For example, the internal force $N_x + dN_x$ is on the positive surface, it should point to the positive x-direction. The internal force N_x is on the negative surface, it should point to the negative x-direction. The externally applied force $b_x A_x dx$ should always in the positive x-direction. This general rule applies to all the free-body diagrams we use in this book, including the ones used in the future chapters.

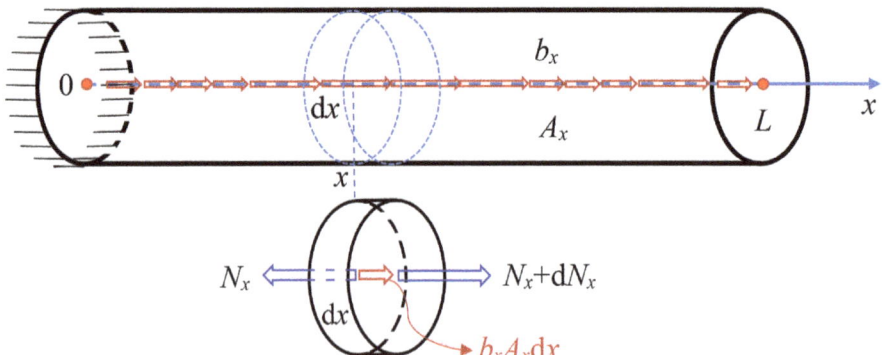

Figure 3.2. A bar fixed at its left-end and subjected to a body force b_x distributed along the bar axial axis. A free-body diagram with length dx is shown with all the forces on it. The normal forces N_x and $N_x + dN_x$ are called internal force because it can only be seen after the free-body is cut out. Force $(b_x A_x dx)$ is the externally applied force. It is there regardless of whether the free-body is cut out or not.

Using the equilibrium condition that all forces in the x-direction must be zero, we have

$$\sum F_x = 0 : -N_x + (N_x + dN_x) + b_x A_x dx = 0 \quad (3.3)$$

In the summation above, forces in the positive x-direction are positive. A minus sign is given to any force pointing to the negative direction of x. Equation (3.3) can be rewritten as

$$\frac{dN_x}{dx} + b_x A_x = 0 \quad (3.4)$$

The stress σ_{xx} on the cross-section should be

$$\sigma_{xx} = \frac{N_x}{A_x} \quad (3.5)$$

Substituting Eq. (3.5) into Eq. (3.4) gives

$$\boxed{\frac{d\sigma_{xx}}{dx} + b_x = 0} \quad (3.6)$$

This is the **equilibrium equation** in terms of stress for 1D bars. Equation (3.6) can be re-written as

$$\frac{d\sigma_{xx}}{dx} = -b_x \quad (3.7)$$

This means that $\frac{d\sigma_{xx}}{dx}$ is the negated externally applied body force. Thus, for 1D cases, the derivative of the stress is called **divergence** of the stress. Equilibrium equation (3.6) states that the **stress divergence** is compensated by the externally **body force**. If there is no body force, the stress will not diverge, its derivative will be zero, as seen in Eq. (3.2) for the previous example where no body force is involved.

3.3.2 Stress solution to 1D equilibrium equation

Equation (3.7) is a differential equation (DE) in stress. If the body force is an integrable function of x, we can find the stress solution analytically. As a simple example, if b_x is constant, we have the general stress solution:

$$\sigma_{xx} = -b_x x + c \quad (3.8)$$

where c is an integral constant[2]. It can be determined using boundary condition (BC). For our problem defined in Fig. 3.2, the stress BC is

$$\text{At} \quad x = L : \sigma_{xx} = 0 \quad (3.9)$$

Using BC Eq. (3.9) in Eq. (3.8), we found
$$c = b_x L \qquad (3.10)$$
The final solution becomes
$$\sigma_{xx} = -b_x x + b_x L = b_x (L - x) \qquad (3.11)$$
It is a linear function of x. It is zero at $x = L$ and has a maximum of $\sigma_{xx} = b_x L$ at $x = 0$.

Readers may set body force b_x to another integrable function, such as a polynomial or trigonometric function and then find the stress solution along the bar. Readers may consider cases with concentrated force f_x at $x = L$, in addition to the body force.

3.4 General measure of force vectors in 3D space

Forces are vectors. Thus, in higher dimensions we need a proper description for it.

3.4.1 *Graphic representation of force vectors*

Consider an arbitrary force acting at the center of a rigid block denoted as **f** in $\mathbb{R}^3$, as shown in Fig. 3.3. It can be decomposed into three components, as discussed Section 2.8.2.

3.4.2 *Components of a force, orthogonal projection*

We first decompose the force vector to two vectors: a vertical component $f_3 \mathbf{i}_3$ and $\mathbf{f}_s$ following the **orthogonal projection rule**. The $\mathbf{f}_s$ is then further decomposed into two vectors: $f_1 \mathbf{i}_1$ and $f_2 \mathbf{i}_2$ following also the orthogonal projection rule. This process results in three components of **f**: f_1, f_2, and f_3, respectively, on x_1-, x_2-, and x_3-axis. The force vector can be written uniquely as

$$\mathbf{f} = \begin{bmatrix} f_1 \\ f_2 \\ f_3 \end{bmatrix} \qquad (3.12)$$

3.4.3 *Equivalence of force vector and its components*

It is clear that this representation of a vector is **unique** under a fixed coordinate, implying that any change, in length or in direction, in **f** will result in unique changes in values of f_1, f_2 and f_3, and vice versa, as long as the **orthogonal projection** rule is followed, and an orthogonal coordinate system is used. This mutual unique representation ensures that force vector **f** are equivalent to its components $f_1, f_2,$ and f_3.

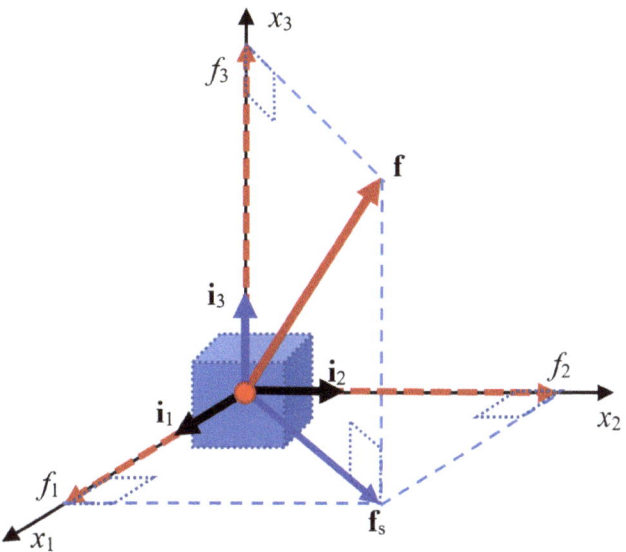

Figure 3.3. A force **f** acting at the center of a rigid block. It can be decomposed into three components f_1, f_2 and f_3 on three axes of an orthogonal coordinate system, respectively, following the orthogonal projection rule.

3.4.4 *Force vector in transformed coordinates*

Note that when the coordinate system transforms (rotates), the values of $f_1, f_2,$ and f_3 change according to the orthogonal projection rule, but the basic property, amplitude and direction, of vector **f** do not change. Therefore, the force vector is a tensor. It is of order one because its components have only one index.

Consider a force vector **f** in coordinate **x**. Using the transformation tensor **T** defined in the previous chapter, the values of force vector **f'** in the rotated coordinate **x'** become

$$f'_i = a_{ij} f_j \tag{3.13}$$

where f'_i is obtained following the orthogonal projection rule. On the right-hand side of Eq. (3.13), we see that the **2nd index** j of **T** is **contracted** with the index of **f**. This leaves one index i, implying that the outcome is a new vector.

In matrix notation, we have

$$\mathbf{f'} = \mathbf{T}\mathbf{f} \tag{3.14}$$

which achieves the correct contraction through the standard matrix multiplication rule.

As proven in the previous chapter, **T** is unitary. Thus, the inverse transformation can be simply done using the transpose of **T**:

$$\mathbf{f} = \mathbf{T}^\top \mathbf{f}' \tag{3.15}$$

Or in the indicial notation:

$$f_i = a_{ji} f'_j \tag{3.16}$$

All we need is to **swap the order of the two indexes** of a_{ij}, which is equivalent to using $\mathbf{T}^\top$ in matrix multiplication.

3.5 General measure of stresses in 3D solids

3.5.1 *General definition of stress*

Consider a 3D solid with arbitrary geometry. It is constrained so that it cannot have any rigid-body displacements, and is loaded by point forces and/or distributed forces like pressure, resulting in stresses distributed over the solid. A schematic drawing is given in Fig. 3.4.

To view the stresses inside the solids, we use an imaginary plane-P to cut off a part of the solid and expose a surface with unit normal direction of **n**. Since any surface can be identified by its normal, let us call it surface-n for convenience in our analysis. The effect of the cut-off part is replaced by a force vector **f** that can be in arbitrary direction changing from location to location. This force results in stresses distributed over the surface-n, implying that the value and direction of the stresses vary also from location to location.

Now, look at a small area ΔA_n on the surface-n. The force on ΔA_n becomes $\Delta \mathbf{f}$, and a stress vector σ_n on it caused by $\Delta \mathbf{f}$ can be obtained using

$$\sigma_n = \lim_{\Delta A_n \to 0} \frac{\Delta \mathbf{f}}{\Delta A_n} \tag{3.17}$$

The direction of stress vector σ_n is the same as **f** at the limit. Here, **n** stands for the surface of the stress vector is on (not the direction of the stress!). If **n** is taken as the direction of the x-axis of the Cartesian coordinate system, we shall have σ_x defined, which is the stress vector is on surface-x. The same argument applies to σ_j on surface-y, and σ_k on surface-z.

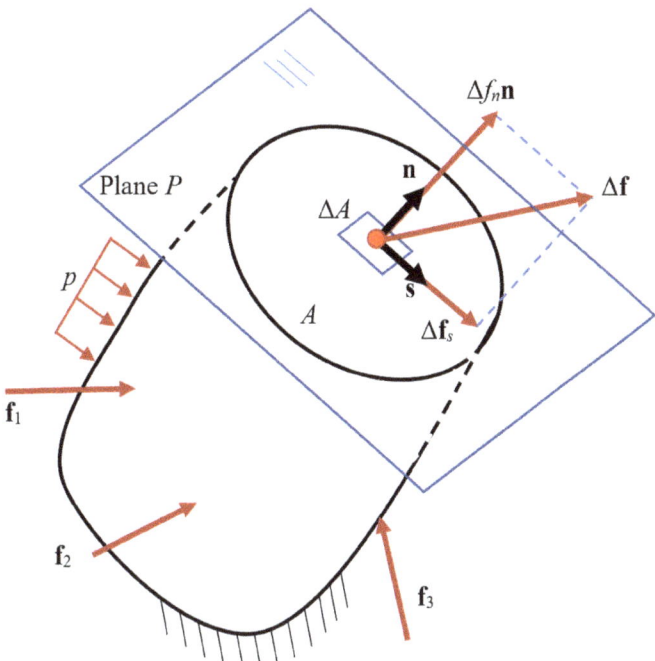

Figure 3.4. A (3D) solid with arbitrary geometry. It is constrained and loaded by forces and/or distributed forces like the pressure. An imaginary plane-P cuts off a part of the solid and exposes the surface-n.

3.5.2 Shear stresses on a surface

The shear force vector $\Delta \mathbf{f}_s$ results in a stress vector expressed as

$$\boldsymbol{\sigma}_s = \lim_{\Delta A_n \to 0} \frac{\Delta \mathbf{f}_s}{\Delta A_n} \qquad (3.18)$$

The direction of stress vector $\boldsymbol{\sigma}_s$ is the same as $\mathbf{f}_s$ at the limit.

Note that $\boldsymbol{\sigma}_n$ and $\boldsymbol{\sigma}_s$ are at the same point in the same surface, following the orthogonal projection rule.

Next, to quantify the shear stress vector $\boldsymbol{\sigma}_s$, we introduce a 2D orthogonal coordinate system (r, q) on the surface, as shown in Fig. 3.5.

The force component Δf_r results in the following shear stress component on surface-n:

$$\sigma_{nr} = \lim_{\Delta A_n \to 0} \frac{\Delta f_r}{\Delta A_n} \qquad (3.19)$$

where σ_{nr} is the stress component on surface-n and in the r-direction.

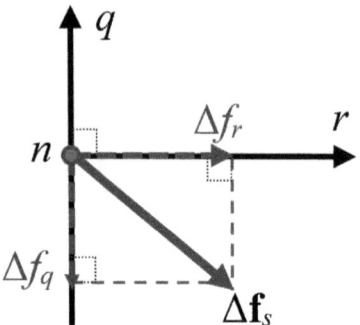

Figure 3.5. The shear force on surface-n (plane P) and its components on the $r - q$ coordinates that are orthogonal. We require that **r** crosses **q** gives **n**, leading to a local orthogonal coordinates obeying the right-hand rule.

The force component Δf_q results in:

$$\sigma_{nq} = \lim_{\Delta A_n \to 0} \frac{\Delta f_q}{\Delta A_n} \qquad (3.20)$$

where σ_{nq} is the stress component on surface-n and in the q-direction.

The choice of coordinates r and q can be arbitrary, as long as they both are orthogonal, and follow the right-hand-rule with **n**: $\mathbf{n} = \mathbf{i}_r \times \mathbf{i}_q$, where $\mathbf{i}_r$ and $\mathbf{i}_q$ are unit vectors, respectively, on the r- and q-axes.

In summary, we obtain three stress components on surface-n at the center of ΔA_n: normal stress σ_{nn}, shear stresses σ_{nr}, and σ_{nq}. The directions of these three stress components are orthogonal. These three stress components quantify completely and uniquely the stress vector $\boldsymbol{\sigma}_n$ on surface-n at the center of ΔA_n for a given local Cartesian coordinate system.

Now, suppose **n** is taken as the direction of the z-axis, r is taken as the x-axis, and q as the y-axis, we shall have normal stress σ_{zz}, shear stress σ_{zx} and shear stress σ_{zy}. These stresses all on the same surface-z pointing, respectively, to x, y, and z directions.

By the same argument, when **n** is taken as x-axis, we shall have σ_{xx}, σ_{xz} and σ_{xy}. When **n** is taken as y-axis, it gives σ_{yy}, σ_{yz}, and σ_{yx}.

Let us state some important **remarks:**

- All stress measure depends on the area of the surface, in addition to its direction. Thus, **there must be two indexes for a stress component.** This is the root reason for stress to be a second-order tensor.
- At any point on a surface in 3D solid, there are in general **three stress components**: one normal and two shears. **The directions of these three stress components are orthogonal**, as long as the coordinates used are orthogonal.

3.5.3 Stress representation in 3D solids

Consider a point in a stressed solid. To view all the stress components at the point, we must expose the surfaces surrounding the point. From the discussion above, we know that each surface can have a total of three stress components: one normal and two shears. To present these stress components on surfaces that are orthogonal with each other, we cut a small cube from the solid surrounding that point, as shown in Fig. 3.6. These surfaces on the cube are along with the Cartesian coordinates and hence orthogonal.

There are six surfaces on the cube, forming three pairs of surfaces with opposite direction. All these pairs of surfaces are infinitely close to represent the point. A Cartesian coordinate is defined with three axes: x, y, and z, all being perpendicular to these pair of surfaces. On each of these six surfaces, we put three stress components on it. All these stress components are now shown in Fig. 3.6, all following our standard sign convention.

On any pair of opposite surfaces, the stress components should be the same because these surfaces are infinitely close. But they point to opposite directions because of the opposite surface direction. Therefore, there are a total of nine stress components on the cube representing the stress states at the point. The notation convention for a stress component σ_{ij} is as follows:

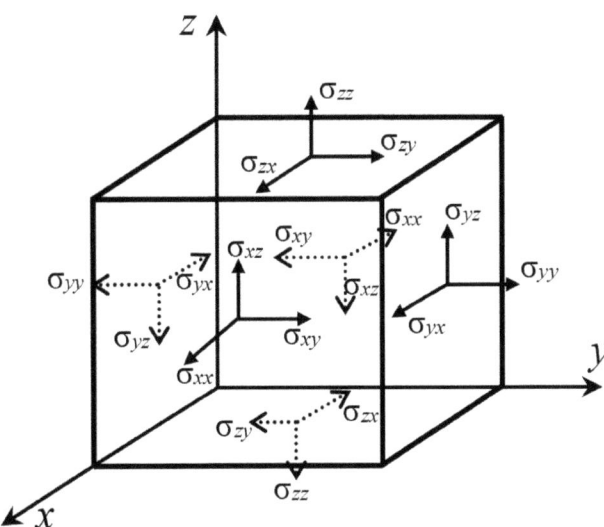

Figure 3.6. Stress components on surfaces of a cube cut along a Cartesian coordinate system. There are three pairs of orthogonal surfaces: surface-x, -y, and -z. Each pair has two opposite normals. The directions of these three stress components on a surface are orthogonal.

$$\sigma \underbrace{i}_{\text{surface that stress is on}} \underbrace{j}_{\text{direction of the stress}} \tag{3.21}$$

in which $i, j = x, y, z$ (or 1, 2, 3). Thus, these indexes covers all the nine stress components uniquely.

3.5.4 Shear stress equivalence, symmetric tensor

Let us examine the relationship between a pair of shear stresses: σ_{yz} and σ_{zy}, by considering the mechanics law of equilibrium conditions, using Fig. 3.7.

We look at the moment equilibrium about the axis of A-A' in the x-direction. The stress components that contribute to such a moment are in red: a pair of σ_{yz} on two surfaces-y, and pair of σ_{zy} on two surfaces-z. Due to the orthogonality, all other stresses contribute nothing to the moment about A-A'. The equilibrium law requires the total moment caused by the forces resulting from these stresses being zero, we have

$$\sum M_{A-A'} = 0 : \underbrace{(\sigma_{yz} dxdz)}_{\text{force}} \underbrace{dy}_{\text{arm}} - \underbrace{(\sigma_{zy} dxdy)}_{\text{force}} \underbrace{dz}_{\text{arm}} = 0 \tag{3.22}$$

which leads to the shear stress equivalence for σ_{yz} and σ_{zy}:

$$\sigma_{yz} = \sigma_{zy} \tag{3.23}$$

Note that the foregoing equation holds regardless of the actual length of dx or dy, or dz, as long as they are small and the coordinates are orthogonal.

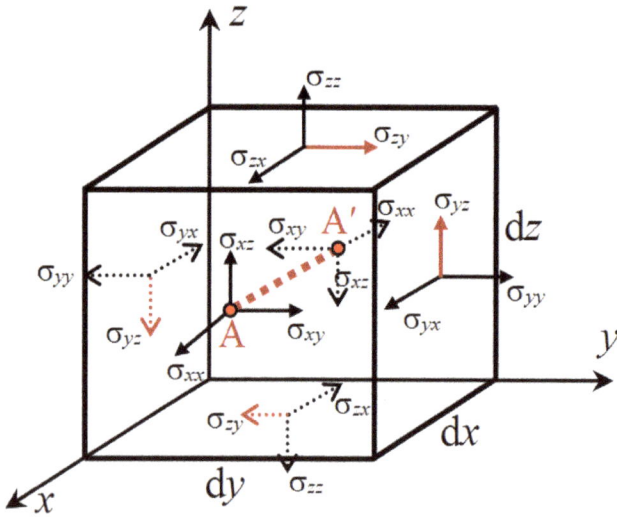

Figure 3.7. Moment equilibrium about the axis of A-A'. All the stress components on the cube that contribute to the moment are in red.

Similarly, examining the moment equilibrium about other two axes in other two directions, we obtain

$$\sigma_{xz} = \sigma_{zx}$$
$$\sigma_{xy} = \sigma_{yx} \qquad (3.24)$$

Readers may try to get one of these equations in Eq. (3.24). Equations (3.23) and (3.24) are known as the **shear stress equivalence**. It can be written in indicial notation as

$$\sigma_{ij} = \sigma_{ji} \quad \forall i, j = 1, 2, 3 \qquad (3.25)$$

Due to these three equivalence conditions, the number of independent stress components become six.

Equations (3.25) is known as the symmetry of stress components, and hence the stress tensor is thus said to be a **symmetric tensor**: their indexes are interchangeable. It is all rooted at the **shear stress equivalence**, as the consequence of the mechanics **equilibrium condition**. As stated in Noether's theorem "every differentiable symmetry of the action of a physical system with conservative forces has a corresponding conservation law". In mechanics of materials, the conservative law is equilibrium.

3.5.5 *Dimensionality and stresses*

Stress in mechanics of materials in general is a symmetric tensor of second order, as discussed above. It can be written in indicial notation simply as σ_{ij}. For 1D problems, i and j have a range of 1, for 2D problems, i and j a range of 2, and for 3D problems, a range of 3.

It has also the following three form of matrix representations:

$$\boldsymbol{\sigma} = [\sigma_{ij}] = \begin{bmatrix} \sigma_{11} & \sigma_{12} & \sigma_{13} \\ \sigma_{21} & \sigma_{22} & \sigma_{23} \\ \sigma_{31} & \sigma_{32} & \sigma_{33} \end{bmatrix} = \begin{bmatrix} \sigma_{xx} & \sigma_{xy} & \sigma_{xz} \\ \sigma_{yx} & \sigma_{yy} & \sigma_{yz} \\ \sigma_{zx} & \sigma_{zy} & \sigma_{zz} \end{bmatrix} \qquad (3.26)$$

Since the stress tensor is symmetric, in space $\mathbb{R}^3$, the matrix is a 3×3 symmetric matrix [1,3,4]. It is clear again that there are only 6 independent components. This book uses all these three notation interchangeably.

In space $\mathbb{R}^2$, it is a 2×2 symmetric matrix (considering symmetry), it has only 3 components. Eq. (3.26) becomes

$$\boldsymbol{\sigma} = [\sigma_{ij}] = \begin{bmatrix} \sigma_{11} & \sigma_{12} \\ \sigma_{21} & \sigma_{22} \end{bmatrix} = \begin{bmatrix} \sigma_{xx} & \sigma_{xy} \\ \sigma_{yx} & \sigma_{yy} \end{bmatrix} \qquad (3.27)$$

In space $\mathbb{R}^1$, it is a scalar, and hence there is only 1 component. One may write it in any of these forms: σ_{11}, σ_{xx}, or simply σ. This is the case we examined at the beginning of this chapter.

The analysis given above shall establish the basic concept about the stresses and stress tensor. It is the right time to reinforce the concept, by writing the following code to create in computer the stress tensor. First, define stress tensor for later symbolic computations.

```
1  # define stress components as symbolic variable
2  s11, s12, s13 = "\u03C3_11", "\u03C3_12", "\u03C3_13"
3  s21, s22, s23 = "\u03C3_21", "\u03C3_22", "\u03C3_23"
4  s31, s32, s33 = "\u03C3_31", "\u03C3_32", "\u03C3_33"
5
6  S = MatrixSymbol('S', 3, 3)
7
8  S = Matrix([[s11, s12, s13],
9              [s21, s22, s23],
10             [s31, s32, s33]])
11 S                                           # a stress tensor
```

$$\begin{bmatrix} \sigma_{11} & \sigma_{12} & \sigma_{13} \\ \sigma_{21} & \sigma_{22} & \sigma_{23} \\ \sigma_{31} & \sigma_{32} & \sigma_{33} \end{bmatrix}$$

Stresses in 2D can be obtained by slicing in Python.

```
1  S2D = S[:2,:2]
2  S2D
```

$$\begin{bmatrix} \sigma_{11} & \sigma_{12} \\ \sigma_{21} & \sigma_{22} \end{bmatrix}$$

Further, the stress in 1D can be obtained by indexing in Python.

```
1  S[0,0]
```

σ_{11}

3.6 Tractions on an arbitrary surface

For a given stress tensor σ at a point in solid, we often need to know the traction $\mathbf{t}_n$ on a pre-specified surface with unit normal direction $\mathbf{n}$. The traction $\mathbf{t}_n$, as shown in Fig. 3.8, on the surface of the solid is the externally applied force (per unit area) there. It carries the same unit of the stress. When the point of interest is inside the solid, it is the stress vector denoted as $\boldsymbol{\sigma}_n$ when the surface at that point is exposed. Positive traction component is in the positive direction of the coordinates, as shown in Fig. 3.8.

We need to derive the formulas for this task.

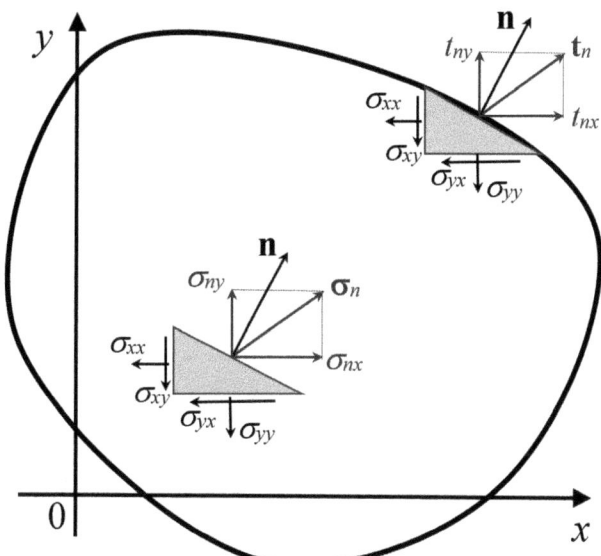

Figure 3.8. As schematic draw of tractions on the surface at a point with normal **n** on the boundary of the solid, or on a surface at a point inside a 2D solid.

3.6.1 *Formulation*

Consider in general a 3D solid, and suppose we know the stress tensor $\boldsymbol{\sigma}$ (in nine components) at a point in solid. We would like to know the traction on a pre-specified surface with unit normal direction of $\mathbf{n} = (n_1, n_2, n_3)$, in which n_1, n_2, n_3 are the direction cosines of the unit normal vector. We shall have

$$n_1 = \cos(x, n)$$
$$n_2 = \cos(y, n) \quad (3.28)$$
$$n_3 = \cos(z, n)$$

The setting is shown in Fig. 3.9. Note in the literature, some authors may use $\mathbf{n} = (l, m, n)$ to denote these direction cosines.

3.6.2 *Stress vector projection*

First, the stress vector $\boldsymbol{\sigma}_n$ on surface-n can be projected orthogonally on the x, y and z-axis, or x_1, x_2 and x_3-axis in indicial notation. It can be expressed as follows:

$$\boldsymbol{\sigma}_n = \sigma_{nx}\mathbf{i}_1 + \sigma_{ny}\mathbf{i}_2 + \sigma_{nz}\mathbf{i}_3 \quad (3.29)$$

where $\mathbf{i}_1, \mathbf{i}_2$, and $\mathbf{i}_3$ are, respectively, the unit vectors on the x_1, x_2 and x_3-axes. These unit vectors are also denoted in literature as $\mathbf{i}, \mathbf{j}$, and $\mathbf{k}$.

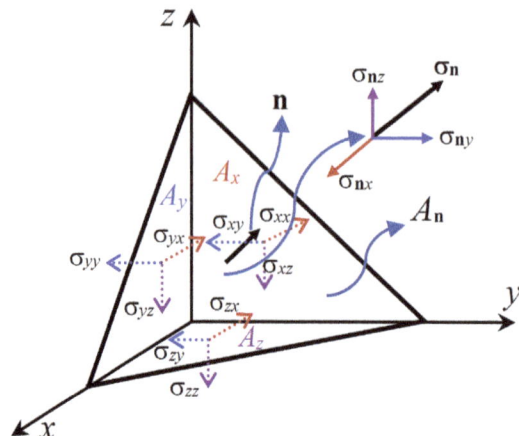

Figure 3.9. Stresses on an arbitrary surface-n for a point in solid with stress tensor given.

Equation (3.29) can be written as

$$\sigma_n = \sigma_{ni} l_i \tag{3.30}$$

in which $\sigma_{n1} = \sigma_{nx}$, $\sigma_{n2} = \sigma_{ny}$, and $\sigma_{n3} = \sigma_{nz}$. Equation (3.30) or (3.29) is a result from stress vector projection.

3.6.3 Area projection

The forces resulting from σ_{ni}, $i = 1, 2, 3$ should be **in equilibrium** with the forces resulting from the given nine stress components. Consider first the components in the x-direction, we have

$$\sigma_{nx} A_n = A_x \sigma_{xx} + A_y \sigma_{yx} + A_z \sigma_{zx} \tag{3.31}$$

where A_n is the area of surface-n, A_x is the area of surface-x, A_y the area of surface-y, and A_z the area of surface-z, as shown in Fig. 3.9. These areas have the following relationship:

$$\begin{aligned} A_x &= n_1 A_n \\ A_y &= n_2 A_n \\ A_z &= n_3 A_n \end{aligned} \tag{3.32}$$

This is because A_x is the **orthogonal projection** of A_n along the x direction, A_y is that along y direction, and A_z is that along z. Equation (3.31) becomes

$$\sigma_{nx} = n_1 \sigma_{xx} + n_2 \sigma_{yx} + n_3 \sigma_{zx} \quad \text{on surface-}n \text{ in } x\text{-direction} \tag{3.33}$$

This is the stress on surface-n in the x-direction. Similarly, we have

$$\sigma_{ny} = n_1\sigma_{xy} + n_2\sigma_{yy} + n_3\sigma_{zy} \quad \text{on surface-}n\text{ in }y\text{-direction}$$
$$\sigma_{nz} = n_1\sigma_{xz} + n_2\sigma_{yz} + n_3\sigma_{zz} \quad \text{on surface-}n\text{ in }z\text{-direction} \tag{3.34}$$

It is clear that Eqs. (3.33) and (3.34) is the consequence of the orthogonal projections of the area, and of the equilibrium condition. Three expressions in Eqs. (3.33) and (3.34) can be written simply in a single equation in indicial notation:

$$\sigma_{nj} = n_i\sigma_{ij} \tag{3.35}$$

Note that the contraction occurs on the **first axis of the stress tensor**, leaving a single index j, producing a vector on the given surface-n. In matrix notation, we have the following formula with dot product to achieve the needed contraction on the first axis of the stress tensor.

$$\boldsymbol{\sigma}_n = \mathbf{n} \cdot \boldsymbol{\sigma} \tag{3.36}$$

Or in matrix form (following matrix-vector multiplication rule):

$$\underbrace{\begin{bmatrix} \sigma_{nx} \\ \sigma_{ny} \\ \sigma_{nz} \end{bmatrix}}_{\boldsymbol{\sigma}_n} = \underbrace{\begin{bmatrix} \sigma_{xx} & \sigma_{yx} & \sigma_{zx} \\ \sigma_{xy} & \sigma_{yy} & \sigma_{zy} \\ \sigma_{xz} & \sigma_{yz} & \sigma_{zz} \end{bmatrix}}_{\boldsymbol{\sigma}^\mathsf{T}} \underbrace{\begin{bmatrix} n_1 \\ n_2 \\ n_3 \end{bmatrix}}_{\mathbf{n}} \tag{3.37}$$

This means the stress vector on surface-n is the orthogonal projection of the transposed stress matrix on the unit normal vector $\mathbf{n}$. Note that the stress vector are written in components in the x-, y-, and z-axes.

3.6.4 *Traction on a boundary surface*

When the surface-n is on a boundary surface of a solid, Eq. (3.35), σ_{ni} shall give the **traction** applied on the boundary surface and is often denoted as t_{ni}. It is written as

$$\boxed{t_{nj} = n_i\sigma_{ij}} \tag{3.38}$$

or explicitly

$$t_{nx} = n_1\sigma_{xx} + n_2\sigma_{yx} + n_3\sigma_{zx}$$
$$t_{ny} = n_1\sigma_{xy} + n_2\sigma_{yy} + n_3\sigma_{zy} \tag{3.39}$$
$$t_{nz} = n_1\sigma_{xz} + n_2\sigma_{yz} + n_3\sigma_{zz}$$

This traction-stress relation bridges the internal stresses σ_{ij} with the externally applied tractions on the boundary of the solid t_{nj}. It is thus frequently used in expressing the **stress boundary conditions** for solid mechanics problems.

Equation (3.39) satisfies the equilibrium conditions, implying that the tractions on the boundary at a point in solid is uniquely determined by stress states at the point. If, for example, surface-n is the surface-x (the surface normal is the x-axis), we have $\mathbf{n} = (1, 0, 0)$, and

$$t_{nx} = \sigma_{xx} \quad t_{ny} = \sigma_{xy} \quad t_{nz} = \sigma_{xz} \tag{3.40}$$

If surface-n is the negative surface-x, we have $\mathbf{n} = (-1, 0, 0)$, and

$$t_{nx} = -\sigma_{xx} \quad t_{ny} = -\sigma_{xy} \quad t_{nz} = -\sigma_{xz} \tag{3.41}$$

Equations (3.40) and (3.41) is a typical relationship between the external stress t_{ni} and internal stress σ_{ij}. Readers may take a moment examine these two equations for better understanding traction. Readers may try to figure out similar relations, if the boundary is on surface-y or on surface-z, both positive and negative.

3.6.5 Stress vector projection, again

3.6.5.1 Normal stress on a surface

Once the stress vector on a surface-n is obtained, one can then compute the normal stress on surface-n, by projecting the stress vector onto the unit normal $\mathbf{n}$ of the surface.

$$\sigma_{nn} = n_i \sigma_{ij} n_j \tag{3.42}$$

This time the contraction occurs on the second axis of the stress tensor. It can be achieved with the dot product of the traction vector on the left with the unit normal $\mathbf{n}$ of the plane on the right. This is the **orthogonal projection** of σ_n on direction $\mathbf{n}$. At the end, both indexes of the stress tensor are contracted, producing just a scalar, which is the normal stress on the normal direction of the surface-n. Its explicit expression can be found using the following code.

```
1  σ11, σ22, σ33, σ23, σ13, σ12 = symbols("σ11, σ22, σ33, σ23, σ13, σ12")
2  n1, n2, n3, σnn = symbols("n1, n2, n3, σ_nn")
3
4  σ = Matrix([[σ11, σ12, σ13],                    # ε: stress tonsor
5              [σ12, σ22, σ23],
6              [σ13, σ23, σ33]])
7
8  n = Matrix([n1, n2, n3])
9  σnn = (n.T@σ@n)[0].simplify()
10 gr.printM(σnn, "The mormal stress on surface-n:")
```

The mormal stress on surface-n:

$$n_1 \left(n_1 \sigma_{11} + n_2 \sigma_{12} + n_3 \sigma_{13} \right) + n_2 \left(n_1 \sigma_{12} + n_2 \sigma_{22} + n_3 \sigma_{23} \right)$$
$$+ n_3 \left(n_1 \sigma_{13} + n_2 \sigma_{23} + n_3 \sigma_{33} \right)$$

Equation (3.42) can also be written in vector operation as

$$\sigma_{nn} = \boldsymbol{\sigma}_n \cdot \mathbf{n} \tag{3.43}$$

Equations (3.42) and (3.43) are identical.

3.6.5.2 Shear stress on a surface

To find the shear stress in a direction on surface-n, all we need to know is the direction cosine of the direction vector. Assume it is given as $\mathbf{q} = (q_1, q_2, q_3)$. Since $\mathbf{q}$ is on the surface-n, $\mathbf{q}$ and $\mathbf{n}$ is orthonormal (orthogonal and all with unit length), $\mathbf{q} \cdot \mathbf{n} = 0$. The shear stress is thus the orthogonal projection the stress vector $\boldsymbol{\sigma}_n$ onto $\mathbf{q}$:

$$\sigma_{nq} = n_i \sigma_{ij} q_j \tag{3.44}$$

Note that there are infinite number of $\mathbf{q}$ on the surface satisfying the orthonormal condition. Among which, there must be a unit vector that is orthonormal to $\mathbf{q}$, satisfying also the right-hand rule: $\mathbf{r} = \mathbf{q} \times \mathbf{n}$. Let us denote it as $\mathbf{r} = (r_1, r_2, r_3)$. The shear stress on the direction of $\mathbf{r}$ becomes

$$\sigma_{nr} = n_i \sigma_{ij} r_j \tag{3.45}$$

In this setting, $\mathbf{r}, \mathbf{q}$, and $\mathbf{n}$ form a Cartesian coordinate system obeying the right-hand-rule.

3.6.5.3 The magnitude of the shear stress on a surface

The magnitude of the shear stress vector on a surface-n can be computed using the Pythagorean theorem (also a type of orthogonal projection),

$$\begin{aligned}|\sigma_{ns}| &= \sqrt{|\boldsymbol{\sigma}_n|^2 - \sigma_{nn}^2} \\ &= \sqrt{\sigma_{n1}^2 + \sigma_{n2}^2 + \sigma_{n3}^2 - \sigma_{nn}^2} \\ &= \sqrt{\sigma_{nq}^2 + \sigma_{nr}^2}\end{aligned} \tag{3.46}$$

Let us write the following Numpy codes to compute stresses on a given surface.

3.6.6 Example: Stresses on a given surface

This example computes stresses on a given surface for a given stress tensor with nine stress components. For the convenience in future possible uses, we write a simple Python function for this kind of task.

```python
def stressesOnN(σ, n, q=False):
    '''Compute the traction (a stress vector), normal stress, and \
    magnitude of the shear stress, and the shear stress on a surface
    with normal vector n.
    Input σ: stress tensor in matrix form with 9 components\
          n: normal vector of the surface
          q: a unit vector on the surface; an vector r orthogonal to q
             will be generated based on the right-hand rule, and
             the shear stresses in both q&r directions will be computed.
             If not provided, the shear stresses are not computed.
    Return: σ_n (traction), σ_nn (normal stress), σ_ns (shear stress)
            σ_nq (shear stress in given q-direction)
            σ_nr(shear stress in r-direction)
    '''
    n = n/np.linalg.norm(n) # normalize vector n -> unit normal vector

    # Compute the traction vector on the surface with normal n
    σ_n = np.dot(n,σ)

    # Compute the normal stress on surface in the normal n direction
    σ_nn = np.dot(σ_n,n)

    # compute the magnitude of the shear stress on the surface
    σ_ns = np.sqrt(σ_n.dot(σ_n)-σ_nn**2)

    # Compute the shear stresses on the surface, if q is given
    if isinstance(q, (list, tuple, np.ndarray)):
        q = q/np.linalg.norm(q)              # normalize vector q
        ndq = np.dot(n,q)

        if ndq > 1e-12:
            print(f'n and q are not orthogonal: {ndq}; σnq & σnr=None.')
            σ_nq = None; σ_nr = None
        else:
            r = np.cross(n, q)      # generate r via right-hand rule
            print(f'Unit vector r on surface-n = {r}\n')

            # Compute the shear stress on the surface direction q
            σ_nq = np.dot(σ_n,q)

            # Compute the shear stress on the surface direction q
            σ_nr = np.dot(σ_n,r)
    elif q == False:
        print(f'Shear stress will not be computed.')
        σ_nq = None; σ_nr = None
    else:
        print(f'q must be array-like')
        σ_nq = None; σ_nr = None

    return σ_n, σ_nn, σ_ns, σ_nq, σ_nr
```

```
1   # Uniaxial loading case studied at the beginning of the chapter.
2
3   # Given stress tensor (matrix) in MPa
4   St = np.array([[1.,0., 0.],[0., 0., 0.],[0.,0.,0.]]) # only σxx.
5   print(f'Stress Tensor (MPa) :\n{St}')
6
7   n = np.array([-1.,0.,0.]) # the surface normal is positive x-direction
8   #n = np.array([1/np.sqrt(3),1/np.sqrt(3),1/np.sqrt(3)])   # Octahedral
9   print(f'A surface with normal direction n = {n}\n')
10
11  q = np.array([0. ,1.,0.])                    # must be orthogonal to n
12  q = q/np.linalg.norm(q)                      # normalize vector q
13  print(f'Normal vector q on surface-n = {q}\n')
14
15  σ_n, σ_nn, σ_ns, σ_nq, σ_nr = stressOnN(St, n, q) # or gr.stressOnN()
16  print(f'Traction (stress vector) on the surface = {σ_n} MPa')
17  print(f'normal stress on the surface = {σ_nn:2.4f} MPa')
18  print(f'Magnitude of the shear stress on the surface = {σ_ns:2.4f} MPa')
19  print(f'shear stress on the surface in q-direction    = {σ_nq} MPa')
20  print(f'shear stress on the surface in r-direction    = {σ_nr} MPa')
```

```
Stress Tensor (MPa) :
[[1. 0. 0.]
 [0. 0. 0.]
 [0. 0. 0.]]
A surface with normal direction n = [-1.  0.  0.]

Normal vector q on surface-n = [0. 1. 0.]

Unit vector r on surface-n = [ 0.  0. -1.]

Traction (stress vector) on the surface = [-1.  0.  0.] MPa
normal stress on the surface = 1.0000 MPa
Magnitude of the shear stress on the surface = 0.0000 MPa
shear stress on the surface in q-direction    = 0.0 MPa
shear stress on the surface in r-direction    = 0.0 MPa
```

In this example, we purposely set the normal vector of the surface in negative x-direction (with $n = $ np.array$([-1., 0., 0.])$), and found that the component of the traction in the x-axis on surface-n is negative. This is because the sign of the traction follows the sign of the coordinates. The normal stress on this negative surface is however positive because it follows the direction of surface-n. This example is ideal in revealing the difference between the traction on a surface and stresses on the same surface. This subtlety confuses many, but these kinds of behavior are correct, and are resulting from the rules of operations for all tensors in solid mechanics: orthogonal coordinates, orthogonal projections, equilibrium laws, and the standard sign convention.

Readers may consider the surface is in the positive x-direction (with $n = \text{np.array}([1., 0., 0.])$) (or any other direction), and run the code. Then, think carefully about the results obtained for this simple problem, until it makes sense. Readers may also create artificial bi-axial stress state say, $\sigma_{xx} = 1$ and $\sigma_{yy} = 2$ with all other components zero, and then perform the tests again, using the code given. Such an exercise may significantly improve the understanding on the behavior of stresses in solids.

The following is a more general stress state case.

```
# stress state used in Example 2.1 in
# Ref: Advanced Mechanics of Materials, by Boresi and Schmidt
np.set_printoptions(precision=4,suppress=True)

# Given stress matrix in MPa
Sv = np.array([[-10.,15., 0.],[15.,30., 0.],[0.,0.,0.]]) #Example 2.1
print(f'Stress Tensor (MPa) :\n{Sv}')

# normal vector of the surface:
n = np.array([1.,1.,0.])   #[1.,1.,2.]         # one may try this also.
n = n/np.linalg.norm(n)
print(f'A surface with normal direction n = {n}\n')

q = np.array([-1., 1., 0])              # must be orthogonal to n
q = q/np.linalg.norm(q)                 # normalize vector q
print(f'Normal vector q on surface n = {q}\n')

o_n, o_nn, o_ns, o_nq, o_nr = stressOnN(Sv, n, q)
print(f'Traction (stress vector) on the surface = {o_n} MPa')
print(f'Normal stress on the surface = {o_nn:2.4f} MPa')
print(f'Magnitude of the shear stress on the surface = {o_ns:2.4f} MPa')
print(f'Shear stress on the surface in q-direction   = {o_nq} MPa')
print(f'Shear stress on the surface in r-direction   = {o_nr} MPa')
```

```
Stress Tensor (MPa) :
[[-10.  15.   0.]
 [ 15.  30.   0.]
 [  0.   0.   0.]]
A surface with normal direction n = [0.7071 0.7071 0.    ]

Normal vector q on surface n = [-0.7071  0.7071  0.    ]

Unit vector r on surface-n = [ 0. -0.  1.]

Traction (stress vector) on the surface = [ 3.5355 31.8198  0.    ] MPa
Normal stress on the surface = 25.0000 MPa
Magnitude of the shear stress on the surface = 20.0000 MPa
Shear stress on the surface in q-direction   = 20.0000 MPa
Shear stress on the surface in r-direction   = 0.0 MPa
```

Note that the code stressOnN() works for problems in any dimension. Following is an example for 2D stress state, which is the same as the previous 3D example.

```
1  # Given stress matrix in MPa
2  Sv2d = Sv[:2,:2]
3  #Sv2d = np.array([[88., 38.], [38.,-35.]])    # one may try this also.
4  print(f'Stress Tensor (MPa) :\n{Sv2d}')
5
6  n = np.array([1., 1.])                # normal vector of the surface
7  n = n/np.linalg.norm(n)
8  print(f'A surface with normal direction n = {n}\n')
9
10 q = np.array([-1.,1.])                # must be orthogonal to n
11 q = q/np.linalg.norm(q)
12 print(f'Unit vector q on surface n = {q}\n')
13
14 σ_n, σ_nn, σ_ns, σ_nq, σ_nr = stressOnN(Sv2d, n, q)
15 print(f'Traction (stress vector) on the surface = {σ_n} MPa')
16 print(f'Normal stress on the surface = {σ_nn:2.4f} MPa')
17 print(f'Magnitude of the shear stress on the surface = {σ_ns:2.4f} MPa')
18 print(f'Shear stress on the surface in q-direction   = {σ_nq} MPa')
19 print(f'Shear stress in r-direction is ignored for 2D cases.')
```

```
Stress Tensor (MPa) :
[[-10.  15.]
 [ 15.  30.]]
A surface with normal direction n = [0.7071 0.7071]

Unit vector q on surface n = [-0.7071  0.7071]

Unit vector r on surface-n = 1.0000000000000002

Traction (stress vector) on the surface = [ 3.5355 31.8198] MPa
Normal stress on the surface = 25.0000 MPa
Magnitude of the shear stress on the surface = 20.0000 MPa
Shear stress on the surface in q-direction   = 20.0000 MPa
Shear stress in r-direction is ignored for 2D cases.
```

Note that in 2D cases, the unit vector **r** is only the third component of the cross product **q** × **n**. The shear in r-direction should be ignored.

With stressOnN(), readers can try more cases, 2D and 3D ones.

3.7 Coordinate transformation of second order tensor (stress)

Consider a coordinate transformation from coordinate system of x_1, x_2, and x_3 to a new coordinate x'_1, x'_2, and x'_3. We can now readily derive the rule for coordinate transformation of stress tensor. It is nothing but a generalization of Eq. (3.44).

3.7.1 *Formulation by simple inspection*

Let us now re-write Eq. (3.44) to include **very carefully** the precise meaning of each of the indexes.

$$\sigma_{nq} = n_i \sigma_{ij} q_j$$

$$\underbrace{\sigma}_{} \quad \underbrace{n}_{\substack{\text{arbitrary-surface} \\ \text{denoted by } r}} \quad \underbrace{q}_{\substack{\text{arbitrary-direction} \\ \text{denoted by } s}} = \underbrace{n_i}_{\text{a surface comp.}} \quad \underbrace{\sigma}_{} \quad \underbrace{i}_{\substack{\text{contract} \\ \text{with surface}}} \quad \underbrace{j}_{\substack{\text{contract with a direction} \\ \text{direction}}} \quad \underbrace{q_j}_{\substack{\text{contract with a direction} \\ \text{comp.}}}$$

$$\underbrace{\begin{bmatrix} x_1' \\ x_2' \\ x_3' \end{bmatrix}}_{r} \quad \underbrace{\begin{bmatrix} x_1' \\ x_2' \\ x_3' \end{bmatrix}}_{s} \quad a_{ri} \quad \underbrace{\begin{bmatrix} x_1 \\ x_2 \\ x_3 \end{bmatrix}}_{i} \quad \underbrace{\begin{bmatrix} x_1 \\ x_2 \\ x_3 \end{bmatrix}}_{j} \quad a_{sj}$$

(3.47)

$$\mathbf{T} = \begin{bmatrix} \underbrace{\cos(x_1', x_1)}_{a_{11}} & \underbrace{\cos(x_1', x_2)}_{a_{12}} & \underbrace{\cos(x_1', x_3)}_{a_{13}} \\ \underbrace{\cos(x_2', x_1)}_{a_{21}} & \underbrace{\cos(x_2', x_2)}_{a_{22}} & \underbrace{\cos(x_2', x_3)}_{a_{23}} \\ \underbrace{\cos(x_3', x_1)}_{a_{31}} & \underbrace{\cos(x_3', x_2)}_{a_{32}} & \underbrace{\cos(x_3', x_3)}_{a_{33}} \end{bmatrix} \quad (3.48)$$

After the coordinate transformation from the base coordinates x_1, x_2, and x_3 to the new coordinates x_1', x_2', and x_3', the n_i should be replaced by a_{ri} because the surface could be one of the three axes x_1', x_2', and x_3'. The q_j should be replaced by a_{sj} because the direction could also be one of the three directions x_1', x_2', and x_3'. The surface-index can use free index r, and the direction-index can use free index s. All the contraction operations stay unchanged, leaving with 2 indexes, resulting in a new description of the same stress tensor in the new coordinate system. After this indexation matching, the new components of the stress tensor in the new coordinate are computed using

$$\boxed{\sigma_{rs}' = a_{ri} \sigma_{ij} a_{sj}} \qquad (3.49)$$

This is the formulation for **stress tensor transformation rule** using indicial notation. This is because the transformation is in reference to the original base coordinates x_1, x_2 and x_3 as defined in Eq. (3.48), which must be contracted so as to get into the new coordinate system x_1', x_2' and x_3'.

Alternatively, we may use the matrix notation and matrix multiplication rule that achieves the same contraction effects, and the formula should be:

$$\boxed{\boldsymbol{\sigma}' = \mathbf{T}\boldsymbol{\sigma}\mathbf{T}^\top} \tag{3.50}$$

The $\mathbf{T}$ on the left ensures the contraction is on the first index of $\boldsymbol{\sigma}$ (with the 2nd index of $\mathbf{T}$), and $\mathbf{T}$ on the right ensures the contraction is on the 2nd index of $\boldsymbol{\sigma}$ (also with the 2nd index of $\mathbf{T}$), and thus the transpose is needed.

3.7.2 Inverse transformation

Since $\mathbf{T}$ is unitary, the inverse transformation can be simply done using the transpose of $\mathbf{T}$:

$$\boldsymbol{\sigma} = \mathbf{T}^\top \boldsymbol{\sigma}' \mathbf{T} \tag{3.51}$$

Or in the indicial notation:

$$\sigma_{rs} = a_{ir}\sigma'_{ij}a_{js} \tag{3.52}$$

All we need is to swap the order of the two indexes of a_{ij} for reverse transformation. This is because the transformation is in reference to x'_1, x'_2 and x'_3 as defined in Eq. (3.48), which must be contracted so as to get back to the coordinate system x_1, x_2 and x_3.

It is seen that transformation of a second-order stress tensor requires two times of operation by the transformation matrix $\mathbf{T}$.

3.7.3 Python example: Stress tensor under coordinate transformation

Consider a stress tensor in 3D solids represented originally in a coordinate system $\mathbf{x}$. When the coordinates is transformed (rotated) to a new coordinate $\mathbf{x}'$, we compute the new components of the same stress tensor in the new coordinate system.

Let us write Numpy codes to compute stresses after a coordinate transformation, using both matrix and indicial formulations.

First, an arbitrary stress tensor in the coordinate system $\mathbf{x}$ can be written using the following code. In coordinate transformation operations, it is more convenient to use the indicial notation: x:1, y:2, and z:3. We shall also make use of the symmetry properties of the stress tensor.

3.7.3.1 Use of matrix formulation

```
1  T = gr.transferMs()
2  T
```

$$\begin{bmatrix} a_{11} & a_{12} & a_{13} \\ a_{21} & a_{22} & a_{23} \\ a_{31} & a_{32} & a_{33} \end{bmatrix}$$

```
1  Sp=T@S@T.T        # S: stress tensor; @: the shorthand of np.matmul
2  Sp[:,0]           # print out the first column of S-prime
```

$$\begin{bmatrix} a_{11}\left(a_{11}\sigma_{11} + a_{12}\sigma_{21} + a_{13}\sigma_{31}\right) + a_{12}\left(a_{11}\sigma_{12} + a_{12}\sigma_{22} + a_{13}\sigma_{32}\right) \\ + a_{13}\left(a_{11}\sigma_{13} + a_{12}\sigma_{23} + a_{13}\sigma_{33}\right) \\ a_{11}\left(a_{21}\sigma_{11} + a_{22}\sigma_{21} + a_{23}\sigma_{31}\right) + a_{12}\left(a_{21}\sigma_{12} + a_{22}\sigma_{22} + a_{23}\sigma_{32}\right) \\ + a_{13}\left(a_{21}\sigma_{13} + a_{22}\sigma_{23} + a_{23}\sigma_{33}\right) \\ a_{11}\left(a_{31}\sigma_{11} + a_{32}\sigma_{21} + a_{33}\sigma_{31}\right) + a_{12}\left(a_{31}\sigma_{12} + a_{32}\sigma_{22} + a_{33}\sigma_{32}\right) \\ + a_{13}\left(a_{31}\sigma_{13} + a_{32}\sigma_{23} + a_{33}\sigma_{33}\right) \end{bmatrix}$$

This is the general formula for the first column stress components after the coordinate transformation with **T**. Others can be easily printed out as follows. Note again that **T** is applied from both sides because **S** is a second-order tensor.

```
1  Sp[:,1]           # print out the 2nd column
```

$$\begin{bmatrix} a_{21}\left(a_{11}\sigma_{11} + a_{12}\sigma_{21} + a_{13}\sigma_{31}\right) + a_{22}\left(a_{11}\sigma_{12} + a_{12}\sigma_{22} + a_{13}\sigma_{32}\right) \\ + a_{23}\left(a_{11}\sigma_{13} + a_{12}\sigma_{23} + a_{13}\sigma_{33}\right) \\ a_{21}\left(a_{21}\sigma_{11} + a_{22}\sigma_{21} + a_{23}\sigma_{31}\right) + a_{22}\left(a_{21}\sigma_{12} + a_{22}\sigma_{22} + a_{23}\sigma_{32}\right) \\ + a_{23}\left(a_{21}\sigma_{13} + a_{22}\sigma_{23} + a_{23}\sigma_{33}\right) \\ a_{21}\left(a_{31}\sigma_{11} + a_{32}\sigma_{21} + a_{33}\sigma_{31}\right) + a_{22}\left(a_{31}\sigma_{12} + a_{32}\sigma_{22} + a_{33}\sigma_{32}\right) \\ + a_{23}\left(a_{31}\sigma_{13} + a_{32}\sigma_{23} + a_{33}\sigma_{33}\right) \end{bmatrix}$$

```
1  Sp[:,2]           # print out the 3rd column
```

$$\begin{bmatrix} a_{31}\left(a_{11}\sigma_{11} + a_{12}\sigma_{21} + a_{13}\sigma_{31}\right) + a_{32}\left(a_{11}\sigma_{12} + a_{12}\sigma_{22} + a_{13}\sigma_{32}\right) \\ + a_{33}\left(a_{11}\sigma_{13} + a_{12}\sigma_{23} + a_{13}\sigma_{33}\right) \\ a_{31}\left(a_{21}\sigma_{11} + a_{22}\sigma_{21} + a_{23}\sigma_{31}\right) + a_{32}\left(a_{21}\sigma_{12} + a_{22}\sigma_{22} + a_{23}\sigma_{32}\right) \\ + a_{33}\left(a_{21}\sigma_{13} + a_{22}\sigma_{23} + a_{23}\sigma_{33}\right) \\ a_{31}\left(a_{31}\sigma_{11} + a_{32}\sigma_{21} + a_{33}\sigma_{31}\right) + a_{32}\left(a_{31}\sigma_{12} + a_{32}\sigma_{22} + a_{33}\sigma_{32}\right) \\ + a_{33}\left(a_{31}\sigma_{13} + a_{32}\sigma_{23} + a_{33}\sigma_{33}\right) \end{bmatrix}$$

3.7.3.2 Use of tensor operation

Alternatively, we can use indicial notation, and np.tensordot() to have the same transformation done.

```
1  def Tensor2_transfer(T,S):
2      '''Coordinate transformation for 2nd order tensors
3      '''
4      S_ = np.tensordot(T , S, axes=([1],[0])) # specify contracting exes
5      S_ = np.tensordot(S_, T, axes=([1],[1]))
6      return S_
```

```
1  TS = Tensor2_transfer(T, S)
2  TS = Matrix(TS)
3  print(f"Transformed stresses (first column only):")
4  TS[:,0]
```

Transformed stresses (first column only):

$$\begin{bmatrix} a_{11}\left(a_{11}\sigma_{11}+a_{12}\sigma_{21}+a_{13}\sigma_{31}\right)+a_{12}\left(a_{11}\sigma_{12}+a_{12}\sigma_{22}+a_{13}\sigma_{32}\right) \\ +a_{13}\left(a_{11}\sigma_{13}+a_{12}\sigma_{23}+a_{13}\sigma_{33}\right) \\ a_{11}\left(a_{21}\sigma_{11}+a_{22}\sigma_{21}+a_{23}\sigma_{31}\right)+a_{12}\left(a_{21}\sigma_{12}+a_{22}\sigma_{22}+a_{23}\sigma_{32}\right) \\ +a_{13}\left(a_{21}\sigma_{13}+a_{22}\sigma_{23}+a_{23}\sigma_{33}\right) \\ a_{11}\left(a_{31}\sigma_{11}+a_{32}\sigma_{21}+a_{33}\sigma_{31}\right)+a_{12}\left(a_{31}\sigma_{12}+a_{32}\sigma_{22}+a_{33}\sigma_{32}\right) \\ +a_{13}\left(a_{31}\sigma_{13}+a_{32}\sigma_{23}+a_{33}\sigma_{33}\right) \end{bmatrix}$$

3.7.4 *Transformation of stress tensor in 2D*

For stress states at any point in 2D, the expression is more manageable, and can be derived explicitly. The transformation rule is the same as Eq. (3.50). Assuming the new coordinates are X-Y rotated from the original x-y coordinates by θ, the Sympy derivation for the expressions for stresses in X-Y is as follows:

```
1  # define stress components as symbolic variable for 2D stresses
2  sxx, syy, sxy, syx = "\u03C3_xx", "\u03C3_yy", "\u03C3_xy", "\u03C3_yx"
3
4  S2D = Matrix([[sxx, sxy],
5               [syx, syy]])
6  S2D                                                    # a 2D stress tensor
```

$$\begin{bmatrix} \sigma_{xx} & \sigma_{xy} \\ \sigma_{yx} & \sigma_{yy} \end{bmatrix}$$

```
1  θ = symbols("θ")                              # rotation angle from x-y
2  T2D = gr.transferMts(θ)[0:2,0:2]
3  T2D
```

$$\begin{bmatrix} \cos(\theta) & \sin(\theta) \\ -\sin(\theta) & \cos(\theta) \end{bmatrix}$$

```
1  S2Dp = T2D @ S2D @ T2D.T    # S2D: stress tensor in 2D, obtained earlier
2  sXX  = S2Dp[0,0].expand()
3  gr.printM(sXX, 'σ_XX in the rotated coordinate X-Y:')
```

σ_XX in the rotated coordinate X-Y:

$$\sigma_{xx}\cos^2(\theta) + \sigma_{xy}\sin(\theta)\cos(\theta) + \sigma_{yx}\sin(\theta)\cos(\theta) + \sigma_{yy}\sin^2(\theta)$$

```
1  sYY = S2Dp[1,1].expand()
2  gr.printM(sYY, 'σ_YY in the rotated coordinate X-Y:')
```

σ_YY in the rotated coordinate X-Y:

$$\sigma_{xx}\sin^2(\sigma) - \sigma_{xy}\sin(\theta)\cos(\theta) - \sigma_{yx}\sin(\theta)\cos(\theta) + \sigma_{yy}\cos^2(\theta)$$

```
1  sXY = S2Dp[0,1].expand()
2  gr.printM(sXY, 'σ_XY in the rotated coordinate X-Y:')
```

σ_XY in the rotated coordinate X-Y:

$$-\sigma_{xx}\sin(\theta)\cos(\theta) + \sigma_{xy}\cos^2(\theta) - \sigma_{yx}\sin^2(\theta) + \sigma_{yy}\sin(\theta)\cos(\theta)$$

The formulas are summarized, with regrouping, as follows:

$$\sigma_{XX} = \sigma_{xx}\cos^2\theta + \sigma_{yy}\sin^2\theta + 2\sigma_{xy}\sin\theta\cos\theta$$
$$\sigma_{YY} = \sigma_{xx}\sin^2\theta + \sigma_{yy}\cos^2\theta - 2\sigma_{xy}\sin\theta\cos\theta \tag{3.53}$$
$$\sigma_{XY} = \sigma_{xy}\cos^2\theta - \sigma_{xy}\sin^2\theta + (\sigma_{yy} - \sigma_{xx})\sin\theta\cos\theta$$

Readers may find these formulas frequently in the literature. It will be used to derive the Mohr circle later. Note that the 2D stresses can also computed easily by treating 2D stresses as a special case of 3D ones.

Let us look at some specific examples.

3.7.4.1 *Example: Uniaxial stress after a coordinate transformation*

Consider a simple uniaxial loading case studied at the beginning of the chapter.

```
1  # Given stress matrix in MPa
2  S_xx = Matrix([[1.,0., 0.],[0., 0., 0.],[0.,0.,0.]]) # It has only σxx.
3  printM(S_xx, 'Given stress matrix', n_dgt=4)
```

Given stress matrix

$$\begin{bmatrix} 1.0 & 0 & 0 \\ 0 & 0 & 0 \\ 0 & 0 & 0 \end{bmatrix}$$

Let us rotate the coordinate by 45 degrees about the z-axis. The transformation matrix becomes

```
1  # Create transformation matrix.
2  about, θ = 'z', 45.                    # rotation axis and angle
3  Tr, _ = gr.transferM(θ, about = about)
4  print(f'Transformation tensor {θ:3.2f}°, w.r.t. {about}:\n{Tr}')
```

```
Transformation tensor 45.00°, w.r.t. z:
[[ 0.7071  0.7071  0.  ]
 [-0.7071  0.7071  0.  ]
 [ 0.      0.      1.  ]]
```

```
1  TTS = Matrix(Tensor2_transfer(Tr, S_xx))
2  printM(TTS, 'Stress after forward transfer', n_dgt=4)
```

Stress after forward transfer

$$\begin{bmatrix} 0.5 & -0.5 & 0 \\ -0.5 & 0.5 & 0 \\ 0 & 0 & 0 \end{bmatrix}$$

This means that uniaxial stress tensor with only one nonzero component of $\sigma_{xx} = 1$ in (x, y, z) has a shear stress components of $\sigma_{XY} = -0.5$ in 45-degree rotated coordinate (X, Y, Z) with respect to z, shown in Fig. 3.10.

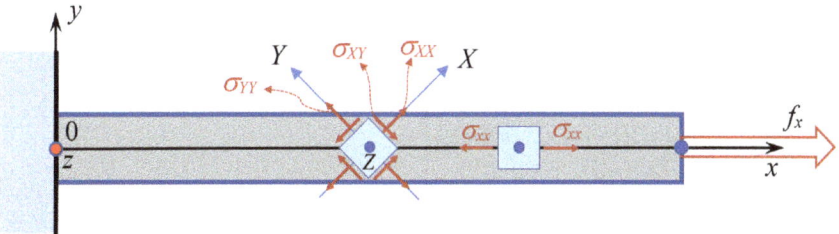

Figure 3.10. Stresses in (X, Y, Z) in a bar subjected to a uniaxial force in the x-direction. Rotation angle: 45°. The arrows for the stress is the **actual** direction of the stresses (not following the standard sign convention for free-body diagrams).

The failure of many ductile materials is controlled by shear stress. Thus, such materials fail at an angle close to 45 degrees when they are subjected to uniaxial tension or compression. Readers may pay attention to the negative sign of σ_{XY}, which explains why the shear stress needs to carry a negative sign in the Mohr circle (see later).

We can now transform the coordinate back:

```
1  TTSb = Matrix(Tensor2_transfer(Matrix(Tr).T, TTS))
2  printM(TTSb, 'Stress after forward & inverse transfer', n_dgt=4)
```

Stress after forward & inverse transfer

$$\begin{bmatrix} 1.0 & 0 & 0 \\ 0 & 0 & 0 \\ 0 & 0 & 0 \end{bmatrix}$$

Let us generate a random transformation matrix, and use it do a coordinate transformation.

```
1  # Create random transformation matrix.
2  about, θ = 'random', 0.        # rotation angle is not used
3  Tr, _ = gr.transferM(θ, about = about)
4  print(f'Transformation tensor {θ:3.2f}°, w.r.t. {about}:\n{Tr}')
```

Transformation tensor 0.00°, w.r.t. random:
[[-0.2402   0.9696 -0.0463]
 [-0.0036   0.0468   0.9989]
 [0.9707   0.2401 -0.0078]]

```
1  TTS = Matrix(Tensor2_transfer(Tr, S_xx))
2  printM(TTS, 'Stress after forward transfer', n_dgt=4)
```

Stress after forward transfer

$$\begin{bmatrix} 0.05768 & 0.0008532 & -0.2331 \\ 0.0008532 & 1.262 \cdot 10^{-5} & -0.003449 \\ -0.2331 & -0.003449 & 0.9423 \end{bmatrix}$$

We see that the simple uniaxial stress state becomes a full matrix, and all the components become nonzero, in some rotated coordinates. Since the transformation is invertible, one can also expect that a complicated stress state can have a simpler representation in a rotated coordinates.

Let us rotate the coordinates back.

```
1  TTSb = Matrix(Tensor2_transfer(Matrix(Tr).T, TTS))
2  printM(TTSb, 'Stress after forward & inverse transfer', n_dgt=4)
```

Stress after forward & inverse transfer

$$\begin{bmatrix} 1.0 & 1.388 \cdot 10^{-16} & -8.327 \cdot 10^{-17} \\ 1.397 \cdot 10^{-16} & 1.311 \cdot 10^{-17} & -5.517 \cdot 10^{-19} \\ -8.383 \cdot 10^{-17} & -6.37 \cdot 10^{-19} & 2.679 \cdot 10^{-20} \end{bmatrix}$$

We recovered the original uniaxial stress state (to machine accuracy).

3.7.4.2 Example: A general stress state under coordinate transformation

Let us look at a more general stress state.

```
1  # define the stress state:
2
3  S_value=Matrix([[8, 2, 3],
4                  [2, 7, 1],
5                  [3, 1, 5]])   # stress values are given arbitrarily
6  print(f'Stress state in values:')
7  S.subs(list(zip(S,S_value)))
```

Stress state in values:

$$\begin{bmatrix} 8 & 2 & 3 \\ 2 & 7 & 1 \\ 3 & 1 & 5 \end{bmatrix}$$

Let us rotate the coordinates by 60 degrees about the z axis. The transformation matrix becomes

```
1  # Create transformation matrix.
2  about, θ = 'z', 60                      # rotation axis and angle
3  Tr, _ = gr.transferM(θ, about = about)
4  print(f'Transformation tensor {θ:3.2f}°, w.r.t. {about}:\n{Tr}')
```

```
Transformation tensor 60.00°, w.r.t. z:
[[ 0.5     0.866  0.   ]
 [-0.866   0.5    0.   ]
 [ 0.      0.     1.   ]]
```

```
1  # This time use the formula derived earlier for transformed stresses.
2  # The results can be obtained by simple substitutions.
3  TS = Matrix(Tensor2_transfer(T, S))         # re-generate the formula
4  T = gr.transferMs()                         # re-generate the transformation matrix
5
6  # Substitute matrix T with values in Tr. We use zip to create a
7  # list of tuples, which tuple pairs T_ij with Tr_ij for substitution:
8  TTS = TS.subs(list(zip(T, Matrix(Tr))))
9  gr.printM(TTS[:,0], 'Stress (1st column) formulas in new coordinates')
```

Stress (1st column) formulas in new coordinates

$$\begin{bmatrix} 0.25\sigma_{11} + 0.433\sigma_{12} + 0.433\sigma_{21} + 0.75\sigma_{22} \\ -0.433\sigma_{11} - 0.75\sigma_{12} + 0.25\sigma_{21} + 0.433\sigma_{22} \\ 0.5\sigma_{31} + 0.866\sigma_{32} \end{bmatrix}$$

```
1  TTSv=Matrix(TTS.subs(list(zip(S, Matrix(S_value)))))    # Subs. stresses
2  gr.printM(TTSv, 'Stresses in the new coordinates')
```

Stresses in the new coordinates

$$\begin{bmatrix} 8.982 & -1.433 & 2.366 \\ -1.433 & 6.018 & -2.098 \\ 2.366 & -2.098 & 5.0 \end{bmatrix}$$

These are the stress components after transformation. We can now transform the coordinate back:

```
1  TTSvb = Matrix(Tensor2_transfer(Tr.T, TTSv))
2  printM(TTSvb, 'Stress after forward & inverse transfer', n_dgt=4)
```

Stress after forward & inverse transfer

$$\begin{bmatrix} 8.0 & 2.0 & 3.0 \\ 2.0 & 7.0 & 1.0 \\ 3.0 & 1.0 & 5.0 \end{bmatrix}$$

The stresses are again recovered.

Let us now look a randomly generated matrix that may or may not be a stress state, and subject it to coordinate transformation, using a randomly generated transformation matrix.

```
1  np.random.seed(8)                        # set a seed for reproducibility
2  M = np.random.randn(3,3)                 # Generate a random matrix
3
4  about, theta = 'random', 0.              # angle theta is not used
5  Tr, _ = gr.transferM(theta, about = about)   # random T.
6
7  print(f'Randomly generated matrix M (asymmetric): \n{M}')
```

```
Randomly generated matrix M (asymmetric):
[[ 0.0912  1.0913 -1.947 ]
 [-1.3863 -2.2965  2.4098]
 [ 1.7278  2.2046  0.7948]]
```

```
1  TM = Matrix(gr.Tensor2_transfer(Tr,M))   # use random transformation
2  printM(TM, 'After forward coordinate transformation', n_dgt=4)
```

After forward coordinate transformation

$$\begin{bmatrix} -1.886 & -3.557 & 1.229 \\ -0.2874 & -1.248 & 0.507 \\ -1.742 & 0.9192 & 1.724 \end{bmatrix}$$

```
1  TMb = Matrix(Tensor2_transfer(Tr.T,TM))
2  printM(TMb, 'After forward & inverse transformation', n_dgt=4)
```

After forward & inverse transformation

$$\begin{bmatrix} 0.0912 & 1.091 & -1.947 \\ -1.386 & -2.296 & 2.41 \\ 1.728 & 2.205 & 0.7948 \end{bmatrix}$$

We obtained the original stress state.

Remark: To perform an **inverse** transformation, all we need is to use the **transposed** transformation matrix, or reverse the indexes of the transformation tensor. This applies to arbitrary matrix. This is because of the unitary property of the transformation matrix.

This example shows that any matrix can be a second-order tensor, provided the rule of transformation defined in Eq. (3.49) is attached to it. The content of the matrix is not the point. The behavior of the content under a coordinate transformation is.

3.8 Principal stresses

For a given stress state, one can find a set of principal stresses, by rotating the coordinates to a particular angle, where all the shear stress components vanish and only normal stresses exist. Such a normal stress is called the **principal stress**. Assuming such a plane is found with a unit normal denoted as

$\mathbf{n} = [n_1 \ n_2 \ n_3]^\mathsf{T}$. Such a plane is called **principal plane**, and $\mathbf{n}$ is called a **principal axis**. On a principal plane, the stress vector $\boldsymbol{\sigma}_n$ must be in the same direction of $\mathbf{n}$ because there is only a normal stress on that plane. It can be written as

$$\boldsymbol{\sigma}_n = \sigma \mathbf{n} \tag{3.54}$$

where σ is the amplitude of the normal stress. It is yet unknown. The components of the normal stress vector $\boldsymbol{\sigma}_n$ should be

$$\sigma_{nx} = \sigma n_1; \quad \sigma_{ny} = \sigma n_2; \quad \sigma_{nz} = \sigma n_3 \tag{3.55}$$

Substituting these into Eqs. (3.33) and (3.34), we obtain:

$$\begin{aligned}\sigma n_1 &= n_1 \sigma_{xx} + n_2 \sigma_{yx} + n_3 \sigma_{zx} \\ \sigma n_2 &= n_1 \sigma_{xy} + n_2 \sigma_{yy} + n_3 \sigma_{zy} \\ \sigma n_3 &= n_1 \sigma_{xz} + n_2 \sigma_{yz} + n_3 \sigma_{zz}\end{aligned} \tag{3.56}$$

which can be re-arranged to

$$\begin{aligned} n_1(\sigma_{xx} - \sigma) + n_2 \sigma_{yx} + n_3 \sigma_{zx} &= 0 \\ n_1 \sigma_{xy} + n_2(\sigma_{yy} - \sigma) + n_3 \sigma_{zy} &= 0 \\ n_1 \sigma_{xz} + n_2 \sigma_{yz} + n_3(\sigma_{zz} - \sigma) &= 0 \end{aligned} \tag{3.57}$$

In matrix form:

$$\left(\underbrace{\begin{bmatrix} \sigma_{xx} & \sigma_{yx} & \sigma_{zx} \\ \sigma_{xy} & \sigma_{yy} & \sigma_{zy} \\ \sigma_{xz} & \sigma_{yz} & \sigma_{zz} \end{bmatrix}}_{\boldsymbol{\sigma}^\mathsf{T} = \boldsymbol{\sigma}} - \sigma \underbrace{\begin{bmatrix} 1 & 0 & 0 \\ 0 & 1 & 0 \\ 0 & 0 & 1 \end{bmatrix}}_{\mathbf{I}} \right) \underbrace{\begin{bmatrix} n_1 \\ n_2 \\ n_3 \end{bmatrix}}_{\mathbf{n}} = \underbrace{\begin{bmatrix} 0 \\ 0 \\ 0 \end{bmatrix}}_{\mathbf{0}} \tag{3.58}$$

Here, we used the symmetry property of the stress tensor.

3.8.1 Eigenvalue problem

Equation (3.58) is a typical eigenvalue equation. It can be written in a concise form of

$$\boxed{[\boldsymbol{\sigma} - \sigma \mathbf{I}]\mathbf{n} = \mathbf{0}} \tag{3.59}$$

where σ is an unknown stress called **eigenvalue** in mathematical term, $\mathbf{I}$ is the identity matrix that has the same dimension as stress tensor $\boldsymbol{\sigma}$, and $\mathbf{n}$ is called **eigenvector** related to the vector of the direction cosines of $\mathbf{n}$. Equation (3.59) is in fact a linear algebraic system equation with $\mathbf{n}$ being the unknown vector, but the system matrix contains a unknown σ. We need to solve the eigenvalue equation to find σ and its corresponding vector $\mathbf{n}$.

Since this linear algebraic system is homogeneous (the right-hand side is zero). The solution exists only if
$$\det(\boldsymbol{\sigma} - \sigma \mathbf{I}) = 0 \quad \text{or} \quad |\boldsymbol{\sigma} - \sigma \mathbf{I}| = 0 \qquad (3.60)$$
Equation (3.60) can then be used to find the eigenvalue σ.

3.8.2 *Characteristic equation, polynomial*

For stresses in 3D solids, the determinate in Eq. (3.60) results in a polynomial equation of order of 3 in σ (implying the problem becomes a nonlinear one). The polynomial equation is known as the **characteristic equation**, for which there will always be 3 roots (complex in general, real if $\boldsymbol{\sigma}$ is symmetric) of σ. Therefore, σ is also called **characteristic value**. These three characteristic values are the principal stresses denoted as σ_1, σ_2, and σ_3. This essentially proves the existence of 3 principal stresses.

Substituting each of these eigenvalues back to Eq. (3.59), **n** can be solved with the help of the normalization condition:
$$\boxed{n_1^2 + n_2^2 + n_3^2 = 1 \quad \text{or} \quad n_i n_i = 1} \qquad (3.61)$$
Note that this equation is nonlinear in n_i.

The eigenvector corresponds to the principal axes. Since Eq. (3.59) is homogeneous and $\boldsymbol{\sigma}$ is a square matrix, the solutions for eigenvectors always exist. In fact, the eigenvector is not unique up to a scalar. This is simply because an eigenvector that satisfies the homogeneous Eq. (3.59), any its scalar multiples will also satisfy it.

Since the problem becomes nonlinear, we are expecting multiple solutions for the eigenvectors (in n_i). We shall choose these eigenvectors relevant to mechanics of materials problems following the right-hand rule. This can be ensured by, for example, computing $\mathbf{n}_3$ using the cross product of $\mathbf{n}_1$ toward $\mathbf{n}_2$:
$$\mathbf{n}_3 = \mathbf{n}_1 \times \mathbf{n}_2 \qquad (3.62)$$

These eigenvectors correspond to the principal stresses, and give the direction cosines of the principal axes.

Note that the cosine function is multivalued. For example, $\cos(\theta) = \cos(-\theta)$. When a direction cosine value, say n_1, is used to find the angle θ_1, the actual angle could be either $-\theta_1$ or θ_1. In addition because both **n** and $-\mathbf{n}$ are all eigenvectors, n_1 and $-n_1$ can also be used to compute the angle θ_1. Since $\cos(\theta_1) = -\cos(180° - \theta_1)$, $(180° - \theta_1)$ may also be the angle. Therefore, one needs to find the correct angle from the possible ones. This can be easily done by using the possible angles to perform coordinate transformations to the stress tensor. The correct angle should produce zero shear stress. We will do so multiple time in later examples.

3.8.3 Python code for finding principal stresses and their directions

Eigenvalue problems can be solved routinely. We write the following code function for solving eigenvalue problems for stresses in 3D solid materials.

```python
def principalS(S):
    '''Compute the principal stresses/strains & their direction cosines.
    inputs:
        S: given stress/strain tensor, numpy array
    return:
        principal stresses/strain (eigenValues), their direction cosines
        (eigenVectors) ranked by its values. Right-hand-rule is enforced
    '''
    eigenValues, eigenVectors = lg.eig(S)

    #Sort in order
    idx = eigenValues.argsort()[::-1]
    eigenValues = eigenValues[idx]
    eigenVectors = eigenVectors[:,idx]
    print('Principal stress/strain (Eigenvalues):\n',eigenValues,'\n')

    # make the first element in the first vector positive (optional):
    #eigenVectors[0,:] = eigenVectors[0,:]/np.sign(eigenVectors[0,0])

    # Determine the sign for given eigenVector-1 and eigenVector-3
    eigenVectors[:,2] = np.cross(eigenVectors[:,0], eigenVectors[:,1])

    angle = np.arccos(eigenVectors[0,0])*180/np.pi    # in degree
    print(f'Principal stress/strain directions:\n{eigenVectors}\n')
    print(f"Possible angles (n1,x)={angle}° or {180-angle}°")

    return eigenValues, eigenVectors
```

3.8.4 Example: Computing the principal stresses and principal directions

This example is studied in Ref. [1]. We write the following code for the same problem, using our code function principalS().

```python
# Advanced Mechanics of Materials, by Boresi and Schmidt
# Example 2.1 Principal Stresses and Principal Directions

St = np.array([[-10, 15, 0], [15, 30, 0], [0, 0, 0]])    # stress tensor

print(f'Stress state:\n{St}\n')

eigenValues, eigenVectors = principalS(St)        # or gr.principalS(St)
```

```
Stress state:
[[-10  15   0]
 [ 15  30   0]
 [  0   0   0]]

Principal stress/strain (Eigenvalues):
[ 35.   0. -15.]

Principal stress/strain directions:
[[-0.3162  0.     -0.9487]
 [-0.9487  0.      0.3162]
 [ 0.      1.      0.    ]]

Possible angles (n1,x)=108.43494882292201° or 71.56505117707799°
```

The solution is found with a single line of code. The computation takes microseconds.

We got two angles because of the non-uniqueness of the eigenvectors and the multivalued feature of cosine functions. To find out which one is correct, we use these two angles, form the transformation matrices, and perform the transformations. The correct one gives zero shear stresses.

```
1  # Form the transformation matrix using the first angle:
2  T0 = gr.transferM(108.43494882292201, about = 'z')[0]
3
4  # Perform the transformation to the stress tensor
5  T0@St@T0.T
```

```
array([[ 17., -24.,  0.],
       [-24.,   3.,  0.],
       [  0.,   0.,  0.]])
```

This angle is not correct. Let us try the second one:

```
1  # Form the transformation matrix using the 2nd angle:
2  T0 = gr.transferM(71.56505117707799, about = 'z')[0]
3
4  # Perform the transformation to the stress tensor
5  T0@St@T0.T
```

```
array([[ 35.,  -0.,   0.],
       [  0., -15.,   0.],
       [  0.,   0.,   0.]])
```

This time we got the correct principal stresses and the shear stress is zero. This θ is the correct one.

Alternative and more systematic means and code to find the angles for principal axes are given in the last section of this chapter.

3.9 Stress invariants

Nine stress components at any point in solid can form three stress invariants that do not change with the coordinates used. For example, these three principal stresses do not change with coordinate transformations. The invariant property is essentially the feature of any 3×3 matrix, including the stress tensor matrix.

In mechanics of materials, these three stress invariants are defined as follows, based on nine stress components or these three principal stresses.

First, the sum of the diagonal terms in $\boldsymbol{\sigma}$, also known as the trace of $\boldsymbol{\sigma}$, gives the first stress invariant:

$$I_1 = \sigma_{11} + \sigma_{22} + \sigma_{33} = \sigma_1 + \sigma_2 + \sigma_3 \qquad (3.63)$$

Second, the sum of the three minor determinants of $\boldsymbol{\sigma}$ gives the second stress invariant:

$$\begin{aligned} I_2 &= \begin{vmatrix} \sigma_{11} & \sigma_{12} \\ \sigma_{12} & \sigma_{22} \end{vmatrix} + \begin{vmatrix} \sigma_{11} & \sigma_{13} \\ \sigma_{13} & \sigma_{33} \end{vmatrix} + \begin{vmatrix} \sigma_{22} & \sigma_{23} \\ \sigma_{23} & \sigma_{33} \end{vmatrix} \\ &= \sigma_{11}\sigma_{22} + \sigma_{11}\sigma_{33} + \sigma_{22}\sigma_{33} - \sigma_{12}^2 - \sigma_{13}^2 - \sigma_{23}^2 \\ &= \sigma_1\sigma_2 + \sigma_1\sigma_3 + \sigma_2\sigma_3 \end{aligned} \qquad (3.64)$$

Third, the determinant of $\boldsymbol{\sigma}$ gives the third stress invariant:

$$I_3 = \begin{vmatrix} \sigma_{11} & \sigma_{12} & \sigma_{13} \\ \sigma_{12} & \sigma_{22} & \sigma_{23} \\ \sigma_{13} & \sigma_{23} & \sigma_{33} \end{vmatrix} = \sigma_1\sigma_2\sigma_3 \qquad (3.65)$$

For 2D cases, say in the x–y plane, we shall have two stress invariants:

$$I_1 = \sigma_{11} + \sigma_{22} = \sigma_1 + \sigma_2 \qquad (3.66)$$

and

$$I_2 = \begin{vmatrix} \sigma_{11} & \sigma_{12} \\ \sigma_{12} & \sigma_{22} \end{vmatrix} = \sigma_{11}\sigma_{22} - \sigma_{12}^2 = \sigma_1\sigma_2 \qquad (3.67)$$

Due to the coordinate independent feature, the stress invariants are useful in establishing failure criteria (see Chapter 14). A solid material failing or not should not depend on the coordinates used to find the stress values. It is a physical property of the material itself.

3.10 Special stresses

3.10.1 *Octahedral stress*

Once the principal stresses are found, we know the principal axes. The so-called octahedral surface is defined in the principal axes with direction

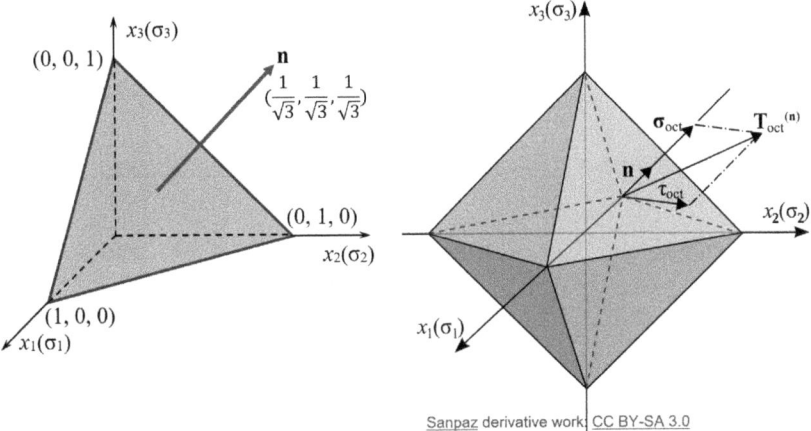

Figure 3.11. The octahedral surface is defined on the principal axes of stresses. Its three normal components are the same: $n_1 = n_2 = n_3 = 1/\sqrt{3}$.

cosines of $\mathbf{n} = (n_1, n_2, n_3) = (\frac{1}{\sqrt{3}}, \frac{1}{\sqrt{3}}, \frac{1}{\sqrt{3}})$, as shown in the figure on the left of Fig. 3.11.

Note that there is a total of eight octahedral surfaces for a point in solid, as shown in the figure on the right in Fig. 3.11.

On the octahedral surface, there is a special stress called octahedral stress. It has two components, octahedral normal stress σ_{oct} and octahedral shear stress τ_{oct}. They can be written as

$$\sigma_{oct} = \frac{1}{3}I_1 = \frac{1}{3}(\sigma_1 + \sigma_2 + \sigma_3)$$
$$\tau_{oct} = \sqrt{\frac{2}{9}I_1^2 - \frac{2}{3}I_2} = \frac{1}{3}\sqrt{(\sigma_1 - \sigma_2)^2 + (\sigma_1 - \sigma_3)^2 + (\sigma_2 - \sigma_3)^2}$$
(3.68)

Note that these octahedral stresses are written in stress invariants, and hence are also invariants. Alternatively, octahedral stress can also be written in nine components.

$$\sigma_{oct} = \frac{1}{3}(\sigma_{11} + \sigma_{22} + \sigma_{33})$$
$$\tau_{oct} = \frac{1}{3}\sqrt{(\sigma_{11} - \sigma_{22})^2 + (\sigma_{22} - \sigma_{33})^2 + (\sigma_{33} - \sigma_{11})^2 + 6(\sigma_{11}^2 - \sigma_{13}^2 - \sigma_{23}^2)}$$
(3.69)

These stresses can be used in failure criteria (see Chapter 14).

3.10.2 *The von Mises stress*

The well-known von Mises stress is widely used in failure criteria for ductile materials. It is related to the distortional energy density in stressed solids and the octahedral shear stress τ_{oct}. The von Mises stress is computed using

$$\sigma_{vm} = \sqrt{\frac{1}{2}[(\sigma_1-\sigma_2)^2 + (\sigma_1-\sigma_3)^2 + (\sigma_2-\sigma_3)^2]} = \frac{3}{\sqrt{2}}\tau_{oct} \qquad (3.70)$$

The von Mises stress relates to the octahedral shear stress by a constant factor.

Alternatively, the von Mises stress can also be written in nine components.

$$\sigma_{vm} = \sqrt{\frac{1}{2}[(\sigma_{11}-\sigma_{22})^2 + (\sigma_{22}-\sigma_{33})^2 + (\sigma_{33}-\sigma_{11})^2 + 6(\sigma_{12}^2 - \sigma_{13}^2 - \sigma_{23}^2)]} \qquad (3.71)$$

3.10.3 *The strain energy stress*

The strain-energy stress is used in the failure criteria based on the strain-energy density (see Chapter 13). It is the effective stress that gives the strain-energy density:

$$U_0 = \frac{1}{2E}\underbrace{[\sigma_1^2 + \sigma_2^2 + \sigma_3^2 - 2\nu(\sigma_2\sigma_3 + \sigma_1\sigma_3 + \sigma_1\sigma_2)]}_{\sigma_{energy}^2} = \frac{1}{2E}\sigma_{energy}^2 \qquad (3.72)$$

where the strain-energy stress is defined as

$$\sigma_{energy} = \sqrt{\sigma_1^2 + \sigma_2^2 + \sigma_3^2 - 2\nu(\sigma_2\sigma_3 + \sigma_1\sigma_3 + \sigma_1\sigma_2)} \qquad (3.73)$$

We will study the strain energy in greater detail in Chapter 13. For now, we use it as defined above.

3.10.4 *Python code for computing various stresses*

We write some useful functions for easy computation of various special stresses, for a given stress state with nine stress components at a point, including the principal stresses, stress invariants, the octahedral stress, mean stress, deviatoric stress (see Section 5.14 for detailed definition), and the von Mises stress.

Our idea is to first conduct eigenvalue analysis for the given stress matrix, and then compute all the other stresses using the eigenvalues. Since the stress matrix is a very small (3×3) symmetric matrix, the time taken for eigenvalue computation is not noticeable. This makes the formulas for all the other types of stress simple and the code is much easier to read.

```
1   def M_eigen(stressMatrix):
2       '''Compute the eigenvalues & eigenvectors for a given stress
3       (or strain) matrix.
4       Input stressMatrix: the stress/strain matrix with 9 components.
5       Return: the eigenvalues are sorted in descending
6       order, and the eigenvectors will be sorted accordingly. The sign
7       for eigenVectors are also adjusted based on the right-hand rule.
8       '''
9       eigenValues, eigenVectors = lg.eig(stressMatrix)
10
11      # Sort in descending order
12      idx = eigenValues.argsort()[::-1]
13      eigenValues = eigenValues[idx]
14      eigenVectors = eigenVectors[:,idx]
15
16      # Determine the sign for given eigenVector-1 and eigenVector-3
17      eigenVectors[2,:] = np.cross(eigenVectors[0,:], eigenVectors[1,:])
18      return eigenValues, eigenVectors
19
20  def von_Mises(e):
21      '''Compute the von Mises stress using principal stresses
22      Input e: eigenvalues of the stress matrix or principal stresses
23      '''
24      vonMises=np.sqrt(((e[0]-e[1])**2+(e[0]-e[2])**2+(e[1]-e[2])**2)/2.)
25      return vonMises
26
27  def Oct_stress(e):
28      '''Compute the octahedral shear stress using principal stresses
29      Input e: eigenvalues of the stress matrix or principal stresses
30      '''
31      sigma_oct= np.sum(e)/3.
32      tau_oct=np.sqrt((e[0]-e[1])**2+(e[0]-e[2])**2+(e[1]-e[2])**2)/3.
33      return sigma_oct, tau_oct
34
35  def energy_stress(e,nu):
36      '''Compute the strain-energy using principal stresses
37      Input e: eigenvalues of the stress matrix or principal stresses
38          nu: Poisson's ratio.
39      '''
40      U_stress=np.sqrt(e[0]**2+e[1]**2+e[2]**2-2*nu*(e[1]*e[2]+
41                      e[0]*e[2]+e[0]*e[1]))
42      return U_stress
43
44  def M_invariant(e):
45      '''Compute the 3 stress/strain invariants using principal stresses
46      Input e: eigenvalues of the stress matrix or principal stresses
47      '''
48      I1 = np.sum(e)
49      I2 = e[0]*e[1]+e[1]*e[2]+e[0]*e[2]
50      I3 = e[0]*e[1]*e[2]
51      return I1, I2, I3
52
53  def meanDeviator(stressM,e):
```

```
54      '''Compute the mean stress, mean stresses, deviatoric stresses.
55      input stressM: given stress (strain or any) matrix with 9 components
56      Input e: eigenvalues of stress/strain matrix or principal stresses
57      Input stressM: The stress/strain matrix
58      '''
59      meanStress = np.mean(e)
60      meanMatrix = np.diag([meanStress,meanStress,meanStress])
61      deviatorM = stressM - meanMatrix
62      return meanStress, meanMatrix, deviatorM
```

3.10.5 Example: Computation of various stresses

Let us do an numerical example using the codes written above.

```
1  # Example 2.5 Stress Invariants:
2  # from Ref: Advanced Mechanics of Materials, by Boresi and Schmidt
3
4  np.set_printoptions(precision=4,suppress=True)
5
6  # Given stress matrix
7  #Sv = np.array([[-10.,15., 0.],[15.,30., 0.],[0.,0.,0.]]) #Example 2.1
8  Sv = np.array([[80.,20, 40],[20., 60, 10],[40., 10, 20]]) #Example 2.5
9  print(f'The given stress tensor =\n{Sv}\n')
10
11 eigenValues, eigenVectors = M_eigen(Sv)           # or gr.M_eigen(Sv)
12 print('Principal stresses (Eigenvalues) = ',eigenValues)
13 print(f'Eigenvectors:\n{eigenVectors}')
14 print(f'Orthogonal:\n{eigenVectors.dot(eigenVectors.T)}\n')
15
16 I1, I2, I3 = M_invariant(eigenValues)
17 print(f'The first  stress invariant I1={I1:9.5f}')
18 print(f'The second stress invariant I2={I2:9.5f}')
19 print(f'The third  stress invariant I3={I3:9.5f}\n')
20
21 sigma_oct, tau_oct = Oct_stress(eigenValues)
22 print(f'The octahedral normal stress is {sigma_oct:9.5f}')
23 print(f'The octahedral shear  stress is {tau_oct:9.5f}\n')
24
25 vonMises = von_Mises(eigenValues)
26 print(f'The von Mises stress is {vonMises:9.5f}\n')
27
28 U_stress = energy_stress(eigenValues, 0.3)
29 print(f'The strain-energy stress is {U_stress:9.5f}\n')
30
31 Tau_max =0.5*(eigenValues[0]-eigenValues[2])
32 print(f'The maximum shear stress is {Tau_max:9.5f}\n')
33
34 meanStress, meanMatrix, deviatorM = meanDeviator(Sv,eigenValues)
35 print(f'The mean stress = {meanStress:9.5f}')
36 print(f'The mean stress tensor:\n{meanMatrix}')
37 print(f'The deviatoric stress tensor:\n{deviatorM}')
```

```
The given stress tensor =
[[80. 20. 40.]
 [20. 60. 10.]
 [40. 10. 20.]]

Principal stresses (Eigenvalues) =  [110.  50.   0.]
Eigenvectors:
[[-0.8165  0.3651 -0.4472]
 [-0.4082 -0.9129 -0.    ]
 [-0.4082  0.1826  0.8944]]
Orthogonal:
[[ 1.  0. -0.]
 [ 0.  1. -0.]
 [-0. -0.  1.]]

The first  stress invariant I1=160.00000
The second stress invariant I2=5500.00000
The third  stress invariant I3=   0.00000

The octahedral normal stress is  53.33333
The octahedral shear stress is  44.96913

The von Mises stress is  95.39392

The strain-energy stress is 106.30146

The maximum shear stress is  55.00000

The mean stress =  53.33333
The mean stress tensor:
[[53.3333  0.      0.    ]
 [ 0.     53.3333  0.    ]
 [ 0.      0.     53.3333]]
The deviatoric stress tensor:
[[ 26.6667  20.      40.    ]
 [ 20.       6.6667  10.    ]
 [ 40.      10.     -33.3333]]
```

3.11 Mohr circles, geometric representation

Mohr circle is a geometric representation of tensors of second order. This section discusses construction of Mohr circles for both 2D and 3D stress states. It is written in reference to the wikipage on Mohr circle (https://en.wikipedia.org/wiki/Mohr%27s_circle) and Ref. [1].

3.11.1 *Formulation for 2D stress states, double angle formula*

To derive the equations for the Mohr circle, we consider first 2D cases, including the plane stress and plane strain, as long as the stress components are given in the x–y coordinate system.

Consider a point in 2D solid. Its state of stress is represented by an infinitely small square shown in Fig. 3.12.

Using Eq. (3.53) and the double angle formulas for sine and cosine, we can derive formulas for stresses in X–Y system that is rotated by θ with respect to the x–y system. The code to derive these formulas is given as follows:

```
# Define the double angle identities in a dictionary:

to2θ = {sp.cos(θ)**2:(1+sp.cos(2*θ))/2, sp.sin(θ)**2:(1-sp.cos(2*θ))/2,
        sp.sin(θ)*cos(θ):sp.sin(2*θ)/2}

# Use sXX, sYY, sXY obtained in code given earlier.
sXX_ = sXX.subs(to2θ).factor().expand()
print(sXX_)
```

σ_xx*cos(θ)**2 + σ_xy*sin(θ)*cos(θ) + σ_yx*sin(θ)*cos(θ) + σ_yy*sin(θ)**2

```
sYY_ = sYY.subs(to2θ).factor().expand()
print(sYY_)
```

σ_xx*sin(θ)**2 - σ_xy*sin(θ)*cos(θ) - σ_yx*sin(θ)*cos(θ) + σ_yy*cos(θ)**2

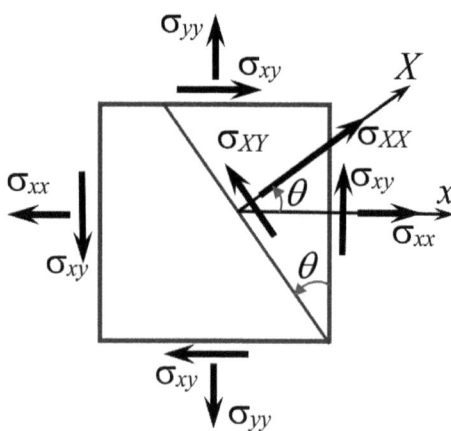

Figure 3.12. 2D stress state on the surfaces of an infinitely small square, and a surface-X with inclined angle θ.

```
1  sXY_ = sXY.subs(to2θ).factor(sp.sin(2*θ)).expand()
2  print(sXY_)
```

-σ_xx*sin(θ)*cos(θ) + σ_xy*cos(θ)**2 - σ_yx*sin(θ)**2 + σ_yy*sin(θ)*cos(θ)

These results are summarized and re-arranged as

$$\boxed{\begin{aligned}\sigma_{XX} &= \frac{1}{2}(\sigma_{xx}+\sigma_{yy}) + \frac{1}{2}(\sigma_{xx}-\sigma_{yy})\cos 2\theta + \sigma_{xy}\sin 2\theta \\ \sigma_{YY} &= \frac{1}{2}(\sigma_{xx}+\sigma_{yy}) - \frac{1}{2}(\sigma_{xx}-\sigma_{yy})\cos 2\theta - \sigma_{xy}\sin 2\theta \\ \sigma_{XY} &= -\frac{1}{2}(\sigma_{xx}-\sigma_{yy})\sin 2\theta + \sigma_{xy}\cos 2\theta\end{aligned}} \quad (3.74)$$

where θ is the rotation angle of the coordinate from x–y to X–Y. We write the following Python function for future use.

3.11.2 Python code for coordinate transformation for 2D stresses, double angle

```
1  def tensorT2angle(σxx, σyy, σxy, θ, py='sympy'):
2      '''
3      Perform coordinate transformation for 2nd order tensors (stress and
4      strain) in 2D, using double angle of θ (rad).
5      inputs: σxx, σyy, σxy in x-y,
6              θ: rad, angle of rotation (counterclockwise) to form X-Y
7              py: if 'sympy', return sympy formulas
8                  if 'numpy', return numpy object
9      return: σXX, σYY, σXY in X-Y
10     '''
11     if py == 'sympy':
12         c2θ = sp.cos(2*θ); s2θ = sp.sin(2*θ)
13     elif py == 'numpy':
14         c2θ = np.cos(2*θ); s2θ = np.sin(2*θ)
15     else:
16         print("Please set py == 'sympy' or 'numpy'")
17
18     σXX = σxx*c2θ/2 + σxx/2 + σxy*s2θ/2 + σxy*s2θ/2 - σyy*c2θ/2 + σyy/2
19     σYY =-σxx*c2θ/2 + σxx/2 - σxy*s2θ/2 - σxy*s2θ/2 + σyy*c2θ/2 + σyy/2
20     σXY =-σxx*s2θ/2 + σxy*c2θ/2 + σxy/2 + σxy*c2θ/2 - σxy/2 + σyy*s2θ/2
21
22     return σXX, σYY, σXY
```

3.11.3 Principal stresses for 2D stress states

For given 2D stress states, one can find the principal stresses by extending it to a 3D stress state, and use principalS() to find the principal stresses.

Alternatively, we can make use of the double angle formula. In this way, we can find the angle of the principal axis with ease.

Since at the principal axes, the shear stress is zero, using the third equation in Eq. (3.74), we obtain

$$\boxed{\tan 2\theta_{p1} = \frac{2\sigma_{xy}}{\sigma_{xx} - \sigma_{yy}}} \qquad (3.75)$$

where $2\theta_{p1}$ is the angle of the principal X-axis with respect to x-axis. This angle found is unique because arctan() is single-valued for arguments in $[-\infty, \infty]$.

After θ_{p1} is found, the principal stresses σ_{XX} and σ_{YY} can be computed using Eq. (3.74). Following is the code.

```
1  def principal2angle(σ2d):
2      '''Compute the principal stresses for a given 2D stress state σ2d.
3      '''
4      θp1 = 0.5*np.arctan(2*σ2d[0,1]/(σ2d[0,0]-σ2d[1,1]))
5      #print(f'Angle (X,x) = {θp1}(rad), {np.rad2deg(θp1)}o')
6
7      σXX,σYY,σXY=tensorT2angle(σ2d[0,0],σ2d[1,1],σ2d[0,1],θp1,py='numpy'
8      #print(f'Principal stresses (MPa) :\n{σXX, σYY, σXY}')
9      return σXX, σYY, σXY, θp1
```

3.11.4 Example: Principal stresses of 2D stress states

In this example, we define a 2D stress state, and compute the principal stresses using the procedure described above. We then compare the results with that obtained using the eigenvalue solver principalS() for 3D stress state extended from the same 2D one.

```
1  # Define a 2D stress state:
2  σ2d = np.array([[88., 38.],            # in MPa
3                  [38.,-35.]])           # one may use other stresses
4  print(f'Stress Tensor (MPa) :\n{σ2d}')
5
6  σXX, σYY, σXY, θp1 = principal2angle(σ2d) # Or: gr.principal2angle(σ2d)
7  print(f'Angle (X,x) = {θp1}(rad), {np.rad2deg(θp1)}o')
8  print(f'Principal stresses (MPa) :\n{σXX, σYY, σXY}')
```

Stress Tensor (MPa) :
[[88. 38.]
 [38. -35.]]
Angle (X,x) = 0.2767336977935464(rad), 15.855672932618990
Principal stresses (MPa) :
(98.79280738773396, -45.792807387733944, 8.881784197001252e-15)

```
1  # Expand σ2d to σ3d, by appending zeros:
2  σ3d = np.zeros((3,3))
3  σ3d[:2,:2] = σ2d
4  print(f'Stress Tensor (MPa) :\n{σ3d} \n Principal stresses:')
5  eigenValues, eigenVectors = principalS(σ3d)
6  print(f' Principal stresses: \n{eigenValues}')
```

```
Stress Tensor (MPa) :
[[ 88.   38.    0.]
 [ 38.  -35.    0.]
 [  0.    0.    0.]]
Principal stresses:
Principal stress/strain (Eigenvalues):
[ 98.7928   0.    -45.7928]

Principal stress/strain directions:
[[ 0.962   0.     0.2732]
 [ 0.2732  0.    -0.962 ]
 [ 0.      1.     0.    ]]

Possible angles (n1,x)=15.85567293261899° or 164.144327067381°
Principal stresses:
[ 98.7928   0.    -45.7928]
```

As shown, we obtained the same principal stress results using these two approaches. The double angle approach principal2angle() gives an unique angle, but the eigenvalue solver principalS() gives two choices.

3.11.5 Mohr circle for 2D and stress states

Using the stress invariants Eqs. (3.66) and (3.67), we shall have

$$\sigma_{XX} + \sigma_{YY} = \sigma_{xx} + \sigma_{yy} (I_1)$$
$$\sigma_{XX}\sigma_{YY} - \sigma_{XY}^2 = \sigma_{xx}\sigma_{yy} - \sigma_{xy}^2 (I_2)$$
(3.76)

Using Eq. (3.74) and with the help of Eq. (3.76), we obtain,

$$\left[\sigma_{XX} - \tfrac{1}{2}(\sigma_{xx} + \sigma_{yy})\right]^2 + \sigma_{XY}^2 = \left[\tfrac{1}{2}(\sigma_{xx} - \sigma_{yy})\right]^2 + \sigma_{xy}^2 \tag{3.77}$$

or

$$(\sigma_{XX} - \sigma_{\text{avg}})^2 + \sigma_{XY}^2 = R^2 \tag{3.78}$$

where $\sigma_{\text{avg}} = \tfrac{1}{2}(\sigma_{xx} + \sigma_{yy})$ is the average or mean stress.

Equation (3.78) is clearly for a circle in the $\sigma_{XX} \sim \sigma_{XY}$ coordinate system with the center at $(\sigma_{\text{avg}}, 0)$ and the radius given by

$$\boxed{R = \sqrt{\left[\tfrac{1}{2}(\sigma_{xx} - \sigma_{yy})\right]^2 + \sigma_{xy}^2}} \tag{3.79}$$

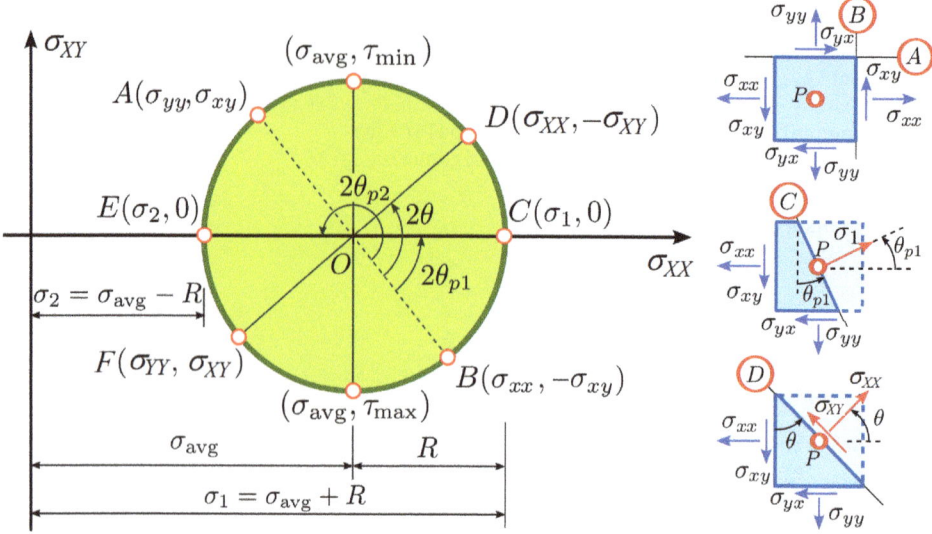

Figure 3.13. Mohr circle for 2D stress states at a point in solid. The shear stress on the surface-x (or surface-X), which is at point B, carries a minus sign. Image modified based on those from Wikimedia Commons, by Sanpaz under the CC BY-SA 3.0 license.

Figure 3.13 shows the detailed relationship between the a Mohr circle and the 2D stress states. It is constructed using the equations given in the previous section. All the states of 2D stresses are located on the circle. The initial stress state before the coordinate transformation is at points B and A. The stress state after the coordinate rotating counterclockwise an arbitrary θ corresponds to point D. Note that the angle θ in the x–y coordinate corresponds to 2θ on the Mohr circle. The principal stress state corresponds to the points C and E on the axis with zero shear stress. The angle of $2\theta_{p1}$ to the principal stress state is computed using Eq. (3.75).

3.11.6 Sign convention for 2D Mohr circles

Note that at any state, there are 2 normal stresses: σ_{XX} and σ_{YY}, while σ_{XX} corresponds to point B (surface-X), and σ_{YY} corresponding to the opposite point A (surface-Y) on the circle, as shown in Fig. 3.13. However, at any state, there is only one shear stress σ_{XY} that must show on the $\sigma_{YY} \smile \sigma_{XY}$ coordinate system with different signs on the upper and lower quadrants. Therefore, we need a **sign convention** to allow the shear stress change signs. There are a few ways to do so, but no standard sign convention for this. Here, we let the shear stress at point A be positive on the Mohr circle: $A(\sigma_{YY}, \sigma_{XY})$, and at point B be negative: $B(\sigma_{XX}, -\sigma_{XY})$. This is true also for the initial state x–y: $A(\sigma_{yy}, \sigma_{xy})$ and $B'(\sigma_{xx}, -\sigma_{xy})$. This allows the shear stress on the surface-X to carry a reversed sign. On the surface-Y, the sign for the shear stress is unchanged. With this **sign reversal**, the shear stress on the Mohr circle gives the same results, as that obtained using the coordinate transformation rule.

One can easily confirm this sign convention by plotting the Mohr circle for the problem shown in Fig. 3.10. When x–y rotates 45° to X–Y, we obtain a negative shear stress. This will be seen in the example later when plotting the Mohr circle for this problem.

More detailed discussions on the sign convention can be found at the wikipage on Mohr circle.

3.11.7 *Python code for generating 2D Mohr circles*

Python codes are made by the author for conveniently producing the Mohr circles for both 2D and 3D stress (or strain) states. The 2D codes are given as follows, and are self-explanatory.

```python
def Mohr_circle2D(σ, lw=1.0):
    '''
    Plot a 2D Mohr circle, for given 2×2 stress/strain matrix (tensor).
    input: σ, numpy array.
           lw, the line width of the circle.
    Example:
    plt.figure(figsize=(6, 6))
    Mohr_circle2D(np.array([[88.,-35], [-35, 38.]]), lw=2.0)
    '''
    σ_avg = (σ[0,0]+σ[1,1])/2                              # mean stress
    center = (σ_avg, 0)                                     # center of the Mohr circle
    R = np.sqrt((σ[0,0]-σ[1,1])**2/4+σ[0,1]**2)             # radius Mohr circle

    plt.scatter(σ_avg, 0, c='r', marker="o", zorder=2) # circle center
    plt.scatter((σ[1,1],σ[0,0]),(σ[0,1],-σ[0,1])) #current state A & B

    plt.plot((σ[1,1],σ[0,0]),(σ[0,1],-σ[0,1]), color='blue', lw=0.5)
    plt.text(σ[1,1]+R/25., σ[0,1]+R/25., "A")              # current state A
    plt.text(σ[0,0]+R/25.,-σ[0,1]+R/25., "B")              # current state B

    circle = plt.Circle(center, R, color='blue', fill=False, lw=lw)
    plt.gca().add_patch(circle) #plt.gca().invert_yaxis()

    plt.axis('scaled')
    plt.xlim(σ_avg-R-R/10, σ_avg+R+R/10)
    plt.xlabel('$σ_{XX}$'); plt.ylabel('$σ_{XY}$')
    #plt.show()
```

3.11.8 Example: Mohr circle of a uniaxial stress state

Consider the uniaxial stress state defined it as a 2D stress state in x–y plane.

$$\begin{bmatrix} \sigma_{xx} & \sigma_{xy} \\ \sigma_{yx} & \sigma_{yy} \end{bmatrix} = \begin{bmatrix} 1 & 0 \\ 0 & 0 \end{bmatrix} \quad (3.80)$$

This is the stress state shown in Fig. 3.10. This stress tensor can be represented in a single Mohr circle, by simply using

```
1  # Consider a bar subject to tension σxx = 1.0:
2  σxx = 1.0; σyy = 0.; σxy=σyx=0.   # stress tensor at initial state
3  σ_tensile = np.array([[σxx, σxy],    # np.array([[1, 0], [0, 0]])
4                         [σyx, σyy]])
5
6  plt.figure(figsize=(3, 3))  # for change the figsize, when needed
7  Mohr_circle2D(σ_tensile, lw=2.0) # or gr.Mohr_circle2D(SM, Lw=2.0)
```

The initial stress state before a coordinate transformation is at points B and A. The center is the mean stress (0.5), and the radius is the absolute value of the maximum shear stress (also 0.5). The principal stress state corresponds to the two points on the axis with zero shear stress.

When the coordinates rotates by $\theta = 45$ degrees, the shear stress found using coordinate transformation formula is -0.5, as shown in Fig. 3.10 (or by the following code). It is the minimum shear stress state.

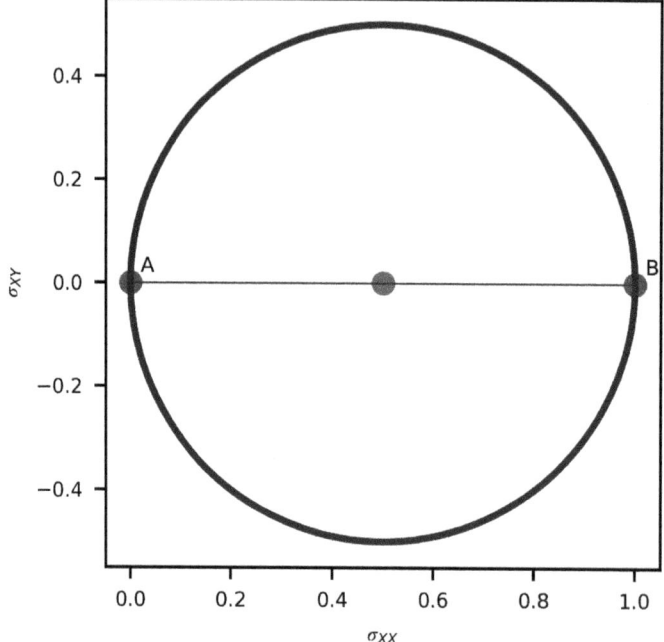

Based on our sign convention for the Mohr circle given earlier, point B is at $(\sigma_{XX}, -\sigma_{XY})$. Therefore, it is at the top of the Mohr circle, due to the sign reversal. The following code computes the transformed stress components and draws the point on the same Mohr circle.

```
1  # Rotate θ_deg:
2  θ_deg = 45.              # this gets to the top point on the Mohr circle
3  θ = np.deg2rad(θ_deg)
4  σXX, σYY, σXY = tensorT2angle(σxx, σyy, σxy, θ, py='numpy')
5
6  σ_45 = np.array([[σXX, σXY],
7                   [σXY, σYY]])
8  print(f" At θ = {θ_deg}◦, stress tensor via coordinate transfer: "
9        f" \n{σ_45} the shear stress should be = {σXY} (the minimum)")
10
11 print(f" Point B(σXX,-σXY) is at the circle top (sign reversal)")
12
13 plt.figure(figsize=(3, 3))      # for change the figsize, when needed
14 Mohr_circle2D(σ_45, lw=2.0)     # or gr.Mohr_circle2D(SM, lw=2.0)
```

At θ = 45.0◦, stress tensor via coordinate transfer:
[[0.5 -0.5]
 [-0.5 0.5]] the shear stress should be = -0.5 (the minimum)
Point B(σXX,-σXY) is at the circle top (sign reversal)

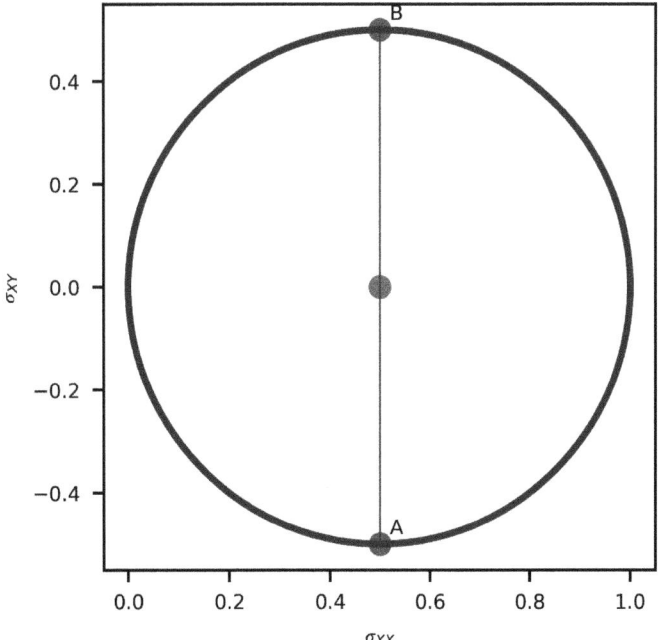

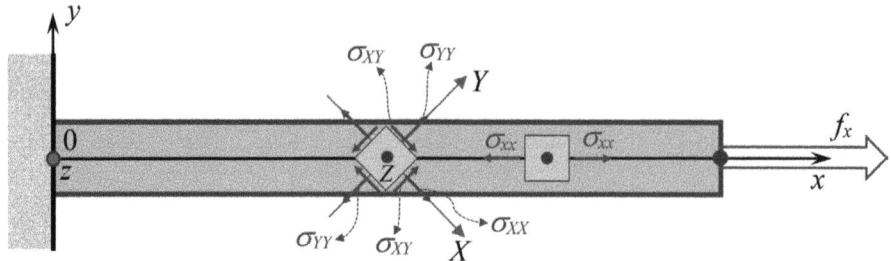

Figure 3.14. Stresses in (X, Y, Z) in a bar subjected to a uniaxial force in the x-direction. Rotation angle: $-45°$. The arrows are the actual direction of the stresses.

If we rotate $\theta = -45$ degrees, the shear stress value found using the coordinate transformation formula should be $\sigma_{XY} = 0.5$, which is the maximum shear stress. On the Mohr circle, point B gets on the bottom at $(\sigma_{XX}, -\sigma_{XY})$, based on our sign convention. The actual directions of all these stress components are drawn in Fig. 3.14.

The following code computes the transformed stress components and draws the Mohr circle.

```
1   # Rotate θ_deg:
2   θ_deg =-45.        # this gets to the bottom point on the Mohr circle
3   θ = np.deg2rad(θ_deg)
4   σXX, σYY, σXY = tensorT2angle(σxx, σyy, σxy, θ, py='numpy')
5
6   σ_135 = np.array([[σXX, σXY],
7                     [σXY, σYY]])
8   print(f" At θ = {θ_deg}o, stress tensor via coordinate transfer: "
9         f" \n{σ_135} the shear stress should be = {σXY} (the maximum)")
10
11  print(f" Point B(σXX,-σXY) is at the circle bottom (sign reversal)")
12
13  plt.figure(figsize=(3, 3))       # for change the figsize, when needed
14  Mohr_circle2D(σ_135, lw=2.0)     # or gr.Mohr_circle2D(SM, lw=2.0)
```

At θ = -45.0o, stress tensor via coordinate transfer:
[[0.5 0.5]
 [0.5 0.5]] the shear stress should be = 0.5 (the maximum)
Point B(σXX,-σXY) is at the circle bottom (sign reversal)

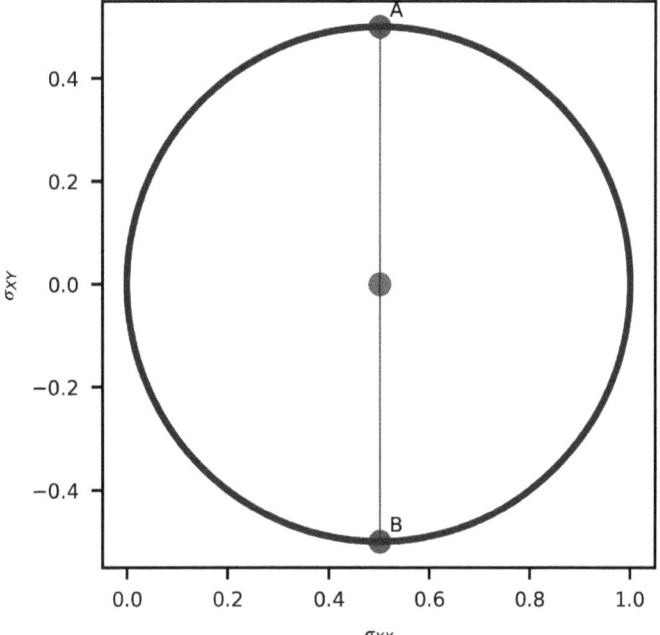

This example is simple, but ideal for revealing the differences of the values of stresses obtained by the coordinate transformation rule and their presentations on the point on the same Mohr circle.

3.11.9 Example: Mohr circle of a general 2D stress state

Consider 2D solid in x–y plane with general stress state given as

$$\begin{bmatrix} \sigma_{xx} & \sigma_{xy} \\ \sigma_{yx} & \sigma_{yy} \end{bmatrix} = \begin{bmatrix} 3 & -1 \\ -1 & 2 \end{bmatrix} \tag{3.81}$$

The Mohr circle can be produced by simply using

```
1  SM = np.array([[3, -1],
2                [-1, 2]])
3  #plt.figure(figsize=(6, 6))      # for change the figsize, when needed
4  Mohr_circle2D(SM, lw=2.0)
```

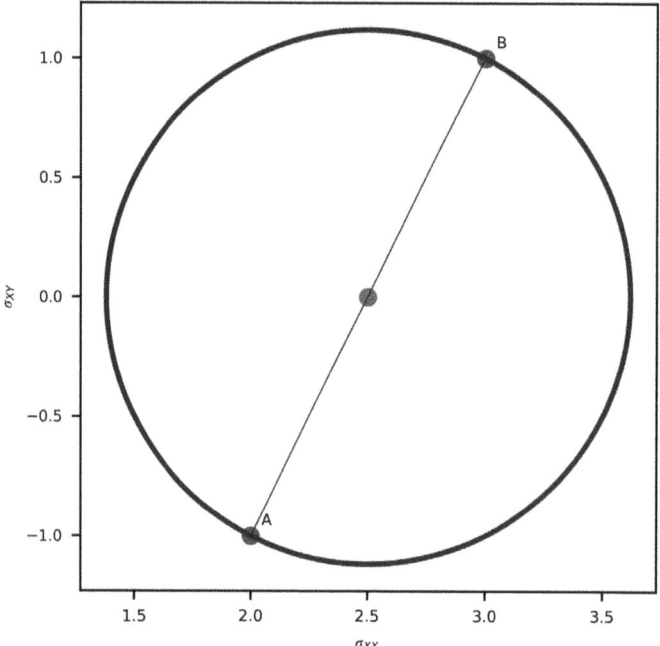

We observe again that the initial stress state before the coordinate transformation is at points B and A. The center is the mean stress, and the radius is the absolute value of the maximum shear stress. The principal stress state corresponds to the two points on the axis of zero shear stress.

3.11.10 *Mohr circle for 3D stress states**

To construct the Mohr circle for stress states at a point in a 3D solid, we assume that the values of the principal stresses $\sigma_1, \sigma_2, \sigma_3$ and their principal directions n_1, n_2, n_3 are known. Those can be found via an eigenvalue analysis, which is easy to do, and we have done multiple times. These principal stresses are ranked so that $\sigma_1 > \sigma_2 > \sigma_3$.

We can then consider the principal stress state with principal axes as the initial state. Using Eq. (3.42), the normal stress on the plane with normal $\mathbf{n}$ with respect to the principal axes is given by

$$\sigma_{nn} = n_i \sigma_{ij} n_j = n_1^2 \sigma_1 + n_2^2 \sigma_2 + n_3^2 \sigma_3 \tag{3.82}$$

This is because all $\sigma_{ij} = 0$ for $i \neq j$ and $\sigma_{ij} = \sigma_i$ for $i = j$ on the principal axes.

Using Eq. (3.46) the share stress on the plane with $\mathbf{n}$ is given by

$$\sigma_{ns}^2 = \sigma_1^2 + \sigma_2^2 + \sigma_3^2 - \sigma_{nn}^2 \tag{3.83}$$

In addition, we have (from the orthogonality of the coordinates):
$$n_1^2 + n_2^2 + n_3^2 = 1 \tag{3.84}$$

We can solve the system of three Eqs. (3.82), (3.83), (3.84) for three unknowns n_1^2, n_2^2, and n_3^2. The results are as follows:

$$n_1^2 = \frac{\sigma_{ns}^2 + (\sigma_{nn} - \sigma_2)(\sigma_{nn} - \sigma_3)}{(\sigma_1 - \sigma_2)(\sigma_1 - \sigma_3)} \geq 0$$

$$n_2^2 = \frac{\sigma_{ns}^2 + (\sigma_{nn} - \sigma_3)(\sigma_{nn} - \sigma_1)}{(\sigma_2 - \sigma_3)(\sigma_2 - \sigma_1)} \geq 0 \tag{3.85}$$

$$n_3^2 = \frac{\sigma_{ns}^2 + (\sigma_{nn} - \sigma_1)(\sigma_{nn} - \sigma_2)}{(\sigma_3 - \sigma_1)(\sigma_3 - \sigma_2)} \geq 0$$

The "≥ 0" on the right-hand side of the equation is because of the square of n_i.

Next because $\sigma_1 > \sigma_2 > \sigma_3$, the numerators in Eq. (3.85) satisfy

$$\sigma_{ns}^2 + (\sigma_{nn} - \sigma_2)(\sigma_{nn} - \sigma_3) \geq 0, \because \sigma_1 - \sigma_2 > 0, \sigma_1 - \sigma_3 > 0$$
$$\sigma_{ns}^2 + (\sigma_{nn} - \sigma_3)(\sigma_{nn} - \sigma_1) \leq 0, \because \sigma_2 - \sigma_3 > 0, \sigma_2 - \sigma_1 < 0 \tag{3.86}$$
$$\sigma_{ns}^2 + (\sigma_{nn} - \sigma_1)(\sigma_{nn} - \sigma_2) \geq 0, \because \sigma_3 - \sigma_1 < 0, \sigma_3 - \sigma_2 < 0$$

The foregoing equations can be rewritten as

$$\sigma_{ns}^2 + \left[\sigma_{nn} - \tfrac{1}{2}(\sigma_2 + \sigma_3)\right]^2 \geq \left(\tfrac{1}{2}(\sigma_2 - \sigma_3)\right)^2$$
$$\sigma_{ns}^2 + \left[\sigma_{nn} - \tfrac{1}{2}(\sigma_1 + \sigma_3)\right]^2 \leq \left(\tfrac{1}{2}(\sigma_1 - \sigma_3)\right)^2 \tag{3.87}$$
$$\sigma_{ns}^2 + \left[\sigma_{nn} - \tfrac{1}{2}(\sigma_1 + \sigma_2)\right]^2 \geq \left(\tfrac{1}{2}(\sigma_1 - \sigma_2)\right)^2$$

These are the equations of three circles in the coordinate $\sigma_{nn}-\sigma_{ns}$ denoted as C_1, C_2, and C_3. Their radii are given respectively as

$$R_1 = \tfrac{1}{2}(\sigma_2 - \sigma_3)$$
$$R_2 = \tfrac{1}{2}(\sigma_1 - \sigma_3) \tag{3.88}$$
$$R_3 = \tfrac{1}{2}(\sigma_1 - \sigma_2)$$

The corresponding centers are at

$$\left(\tfrac{1}{2}(\sigma_2 + \sigma_3), 0\right)$$
$$\left(\tfrac{1}{2}(\sigma_1 + \sigma_3), 0\right) \tag{3.89}$$
$$\left(\tfrac{1}{2}(\sigma_1 + \sigma_2), 0\right)$$

128   *Mechanics of Materials: Formulations and Solutions with Python*

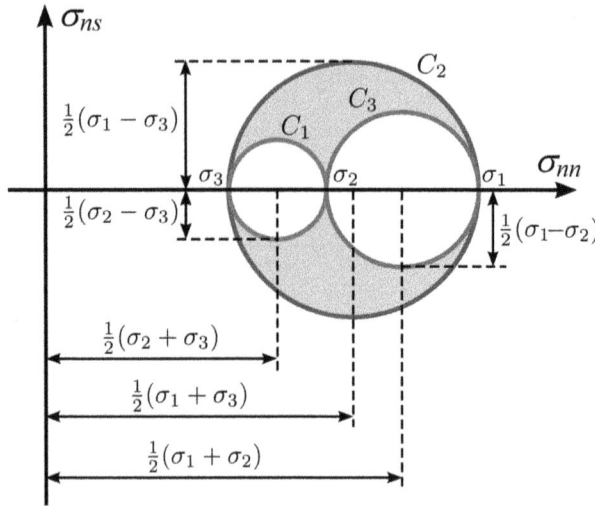

Figure 3.15.   Mohr circle for 3D stress states at a point in solid. Image modified base on that from the Wikimedia Commons in public domain, by Sanpaz.

Equations (3.87) are the Mohr circles for 3D stress states. The inequality means that the stress states at any point in solid (σ_{nn}, σ_{ns}) shall be on these circles or within the shaded green area enclosed by these circles shown in Fig. 3.15.

3.11.11   *Python code for generating 3D Mohr circles*

```
 1  def Mohr_circle3D(σ, lw=1.0):
 2      '''
 3      Plot a 3D Mohr circle, for given 3×3 stress/strain matrix (tensor)
 4      σ in numpy array. lw: the line width of the circle.
 5      Example:
 6      plt.figure(figsize=(6, 6))
 7      Mohr_circle3D([100., 20., -50.], lw=2.0)
 8      '''
 9      pσ, _ = gr.principalS(σ)                    # ranked principal stresses
10
11      σ_avg12 = (pσ[0]+pσ[1])/2                   # 3 average stresses
12      σ_avg13 = (pσ[0]+pσ[2])/2
13      σ_avg23 = (pσ[1]+pσ[2])/2
14
15      center1 = (σ_avg12, 0)                      # 3 centers
16      center2 = (σ_avg13, 0)
17      center3 = (σ_avg23, 0)
18
```

```
19      R1 = (pσ[0]-pσ[1])/2                                    # 3 radii
20      R2 = (pσ[0]-pσ[2])/2
21      R3 = (pσ[1]-pσ[2])/2
22
23      # centers of 3 circles:
24      circle2 = plt.Circle(center2, R2, edgecolor='blue', \
25                          facecolor='lightgreen', lw=lw)
26      circle1 = plt.Circle(center1, R1, edgecolor='c', \
27                          facecolor='white', lw=lw)
28      circle3 = plt.Circle(center3, R3, edgecolor='orange', \
29                          facecolor='white', lw=lw)
30
31      plt.gca().add_patch(circle2)           # plot on the same plot
32      plt.gca().add_patch(circle1)
33      plt.gca().add_patch(circle3)
34
35      plt.axis('scaled')
36      plt.xlabel('$σ_{nn}$'); plt.ylabel('$σ_{ns}$')
37
38      # three principal stresses:
39      plt.scatter(pσ, (0,0,0), c='r', marker="o", s=6, zorder=2)
40
41      # three centers:
42      centers = (σ_avg12,σ_avg13,σ_avg23)
43      plt.scatter(centers, (0,0,0), c='k', \
44                 marker="o", facecolor='none', s=4, zorder=2)
45
46      plt.scatter(centers,( R1, R2, R3), c='r', marker="o",\
47                 s=6, zorder=2)       # maximum shear stress (positive)
48      plt.scatter(centers,(-R1,-R2,-R3), c='r', marker="o",\
49                 s=6, zorder=2)       # maximum shear stress (negative)
50      plt.show()
```

3.11.12 Example: Mohr circle of a three-dimensional (3D) stress state

Consider general problems in 3D. The stress tensor will be 3D, and can be represented in Mohr circles, and there are in general three of them connected together. The 3D codes are given as follows, and are self-explanatory.

Following is an example.

```
1   σmatrix = np.array([[2, 1, 3],      # stresses at a point in 3D solid
2                       [1, 9, 2],
3                       [3, 2, 7]])
4
5   #plt.figure(figsize=(6, 6))                    # for adjusting figsize
6   Mohr_circle3D(σmatrix, lw=2.0)    # or gr.Mohr_circle3D(σmatrix, lw=2.0)
```

```
Principal stress/strain (Eigenvalues):
[10.9578  6.4478  0.5945]

Principal stress/strain directions:
[[-0.2832  0.3133 -0.9065]
 [-0.7523 -0.6588  0.0074]
 [-0.5948  0.684   0.4222]]

Possible angles (n1,x)=106.4510639474349° or 73.5489360525651°
```

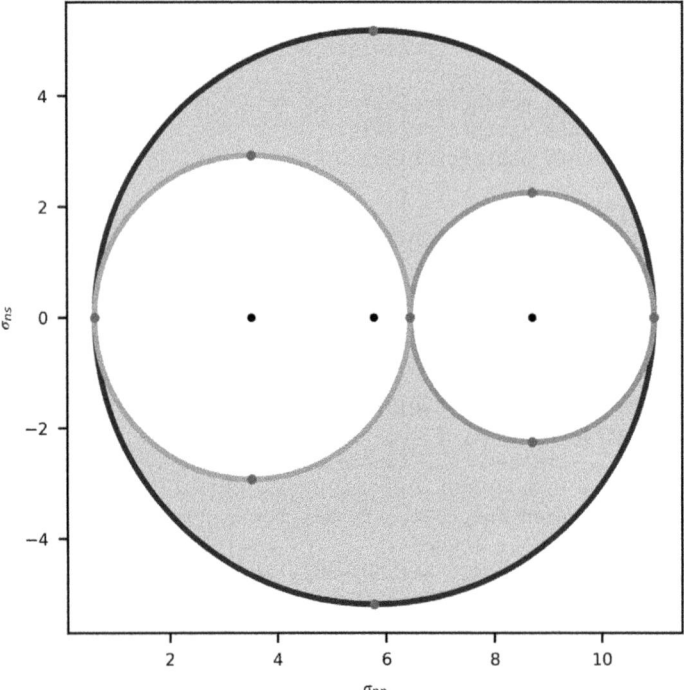

The three principal stresses (the initial stress state) are at the "kissing" points of these three circles, corresponding to zero shear stress. The centers of these three circles are the average values of a pair of the principal stresses. The radius of each circle is the absolute value of a maximum shear stress. The maximum shear stress is the radius of the largest circle.

Alternatively, readers may use the following external modules continuum_mechanics to plot Mohr circles. The sample codes are given as follows:

```
1  #from continuum_mechanics.visualization import mohr2d, mohr3d
2  #mohr3d(Matrix([[2, 1, 3], [1, 9, 2], [3, 2, 7]]));
3  #mohr3d(Matrix([[2, 1, 3], [1, 9, 2], [3, 2, 7]]))
4  #SM = Matrix([[1, 0], [0, 0]]); mohr2d(SM);
```

3.12 Determination of the principal axis angle*

Solving the eigenvalue problem given in Eq. (3.59) gives only the possible direction cosines of the principal axis. Also, the eigenvector is not unique. For example, an **n** satisfies Eq. (3.59), −**n** also satisfies it. Therefore, a direction cosine, say n_1, can be set as both negative and positive, as long as n_2 and n_3 change their signs accordingly. This creates problems if one would want to find precise angle of the principal axis.

For 2D cases, we can use Eq. (3.75), the angle found is unique because $\arctan(x)$ is single-valued for x in $[-\infty, \infty]$. Alternatively, we can perform tests as done in Section 3.8.4 to find the angle.

For stresses in 3D, there will be three angles, and hence finding those can be a challenge. This section discusses techniques for 3D cases.

3.12.1 *Transformation matrices for Yaw, Pitch, and Roll*

Let us now consider the general cases of 3D stress states. In such a case, all these 6 stress components may not be zero. We need to find the principal stress and the corresponding precise angles. This problem becomes quite complicated, and the solution cannot be found in existing textbooks and open literature, to the best knowledge of the author. The author has found an effective method to get this done, and the strategy is given as follows:

1. Create the transformation matrix using the Yaw, Pitch, and Roll formulation given in the previous chapter in, for example, Sympy formulas.
2. Perform the coordinate transformation to any given general stress state, which leads to three expressions for shear stresses on all the off-diagonal terms. These expressions will be in the Yaw angle θ, Pitch angle β, and Roll angle γ, and are in general quite complicated full-page long formulas.
3. Force these three expressions to zero to find the angles, θ, β, γ. This can be done using an equation solver.

Due to the tensor properties, we know that the solution for θ, β, γ exists.

This procedure is essentially a diagonalization procedure. It produces a diagonal stress matrix. It not only finds the angles, θ, β, γ, but also these principal stresses that are on the diagonal.

The following are the codes to get it done.

```
1  # define variables for angle of rotation
2  θ, β, γ = symbols("θ, β, γ")
3
4  # Create the matrices for Yaw, Pitch, and Roll for transformations:
5  Tz = gr.transf_YPRs(θ, about = 'z')[0]
6  Ty = gr.transf_YPRs(β, about = 'y')[0]
7  Tx = gr.transf_YPRs(γ, about = 'x')[0]
8
9  print(f'Ty = {Ty}, \nTx = {Tx}')
10 Tz         # take a look at one of these
```

Ty = Matrix([[cos(β), 0, -sin(β)], [0, 1, 0], [sin(β), 0, cos(β)]]),
Tx = Matrix([[1, 0, 0], [0, cos(γ), sin(γ)], [0, -sin(γ), cos(γ)]])

$$\begin{bmatrix} \cos(\theta) & \sin(\theta) & 0 \\ -\sin(\theta) & \cos(\theta) & 0 \\ 0 & 0 & 1 \end{bmatrix}$$

```
1  # Construct the transformation matrix for Yaw, Pitch and Roll together.
2  T_ypr = Tz@Ty@Tx
3  T_ypr
```

$$\begin{bmatrix} \cos(\beta)\cos(\theta) & \sin(\beta)\sin(\gamma)\cos(\theta) + \sin(\theta)\cos(\gamma) \\ -\sin(\theta)\cos(\beta) & -\sin(\beta)\sin(\gamma)\sin(\theta) + \cos(\gamma)\cos(\theta) \\ \sin(\beta) & -\sin(\gamma)\cos(\beta) \end{bmatrix}$$

$$\begin{matrix} -\sin(\beta)\cos(\gamma)\cos(\theta) + \sin(\gamma)\sin(\theta) \\ \sin(\beta)\sin(\theta)\cos(\gamma) + \sin(\gamma)\cos(\theta) \\ \cos(\beta)\cos(\gamma) \end{matrix}$$

3.12.2 Stress transformation via Yaw, Pitch, and Roll

We now perform stress transformation using all three Yaw, Pitch, and Roll angles.

```
1  # Define a stress tensor:
2  St = np.array([[10, 15, 8], [15, -30, 5], [8, 5, 9]]) # stress matrix
3
4  # Perform coordinate transformation to the given stress tensor.
5  S_ypr = T_ypr@St@T_ypr.T
6
7  #S_ypr[0,1]    # reader may take a look at one shear stress (1/3 page)
```

These three shear stress components will be in $S_{ypr}[0,1]$, $S_{ypr}[0,2]$, $S_{ypr}[1,2]$. These are all quite lengthy.

3.12.3 *Diagonalization of a stress tensor in 3D*

To obtain the principal stresses, we can find a set of Yaw, Pitch, and Roll angles, under which the stress tensor becomes diagonal. This means that all these off-diagonal shear stresses must be zero. This provides three equations. We shall solve these equations for these three angles: θ, β, γ.

This set of equations are nonlinear. There are many Python methods to solve this type of equations. Here, we use scipy.optimize.fsolve() for this task.

```
1  # Import the external solver: fsolve:
2  from scipy.optimize import fsolve
3
4  # Create these three equations:
5  eqns = lambda x: [S_ypr[0,1].subs({θ:x[0], β:x[1], γ:x[2]}),
6                    S_ypr[0,2].subs({θ:x[0], β:x[1], γ:x[2]}),
7                    S_ypr[1,2].subs({θ:x[0], β:x[1], γ:x[2]})]
8
9  eqns([0.1, 0.1, 0.1])    # test the lambda function defined
```

[10.7685926296899, 7.56582183504705, 9.59290586151537]

For solving an nonlinear equations, we often need to give an initial guess, which can be essentially arbitrary. We set it with random values, but make these values small so that the solutions found can be near zero. This is because our functions are nonlinear, sinusoidal and periodical. There are infinite number of solutions in $(-\infty, \infty)$. We would like to have the primary solutions within $(-\pi, \pi)$.

```
1  init_guess = np.random.randn(3)/999.           # make it small
2
3  # Solve the set of three equations numerically:
4  sln = fsolve(eqns, init_guess)
5
6  print(f'Solution, θ, β, γ = {sln} (rad)')
7  print(f'Solution, θ, β, γ = {sln*180/np.pi} (degree)')
```

Solution, θ, β, γ = [0.3839 -0.6111 0.173] (rad)
Solution, θ, β, γ = [21.9951 -35.0143 9.9117] (degree)

We find the solutions in a few seconds.

Finally, we can check the results, by using these θ, β, γ angles found to perform coordinate transformation:

```
1  T_ypr_ = T_ypr.subs({θ:sln[0], β:sln[1], γ:sln[2]})
2  S_ypr_ = T_ypr_@St@T_ypr_.T
3  gr.printM(S_ypr_, 'Principal stresses found',n_dgt=5)
```

Principal stresses found

$$\begin{bmatrix} 21.679 & -3.4318 \cdot 10^{-10} & -1.8484 \cdot 10^{-10} \\ -3.4318 \cdot 10^{-10} & -35.115 & 2.3496 \cdot 10^{-10} \\ -1.8483 \cdot 10^{-10} & 2.3496 \cdot 10^{-10} & 2.4368 \end{bmatrix}$$

We obtained all the expected results. This process is summarized into a code function for future use.

3.12.4 *Python code to find Yaw, Pitch, and Roll angles to diagonalize stress tensors*

```python
def principalS_ypr(St):
    '''Diagonalization of a stress tensor in 3D, through coordinate
    transformation via three Yaw, Pitch and Roll angles.
    Input: St: Stress tensor, array like, 3 by 3
    return: θ, β, γ: Yaw, Pitch and Roll angles, in rad.
            TS_ypr: Diagonalized stress matrix array like, 3 by 3
    '''
    from scipy.optimize import fsolve

    θ, β, γ = sp.symbols("θ, β, γ")   # Angles for Yaw, Pitch and Roll

    # Create the matrices for Yaw, Pitch, and Roll for transformations:
    Tz = gr.transf_YPRs(θ, about = 'z')[0]
    Ty = gr.transf_YPRs(β, about = 'y')[0]
    Tx = gr.transf_YPRs(γ, about = 'x')[0]

    # Construct the transformation matrix for Yaw, Pitch and Roll.
    T_ypr = Tz@Ty@Tx

    # Perform coordinate transformation to the given stress tensor.
    S_ypr = T_ypr@St@T_ypr.T

    # Create these three equations:
    eqns = lambda x: [S_ypr[0,1].subs({θ:x[0], β:x[1], γ:x[2]}),
                      S_ypr[0,2].subs({θ:x[0], β:x[1], γ:x[2]}),
                      S_ypr[1,2].subs({θ:x[0], β:x[1], γ:x[2]})]

    # Give an initial guess randomly:
    init_guess = np.random.randn(3)/999.         # make it small

    # Solve the set of three equations numerically:
    sln = fsolve(eqns, init_guess)
    print(f'Solution, θ, β, γ = {sln} (rad)')
    print(f'Solution, θ, β, γ = {sln*180/np.pi} (degree)')

    # Finally, we can check the results, use θ, β, γ found to
    # perform coordinate trans formation:
    T_ypr_ = T_ypr.subs({θ:sln[0], β:sln[1], γ:sln[2]})
    S_ypr_ = T_ypr_@St@T_ypr_.T

    return θ, β, γ, S_ypr_
```

3.12.5 Example: Diagonalize a stress tensor via Yaw, Pitch, and Roll

```
1  St = np.array([[10, 15, 8], [15, -30, 5], [8, 5, 9]])  # stress matrix
2  θ, β, γ, S_ypr_ = principalS_ypr(St)
3
4  gr.printM(S_ypr_, 'Principal stresses found',n_dgt=5)
```

Solution, θ, β, γ = [0.3839 -0.6111 0.173] (rad)
Solution, θ, β, γ = [21.9951 -35.0143   9.9117] (degree)
Principal stresses found

$$\begin{bmatrix} 21.679 & -5.7027 \cdot 10^{-11} & -4.9118 \cdot 10^{-11} \\ -5.703 \cdot 10^{-11} & -35.115 & 2.4794 \cdot 10^{-11} \\ -4.9117 \cdot 10^{-11} & 2.4795 \cdot 10^{-11} & 2.4368 \end{bmatrix}$$

The code function can, of course, be used for finding the angles for these 2D examples studied earlier:

```
1  St = np.array([[-10, 15, 0], [15, 30, 0], [0, 0, 0]])  # stress tensor
2  θ, β, γ, S_ypr_ = principalS_ypr(St)
3
4  gr.printM(S_ypr_, 'Principal stresses found',n_dgt=5)
```

Solution, θ, β, γ = [-0.3218 -0.   -0.   ] (rad)
Solution, θ, β, γ = [-18.4349 -0.   -0.   ] (degree)
Principal stresses found

$$\begin{bmatrix} -15.0 & 9.77 \cdot 10^{-15} & 2.1503 \cdot 10^{-12} \\ 8.8818 \cdot 10^{-15} & 35.0 & 6.3698 \cdot 10^{-13} \\ 2.1503 \cdot 10^{-12} & 6.3698 \cdot 10^{-13} & -2.9666 \cdot 10^{-25} \end{bmatrix}$$

Since this example is a 2D stress state, the pitch and roll angles are zero.

We can also test it for stress tensor generated randomly:

```
1  n = 3
2  St = np.random.randn(n,n)           # Generate a random matrix
3  St = (St + St.T)/2                  # random symmetric matrix
4  print(f'Stress state:\n{St}\n')
5
6  θ, β, γ, S_ypr_ = gr.principalS_ypr(St)
7  gr.printM(S_ypr_, 'Principal stresses found',n_dgt=5)
```

Stress state:
```
[[ 1.1661  0.8021 -0.2455]
 [ 0.8021 -0.2215 -0.0158]
 [-0.2455 -0.0158 -0.5416]]
```

Solution, θ, β, γ = [0.3917 -0.1998 -0.7082] (rad)
Solution, θ, β, γ = [22.4422 -11.4482 -40.5775] (degree)
Principal stresses found

$$\begin{bmatrix} 1.5581 & 9.3682 \cdot 10^{-13} & -4.8572 \cdot 10^{-15} \\ 9.369 \cdot 10^{-13} & -0.66529 & 1.2548 \cdot 10^{-12} \\ -4.8295 \cdot 10^{-15} & 1.2549 \cdot 10^{-12} & -0.48968 \end{bmatrix}$$

We obtained all these yaw, pitch, and roll angles, and the principal stresses in the diagonal terms in the output matrix.

3.13 Remarks

This busy chapter discusses first how the stress tensor (order 2) result from the external force tensor (order 1) all defined in an orthogonal coordinate system, and then the states of stress and its change when the coordinate system rotates.

1. The value of the stress tensor depends on both the force tensor and the surface of the stress is on. This is the root reason why a stress tensor has two indexes and hence being a second-order tensor.
2. In all the formulas derived for the stress tensor, orthogonal projections for both forces and areas are strictly followed.
3. When coordinate rotates, the presentation of the stress tensor changes. It uses twice the transformation matrix because it is a second-order tensor.
4. The basic properties of a stress tensor (three stress invariants, principal stresses, and the Mohr circle) do not change with the coordinate transformation. Special stresses are defined using these invariant property of stresses. These special stress will be used in establishing failure criteria.
5. At the principal axes, the shear stress components vanish, and only the normal stresses present, which are called principal stresses. These are invariant.
6. Principal stresses and the axes can be found using the eigenvalue solver provided in this chapter.
7. Mohr circle is a geometric representation of stresses in 2D and 3D. For a given stress tensor, the circle is determined. Stress states of all possible rotated coordinates stay on the circle (for 2D) or in the region on and bounded by these circles (for 3D). It is useful to intuitively locate the principal stresses and maximum shear stresses.

8. The Mohr circle applies exactly the same way to the strain tensor to be discussed in the following chapter.
9. A code principalS_ypr() has been developed to find the three angles: Yaw, Pitch and Roll, for the principal axes for 3D stress states. It produces also a diagonalized stress tensor.

With the detailed formulations, discussions, examples, and code provided to play with in the chapter, we shall have a good understanding on forces, stresses, stress states in the solids, on how to perform coordinate transformation with them, and how to find the special stresses for given a stress state.

The following chapter will discuss another important pair of tensors: displacement vector (first order) and strain tensor (second order).

References

[1] A.P. Boresi, R.J. Schmidt and O.M. Sidebottom, *Advanced Mechanics of Materials*. John Wiley & Sons, New York, 1985.
[2] G.R. Liu, *Calculus: A Practical Course with Python*, World Scientific, New Jersey, 2025.
[3] G.R. Liu and S.S. Quek, *The Finite Element Method: A Practical Course*. Butterworth-Heinemann, Amsterdam, 2013.
[4] G.R. Liu and T.T. Nguyen, *Smoothed Finite Element Methods*. Taylor and Francis Group, New York, 2010.

Chapter 4

Displacements and Strains

Let us import the necessary external modules or dependencies for later use in this chapter.

```
1  # Place cursor in this cell, and press Ctrl+Enter to import dependences.
2  import sys                         # for accessing the computer system
3  sys.path.append('../grbin/')   # Change to the directory in your system
4
5  from commonImports import *    # Import dependences from '../grbin/'
6  import grcodes as gr           # Import the module of the author
7  importlib.reload(gr)           # When grcodes is modified, reload it
8
9  from continuum_mechanics import vector
10 init_printing(use_unicode=True)      # For Latex-like quality printing
11 np.set_printoptions(precision=4,suppress=True)   # Digits in print-outs
```

4.1 Introduction

When a solid material is loaded, points in the solid are displaced (in a continuous form). This results in a field of displacements over the solid body. The displacement at a point has three components, each representing the displacement in a direction. Hence, displacement is a vector, a tensor of first order. When the coordinates transform (rotate), the displacement transforms in the same way as the force vector discussed in the previous chapter.

If the field of displacements is not uniform (otherwise, it is trivial and the solid undergoes a rigid body movement), it results in a field of strains over the solid body. In general, displacement components are functions of coordinates, forming a vector of functions. In this book, we assume that all the displacement functions are continuous. This means that there is not breakage in the material under loading. The gradient of the displacement vector function leads to the strain field in the solid.

Strains are tensors of second order. The components of the strains change when the coordinate transforms. The transformation rule for strains is the same as the stress tensor discussed in the previous chapter. We will find out why this is true.

This chapter discusses the details about these important displacement-strain relations, including the representation of displacement vector in 3D space, coordinate transformation rule, how the strains are resulted from the derivatives of the displacement components, derivation of the strain transformation rule, and strain invariants. This chapter is written in reference to [1–6].

As usual, our discussion is accompanied with Python codes written in reference to the documentations of Python, Sympy, and other Python modules with assistance from the AI tools, especially ChatGPT.

4.2 Displacement vector, representations

Forces on a solid can result in displacements. Displacement is a vector because it has two basic properties: magnitude and direction. Consider an infinitely small block representing a point in a solid defined in a 3D Cartesian coordinate system. The block is somehow moved (displaced) to a new location, as shown in Fig. 4.1.

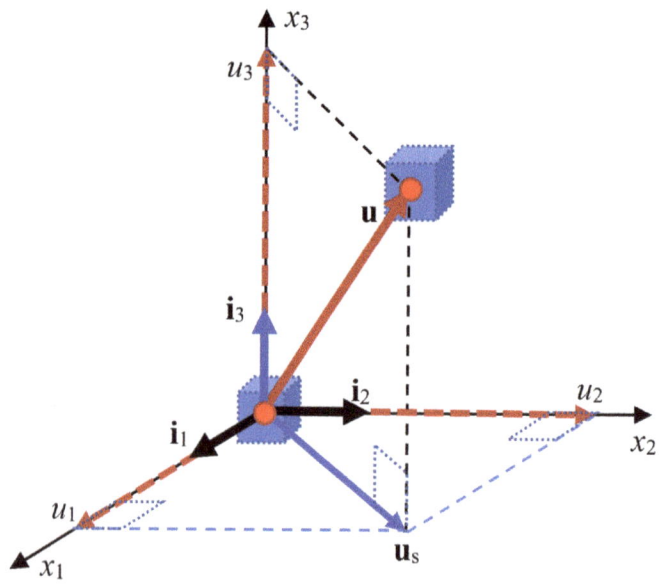

Figure 4.1. An infinitely small block is displaced in space. A vector **u** is used to represent the displacements. The displacement vector can be decomposed to three components following the orthogonal projection rule.

We now present the displacements of the block using a vector **u** under a Cartesian coordinate system, in the same way for a force vector. The displacement vector is first decomposed to two vectors: a vertical component $u_3\mathbf{i}_3$ and $\mathbf{u}_s$ following the orthogonal projection rule. The $\mathbf{u}_s$ is then further decomposed into two vectors: $u_1\mathbf{i}_1$ and $u_2\mathbf{i}_2$ following also the orthogonal projection rule. This process results in three components of **u**: u_1, u_2, and u_3, respectively, on x_1-, x_2-, and x_3-axis. The displacement vector can be written as

$$\mathbf{u} = \begin{bmatrix} u_1 \\ u_2 \\ u_3 \end{bmatrix} \qquad (4.1)$$

It is clear that this representation of a vector is complete and unique under the Cartesian coordinate system, implying that any change, in length or direction, in **u** will result in unique changes in u_1, u_2, and u_3, and vice versa, as long as the orthogonal projection rule is followed.

When we denote the coordinates as (x, y, z), the corresponding displacement components are often denoted as u_x, u_y, u_z or in shorthand u, v, w.

In solid mechanics, we deal with deformable solids. Therefore, any point in a solid body can have displacements, and they often change with the coordinates. Thus, we have in general displacement **vector functions**.

Note also that there are many other vectors, such as velocity and acceleration, rooted at the displacement vector. The same rules of representation apply.

4.3 Displacement vector in transformed coordinates

4.3.1 *Forward transformation*

Consider a displacement vector **u** in coordinate **x**, and the transformed (rotated) coordinate $\mathbf{x}'$. Using the transformation tensor **T** defined in Section 2.10, the values of vector $\mathbf{u}'$ becomes

$$\boxed{u'_i = a_{ij} u_j} \qquad (4.2)$$

where u'_i is obtained following the orthogonal projection rule. On the right-hand side of Eq. (4.2), we see that the 2nd index j of **T** is **contracted** with the index of **u**. This leaves one index i, implying that the outcome is a vector.

In matrix notation, we have

$$\boxed{\mathbf{u}' = \mathbf{T}\mathbf{u}} \qquad (4.3)$$

which achieves the correct contraction through the standard matrix multiplication rule.

4.3.2 Inverse transformation

Since **T** is unitary (see Section 2.10), the inverse transformation can be simply done using the transpose of **T**:

$$\mathbf{u} = \mathbf{T}^\top \mathbf{u}' \tag{4.4}$$

Or in the indicial notation:

$$u_i = a_{ji} u'_j \tag{4.5}$$

All we need is to swap the order of the two indexes of a_{ij}. This can also be achieved by contracting the first index of a_{ij}.

4.4 Vector of displacement functions

4.4.1 Displacement functions, a vector function

Since each displacement component is a function of coordinates, and the displacement components form a vector, we have displacement vector functions. Let us define symbolic functions for each displacement components, and then form the displacement vector function in a Sympy code.

```
# Define each component of the displacement vector as a function of
# coordinates
x1, x2, x3 = sp.symbols('x1, x2, x3')        # symbolic coordinates
u1 = Function('u1')(x1, x2, x3)    # symbolic function of coordinates
u2 = Function('u2')(x1, x2, x3)
u3 = Function('u3')(x1, x2, x3)

u = Matrix([u1, u2, u3])           # form a vector of symbolic functions
u                                                           # print out
```

$$\begin{bmatrix} u_1(x_1, x_2, x_3) \\ u_2(x_1, x_2, x_3) \\ u_3(x_1, x_2, x_3) \end{bmatrix}$$

In addition, at any point in solid, the components of the displacement vector may change when the coordinate at that point transforms. Thus, we need to evaluate such a change due to the coordinate transformation.

Let us create a coordinate transformation matrix and compute the new displacement vector function symbolically.

```
T = gr.transferMs()             # form a symbolic transformation matrix
T                                                           # print out
```

$$\begin{bmatrix} a_{11} & a_{12} & a_{13} \\ a_{21} & a_{22} & a_{23} \\ a_{31} & a_{32} & a_{33} \end{bmatrix}$$

```
1  up=T@u   # transformed displacements; @ is shorthand for multiplication
2  printM(up, 'Formulas for vector components of u',n_dgt=4)
```

Formulas for vector components of u

$$\begin{bmatrix} a_{11}u_1(x_1,x_2,x_3) + a_{12}u_2(x_1,x_2,x_3) + a_{13}u_3(x_1,x_2,x_3) \\ a_{21}u_1(x_1,x_2,x_3) + a_{22}u_2(x_1,x_2,x_3) + a_{23}u_3(x_1,x_2,x_3) \\ a_{31}u_1(x_1,x_2,x_3) + a_{32}u_2(x_1,x_2,x_3) + a_{33}u_3(x_1,x_2,x_3) \end{bmatrix}$$

This gives closed-form formulas for computing the new displacement component functions. If we substitute the displacement vector with a vector of fixed values for the transformation matrix, we have simplified formulas. Let us show this by first generating randomly such a vector with fixed numerical values.

```
1  # A test on a random vector with fixed numerical values.
2  np.random.seed(8)                    # use a seed for re-producibility
3  n = 3
4  Vr = np.random.randn(n,1)            # Generate a random vector at a point
5  print(Vr)
```

```
[[ 0.0912]
 [ 1.0913]
 [-1.947 ]]
```

Now, compute the transformed displacement component functions at the point.

```
1  upT = up.subs(list(zip(u, Matrix(Vr))))
2  printM(upT, 'Reduced formulas for vector components of u', n_dgt=4)
```

Reduced formulas for vector components of u

$$\begin{bmatrix} 0.0912a_{11} + 1.091a_{12} - 1.947a_{13} \\ 0.0912a_{21} + 1.091a_{22} - 1.947a_{23} \\ 0.0912a_{31} + 1.091a_{32} - 1.947a_{33} \end{bmatrix}$$

These are reduced closed-form formulas for computing new displacement components for any given transformation matrix T. If we further substitute T with fixed numerical values, we obtain the displacement vector in numerical values. Let us generate a T with random numbers.

```
1  theta = 0.
2  Tr, about = gr.transferM(theta, about = "random")
3  print(Tr)
```

```
[[-0.5727 -0.3829  0.7249]
 [ 0.7137  0.2022  0.6706]
 [-0.4033  0.9014  0.1575]]
```

```
1  upv = upT.subs(list(zip(T, Matrix(Tr))))
2  printM(upv, 'Transformed random vector by a random T:', n_dgt=4)
```

Transformed random vector by a random T:

$$\begin{bmatrix} -1.881 \\ -1.02 \\ 0.6404 \end{bmatrix}$$

The above example shows that one can perform coordinate transformation for displacement vector function at any point in the solid, and at the same time evaluate its variation with the coordinates.

4.4.2 Code using tensor operation for transformation

Alternatively, the same transformation can be done using tensor notation. The following is a Python code for transformation for a first-order tensor, such as a displacement vector.

```
1  def Tensor1_transfer(T,u):
2      '''Transformation for 1st order tensor u using transformation
3      matrix T.
4      u and T are array alike.
5      '''
6      S = np.tensordot(T, u, axes=([1],[0]))    # use numpy tensordot()
7                  # contracting axis 1 of T with axis 0 of u
8      return S
```

```
1  Tu = Tensor1_transfer(T, u)         # or: gr.Tensor1_transfer(T, u)
2  TuS = Matrix(Tu)                                #Covert to sympy matrix
3  print(f"Transformed vector:")
4  TuS                                                              # print out
```

Transformed vector:

$$\begin{bmatrix} a_{11}u_1(x_1,x_2,x_3) + a_{12}u_2(x_1,x_2,x_3) + a_{13}u_3(x_1,x_2,x_3) \\ a_{21}u_1(x_1,x_2,x_3) + a_{22}u_2(x_1,x_2,x_3) + a_{23}u_3(x_1,x_2,x_3) \\ a_{31}u_1(x_1,x_2,x_3) + a_{32}u_2(x_1,x_2,x_3) + a_{33}u_3(x_1,x_2,x_3) \end{bmatrix}$$

These formulas are the same as those obtained earlier using matrix multiplication.

We can also perform numerically the transformation using the same function. Following is an example for randomly generated transformation matrix and vector:

```
1  TrV = Tensor1_transfer(Tr, Vr)
2  p_text = 'Random vector in a transformed coordinates with random T:'
3  printM(Matrix(TrV), p_text, n_dgt=4)
```

Random vector in a transformed coordinates with random T:

$$\begin{bmatrix} -1.881 \\ -1.02 \\ 0.6404 \end{bmatrix}$$

We obtained the same results as we did when using matrix transformation. We can transform it back:

```
1  print(Tensor1_transfer(Tr.T, TrV))   # one may use Tr with swapping axes
```

```
[[ 0.0912]
 [ 1.0913]
 [-1.947 ]]
```

This is the vector before coordinate transformation.

4.5 Strains derived via kinematic relations

Strains are often defined based on the following mechanics principles [1,5,6]. We do so here starting with a simple case.

4.5.1 *An ideal uniform bar under tension*

Consider a uniform bar with length L and height h. It is constrained at the left-end, and subjected to a uniformly distributed stress at the right-end, as shown in Fig. 4.2.

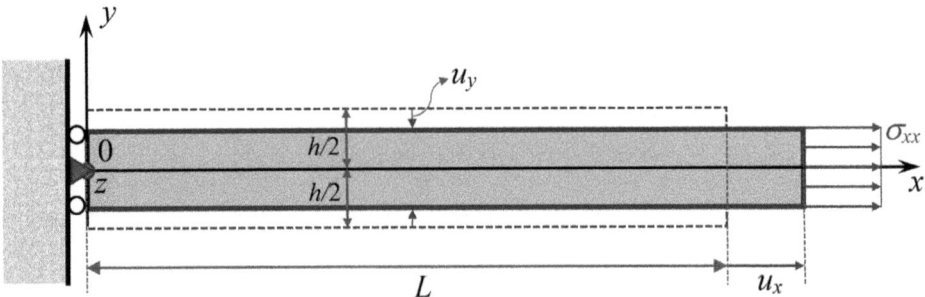

Figure 4.2. A uniform bar constrained at the left-end: the middle point is fixed, but other points on the cross-section is allowed to move freely vertically. The bar with dashed line is the undeformed bar. It is subjected to a uniformly distributed stress at the right-end. Physical observation finds that the bar is elongated in the x-direction and shrunk in the y-direction. Such deformation results in strains in both x- and y-directions.

The particles at any point in the bar are displaced by the applied stress. Assume the displacement changes linearly in both x- and y-directions. The length of the bar becomes L' and the height of the bar becomes h'. Thus, the right-end of the bar has a displacement of u_x in the x-direction, and the top (or bottom) surface has a displaced u_y in the y-direction. The strains in the bar are defined as

$$\varepsilon_{xx} = \frac{L' - L}{L} = \frac{u_x}{L} \qquad (4.6)$$

which is the rate of length change of the bar in x-direction when its right-end has a displacement u_x in the x-direction. Note that the strain measure here is in reference to the original length of the bar. Similarly,

$$\varepsilon_{yy} = \frac{h'/2 - h/2}{h/2} = \frac{u_y}{h/2} \qquad (4.7)$$

where u_y is the displacement in the y-direction, resulting from u_x at the right-end. Strain ε_{yy} is the rate of length change of the bar in the y-direction. The strain measure is in reference to the original half-height of the bar.

Note that strain is dimensionless because the displacement carries a length dimension.

In a general problem of mechanics of materials, the displacements are usually not linearly distributed, and have a gradient field. Thus, the general definition for a strain is defined for a local point in the deformed material as

the rate of change in length of an infinitely small fiber at a point in the material.

Therefore, the strains will be written in the derivatives of the displacements. Since the tip of the fiber can move in x-, y-, and z-directions, we shall have normal and shear strains. Also, the fiber can have x-, y-, and z-orientations; we thus shall have a tensor of second order (two indexes). These expressions are derived as follows.

4.5.2 Normal strains

Consider in general a small line segment (fiber) in the s direction has length of Δs. It is elongated in a deformed solid. Assume the two tips of Δs has a relative displacement Δu_s also in the s direction. The strain along that direction is defined as

$$\varepsilon_{ss} = \lim_{\Delta s \to 0} \frac{\Delta u_s}{\Delta s} = \frac{\partial u_s}{\partial s} \tag{4.8}$$

In particular, if the fiber is Δx in the x-direction, we shall have the normal strain as

$$\varepsilon_{xx} = \lim_{\Delta x \to 0} \frac{\Delta u_x}{\Delta x} = \frac{\partial u_x}{\partial x} \tag{4.9}$$

Similarly, if the fiber is in the y- and z-direction, respectively, we shall have the normal strains ε_{yy} and ε_{zz}:

$$\varepsilon_{yy} = \lim_{\Delta y \to 0} \frac{\Delta u_y}{\Delta y} = \frac{\partial u_y}{\partial y}$$

$$\varepsilon_{zz} = \lim_{\Delta z \to 0} \frac{\Delta u_z}{\Delta z} = \frac{\partial u_z}{\partial z} \tag{4.10}$$

4.5.3 Shear strains

Consider a small square solid abdc experiencing a small deformation to ab'd'c', as shown in Fig. 4.3. The angles α_{xy} and α_{yx} can be calculated using

$$\alpha_{xy} = \frac{\partial u_y}{\partial x}; \quad \alpha_{yx} = \frac{\partial u_x}{\partial y} \tag{4.11}$$

The total angular change becomes

$$\gamma_{xy} = \alpha_{xy} + \alpha_{yx} = \frac{\partial u_y}{\partial x} + \frac{\partial u_x}{\partial y} \tag{4.12}$$

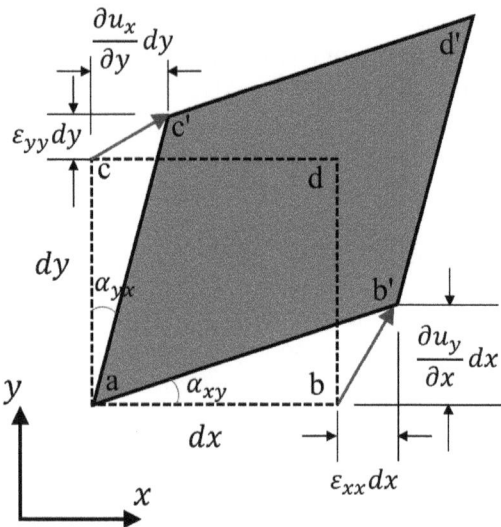

Figure 4.3. A small square solid abdc is deformed to ab'd'c'.

where γ_{xy} is the so-called engineering shear strain. The shear strain ε_{xy} is defined as

$$\varepsilon_{xy} := \frac{1}{2}\gamma_{xy} = \frac{1}{2}\left(\frac{\partial u_y}{\partial x} + \frac{\partial u_x}{\partial y}\right) \tag{4.13}$$

It is obvious that the strains are symmetric. Similarly, we can obtain shear strains ε_{yz} and ε_{xz}.

In summary, all the strains (normal or shear) can be written in a unified form of

$$\boxed{\varepsilon_{ij} = \frac{1}{2}\left(\frac{\partial u_i}{\partial x_j} + \frac{\partial u_j}{\partial x_i}\right) = \frac{1}{2}\left(u_{i,j} + u_{j,i}\right)} \tag{4.14}$$

Here, we used "," in the subscript to represent a partial differentiation. As shown, the small strain is linear in displacement derivatives. The foregoing equation is also known as the **strain–displacement relation** for small strains. It will be used frequently in this book.

4.6 Strains, derivatives of displacement functions

4.6.1 *Displacement gradient*

The displacement vector in solid mechanics has three components, each of which is a function of coordinates. It can be defined symbolically in Python code as follows:

```
1  x, y, z = symbols("x, y, z")          # define symbolic coordinates
2  X = Matrix([x, y, z])
3
4  ux = Function('u_x')(x, y, z) # define symbolic displacement functions
5  uy = Function('u_y')(x, y, z)
6  uz = Function('u_z')(x, y, z)
7
8  U = Matrix([ux, uy, uz])    # displacement vector of symbolic functions
9  U
```

$$\begin{bmatrix} u_x(x,y,z) \\ u_y(x,y,z) \\ u_z(x,y,z) \end{bmatrix}$$

The displacement **U** is a vector, and each of its components is a function of coordinates. When we get the gradient of the displacement vector, we are getting the gradient of each of the 3 component functions. Since the gradient of one component function gives a vector, we shall now obtain a matrix as follows (see Section 2.14), using the **outer product** of ∇ (vector of differential operators) and $\mathbf{U}(\mathbf{x})$:

$$\nabla \mathbf{u}(\mathbf{x}) := \underbrace{\begin{bmatrix} \frac{\partial}{\partial x} \\ \frac{\partial}{\partial y} \\ \frac{\partial}{\partial z} \end{bmatrix}}_{\nabla} \otimes \underbrace{\begin{bmatrix} u_x(\mathbf{x}) \\ u_y(\mathbf{x}) \\ u_z(\mathbf{x}) \end{bmatrix}}_{\mathbf{U}(\mathbf{x})}$$

$$= \begin{bmatrix} \frac{\partial}{\partial x} u_x(x,y,z) & \frac{\partial}{\partial x} u_y(x,y,z) & \frac{\partial}{\partial x} u_z(x,y,z) \\ \frac{\partial}{\partial y} u_x(x,y,z) & \frac{\partial}{\partial y} u_y(x,y,z) & \frac{\partial}{\partial y} u_z(x,y,z) \\ \frac{\partial}{\partial z} u_x(x,y,z) & \frac{\partial}{\partial z} u_y(x,y,z) & \frac{\partial}{\partial z} u_z(x,y,z) \end{bmatrix} \qquad (4.15)$$

It is a matrix of differentials of the displacements, known as the **displacement gradient**. We can use the following code to obtain it.

```
1  def grad_vf(vf, X):
2      '''Compute the gradient of a given vector function, vf
3         input: vf, sp.Matrix, a column vector of 2 or 3.
4                X, coordinates, array-like of symbols
5         return: grad_f, sp.Matrix, a matrix of 2 by 2 or 3 by 3.
6      '''
7      grad_f = sp.Matrix([sp.diff(vf,xi).T for xi in X])
8      return grad_f
```

```
1  U_g = grad_vf(U, X)  # displacement gradient. Or use vector.grad_vec(U)
2  U_g                   # Or gr.grad_vf(U, X)
```

$$\begin{bmatrix} \frac{\partial}{\partial x} u_x(x,y,z) & \frac{\partial}{\partial x} u_y(x,y,z) & \frac{\partial}{\partial x} u_z(x,y,z) \\ \frac{\partial}{\partial y} u_x(x,y,z) & \frac{\partial}{\partial y} u_y(x,y,z) & \frac{\partial}{\partial y} u_z(x,y,z) \\ \frac{\partial}{\partial z} u_x(x,y,z) & \frac{\partial}{\partial z} u_y(x,y,z) & \frac{\partial}{\partial z} u_z(x,y,z) \end{bmatrix}$$

4.6.2 Small strains, derived via displacement gradient

The displacement gradient matrix given in Eq. (4.15) is in general asymmetric. It can be used to produce symmetric tensors. There are two approaches to do this. One is to decompose $\nabla \mathbf{u}(\mathbf{x})$ into two matrices: one symmetric and another antisymmetric. The symmetric part is as follows:

$$\frac{1}{2}\left[\nabla \mathbf{u}(\mathbf{x}) + [\nabla \mathbf{u}(\mathbf{x})]^\top\right] \tag{4.16}$$

This leads to a symmetric tensor that will be the small strains in solids. In indicial notation, it is given in Eq. (4.14).

We write the following simple code to derive this relation.

```
1  E_1 = (U_g + U_g.T)/2      # Or use vector.sym_grad(U)
2  E_1                         # strain tensor, symmetric
```

$$\begin{bmatrix} \frac{\partial}{\partial x} u_x(x,y,z) & \frac{\frac{\partial}{\partial y} u_x(x,y,z)}{2} + \frac{\frac{\partial}{\partial x} u_y(x,y,z)}{2} & \frac{\frac{\partial}{\partial z} u_x(x,y,z)}{2} + \frac{\frac{\partial}{\partial x} u_z(x,y,z)}{2} \\ \frac{\frac{\partial}{\partial y} u_x(x,y,z)}{2} + \frac{\frac{\partial}{\partial x} u_y(x,y,z)}{2} & \frac{\partial}{\partial y} u_y(x,y,z) & \frac{\frac{\partial}{\partial z} u_y(x,y,z)}{2} + \frac{\frac{\partial}{\partial y} u_z(x,y,z)}{2} \\ \frac{\frac{\partial}{\partial z} u_x(x,y,z)}{2} + \frac{\frac{\partial}{\partial x} u_z(x,y,z)}{2} & \frac{\frac{\partial}{\partial z} u_y(x,y,z)}{2} + \frac{\frac{\partial}{\partial y} u_z(x,y,z)}{2} & \frac{\partial}{\partial z} u_z(x,y,z) \end{bmatrix}$$

This gives the analytic formulas for the linear strain tensor in terms of the derivatives of the displacement components. It is valid only when strain is small.

The strain tensor given in the above matrix is of second order and symmetric. We now write the strain tensor as symbolic variables.

Displacements and Strains

```
1  Exx, Exy, Exz = "\u03B5_xx", "\u03B5_xy", "\u03B5_xz"
2  Eyx, Eyy, Eyz = "\u03B5_yx", "\u03B5_yy", "\u03B5_yz"
3  Ezx, Ezy, Ezz = "\u03B5_zx", "\u03B5_zy", "\u03B5_zz"
4
5  E = Matrix([[Exx, Exy, Exz],      # E: stress tensor
6               [Eyx, Eyy, Eyz],
7               [Ezx, Ezy, Ezz]])
8  E
```

$$\begin{bmatrix} \varepsilon_{xx} & \varepsilon_{xy} & \varepsilon_{xz} \\ \varepsilon_{yx} & \varepsilon_{yy} & \varepsilon_{yz} \\ \varepsilon_{zx} & \varepsilon_{zy} & \varepsilon_{zz} \end{bmatrix}$$

We can write the strain–displacement relations (small strains) in matrix form as follows [1, 5, 6].

$$\begin{bmatrix} \varepsilon_{xx} & \varepsilon_{xy} & \varepsilon_{xz} \\ \varepsilon_{yx} & \varepsilon_{yy} & \varepsilon_{yz} \\ \varepsilon_{zx} & \varepsilon_{zy} & \varepsilon_{zz} \end{bmatrix} = \frac{1}{2} \begin{bmatrix} 2\frac{\partial u_x}{\partial x} & \frac{\partial u_x}{\partial y}+\frac{\partial u_y}{\partial x} & \frac{\partial u_x}{\partial z}+\frac{\partial u_z}{\partial x} \\ \frac{\partial u_x}{\partial y}+\frac{\partial u_y}{\partial x} & 2\frac{\partial u_y}{\partial y} & \frac{\partial u_y}{\partial z}+\frac{\partial u_z}{\partial y} \\ \frac{\partial u_x}{\partial z}+\frac{\partial u_z}{\partial x} & \frac{\partial u_y}{\partial z}+\frac{\partial u_z}{\partial y} & 2\frac{\partial u_z}{\partial z} \end{bmatrix} \qquad (4.17)$$

In indicial notation, it is the same as that given in Eq. (4.14). We obtained the displacement-strain relations via purely mathematic operations.

4.6.3 Example: Computation of small strains

Following is an example of computing small strains for given functions for the three displacement components.

```
1  # Define displacement vector function:
2  #uv=Matrix([.058*x-.025*y+.065*x*y,.016*x+.025*y-.03*x*y,.005*z])#Large
3  uv=Matrix([.0058*x-.0025*y+.0065*x*y, .0016*x+.0025*y-.003*x*y,.005*z])
4  uv             # the given functions for three displacement components
```

$$\begin{bmatrix} 0.0065xy + 0.0058x - 0.0025y \\ -0.003xy + 0.0016x + 0.0025y \\ 0.005z \end{bmatrix}$$

Compute the strains, using the formula E_1 obtained earlier and the subs() function.

```
1  e_small = E_l.subs(list(zip(U,uv))).doit().expand()
2  gr.printM(e_small*1000., 'Strains (e^-3)', n_dgt=4)
```

Strains (e^-3)

$$\begin{bmatrix} 6.5y + 5.8 & 3.25x - 1.5y - 0.45 & 0 \\ 3.25x - 1.5y - 0.45 & 2.5 - 3.0x & 0 \\ 0 & 0 & 5.0 \end{bmatrix}$$

Further, we find the strains at a point with coordinates, say at $(1., 1., 1.)$.

```
1  e_v = e_small.subs({x:1.,y:1,z:1.})
2  gr.printM(e_v*1000., 'Strains (e^-3)', n_dgt=4)
```

Strains (e^-3)

$$\begin{bmatrix} 12.3 & 1.3 & 0 \\ 1.3 & -0.5 & 0 \\ 0 & 0 & 5.0 \end{bmatrix}$$

Remark: The gradient of the displacement vector at a point in a solid gives the strain tensor/matrix at that point.

4.6.4 Rotation-displacement relations*

An antisymmetric matrix can be obtained using the displacement gradient matrix:

$$\frac{1}{2}\left[\nabla \mathbf{u}(\mathbf{x}) - [\nabla \mathbf{u}(\mathbf{x})]^\top\right] \qquad (4.18)$$

This will be the rotation in solids. The following Python code creates the matrix of rotations.

```
1  eps = (U_g-U_g.T)/2          # compute rotations, anti-symmetric
2  eps
```

$$\begin{bmatrix} 0 & -\frac{\frac{\partial}{\partial y}u_x(x,y,z)}{2} + \frac{\frac{\partial}{\partial x}u_y(x,y,z)}{2} & -\frac{\frac{\partial}{\partial z}u_x(x,y,z)}{2} + \frac{\frac{\partial}{\partial x}u_z(x,y,z)}{2} \\ \frac{\frac{\partial}{\partial y}u_x(x,y,z)}{2} - \frac{\frac{\partial}{\partial x}u_y(x,y,z)}{2} & 0 & -\frac{\frac{\partial}{\partial z}u_y(x,y,z)}{2} + \frac{\frac{\partial}{\partial y}u_z(x,y,z)}{2} \\ \frac{\frac{\partial}{\partial z}u_x(x,y,z)}{2} - \frac{\frac{\partial}{\partial x}u_z(x,y,z)}{2} & \frac{\frac{\partial}{\partial z}u_y(x,y,z)}{2} - \frac{\frac{\partial}{\partial y}u_z(x,y,z)}{2} & 0 \end{bmatrix}$$

We see that the matrix of rotations is anti-symmetric. It is obvious that the rotation components do not contribute anything to the strains defined earlier. Therefore, anti-symmetric part of the gradient of the displacement vector must be excluded in the measure of strains. This will be done in the formulations for both small and large strains. The matrix of rotations can also be obtained using the vector.antisym_grad().

4.6.5 Large strains, derived via displacement gradient*

Using the displacement gradient, one can also construct a symmetric matrix as follows:

$$\frac{1}{2}\nabla \mathbf{u}(\mathbf{x})[\nabla \mathbf{u}]^\top \tag{4.19}$$

This gives a symmetric tensor, the nonlinear part of the **large strains**. The formula for the large strain, known as the so-called Green–Lagrange Strain tensor denoted as ε_{GL}, can then be obtained by adding up the linear and nonlinear strains:

$$\boxed{\varepsilon_{GL} = \frac{1}{2}\left[\nabla \mathbf{u}(\mathbf{x}) + [\nabla \mathbf{u}(\mathbf{x})]^\top + \nabla \mathbf{u}(\mathbf{x})[\nabla \mathbf{u}]^\top\right]} \tag{4.20}$$

In indicial notation, it can be written as

$$\boxed{\varepsilon_{ij} = \frac{1}{2}\left(\frac{\partial u_i}{\partial x_j} + \frac{\partial u_j}{\partial x_i} + \frac{\partial u_k}{\partial x_i}\frac{\partial u_k}{\partial x_j}\right) = \frac{1}{2}\left(u_{i,j} + u_{j,i} + u_{k,i}u_{k,j}\right)} \tag{4.21}$$

Note the contraction over k. This means that the nonlinear part is a sum of three nonlinear terms.

```
1  E_GL = (U_g+U_g.T + U_g@U_g.T)/2
2  gr.printM(E_GL[:,0], ' Formulas for large strains (1st column):')
```

Formulas for large strains (1st column):

$$\begin{bmatrix} \frac{\left(\frac{\partial u_x}{\partial x}\right)^2}{2} + \frac{\partial u_x}{\partial x} + \frac{\left(\frac{\partial u_y}{\partial x}\right)^2}{2} + \frac{\left(\frac{\partial u_z}{\partial x}\right)^2}{2} \\ \frac{\frac{\partial u_x}{\partial x}\frac{\partial u_x}{\partial y}}{2} + \frac{\frac{\partial u_x}{\partial y}}{2} + \frac{\frac{\partial u_y}{\partial x}\frac{\partial u_y}{\partial y}}{2} + \frac{\frac{\partial u_y}{\partial x}}{2} + \frac{\frac{\partial u_z}{\partial x}\frac{\partial u_z}{\partial y}}{2} \\ \frac{\frac{\partial u_x}{\partial x}\frac{\partial u_x}{\partial z}}{2} + \frac{\frac{\partial u_x}{\partial z}}{2} + \frac{\frac{\partial u_y}{\partial x}\frac{\partial u_y}{\partial z}}{2} + \frac{\frac{\partial u_z}{\partial x}\frac{\partial u_z}{\partial z}}{2} + \frac{\frac{\partial u_z}{\partial x}}{2} \end{bmatrix}$$

Strains given in Eq. (4.21) depend purely on the derivatives of the displacement components. Thus, it should be some kind of strain measure. Since it has terms of products of the displacement derivatives, it is clearly nonlinear. Equation (4.21) is a measure of large strains containing both linear and nonlinear parts. In addition, if the derivatives of the displacements are small, the higher-order 3rd term in Eq. (4.21) can be omitted. In that case, Eq. (4.21) is the small strain given in Eq. (4.14).

The nonlinear part can be obtained easily using the following code.

```
1 E_nl = (U_g@U_g.T)/2
2 gr.printM(E_nl[:,0],'Formulas for nonlinear strains only (1st column):')
```

Formulas for nonlinear strains only (1st column):

$$
\begin{bmatrix}
\frac{\left(\frac{\partial}{\partial x} u_x(x,y,z)\right)^2}{2} + \frac{\left(\frac{\partial}{\partial x} u_y(x,y,z)\right)^2}{2} + \frac{\left(\frac{\partial}{\partial x} u_z(x,y,z)\right)^2}{2} \\
\frac{\frac{\partial}{\partial x} u_x(x,y,z) \frac{\partial}{\partial y} u_x(x,y,z)}{2} + \frac{\frac{\partial}{\partial x} u_y(x,y,z) \frac{\partial}{\partial y} u_y(x,y,z)}{2} + \frac{\frac{\partial}{\partial x} u_z(x,y,z) \frac{\partial}{\partial y} u_z(x,y,z)}{2} \\
\frac{\frac{\partial}{\partial x} u_x(x,y,z) \frac{\partial}{\partial z} u_x(x,y,z)}{2} + \frac{\frac{\partial}{\partial x} u_y(x,y,z) \frac{\partial}{\partial z} u_y(x,y,z)}{2} + \frac{\frac{\partial}{\partial x} u_z(x,y,z) \frac{\partial}{\partial z} u_z(x,y,z)}{2}
\end{bmatrix}
$$

The nonlinear part is still too large to print out in whole, and hence we printed out only the first column. To write it out in whole, we simplify the expressions, which leads to the same formulas given in Refs. [1,3]:

$$
\mathbf{E}_{nl} = \frac{1}{2}
\begin{bmatrix}
\left(\frac{\partial u}{\partial x}\right)^2 + \left(\frac{\partial v}{\partial x}\right)^2 + \left(\frac{\partial w}{\partial x}\right)^2 & \frac{\partial u}{\partial x}\frac{\partial u}{\partial y} + \frac{\partial v}{\partial x}\frac{\partial v}{\partial y} + \frac{\partial w}{\partial x}\frac{\partial w}{\partial y} \\
\frac{\partial u}{\partial x}\frac{\partial u}{\partial y} + \frac{\partial v}{\partial x}\frac{\partial v}{\partial y} + \frac{\partial w}{\partial x}\frac{\partial w}{\partial y} & \left(\frac{\partial u}{\partial y}\right)^2 + \left(\frac{\partial v}{\partial y}\right)^2 + \left(\frac{\partial w}{\partial y}\right)^2 \\
\frac{\partial u}{\partial x}\frac{\partial u}{\partial z} + \frac{\partial v}{\partial x}\frac{\partial v}{\partial z} + \frac{\partial w}{\partial x}\frac{\partial w}{\partial z} & \frac{\partial u}{\partial y}\frac{\partial u}{\partial z} + \frac{\partial v}{\partial y}\frac{\partial v}{\partial z} + \frac{\partial w}{\partial y}\frac{\partial w}{\partial z} \\
\frac{\partial u}{\partial x}\frac{\partial u}{\partial z} + \frac{\partial v}{\partial x}\frac{\partial v}{\partial z} + \frac{\partial w}{\partial x}\frac{\partial w}{\partial z} \\
\frac{\partial u}{\partial y}\frac{\partial u}{\partial z} + \frac{\partial v}{\partial y}\frac{\partial v}{\partial z} + \frac{\partial w}{\partial y}\frac{\partial w}{\partial z} \\
\left(\frac{\partial u}{\partial z}\right)^2 + \left(\frac{\partial v}{\partial z}\right)^2 + \left(\frac{\partial w}{\partial z}\right)^2
\end{bmatrix}
\quad (4.22)
$$

Note that the nonlinear strains given in Eq. (4.21) is an exact expression, implying that it is derived without any assumption or approximation. It is not obtained from a Taylor series expansion by cutting off the third-order terms and beyond.

Remark: The strains at a point in a solid are determined by the displacements derivatives at that point.

4.6.6 Example: Computation of large strains*

Following is an example for computing large strains for displacement functions given below, simply using subs().

```
1  # Define displacement vector function:
2  #df=Matrix([.058*x-.025*y+.065*x*y,.016*x+.025*y-.03*x*y,.005*z])#Large
3  df=Matrix([.0058*x-.0025*y+.0065*x*y, .0016*x+.0025*y-.003*x*y,.005*z])
4  df
```

$$\begin{bmatrix} 0.0065xy + 0.0058x - 0.0025y \\ -0.003xy + 0.0016x + 0.0025y \\ 0.005z \end{bmatrix}$$

```
1  egl = E_GL.subs(list(zip(U,df))).doit().expand()
2  gr.printM(egl*1000., 'Green-Lagrange Strains (e^-3)', n_dgt=4)
```

Green-Lagrange Strains (e^-3)

$$\begin{bmatrix} 0.02562y^2 + 6.533y + 5.818 & \begin{array}{c} 0.02562xy + 3.266x \\ -1.512y - 0.4552 \end{array} & 0 \\ \begin{array}{c} 0.02562xy + 3.266x \\ -1.512y - 0.4552 \end{array} & 0.02562x^2 - 3.024x + 2.506 & 0 \\ 0 & 0 & 5.013 \end{bmatrix}$$

Further, we find the solution at a given coordinate, say at $(1., 1., 1.)$.

```
1  egln = egl.subs({x:1.,y:1,z:1.})
2  gr.printM(egln*1000., 'Values of the GL Strains (e^-3)', n_dgt=4)
```

Values of the GL Strains (e^-3)

$$\begin{bmatrix} 12.38 & 1.325 & 0 \\ 1.325 & -0.4919 & 0 \\ 0 & 0 & 5.013 \end{bmatrix}$$

It is found that the Green–Lagrange strain (large strain) is very close to the small strain obtained in the previous section. Readers may make the displacements larger (by 10 times for example), and compute both small and large strains again to see the differences.

4.6.7 Large strains derived via conventional deformation gradient**

For problems with large displacements and hence large strains, the strain–displacement relation will be nonlinear [1,6]. Traditionally, one often derive large strains using the deformation gradient. Let us do so using symbolic computation.

Assume a solid is initially at state 0 and is described using coordinates $x, y,$ and z. We define the initial coordinate vector $\mathbf{x}_0$ as follows:

```
1  x, y, z = symbols("x, y, z")           # define symbolic coordinates
2  X = Matrix([x, y, z])
3  x0 = Matrix([x, y, z])                 # define initial coordinates (symbolic)
4  x0
```

$$\begin{bmatrix} x \\ y \\ z \end{bmatrix}$$

Vector $\mathbf{x}_0$ can be treated as a vector of (simple) functions. The gradient of $\mathbf{x}_0$ with respect to $x, y,$ and z becomes an identity matrix as expected:

```
1  x0_g = gr.grad_vf(x0, X)               # gradient of the initial coordinates
2  x0_g
```

$$\begin{bmatrix} 1 & 0 & 0 \\ 0 & 1 & 0 \\ 0 & 0 & 1 \end{bmatrix}$$

Assume now the solid is stressed and deformed heavily to a new state t. At any point in the solid, the displacements are denoted as vector $\mathbf{U}$ with components not necessarily small. The coordinates of the point now become $\mathbf{x}_t$.

```
1  x_t = Function('x_t')(x,y,z)           # define symbolic functions
2  y_t = Function('y_t')(x,y,z)
3  z_t = Function('z_t')(x,y,z)
```

```
1  xt = Matrix([x_t, y_t, z_t])           # form a vector of symbolic functions
2  xt
```

$$\begin{bmatrix} x_t(x, y, z) \\ y_t(x, y, z) \\ z_t(x, y, z) \end{bmatrix}$$

4.6.7.1 Deformation gradient, definition

Note that $\mathbf{x}_t$ is the vector of functions of x, y, and z. The **deformation gradient**, often denoted as $\mathbf{F}$, can be defined as the gradient of $\mathbf{x}_t$, that is a vector function, with respect to x, y, and z. It becomes a matrix as follows:

$$\mathbf{F} = \left[\frac{\partial x_i^t}{\partial x_j^0}\right] = \begin{bmatrix} \frac{\partial x_1^t}{\partial x_1^0} & \frac{\partial x_1^t}{\partial x_2^0} & \frac{\partial x_1^t}{\partial x_3^0} \\ \frac{\partial x_2^t}{\partial x_1^0} & \frac{\partial x_2^t}{\partial x_2^0} & \frac{\partial x_2^t}{\partial x_3^0} \\ \frac{\partial x_3^t}{\partial x_1^0} & \frac{\partial x_3^t}{\partial x_2^0} & \frac{\partial x_3^t}{\partial x_3^0} \end{bmatrix} \qquad (4.23)$$

The code to compute it is given as follows:

```
1  F = vector.grad_vec(xt).T
2  F
```

$$\begin{bmatrix} \frac{\partial}{\partial x}x_t(x,y,z) & \frac{\partial}{\partial y}x_t(x,y,z) & \frac{\partial}{\partial z}x_t(x,y,z) \\ \frac{\partial}{\partial x}y_t(x,y,z) & \frac{\partial}{\partial y}y_t(x,y,z) & \frac{\partial}{\partial z}y_t(x,y,z) \\ \frac{\partial}{\partial x}z_t(x,y,z) & \frac{\partial}{\partial y}z_t(x,y,z) & \frac{\partial}{\partial z}z_t(x,y,z) \end{bmatrix}$$

Note the deformation gradient matrix is in general asymmetric. Since it contains rigid body movements and that may result in false strains. Thus, we need to go through the following process.

4.6.7.2 Deformation gradient, in displacements

First, we write the deformation gradient in terms of the displacement gradient. Because the new coordinates $\mathbf{x}_t$ is the result of the displacement $\mathbf{U}$, we shall have (assume the material is continuous):

$$\mathbf{x}^t = \mathbf{x}^0 + \mathbf{U} \qquad (4.24)$$

Let us use a code to do this.

```
1  xt = x0 + U
2  F = gr.grad_vf(xt, X).T                # Or use vector.grad_vec(xt).T
3  gr.printM(F, 'F in displacement gradient:', n_dgt=1)
```

F in displacement gradient:

$$\begin{bmatrix} \frac{\partial}{\partial x}u_x(x,y,z)+1 & \frac{\partial}{\partial y}u_x(x,y,z) & \frac{\partial}{\partial z}u_x(x,y,z) \\ \frac{\partial}{\partial x}u_y(x,y,z) & \frac{\partial}{\partial y}u_y(x,y,z)+1 & \frac{\partial}{\partial z}u_y(x,y,z) \\ \frac{\partial}{\partial x}u_z(x,y,z) & \frac{\partial}{\partial y}u_z(x,y,z) & \frac{\partial}{\partial z}u_z(x,y,z)+1 \end{bmatrix}$$

As seen, the deformation gradient is the transpose of the displacement gradient (see Eq. (4.15)) plus an identity matrix. These 1s on the diagonal are the rigid body movements, and hence have nothing to do with the displacements derivatives, and hence needs to be removed in a proper manner.

4.6.7.3 Cauchy–Green deformation tensor

Using the deformation gradient, we can form the so-called Cauchy–Green deformation tensor.

$$\mathbf{F}_{CG} = \mathbf{F}^\top \mathbf{F} \tag{4.25}$$

It is clear that $\mathbf{F}_{CG}$ is a symmetric tensor of second order. We can use the following code to form it.

```
1  F_CG = F.T@F              # the Cauchy-Green deformation tensor
2  F_CG[:,0]                 # the 1st column (too large to print out in whole)
```

$$\begin{bmatrix} \left(\frac{\partial u_x}{\partial x}+1\right)^2 + \left(\frac{\partial u_x}{\partial y}\right)^2 + \left(\frac{\partial u_x}{\partial z}\right)^2 \\ \left(\frac{\partial u_x}{\partial x}+1\right)\frac{\partial u_x}{\partial y} + \left(\frac{\partial u_y}{\partial y}+1\right)\frac{\partial u_x}{\partial y} + \frac{\partial u_x}{\partial z}\frac{\partial u_z}{\partial y} \\ \left(\frac{\partial u_x}{\partial x}+1\right)\frac{\partial u_x}{\partial z} + \left(\frac{\partial u_z}{\partial z}+1\right)\frac{\partial u_x}{\partial z} + \frac{\partial u_x}{\partial y}\frac{\partial u_y}{\partial z} \end{bmatrix}$$

```
1  #F_CG[:,1]                # the 2nd column
2  #F_CG[:,2]                # the 3rd column
```

Note that the 1s are still there.

4.6.7.4 Green–Lagrange strain tensor

Using the Cauchy–Green deformation tensor, we can form the so-called Green–Lagrange strain tensor, denoted as ε_{GL}.

$$\boxed{\varepsilon_{GL} = \frac{1}{2}(\mathbf{F}_{CG} - \mathbf{I})} \tag{4.26}$$

The subtraction of identity matrix $\mathbf{I}$ removes these 1s. We can use the following code to form ε_{GL}.

```
1  E_GL=sp.S.Half*(F_CG-sp.eye(3)).expand()  # E_GL: Green-Lagrange Strain
2  E_GL[:,0]                     #E_GL[:,1]  #E_GL[:,2]
```

$$\left[\begin{array}{c} \frac{\left(\frac{\partial u_x}{\partial x}\right)^2}{2} + \frac{\partial u_x}{\partial x} + \frac{\left(\frac{\partial u_y}{\partial x}\right)^2}{2} + \frac{\left(\frac{\partial u_z}{\partial x}\right)^2}{2} \\ \frac{\frac{\partial u_x}{\partial x}\frac{\partial u_x}{\partial y}}{2} + \frac{\frac{\partial u_x}{\partial y}}{2} + \frac{\frac{\partial u_y}{\partial x}\frac{\partial u_y}{\partial y}}{2} + \frac{\frac{\partial u_y}{\partial x}}{2} + \frac{\frac{\partial u_z}{\partial x}\frac{\partial u_z}{\partial y}}{2} \\ \frac{\frac{\partial u_x}{\partial x}\frac{\partial u_x}{\partial z}}{2} + \frac{\frac{\partial u_x}{\partial z}}{2} + \frac{\frac{\partial u_y}{\partial x}\frac{\partial u_y}{\partial z}}{2} + \frac{\frac{\partial u_z}{\partial x}\frac{\partial u_z}{\partial z}}{2} + \frac{\frac{\partial u_z}{\partial x}}{2} \end{array}\right]$$

In the code above, we used S.Half that is a singleton. It is the Rational (1, 2) in Sympy. Green–Lagrange strain tensor is for large strains. It contains quadratic terms and hence it is nonlinear in derivatives of the displacements.

Note that ε_{GL} given in Eq. (4.26) is the same as that in Eq. (4.20) obtained earlier via a much simpler procedure, via an argument of symmetry.

Recall that we have obtained the formula for linear strain tensor earlier, which is given in Eq. (4.17). We can subtract the linear strain from the Green–Lagrange strain and obtain the nonlinear part of the strain, which will be exactly the same as the one obtained earlier, via also a simple symmetry argument.

4.7 Strain–displacement relation, a general derivation

Similar to the stress tensor, the strain tensor has both indicial and matrix representations. Since the strain tensor is symmetric, the matrix is a symmetric matrix. It is also similar to the stress tensor, in terms of formulation and operations.

- In space $\mathbb{R}^3$, it is a 3×3 matrix, and hence has, in general, a total of 9 components. Due to the symmetry, we have only 6 components.
- In space $\mathbb{R}^2$, it is a 2×2 matrix, and hence has, in general, a total of 4 components. Due to the symmetry, we have only 3 components.
- In space $\mathbb{R}^1$, it is a scalar, and hence there is only 1 component.

We present now a general approach to derive the strain–displacement relation, considering large deformation in 3D solids. This derivation shall lead to the strain tensor transformation rule.

4.7.1 Normal strain along a fiber, magnification factor*

Consider a solid occupying domain Ω in the Cartesian coordinate system with mutual orthogonal axes (x_1, x_2, x_3). It undergoes a large deformation and its domain becomes a deformed one, Ω^*, as shown in Fig. 4.4.

An infinitely small fiber in the undeformed solid is denoted as P-Q. The coordinates of the two ends of the fiber are $P(x_1, x_2, x_3)$ and $Q(x_1+dx_1, x_2+dx_2, x_3+dx_3)$. After the deformation the fiber becomes P*-Q*, and its coordinates become $P^*(x_1^*, x_2^*, x_3^*)$ and $Q^*(x_1^* + dx_1^*, x_2^* + dx_2^*, x_3^* + dx_3^*)$. The displacements at point P is denoted as (u_1, u_2, u_3).

4.7.1.1 Evaluation of the length of a fiber in a deformed solid

Assume the solid is a continuum media and also the deformation is continuous, implying that each of the coordinates (x_1^*, x_2^*, x_3^*) is differentiable with respect to x_1, x_2, and x_3. Hence, we have

$$dx_i^* = \frac{\partial x_i^*}{\partial x_1}dx_1 + \frac{\partial x_i^*}{\partial x_2}dx_2 + \frac{\partial x_i^*}{\partial x_3}dx_3 = \frac{\partial x_i^*}{\partial x_j}dx_j = x_{i,j}^* dx_j \quad (4.27)$$

where we used again the summation convention for repeated indexes in a term, and the subscript "," stands for partial differentiation.

Since the deformation is assumed continuous, the coordinates at a point in the deformed solid should be the sum of the coordinates of the undeformed

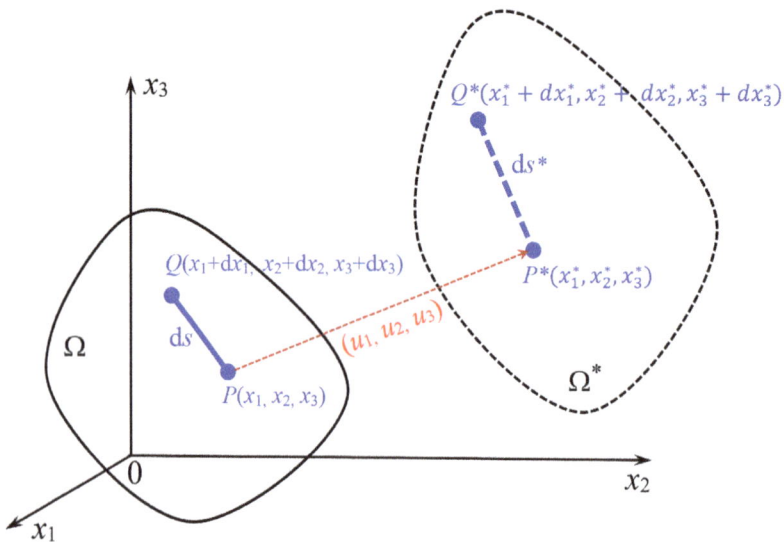

Figure 4.4. A solid continuum media undergoes large deformation.

solid plus the displacements at the same point:

$$x_i^* = x_i + u_i, \quad i = 1, 2, 3 \tag{4.28}$$

In addition, each of the displacement components will also be differentiable, which gives

$$x_{i,j}^* = x_{i,j} + u_{i,j} \tag{4.29}$$

Since the solid is a continuum media, the length of the infinitely small fiber can be measured using

$$(ds)^2 = (dx_1)^2 + (dx_2)^2 + (dx_3)^2 = dx_i dx_i \tag{4.30}$$

The length of the infinitely small fiber in the deformed solid can also be obtained using

$$(ds^*)^2 = (dx_1^*)^2 + (dx_2^*)^2 + (dx_3^*)^2 = dx_i^* dx_i^* \tag{4.31}$$

4.7.1.2 Engineering strain of a given fiber

Assume now the infinitely small fiber has an orientation in the Cartesian coordinate system defined as $\mathbf{n} = [n_1, n_2, n_3]^\top$, where n_i is the direction cosine of the fiber: $n_i = \cos(\mathbf{n}, x_i)$. The engineering strain of the fiber is a measure with respect to the original length of the fiber, and is expressed as,

$$\varepsilon_E = \frac{ds^* - ds}{ds} = \frac{ds^*}{ds} - 1 \tag{4.32}$$

which gives

$$\frac{ds^*}{ds} = \varepsilon_E + 1 \tag{4.33}$$

4.7.1.3 Magnification factor of a fiber, large normal strain

We now define the so-called magnification factor as

$$M := \frac{1}{2} \left[\frac{(ds^*)^2 - (ds)^2}{(ds)^2} \right] = \frac{1}{2} \left[\left(\frac{ds^*}{ds} \right)^2 - 1 \right] \tag{4.34}$$

Readers may take a note of the difference between the expressions of the engineering strain and the magnification factor, which can be expressed as

$$M = \varepsilon_E + \frac{1}{2} \varepsilon_E^2 \tag{4.35}$$

The magnification factor M is known as the Green–Lagrange normal strain for a fiber in a solid undergoing large deformation. We see the nonlinear (squared) term, due to the large strain. When ε_E is small, we have $M \approx \varepsilon_E$.

4.7.1.4 Magnification factor and the strain components

Now, we are ready to derive the relationship between the magnification factor of the given fiber P-Q and the strain components at point P. From Eqs. (4.27) and (4.29), we have

$$dx_i^* = x_{i,j}^* dx_j = \underbrace{(x_{i,j} + u_{i,j})}_{x_{i,j}^*} dx_j \qquad (4.36)$$

Note that

$$x_{i,j} = \frac{\partial x_i}{\partial x_j} = \delta_{ij} \qquad (4.37)$$

Equation (4.36) becomes

$$dx_i^* = x_{i,j}^* dx_j = (\delta_{ij} + u_{i,j}) dx_j \qquad (4.38)$$

Let us evaluate the difference of the squared length of the fiber before and after deformation. Using Eqs. (4.30) and (4.31) we shall have

$$[(ds^*)^2 - (ds)^2] = dx_i^* dx_i^* - dx_i dx_i$$

$$= \left(\underbrace{(\delta_{ij} + u_{i,j})}_{dx_i^*} \underbrace{(\delta_{il} + u_{i,l})}_{dx_i^*} \right) dx_j dx_l - dx_i dx_i$$

$$= \left(\underbrace{\delta_{ij}\delta_{il}}_{\delta_{jl}} + \underbrace{\delta_{ij} u_{i,l}}_{u_{j,l}} + \underbrace{\delta_{il} u_{i,j}}_{u_{l,j}} + u_{i,j} u_{i,l} \right) dx_j dx_l - dx_i dx_i$$

$$= (\delta_{jl} + u_{j,l} + u_{l,j} + u_{i,j} u_{i,l}) dx_j dx_l - dx_i dx_i$$

$$= \underbrace{\delta_{jl} dx_j dx_l}_{dx_j dx_j} + \underbrace{(u_{j,l} + u_{l,j} + u_{i,j} u_{i,l})}_{2\varepsilon_{jl}} dx_j dx_l - dx_i dx_i$$

$$= 2\varepsilon_{jl} dx_j dx_l = 2\varepsilon_{ij} dx_i dx_j \qquad (4.39)$$

In the last line in the above derivation, we used Eq. (4.21).

Finally, using Eq. (4.34), we obtain:

$$M = \frac{1}{2}\left[\frac{(ds^*)^2 - (ds)^2}{(ds)^2}\right] = \frac{1}{2}[(ds^*)^2 - (ds)^2]\frac{1}{ds}\frac{1}{ds}$$

$$= \varepsilon_{ij}\frac{dx_i}{ds}\frac{dx_j}{ds} = n_i \varepsilon_{ij} n_j \qquad (4.40)$$

In the foregoing equation, we used

$$n_i = \frac{dx_i}{ds} \qquad (4.41)$$

Magnification factor M and the strain components ε_{ij} are related as

$$M = n_i \varepsilon_{ij} n_j \qquad (4.42)$$

where n_i is the direction cosine of the fiber: $n_i = \cos(\mathbf{n}, x_i)$, before deformation. The explicit formula for M can be obtained using the following code.

```
1  ε11, ε22, ε33, ε23, ε13, ε12 = symbols("ε11, ε22, ε33, ε23, ε13, ε12")
2  n1, n2, n3, εnn = symbols("n1, n2, n3, ε_nn")
3
4  ε = Matrix([[ε11, ε12, ε13],              # ε: stress tensor
5              [ε12, ε22, ε23],
6              [ε13, ε23, ε33]])
7
8  n = Matrix([n1, n2, n3])
9  M = (n.T@ε@n)[0].simplify()
10 M          # εnn
```

$$n_1 (n_1 \varepsilon_{11} + n_2 \varepsilon_{12} + n_3 \varepsilon_{13}) + n_2 (n_1 \varepsilon_{12} + n_2 \varepsilon_{22} + n_3 \varepsilon_{23})$$
$$+ n_3 (n_1 \varepsilon_{13} + n_2 \varepsilon_{23} + n_3 \varepsilon_{33})$$

Comparing to Eq. (3.42), it is clear that M is the normal strain of the fiber with direction cosine $\mathbf{n}$, ε_{nn}.

4.7.1.5 Direction cosine of a fiber after deformation*

Let us now examine the direction cosine of the fiber after deformation. We start with the definition of the direction cosine for the deformed fiber, which gives

$$n_i^* = \frac{dx_i^*}{ds^*} \qquad (4.43)$$

which can be written as

$$n_i^* = \frac{dx_i^*}{ds} \frac{ds}{ds^*} \qquad (4.44)$$

Using Eq. (4.38), we have

$$\frac{dx_i^*}{ds} = x_{i,j}^* \frac{dx_j}{ds} = (\delta_{ij} + u_{i,j})n_j \qquad (4.45)$$

From Eq. (4.33), we have

$$ds^* = \underbrace{(1+\varepsilon_E)}_{\text{length change rate}} ds \qquad (4.46)$$

where $(\varepsilon_E + 1)$ accounts for the length change rate in a fiber. Using Eqs. (4.44), (4.46), and (4.45), we have

$$(1+\varepsilon_E)n_i^* = (\delta_{ij} + u_{i,j})n_j \qquad (4.47)$$

4.7.2 Shear strains caused by large deformation*

Consider two orthogonal infinitely small fibers in a solid undergoes large deformation, as shown in Fig. 4.5. Fiber P-A has direction cosines of $\mathbf{n} = (n_1, n_2, n_3)$, and Fiber P-B has $\mathbf{m} = (m_1, m_2, m_3)$. These two fibers are deformed, respectively, to P*-A* and P*-B*. The direction cosines become, respectively, $\mathbf{n}^* = (n_1^*, n_2^*, n_3^*)$ and $\mathbf{m}^* = (m_1^*, m_2^*, m_3^*)$. Our task now is to evaluate the shear strain between these two fibers.

Since P-A and P-B are orthogonal, we shall have initially

$$n_i m_i = 0 \qquad (4.48)$$

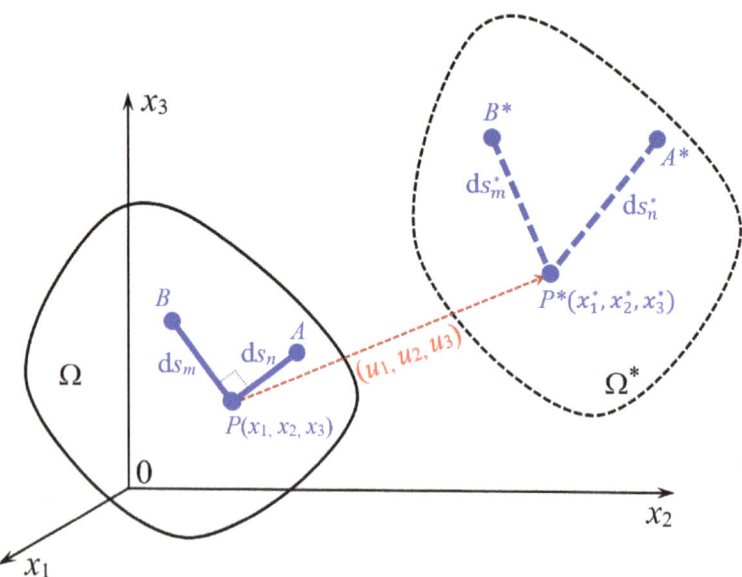

Figure 4.5. Two orthogonal fibers in a solid continuum media undergoes large deformation. The angular change of these two fibers gives the shear strains.

After deformation, P*-A* and P*-B* are not necessarily orthogonal, and their angle changed from $\pi/2$ can be evaluated using

$$\cos\theta^* = n_i^* m_i^* \tag{4.49}$$

The shear strain between fiber **n** and **m** is then defined as

$$\varepsilon_{nm} = \frac{1}{2} \underbrace{(1+\varepsilon_{En})}_{\text{length change rate of } \mathbf{n}} \underbrace{(1+\varepsilon_{Em})}_{\text{length change rate of } \mathbf{m}} \cos\theta^*$$

$$= \frac{1}{2}(1+\varepsilon_{En})(1+\varepsilon_{Em}) n_i^* m_i^*$$

$$= \frac{1}{2}(1+\varepsilon_{En})n_i^* (1+\varepsilon_{Em})m_i^* \tag{4.50}$$

Using Eqs. (4.50), and (4.47), we have

$$\varepsilon_{nm} = \frac{1}{2}(\delta_{ik}+u_{i,k})n_k(\delta_{il}+u_{i,l})m_l$$

$$= \frac{1}{2}(\delta_{ik}\delta_{il} + \delta_{ik}u_{i,l} + u_{i,k}\delta_{il} + u_{i,k}\delta_{il})n_k m_l$$

$$= \frac{1}{2}(\delta_{kl} + u_{k,l} + u_{l,k} + u_{i,k}u_{i,l})n_k m_l$$

$$= \frac{1}{2}\underbrace{\delta_{kl}n_k m_l}_{n_k m_k=0} + \frac{1}{2}\underbrace{(u_{k,l}+u_{l,k}+u_{i,k}u_{i,l})}_{\varepsilon_{kl}} n_k m_l$$

$$= n_k \varepsilon_{kl} m_l \tag{4.51}$$

where ε_{nm} is the shear strain between two fibers with direction cosines **n** and **m**, which are initially orthogonal. The resulting strain expression ε_{kl} in the foregoing equation is the same as that obtained in Eq. (4.21), using gradient of displacement vector functions.

It may be worth to note that when $\mathbf{m} = \mathbf{n}$, ε_{nn} becomes the magnification factor obtained in Eq. (4.42). This is because the length change of the infinitely small fibers are taken into consideration in Eq. (4.50), while considering the angular change. Therefore, Eq. (4.51) is more general than Eq. (4.42). It gives both normal and shear strains.

4.7.3 Example: Symbolic computation of strains

Suppose we know the strain tensor ε (in 9 components) at a point in solid. We want to know the strain along an arbitrary fiber with direction of $\mathbf{n} = (l, m, n)$, which are the direction cosines of the fiber with respect to the coordinates x_1, x_2, x_3. In indicial notation, we denote $\mathbf{n} = (n_1, n_2, n_3)$. The normal strain of this fiber is the magnification factor given in Eq. (4.42), and can be computed using

$$\boxed{\varepsilon_{nn} = n_i \varepsilon_{ij} n_j} \qquad (4.52)$$

In matrix notation, we have

$$\boxed{\varepsilon_{nn} = \mathbf{n}\boldsymbol{\varepsilon}\mathbf{n}^\top} \qquad (4.53)$$

Consider now another arbitrary fiber orthogonal to $\mathbf{n}$ with direction cosines denoted as $\mathbf{m} = (m_1, m_2, m_3)$. The shear strain between fibers $\mathbf{n}$ and $\mathbf{m}$ is then computed using Eq. (4.51):

$$\varepsilon_{nm} = n_i \varepsilon_{ij} m_j \qquad (4.54)$$

In matrix notation, we have

$$\varepsilon_{nm} = \mathbf{n}\boldsymbol{\varepsilon}\mathbf{m}^\top \qquad (4.55)$$

Let us write the following Numpy codes to show this. We first give the symbolic computation code.

```
1  #symbolic computation code
2  n1, n2, n3 = "n1", "n2", "n3"
3  n = Matrix([[n1, n2, n3]])          #Direction cosines for fiber n
4  n
```

$$\begin{bmatrix} n_1 & n_2 & n_3 \end{bmatrix}$$

```
1  εnn = n@ε@n.T                       # ε_nn: normal strain of fiber n
2  εnn    # the magnification factor M
```

$$\begin{bmatrix} n_1 \left(n_1 \varepsilon_{11} + n_2 \varepsilon_{12} + n_3 \varepsilon_{13} \right) + n_2 \left(n_1 \varepsilon_{12} + n_2 \varepsilon_{22} + n_3 \varepsilon_{23} \right) \\ + n_3 \left(n_1 \varepsilon_{13} + n_2 \varepsilon_{23} + n_3 \varepsilon_{33} \right) \end{bmatrix}$$

```
1  εnn.subs({n3:0})                    # Formula of normal strain for n3 = 0 (2D)
```

$$\begin{bmatrix} n_1 \left(n_1 \varepsilon_{11} + n_2 \varepsilon_{12} \right) + n_2 \left(n_1 \varepsilon_{12} + n_2 \varepsilon_{22} \right) \end{bmatrix}$$

```
1  m1, m2, m3 = "m1", "m2", "m3"
2  m = Matrix([[m1, m2, m3]])          # Direction cosines for fiber m
3  m
```

$$\begin{bmatrix} m_1 & m_2 & m_3 \end{bmatrix}$$

```
1  εnm = n@ε@m.T              # Shear strain of between fiber n and m
2  εnm
```

$$[m_1(n_1\varepsilon_{11} + n_2\varepsilon_{12} + n_3\varepsilon_{13}) + m_2(n_1\varepsilon_{12} + n_2\varepsilon_{22} + n_3\varepsilon_{23})$$
$$+ m_3(n_1\varepsilon_{13} + n_2\varepsilon_{23} + n_3\varepsilon_{33})]$$

```
1  εnm.subs({n3:0, m3:0})     # Formula of shear strain for n3 = m3 = 0 (2D)
```

$$[m_1(n_1\varepsilon_{11} + n_2\varepsilon_{12}) + m_2(n_1\varepsilon_{12} + n_2\varepsilon_{22})]$$

4.7.4 Example: Computation of displacements, normal and shear strains

Consider a 3D solid brick shown in Fig. 4.6.

The displacements are given in the following formulas [1].

$$u = c_1 xyz, \quad v = c_2 xyz, \quad w = c_3 xyz \quad (4.56)$$

where c_1, c_2 and c_3 are unknown constants. The displacements at point E are found by a measurement as $(0.004, 0.002, -0.004)$ (m).

1. Determine the functions for all the displacement components.
2. Compute the gradient of the displacement vector functions.
3. Compute the strain functions in the solid, and the values of the strains at point E.

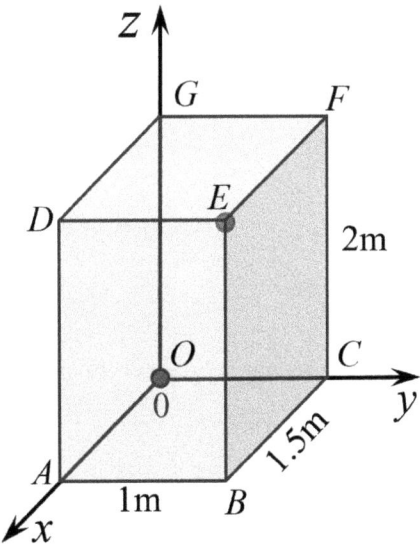

Figure 4.6. A 3D solid brick.

4. Compute the normal strains at point E along the $\overrightarrow{EA}$ direction, and along $\overrightarrow{EA}$ direction.
5. Compute the shear strains at point E between $\overrightarrow{EA}$ and $\overrightarrow{EF}$.

Solution: The formulas for the displacement components contain unknown constants c_1, c_2, c_3. Therefore, we need to determine these constants. This can be done using the conditions given for the displacements at point E, which gives the following three algebraic equations:

$$c_1 x_E y_E z_E = u_E$$
$$c_2 x_E y_E z_E = v_E \qquad (4.57)$$
$$c_3 x_E y_E z_E = w_E$$

where x_E, y_E, z_E are the coordinates at point E, and u_E, v_E, w_E are the displacements at point E. These three equations can be solved easily one-by-one for c_1, c_2, c_3 because these are not coupled. In general, this type equation may be coupled. Thus, we use Sympy solver to obtain the solution, assuming these equations are a coupled system.

After the displacements are determined, we can answer all the remaining questions with ease. We write a code to complete the task.

```
1  # Ref: Advanced Mechanics of Materials, by Boresi and Schmidt
2  # Example 2.8 Three-Dimensional Strain State
3
4  # 1) Determine the functions for all the displacement components.
5  x, y, z = symbols("x, y, z")              # define symbolic coordinates
6  X = Matrix([x, y, z])
7  c1, c2, c3 = symbols("c_1, c_2, c_3")
8
9  u = c1*x*y*z
10 v = c2*x*y*z
11 w = c3*x*y*z
12
13 xE = {x:3/2, y:1, z:2}                    # Coordinates at point E
14 dE = [4, 2, -4]   # x 1e-3                # displacement at point E
15 sln_cs=sp.solve([u.subs(xE)-dE[0],v.subs(xE)-dE[1],w.subs(xE)-dE[2]],
16                 [c1, c2, c3])
17 sln_cs                  # should be: {c1:0.004/3, c2:0.002/3, c3:-0.004/3}
```

$\{c_1 : 1.33333333333333, \ c_2 : 0.666666666666667, \ c_3 : -1.33333333333333\}$

```
1  # substuting the constants back to the formulas for the displacements:
2  U = Matrix([u.subs(sln_cs), v.subs(sln_cs), w.subs(sln_cs)])
3  gr.printM(U.T, 'The displacement vector functions are found as:')
```

The displacement vector functions are found as

$$\begin{bmatrix} 1.333xyz & 0.6667xyz & -1.333xyz \end{bmatrix}$$

The displacement functions are found as

$$\begin{bmatrix} u \\ v \\ w \end{bmatrix} = \begin{bmatrix} \frac{0.004}{3}xyz \\ \frac{0.002}{3}xyz \\ -\frac{0.004}{3}xyz \end{bmatrix} \tag{4.58}$$

```
1  # 2) Compute the gradient of the displacement vector functions.
2  np.set_printoptions(precision=4, suppress=True)
3
4  # the gradient of the displacement vector functions
5  U_g = gr.grad_vf(U, X)                # Or use: vector.grad_vec(U)
6  printM(U_g, 'Displacement gradient (matrix):', n_dgt=4)
```

Displacement gradient (matrix):

$$\begin{bmatrix} 1.333yz & 0.6667yz & -1.333yz \\ 1.333xz & 0.6667xz & -1.333xz \\ 1.333xy & 0.6667xy & -1.333xy \end{bmatrix}$$

```
1  # 3) Compute the strain functions in the solid
2  strains = 0.5*(U_g + U_g.T)    # symmetric part of displacement gradient
3  printM(strains,'Strain tensor (small) functions', n_dgt=4)
```

Strain tensor (small) functions

$$\begin{bmatrix} 1.333yz & 0.6667xz + 0.3333yz & 0.6667xy - 0.6667yz \\ 0.6667xz + 0.3333yz & 0.6667xz & 0.3333xy - 0.6667xz \\ 0.6667xy - 0.6667yz & 0.3333xy - 0.6667xz & -1.333xy \end{bmatrix}$$

```
1  # 3) the values of the strains at point E.
2
3  ε_val = strains.subs(xE)              # values of the strains at point E
4  printM(ε_val, 'Strain tensor values (e-3):', n_dgt=4)
```

Strain tensor values (e-3):

$$\begin{bmatrix} 2.667 & 2.667 & -0.3333 \\ 2.667 & 2.0 & -1.5 \\ -0.3333 & -1.5 & -2.0 \end{bmatrix}$$

```
1  # 4) Compute the normal strains at point E along the fiber EA direction:
2
3  ε_val = np.array(ε_val)
4
5  # compute the direction cosine for EA using Eq.(2.28):
6  # E(1.5, 1.0, 2.0); A(1.5, 0.0, 0.0);
7  length_EA = sp.sqrt((1.5-1.5)**2+(0.0-1.0)**2+(0.0-2.0)**2)
8  n_EA = np.array([(1.5-1.5),(0.0-1.0),(0.0-2.0)])/length_EA   # ni, E->A
9  print(f'The fiber direction N = {n_EA}')
10
11 εnn_EA = n_EA@ε_val@n_EA                # Normal strain of fiber E->A
12 print(f'Normal strain on fiber N  = {εnn_EA:.4f}e-3')
13
14 # normal strain at point E along EF direction:
15 n_EF = np.array([-1., 0, 0])            # fiber along E->F
16 print(f'The fiber direction M = {n_EF}')
17
18 εnn_EF = n_EF@ε_val@n_EF                #Normal strain of fiber E->F
19 print(f'Normal strain on fiber M = {εnn_EF:.4f}e-3')
```

The fiber direction N = [0 -0.447213595499958 -0.894427190999916]
Normal strain on fiber N = -2.4000e-3
The fiber direction M = [-1. 0. 0.]
Normal strain on fiber M = 2.6667e-3

```
1  #5) Compute the shear strains at point E between fibers EA and EF.
2
3  ε_NM = n_EA@ε_val@n_EF          #Shear strain between fiber E->A and E->F
4  print(f'Shear strain between two fibers = {ε_NM:.4f}e-3')
5  print(f'Engineering shear strain between two fibers = {2.*ε_NM:.4f}e-3')
```

Shear strain between two fibers = 0.8944e-3
Engineering shear strain between two fibers = 1.7889e-3

4.8 Coordinate transformation of strain tensor

Consider a coordinate transformation from base coordinate of x_1, x_2, and x_3 to a new coordinate x'_1, x'_2, and x'_3. We can now readily derive the rule for coordinate transformation of strain tensor. It is a generalization of Eq. (4.54).

After the coordinate transformation from the coordinate system of x_1, x_2, and x_3 to a new coordinate x'_1, x'_2, and x'_3, the n_i should be replaced by a_{ri} (with r range from 1 to 3). This is because the fiber direction could be one of the three axes x'_1, x'_2, and x'_3. The m_j should be replaced by a_{sj} (with s range from 1 to 3) because the fiber-tip-moving direction could also be one of the three directions x'_1, x'_2, and x'_3. The fiber directions can be indexed by r, and the tip-moving directions can be indexed by s. All the contraction operations stay unchanged, leaving with 2 indexes, resulting in a new description of the same strain tensor. After this indexation matching, we re-write Eq. (4.54) to include the precise meaning of each of the indexes.

$$\varepsilon_{nm} = n_i \varepsilon_{ij} m_j$$

ε — $\underbrace{n}_{\text{fiber-dir denoted by } r}$ $\underbrace{m}_{\text{tip-move-dir denoted by } s}$ $=$ $\underbrace{n_i}_{\text{controls fiber-dir}}$ ε $\underbrace{i}_{\substack{\text{contract} \\ \text{with fiber-dir}}}$ $\underbrace{j}_{\substack{\text{contract with} \\ \text{tip-move-dir}}}$ $\underbrace{m_j}_{\text{controls tip-move-dir}}$

$\underbrace{\begin{bmatrix} x'_1 \\ x'_2 \\ x'_3 \end{bmatrix}}_{r}$ $\underbrace{\begin{bmatrix} x'_1 \\ x'_2 \\ x'_3 \end{bmatrix}}_{s}$ a_{ri} $\underbrace{\begin{bmatrix} x_1 \\ x_2 \\ x_3 \end{bmatrix}}_{i}$ $\underbrace{\begin{bmatrix} x_1 \\ x_2 \\ x_3 \end{bmatrix}}_{j}$ a_{sj}

(4.59)

$$\mathbf{T} = \begin{bmatrix} \underbrace{\cos(x'_1, x_1)}_{a_{11}} & \underbrace{\cos(x'_1, x_2)}_{a_{12}} & \underbrace{\cos(x'_1, x_3)}_{a_{13}} \\ \underbrace{\cos(x'_2, x_1)}_{a_{21}} & \underbrace{\cos(x'_2, x_2)}_{a_{22}} & \underbrace{\cos(x'_2, x_3)}_{a_{23}} \\ \underbrace{\cos(x'_3, x_1)}_{a_{31}} & \underbrace{\cos(x'_3, x_2)}_{a_{32}} & \underbrace{\cos(x'_3, x_3)}_{a_{33}} \end{bmatrix} \quad (4.60)$$

We obtain:

$$\boxed{\varepsilon'_{rs} = a_{ri} \varepsilon_{ij} a_{sj}} \quad (4.61)$$

This is the formula for strain tensor transformation using indicial notation. Note the indexes i and j are all contracted, leaving with the free indexes r and s, resulting in a second-order tensor, the transformed strain tensor.

Alternatively, we may use the matrix notation and matrix multiplication rule, and the formula should be:

$$\boxed{\varepsilon' = T\varepsilon T^\top} \qquad (4.62)$$

The T on the left ensures the contraction is on the first index of ε (with the 2nd index of T), and T on the right ensures the contraction is on the 2nd index of ε and the transpose is needed so that the contraction is always with the 2nd index of T, related to the base coordinate axes.

Since T is unitary, the inverse transformation can be simply done using the transpose of T:

$$\varepsilon = T^\top \varepsilon' T \qquad (4.63)$$

Or in the indicial notation:

$$\varepsilon_{rs} = a_{ir}\varepsilon'_{ij}a_{js} \qquad (4.64)$$

All we need is to swap the order of the two indexes of a_{ij}.

It is seen again that transformation of a second order strain tensor requires two times of operation by the rotation matrix T.

It is clear now that the coordinate transformation rule of strain tensor is exactly the same as that for the stress tensor discussed in the previous chapter. After all, these are all second-order tensors. Thus, the same codes given in the previous chapter can be used. In addition, the Mohr circle presentation applies also to strain tensors in exactly the same manner.

Let us write the following Python codes to perform coordinate transformation for strains, using both matrix notation and indicial notation.

```
1  T = gr.transferMs()           # form a symbolic transformation matrix
2  T                             # print out
```

$$\begin{bmatrix} a_{11} & a_{12} & a_{13} \\ a_{21} & a_{22} & a_{23} \\ a_{31} & a_{32} & a_{33} \end{bmatrix}$$

```
1  εp = T@ε@T.T                  # ε: Strain tensor defined earlier
2  εp[:,0]                       # print out the first column
```

$$\begin{bmatrix}
a_{11}\left(a_{11}\varepsilon_{11}+a_{12}\varepsilon_{12}+a_{13}\varepsilon_{13}\right)+a_{12}\left(a_{11}\varepsilon_{12}+a_{12}\varepsilon_{22}+a_{13}\varepsilon_{23}\right) \\
\quad+a_{13}\left(a_{11}\varepsilon_{13}+a_{12}\varepsilon_{23}+a_{13}\varepsilon_{33}\right) \\
a_{11}\left(a_{21}\varepsilon_{11}+a_{22}\varepsilon_{12}+a_{23}\varepsilon_{13}\right)+a_{12}\left(a_{21}\varepsilon_{12}+a_{22}\varepsilon_{22}+a_{23}\varepsilon_{23}\right) \\
\quad+a_{13}\left(a_{21}\varepsilon_{13}+a_{22}\varepsilon_{23}+a_{23}\varepsilon_{33}\right) \\
a_{11}\left(a_{31}\varepsilon_{11}+a_{32}\varepsilon_{12}+a_{33}\varepsilon_{13}\right)+a_{12}\left(a_{31}\varepsilon_{12}+a_{32}\varepsilon_{22}+a_{33}\varepsilon_{23}\right) \\
\quad+a_{13}\left(a_{31}\varepsilon_{13}+a_{32}\varepsilon_{23}+a_{33}\varepsilon_{33}\right)
\end{bmatrix}$$

The computation can be done using the same function Tensor2_transfer() defined earlier.

```
1  εp_ = gr.Tensor2_transfer(T,ε)    # use the same code function for stress
2  εp_ = Matrix(εp_)
3
4  print(f"Transformed strain via tensor operation (first column only):")
5  εp_[:,0]               # print first column only; εp_[:,1] # εp_[:,12]
```

Transformed strain via tensor operation (first column only):

$$\begin{bmatrix}
a_{11}\left(a_{11}\varepsilon_{11}+a_{12}\varepsilon_{12}+a_{13}\varepsilon_{13}\right)+a_{12}\left(a_{11}\varepsilon_{12}+a_{12}\varepsilon_{22}+a_{13}\varepsilon_{23}\right) \\
\quad+a_{13}\left(a_{11}\varepsilon_{13}+a_{12}\varepsilon_{23}+a_{13}\varepsilon_{33}\right) \\
a_{11}\left(a_{21}\varepsilon_{11}+a_{22}\varepsilon_{12}+a_{23}\varepsilon_{13}\right)+a_{12}\left(a_{21}\varepsilon_{12}+a_{22}\varepsilon_{22}+a_{23}\varepsilon_{23}\right) \\
\quad+a_{13}\left(a_{21}\varepsilon_{13}+a_{22}\varepsilon_{23}+a_{23}\varepsilon_{33}\right) \\
a_{11}\left(a_{31}\varepsilon_{11}+a_{32}\varepsilon_{12}+a_{33}\varepsilon_{13}\right)+a_{12}\left(a_{31}\varepsilon_{12}+a_{32}\varepsilon_{22}+a_{33}\varepsilon_{23}\right) \\
\quad+a_{13}\left(a_{31}\varepsilon_{13}+a_{32}\varepsilon_{23}+a_{33}\varepsilon_{33}\right)
\end{bmatrix}$$

4.9 Principal strains

Similar to stress states, for a given strain state, one can find a set of principal strains, by rotating the coordinates to a particular angle, where all the shear strain components vanish. Obtaining such a coordinate direction and the principal strains can also be casted to an eigenvalue problem, which can be written in the exact form by replacing the stress tensor with strain tensor.

$$\boxed{[\varepsilon - \varepsilon I]\mathbf{n} = \mathbf{0}} \tag{4.65}$$

All the formulations and theory for the strain tensor is the same as that for the stress tensor. Thus, the same codes given in the previous chapter can be used.

4.10 Strain invariants

Similar to stress tensor, the strain tensor has also nine strain components. Thus, it can also form three strain invariants that do not change with the coordinate change. The invariant property is essentially the feature of any 3×3 matrix, including the stress and strain tensor matrix.

These three strain invariants are defined as follows, based on nine strain components or these three principal strains.

First, the trace of the strain tensor gives the first strain invariant:

$$\bar{I}_1 = \varepsilon_{11} + \varepsilon_{22} + \varepsilon_{33} = \varepsilon_1 + \varepsilon_2 + \varepsilon_3 \tag{4.66}$$

Second, the sum of the three minor determinants of the strain tensor gives the second strain invariant:

$$\bar{I}_2 = \begin{vmatrix} \varepsilon_{11} & \varepsilon_{12} \\ \varepsilon_{12} & \varepsilon_{22} \end{vmatrix} + \begin{vmatrix} \varepsilon_{11} & \varepsilon_{13} \\ \varepsilon_{13} & \varepsilon_{33} \end{vmatrix} + \begin{vmatrix} \varepsilon_{22} & \varepsilon_{23} \\ \varepsilon_{23} & \varepsilon_{33} \end{vmatrix}$$
$$= \varepsilon_{11}\varepsilon_{22} + \varepsilon_{11}\varepsilon_{33} + \varepsilon_{22}\varepsilon_{33} - \varepsilon_{12}^2 - \varepsilon_{13}^2 - \varepsilon_{23}^2 \tag{4.67}$$
$$= \varepsilon_1\varepsilon_2 + \varepsilon_1\varepsilon_3 + \varepsilon_2\varepsilon_3$$

Third, the determinant of the strain tensor gives the third strain invariant:

$$\bar{I}_3 = \begin{vmatrix} \varepsilon_{11} & \varepsilon_{12} & \varepsilon_{13} \\ \varepsilon_{12} & \varepsilon_{22} & \varepsilon_{23} \\ \varepsilon_{12} & \varepsilon_{23} & \varepsilon_{33} \end{vmatrix} = \varepsilon_1\varepsilon_2\varepsilon_3 \tag{4.68}$$

For 2D cases, say in the x–y plane, we shall have two strain invariants:

$$\bar{I}_1 = \varepsilon_{11} + \varepsilon_{22} = \varepsilon_1 + \varepsilon_2 \tag{4.69}$$

and

$$\bar{I}_2 = \begin{vmatrix} \varepsilon_{11} & \varepsilon_{12} \\ \varepsilon_{12} & \varepsilon_{22} \end{vmatrix} = \varepsilon_{11}\varepsilon_{22} - \varepsilon_{12}^2 = \varepsilon_1\varepsilon_2 \tag{4.70}$$

Let us look at a specific example.

4.11 Example: Strain components after coordinate rotation

Consider the same matrix used earlier.

```
1  # Define a strain tensor:
2  ev=Matrix([[2.667,2.667,-0.3333],[2.667,2.0,-1.5],[-0.3333,-1.5,-2.0]])
3  printM(ev, 'Strain tensor values (e-3):', n_dgt=4)
```

Strain tensor values (e-3):

$$\begin{bmatrix} 2.667 & 2.667 & -0.3333 \\ 2.667 & 2.0 & -1.5 \\ -0.3333 & -1.5 & -2.0 \end{bmatrix}$$

Let us rotate the coordinate by 60 degrees about the z-axis. The transformation matrix becomes

```
1  about, theta = 'z', 60.                    # rotation angle and axis
2  Tz, about = gr.transferM(theta, about = about)
3  print(f'Transformation tensor {theta:3.2f}°, w.r.t. {about}:\n{Tz}')
```

```
Transformation tensor 60.00°, w.r.t. z:
[[ 0.5    0.866  0.  ]
 [-0.866  0.5    0.  ]
 [ 0.     0.     1.  ]]
```

```
1  ε60 = εv.subs(list(zip(T, Matrix(Tz))))
2  gr.printM(ε60, 'Stress formulas in the new coordinates',n_dgt=4)
```

Stress formulas in the new coordinates

$$\begin{bmatrix} 2.667 & 2.667 & -0.3333 \\ 2.667 & 2.0 & -1.5 \\ -0.3333 & -1.5 & -2.0 \end{bmatrix}$$

These are the strain components after 60 degree of transformation. Note that to compute the engineering shear strains, we simply use:

$$\gamma_{ij} = 2\varepsilon_{ij} \quad i \neq j \tag{4.71}$$

4.12 Example: Computation of various strains

This example computes

- principal strains, strain invariants, mean strain, and deviatoric strains (see Chapter 5 for detailed definition)

for a given strain state with 9 strain components. The definitions and formulas are essentially the same as those for stresses discussed in the previous chapter. Thus, all these are done using the functions made earlier for computing various stresses.

```
1   # Ref: Advanced Mechanics of Materials, by Boresi and Schmidt
2   # Example 2.9 State of Strain in a Torsion-Tension Member
3
4   import numpy as np
5   np.set_printoptions(precision=4,suppress=True)
6
7   #Given strain matrix in e-3 unit
8   #Ev=np.array([[2.67, 1.335,.335],[1.335, 2.,-1.5],[.335,-1.5,-2.]]) #e-3
9   Ev = np.array([[-.832, 0.,    0.],
10                 [0., -.832,.8725],
11                 [0., .8725,1.667]]) #e-3
12  print(f'strain tensor (e-3) :\n{Ev}')
13
14  eigenValues, eigenVectors = gr.M_eigen(Ev)
15  print('Principal strains (Eigenvalues) (e-3) =',eigenValues)
16  print(f'Eigenvectors:\n{eigenVectors}')
17  print(f'Orthogonal:\n{eigenVectors.dot(eigenVectors.T)}')
18
19  I1, I2, I3 = gr.M_invariant(eigenValues)
20  print(f'The first  strain invariant I1={I1:9.5f}e-3')
21  print(f'The second strain invariant I2={I2:9.5f}e-6')
22  print(f'The third  strain invariant I3={I3:9.5f}e-9')
23
24  meanstrain, meanMatrix, deviatorM = gr.meanDeviator(Ev,eigenValues)
25  print(f'The mean strain = {meanstrain:9.5f}')
26  print(f'The mean strain tensor:\n{meanMatrix}')
27  print(f'The deviatoric strain tensor:\n{deviatorM}')
```

```
strain tensor (e-3) :
[[-0.832  0.      0.    ]
 [ 0.    -0.832   0.8725]
 [ 0.     0.8725  1.667 ]]
Principal strains (Eigenvalues) (e-3) = [ 1.9415 -0.832  -1.1065]
Eigenvectors:
[[ 0.     1.      0.    ]
 [ 0.3001 0.      0.9539]
 [ 0.9539 0.     -0.3001]]
Orthogonal:
[[1. 0. 0.]
 [0. 1. 0.]
 [0. 0. 1.]]
The first  strain invariant I1=  0.00300e-3
The second strain invariant I2= -2.84292e-6
The third  strain invariant I3=  1.78730e-9
The mean strain =   0.00100
The mean strain tensor:
[[0.001 0.    0.   ]
 [0.    0.001 0.   ]
 [0.    0.    0.001]]
The deviatoric strain tensor:
[[-0.833  0.      0.    ]
 [ 0.    -0.833   0.8725]
 [ 0.     0.8725  1.666 ]]
```

4.13 Compatibility equations

4.13.1 *2D problems*

For 2D problems, there are two displacement components u and v, but we have three equations for strain–displacement relations. Using Eq. (4.14), these can be written as

$$\varepsilon_{xx} = \frac{\partial u}{\partial x}$$
$$\varepsilon_{yy} = \frac{\partial v}{\partial y} \qquad (4.72)$$
$$\varepsilon_{xy} = \frac{1}{2}\left(\frac{\partial u}{\partial y} + \frac{\partial v}{\partial x}\right)$$

Therefore, these two displacement components cannot in general be determined, if the three strain components are chosen arbitrarily. These strain components must be somehow related. Such relation is called **compatibility equation**.

Based on the Schwarz's theorem which states that if a function is second order continuously differentiable, the mixed partial derivatives must be equal, regardless of the order of differentiation.

To obtain such a relation, we perform twice partial differentiation to the ε_{xx} with respect to y, twice partial differentiation to the ε_{yy} with respect to x, and twice partial differentiation to ε_{xy} with respect to x and y, which gives

$$\frac{\partial^2 \varepsilon_{xx}}{\partial y^2} = \frac{\partial^3 u}{\partial x \partial y^2}$$
$$\frac{\partial^2 \varepsilon_{yy}}{\partial x^2} = \frac{\partial^3 v}{\partial y \partial x^2} \qquad (4.73)$$
$$\frac{\partial^2 \varepsilon_{xy}}{\partial x \partial y} = \frac{1}{2}\left(\frac{\partial^3 u}{\partial x \partial y^2} + \frac{\partial^3 v}{\partial y \partial x^2}\right)$$

Substituting the first two expressions in the 3rd expression gives

$$\boxed{\frac{\partial^2 \varepsilon_{xx}}{\partial y^2} + \frac{\partial^2 \varepsilon_{yy}}{\partial x^2} = 2\frac{\partial^2 \varepsilon_{xy}}{\partial x \partial y}} \qquad (4.74)$$

This is the compatibility equation for 2D problems. All the strains must satisfy this equation for correct displacements.

4.13.2 3D problems*

For 3D problems, there are three displacement components (u, v, w), but we have six equations for strain–displacement relations as given in Eq. (4.14). Therefore, these three displacement components must also be compatible. These compatibility equations can be derived through a similar procedure used for 2D cases, and have been found as [1]

$$\frac{\partial^2 \varepsilon_{xx}}{\partial y^2} + \frac{\partial^2 \varepsilon_{yy}}{\partial x^2} = 2\frac{\partial^2 \varepsilon_{xy}}{\partial x \partial y}$$

$$\frac{\partial^2 \varepsilon_{xx}}{\partial z^2} + \frac{\partial^2 \varepsilon_{zz}}{\partial x^2} = 2\frac{\partial^2 \varepsilon_{xz}}{\partial x \partial z}$$

$$\frac{\partial^2 \varepsilon_{yy}}{\partial z^2} + \frac{\partial^2 \varepsilon_{zz}}{\partial y^2} = 2\frac{\partial^2 \varepsilon_{yz}}{\partial y \partial z}$$

$$\frac{\partial^2 \varepsilon_{xy}}{\partial z^2} + \frac{\partial^2 \varepsilon_{zz}}{\partial x \partial y} = \frac{\partial^2 \varepsilon_{yz}}{\partial x \partial z} + \frac{\partial^2 \varepsilon_{xz}}{\partial y \partial z}$$

$$\frac{\partial^2 \varepsilon_{xz}}{\partial y^2} + \frac{\partial^2 \varepsilon_{yy}}{\partial x \partial z} = \frac{\partial^2 \varepsilon_{xy}}{\partial y \partial z} + \frac{\partial^2 \varepsilon_{yz}}{\partial x \partial y}$$

$$\frac{\partial^2 \varepsilon_{yz}}{\partial x^2} + \frac{\partial^2 \varepsilon_{xx}}{\partial y \partial z} = \frac{\partial^2 \varepsilon_{xz}}{\partial x \partial y} + \frac{\partial^2 \varepsilon_{xy}}{\partial x \partial z}$$

(4.75)

These compatibility equations can also be written in terms of stresses, using the constitutive equations (to be discussed in the next chapter). These compatibility equations need to be satisfied when solving a solid mechanics problem.

4.14 Remarks

This chapter discusses first how the strain tensor (order 2) results from the gradient of displacement tensor (order 1) all defined in a Cartesian coordinate system. The derivation is done in two ways: a mathematical one that used the gradient of the function of the displacement tensor, and mechanics one that uses the kinematics of motion. The states of strain and its change are also discussed, when the coordinate system transforms (rotates).

1. The value of the strain tensor depends on both the gradient of the displacement tensor and the orientation of the fiber for which the strain is measured. This is the root reason why a strain tensor has two indexes and hence being a second-order tensor.
2. In all the formulas derived for the strain tensor, orthogonal projections are strictly followed.

3. When coordinate rotates, the presentation of the strain tensor changes. It uses twice the transformation matrix because it is a second-order tensor. The process is the same as for the stress tensors.
4. The basic properties of a strain tensor (three strain invariants, principal strains, and the Mohr circle) do not change with the coordinate transformation. These invariants can be used in establishing some failure criteria.
5. At the principal axes, the shear strain components vanish, and only the normal strains are present, which are called principal strains. These are invariant.
6. Principal strains and the axes can also be found using the eigenvalue solver provided in this and the previous chapters.
7. Mohr circle can also be used as a geometric representation of strains in 2D and 3D, but less often compared to stresses. The codes for producing the Mohr circle for strains are the same as those for stresses given in the previous chapter.
8. Since the number of strain components is more than the number of displacement components, compatibility equations are derived, which need to be satisfied by the strains to ensure the deformation in solid is compatible (no gaps and overlaps).

References

[1] A.P. Boresi, R.J. Schmidt and O.M. Sidebottom, *Advanced Mechanics of Materials*. John Wiley & Sons, New York, 1985.
[2] J.M. Gere and S.P. Timoshenko, *Mechanics of Materials*, 1972.
[3] S. Timoshenko and J.N. Goodier, *Theory of Elasticity*. McGraw Hill, New York, 1970. http://books.google.com/books?id=yFISAAAAIAAJ&dq=theory+of+elasticity&ei=ICiKSsr3G4jwkQSbxMyPCg
[4] Z.L. Xu, *Elasticity*, Vols. 1&2. People's Publisher, China, 1979.
[5] G.R. Liu and S.S. Quek, *The Finite Element Method: A Practical Course*. Butterworth-Heinemann, New York, 2013.
[6] G.R. Liu and T.T. Nguyen, *Smoothed Finite Element Methods*. Taylor and Francis Group, New York, 2010.

Chapter 5

Elasticity Tensor and Constitutive Equations

We import the necessary external modulus or dependencies for later use in this chapter.

```
1  # Place cursor in this cell, and press Ctrl+Enter to import dependences.
2  import sys                              # for accessing the computer system
3  sys.path.append('../grbin/')   # Change to the directory in your system
4
5  from commonImports import *        # Import dependences from '../grbin/'
6  import grcodes as gr                   # Import the module of the author
7  importlib.reload(gr)                   # When grcodes is modified, reload it
8
9  from continuum_mechanics import vector
10 init_printing(use_unicode=True)        # For latex-like quality printing
11 #np.set_printoptions(precision=4,suppress=True) # Digits in print-outs
12 np.set_printoptions(formatter={'float': '{: 0.3e}'.format})
```

Chapter 3 defines the stresses and their relations with forces, and Chapter 4 defines strains and their relationship with the displacements. We need to know the relationship between the stress and strain tensors. The elasticity tensor plays this role. This chapter is written in reference to [1–5].

As usual, our discussion is accompanied with Python codes written in reference to the documentations of Python, Sympy, and other Python modules.

5.1 Simplest case: 1D tests

To begin with, we look at the simplest uniaxially loaded 1D specimen that often has a rectangular or circular cross-section. The cross-sectional dimension is much smaller than its length.

Its normal strain ε_{xx} can be obtained as

$$\varepsilon_{xx} = \frac{\Delta L}{L} \tag{5.1}$$

where x stands for the x-axis, which is vertical if the specimen is pulled vertically using a tensile machine, as shown in the left figure in Fig. 5.1. The ΔL is the length change in L under stress σ_{xx} applied by the tensile test machine. The length change can be measured during experiment corresponding to the level of stress.

Equation (5.1) shows that the strain is in fact an averaged one over the base length L, which depends on the measurement method. If the displacements between the two points are measured, L will be the distance of these two points. If a strain gauge is used, it is the effective length of the gauge.

Assume the specimen or bar is made of linear elastic isotropic material, implying that the stress and strain are proportional, and related thorough a material constant E called Young's modulus.

$$\sigma_{xx} = E\varepsilon_{xx} \tag{5.2}$$

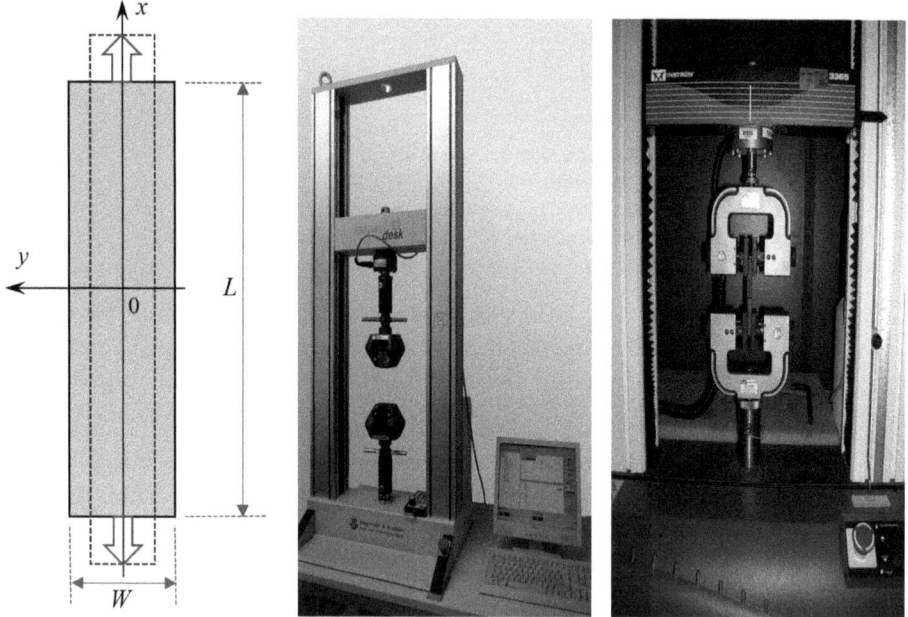

Figure 5.1. Left: A tensile spacemen; Middle: A universal testing machine (Hegewald & Peschke) for measuring the elastic constants and tensile strength for materials. Right: A specimen is mounted in the machine.
Source: (Middle) Image permission by Vom Hofe Kaltstauchdraht GmbH, Altena from en.wikipedia Wikimedia Commons under the CC BY-SA 2.0 de license. (Right) Instron under the CC0 license.

When the specimen is stretched in the x-direction, it may shrink in the y-direction (horizontal). The shrink strain ε_{yy} is calculated using

$$\varepsilon_{yy} = \frac{\Delta W}{W} \tag{5.3}$$

where ΔW is the length change in W under the same stress σ_{xx}. It can also be measured during an experiment. This behavior of material is quantified by Poisson's ratio, denoted often by ν. It gives the relationship between two mutually orthogonal normal strains:

$$\varepsilon_{yy} = -\nu \varepsilon_{xx} \tag{5.4}$$

where ε_{yy} is a normal strain in the y-direction. Clearly Poisson's ratio is dimensionless. The minus sign ensures that Poisson's ratio is positive for usual materials that shrink under tension.

Note that for this 1D bar problem, σ_{yy} will be zero because the width of the specimen is much smaller compared to its length, and no force or pressure is applied in the y-direction.

Using a test machine shown in Fig. 5.1, Young's modulus E, Poisson's ratio ν, yield stress, failure stress, etc. can be measured.

Equation (5.2) is well known as Hooke's Law. Since strain is dimensionless, E carries the same unit as stress: force/area and $[Pa = N/m^2]$ is often used. Young's modulus is the most important elastic constant of a material. For materials that are made through a standardized processes, Young's modulus can be found in material databases. A list of approximate or nominal Young's modulus for commonly used materials can be found at Young modulus (https://en.wikipedia.org/wiki/Young%27s_modulus).

Interested readers may take a look at the online video Tensile Test Experiment (https://www.youtube.com/watch?v=D8U4G5kcpcM).

5.2 Comparison between Young's modulus and spring constants

To make a good sense of Young's modulus, we can compare it with the spring constant that we learned in high school, known also as Hooke's Law. Consider a spring fixed at one end, and it is pulled at another by a force f. The spring is then elongated and the elongation is denoted as e. The relationship between the force and displacement is then expressed as

$$f = ke \tag{5.5}$$

where k is the spring constant. It carries a unit of force/length or $[\frac{N}{m}]$. It is the **stiffness** of the spring.

Consider now an uniform bar with cross-section area A and length L. It is fixed also at one end, and pulled at another by a force f. The bar is elongated and the elongation is denoted as e. Using Eq. (5.2), $\sigma_{xx} = E\varepsilon_{xx}$, we have

$$\underbrace{A\sigma_{xx}}_{f} = \underbrace{\frac{AE}{L}}_{k} \underbrace{L\varepsilon_{xx}}_{e} \tag{5.6}$$

It is clear that $\frac{AE}{L}$ is equivalent to the spring constant k. The unit of $\frac{AE}{L}$ is also $\frac{m^2 \frac{N}{m^2}}{m} = \frac{N}{m}$. It is the stiffness of the bar. Clearly, E and A are proportional, and L is inversely proportional to the bar stiffness.

5.3 General formulation

For a general problem in mechanics of materials, there are usually 9 stress and strain components. Therefore, a general material property law is needed to relate these stresses and strains. Such a law is called the **constitutive equation**. It is assumed that each stress component is related to all the 9 strain components. Since there are 9 stress components, we shall have a total of $9 \times 9 = 81$ material constants. For linear elastic materials, it can be given in the most general form in tensor notation as [6–8]

$$\boxed{\sigma_{ij} = C_{ijkl}\varepsilon_{kl}} \tag{5.7}$$

This is also known as the generalized Hooke's Law [5], where C_{ijkl} is a fourth-order tensor, carrying 4 indexes. It carries the same unit as Young's modulus E: [Pa]. It bridges two second-order tensors, σ_{ij} and ε_{ij}. This ensures that each of the stress components depends on all the strain components with independent elastic constants. Since each of the four indexes of C_{ijkl} has the same range of 3. There are a total of $(3 \times 3 \times 3 \times 3 =)$ 81 elasticity constants for materials in 3D. Thus, Eq. (5.7) is the most general constitutive equation for solid materials.

Due to the symmetry of the stress and strain tensors, and the special features the materials may have, some of these constants may equal, and some of them may be zero. The actual number of the independent elastic constants needed will be drastically reduced, which is discussed in detail later.

5.4 Coordinate transformation of fourth-order tensor*

For many materials, their elastic constants change with directions, such as anisotropic composite materials widely used in various industries and most materials in nature [8, 9]. For analyzing these materials, we must find out the rule of coordinate transformation for fourth-order tensors, so that we can compute these constants when the coordinate rotates. This process is discussed in this section.

Assume Eq. (5.7) is defined in coordinate system $\mathbf{x}$, and now it is rotated to $\mathbf{x}'$. Using the rule of the second-order stress transformation, we obtain:

$$\sigma'_{rs} = a_{ri}a_{sj} \underbrace{\sigma_{ij}}_{C_{ijkl}\varepsilon_{kl}} \implies \sigma'_{rs} = a_{ri}a_{sj}C_{ijkl}\varepsilon_{kl} \tag{5.8}$$

On the other hand, the corresponding strain transformation rule gives

$$\varepsilon'_{kl} = a_{kp}a_{lq}\varepsilon_{pq} \tag{5.9}$$

Its reverse strain transformation becomes

$$\varepsilon_{kl} = a_{pk}a_{ql}\varepsilon'_{pq} \tag{5.10}$$

The swaps of the subscripts in a_{ij} are the result of the inverse transformation that is from coordinate system $\mathbf{x}'$ to $\mathbf{x}$, as discussed in the previous chapter.

Substituting Eq. (5.10) into the 2nd equation in Eq. (5.8), we obtain:

$$\sigma'_{rs} = \underbrace{a_{ri}a_{sj}C_{ijkl}a_{pk}a_{ql}}_{C'_{rspq}} \varepsilon'_{pq} \tag{5.11}$$

It is clear that i, j, k and l are contracted, resulting in free indexes r, s, p, q. We finally obtain

$$\boxed{C'_{rspq} = a_{ri}a_{sj}C_{ijkl}a_{pk}a_{ql}} \tag{5.12}$$

This is the important rule of **transformation of fourth-order tensor** of elastic stiffness, from coordinate system $\mathbf{x}$ to $\mathbf{x}'$.

When inverse transformation is needed, from $\mathbf{x}'$ to $\mathbf{x}$, all we need is to swap the two indexes of a_{ij}. We will demonstrate this in the later examples.

It is seen that transformation of a fourth-order tensor requires **4 times of operation** by a_{ij}.

5.5 Codes for fourth order tensor transformation*

We write, for the first time, a Python code to perform transformations of fourth-order tensors. To the best knowledge of the author, no such code have been written. This is made possible, by using the np.tensordot() function that allows contraction operations by specifying individual axis in a multi-dimensional numpy array.

```
1  def Tensor4_transfer(T,C4):
2      '''Transformation of 4th order tensors
3      input: T, transformation matrix (aij); C4: tensor to be transformed
4      return:C4, transformed tensor of 4th order'''
5
6      C4 = np.tensordot( T, C4, axes=([1],[0]))   # contract i
7      C4 = np.tensordot( T, C4, axes=([1],[1]))   # contract j
8      C4 = np.tensordot(C4,  T, axes=([3],[1]))   # contract l
9      C4 = np.tensordot(C4,  T, axes=([2],[1]))   # contract k
10
11     return C4
```

5.6 Reduction of elastic constants

5.6.1 *Reduction due to stress and strain symmetry, $81 \to 36$*

As discussed earlier, in general, have a total of $(3 \times 3 \times 3 \times 3 =)$ 81 elasticity constants for materials in 3D. This means that we must find these 81 constants via experiments for a material. This is not practical in engineering applications, even if it is possible. This number needs to be and can be drastically reduced. First, we can use the symmetry of the stress and strain tensors to prove the symmetry of the fourth-order tensor.

Due to the symmetry of the stress tensors $\sigma_{ij} = \sigma_{ji}$, implying that ij can be swapped. Thus, these constants satisfy the following symmetry equations.

$$C_{ijkl} = C_{jikl} \quad \text{from stress tensor symmetry} \tag{5.13}$$

This reduces the number by $(3 \times 9) = 27$ constants. Next, because of the symmetry of the strain tensors $\varepsilon_{kl} = \varepsilon_{lk}$, implying that kl can be swapped. These constants satisfy the following additional symmetry equations.

$$C_{ijkl} = C_{ijlk} \quad \text{from strain tensor symmetry} \tag{5.14}$$

This reduces the number by $(3 \times 6) = 18$ constants. The total number of the elasticity constants becomes 36, and Eq. (5.7) can now be written in the following 6×6 matrix form, which indexes are still properly matched.

$$\begin{bmatrix} \sigma_{11} \\ \sigma_{22} \\ \sigma_{33} \\ \sigma_{23} \\ \sigma_{13} \\ \sigma_{12} \end{bmatrix} = \begin{bmatrix} C_{1111} & C_{1122} & C_{1133} & C_{1123} & C_{1113} & C_{1112} \\ C_{2211} & C_{2222} & C_{2233} & C_{2223} & C_{2213} & C_{2212} \\ C_{3311} & C_{3322} & C_{3333} & C_{3323} & C_{3313} & C_{3312} \\ C_{2311} & C_{2322} & C_{2333} & C_{2323} & C_{2313} & C_{2312} \\ C_{1311} & C_{1322} & C_{1333} & C_{1323} & C_{1313} & C_{1312} \\ C_{1211} & C_{1222} & C_{1233} & C_{1223} & C_{1213} & C_{1212} \end{bmatrix} \begin{bmatrix} \varepsilon_{11} \\ \varepsilon_{22} \\ \varepsilon_{33} \\ 2\varepsilon_{23} \\ 2\varepsilon_{13} \\ 2\varepsilon_{12} \end{bmatrix} \quad (5.15)$$

Note that the stress and strain tensors are now written as vectors because of its symmetry and thus the 9 components is reduced to 6.

This number can be further reduced, by making use of the smoothness of the strain energy density in a stressed elastic solid.

5.6.2 Reduction due to the smoothness of the strain energy density

5.6.2.1 Strain energy density in solids

When a solid is stressed, it stores some energy known as the **strain energy**. The strain energy per unit volume is called strain energy density, carrying a unit of $\frac{N}{m^2}\frac{m}{m} = \frac{J}{m^3}$. In general, the strain energy density will be different at different locations in solid, and hence is a function of coordinates. Here, let us evaluate the strain energy density at a point in solid.

Consider first a simple case, where at a point in solid there is only a single component of stress σ and its corresponding strain ε all in the same direction. The stress-strain relationship can be schematically shown Fig. 5.2.

The stress-strain relation is the function represented by the thick black curve, which gives the stress as a function of strain: $\sigma(\varepsilon)$. Assume the strain compatible for variation, implying the operation of $d\varepsilon'$ is possible, and thus the integration of stress-strain function can be carried out. The strain energy density U_0 is then defined by the area under the curve:

$$U_0 = \int_0^\varepsilon \sigma(\varepsilon') d\varepsilon' \quad (5.16)$$

Here we use prime to indicate the integration variable.

Note that the complementary energy density C_0 is the area above the thick black curve shown Fig. 5.2. We will make use of it in developing an energy method in Chapter 13. For now, our focus is on the strain energy density U_0.

Assume U_0 is second order continuously differentiable with respect to the strains, we can then perform this following analysis.

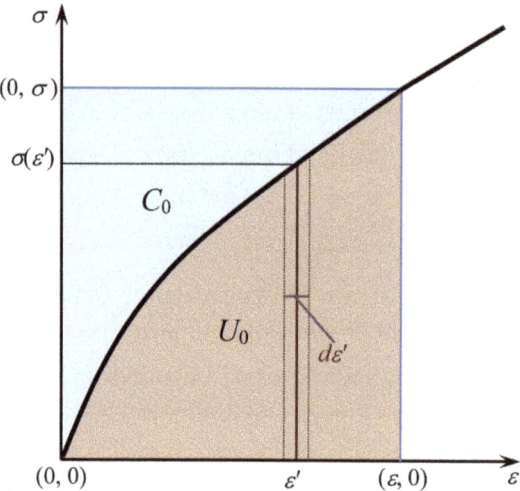

Figure 5.2. Strain energy density U_0 and complementary energy density C_0 at a point in a stressed solid.

First, for 1D solids, we shall have

$$\boxed{\frac{\partial U_0}{\partial \varepsilon} = \sigma} \qquad (5.17)$$

In a general 3D solid, we shall have a total of 9 mutually orthogonal stress components, and corresponding 9 **mutually orthogonal** strain components. Each pair of these 9 stress-strain component pairs is in the same direction. Therefore, the strain energy density should be the sum of all the strain energy densities of the corresponding stress-strain component pairs, which can be expressed as

$$\boxed{\begin{aligned} U_0 = &\int_0^{\varepsilon_{11}} \sigma(\varepsilon'_{11})d\varepsilon'_{11} + \int_0^{\varepsilon_{22}} \sigma(\varepsilon'_{22})d\varepsilon'_{22} + \int_0^{\varepsilon_{33}} \sigma(\varepsilon'_{33})d\varepsilon'_{33} \\ &+ 2\int_0^{\varepsilon_{23}} \sigma(\varepsilon'_{23})d\varepsilon'_{23} + 2\int_0^{\varepsilon_{13}} \sigma(\varepsilon'_{13})d\varepsilon'_{13} + 2\int_0^{\varepsilon_{12}} \sigma(\varepsilon'_{12})d\varepsilon'_{12} \end{aligned}} \qquad (5.18)$$

This is the most **general formula** for evaluating the **strain energy density** in a 3D solid. Note that this is possible because of the use of orthogonal coordinates, over which the stress and strain tensors are defined. This formula is used not only in this chapter, but also in Chapter 13 when discussing energy methods. It is a widely used fundamental formula in solid mechanics.

5.6.2.2 Symmetry from the smoothness of the strain energy density*

In Eq. (5.18), we used the shear stress equivalence, and hence 9 terms is reduced to 6. Partial differentiations of the strain energy density with respect to each of these 6 strains give

$$\frac{\partial U_0}{\partial \varepsilon_{11}} = \sigma_{11}; \quad \frac{\partial U_0}{\partial \varepsilon_{22}} = \sigma_{22}; \quad \frac{\partial U_0}{\partial \varepsilon_{33}} = \sigma_{33}$$

$$\frac{\partial U_0}{\partial \varepsilon_{23}} = 2\sigma_{23}; \quad \frac{\partial U_0}{\partial \varepsilon_{13}} = 2\sigma_{13}; \quad \frac{\partial U_0}{\partial \varepsilon_{12}} = 2\sigma_{12}$$

(5.19)

Using the first two equations in Eq. (5.15), we obtain

$$\frac{\partial U_0}{\partial \varepsilon_{11}} = \sigma_{11} = C_{1111}\varepsilon_{11} + C_{1122}\varepsilon_{22} + C_{1133}\varepsilon_{33} + 2C_{1123}\varepsilon_{23}$$

$$+ 2C_{1113}\varepsilon_{13} + 2C_{1112}\varepsilon_{12}$$

$$\frac{\partial U_0}{\partial \varepsilon_{22}} = \sigma_{22} = C_{2211}\varepsilon_{11} + C_{2222}\varepsilon_{22} + C_{2233}\varepsilon_{33} + 2C_{2223}\varepsilon_{23}$$

$$+ 2C_{2213}\varepsilon_{13} + 2C_{2212}\varepsilon_{12} \quad (5.20)$$

Next, we perform the following 2nd derivatives to U_0, which leads to

$$\frac{\partial^2 U_0}{\partial \varepsilon_{11} \partial \varepsilon_{22}} = C_{1122}; \quad \frac{\partial^2 U_0}{\partial \varepsilon_{22} \partial \varepsilon_{11}} = C_{2211} \quad (5.21)$$

Based on Schwarz's theorem, if a function is second order continuously differentiable, the mixed partial derivatives must be equal, regardless of the sequential order of differentiation. This gives

$$C_{1122} = C_{2211} \quad (5.22)$$

Similarly, using the first and fourth equations in Eq. (5.15), we obtain

$$C_{1123} = C_{2311} \quad (5.23)$$

Repeating the same, for other pairs of equations in Eq. (5.15), we shall find the following additional symmetry condition for the fourth-order elasticity tensor:

$$C_{ijkl} = C_{klij} \quad (5.24)$$

This means that ij and kl can be swapped.

5.6.2.3 Reduction due to strain energy density symmetry, 36 → 21*

Condition Eq. (5.24) makes the **C** in Eq. (5.15) symmetric, and hence reduces number of elasticity constants by 15. The independent constants become 21. The stress-strain relation can be written in the following

symmetric matrix of 6×6.

$$\begin{bmatrix} \sigma_{11} \\ \sigma_{22} \\ \sigma_{33} \\ \sigma_{23} \\ \sigma_{13} \\ \sigma_{12} \end{bmatrix} = \underbrace{\begin{bmatrix} C_{1111} & C_{1122} & C_{1133} & C_{1123} & C_{1113} & C_{1112} \\ & C_{2222} & C_{2233} & C_{2223} & C_{2213} & C_{2212} \\ & & C_{3333} & C_{3323} & C_{3313} & C_{3312} \\ & & & C_{2323} & C_{2313} & C_{2312} \\ & sym. & & & C_{1313} & C_{1312} \\ & & & & & C_{1212} \end{bmatrix}}_{\mathbf{C}} \underbrace{\begin{bmatrix} \varepsilon_{11} \\ \varepsilon_{22} \\ \varepsilon_{33} \\ 2\varepsilon_{23} \\ 2\varepsilon_{13} \\ 2\varepsilon_{12} \end{bmatrix}}_{\vec{\varepsilon}} \quad (5.25)$$

5.6.3 Tensor vs. matrix, a discussion*

Note in Eq. (5.25) that the stress and strain tensors are now written as vectors. Due to symmetry the 9 components in the tensor is reduced to 6. In this form of vector stress or strain, it is no longer a tensor because our transformation rule does not apply any more. Therefore, when we want to perform coordinate rotation and examine the stress and strain components, we would still need to use their tensor form of stress, Eq. (5.8), or strain Eq. (5.9).

The same situation is for the elasticity constant matrix **C**. It is no longer a tensor. The transformation of its components needs to go back to use the tensor transformation rule given in Eq. (5.12), or use our function Tensor4_Transfer() given earlier. Therefore, conversion between 2D matrix **C** and fourth-order tensor C_{ijkl} is needed.

5.7 The Voigt notation*

By substituting pairs of indexes $11 \to 1$, $22 \to 2$, $33 \to 3$, $23 \to 4$, $13 \to 5$, $12 \to 6$, Eq. (5.25) can be further expressed using the so-called Voigt notation [9, 10]:

$$\begin{bmatrix} \sigma_{11} \\ \sigma_{22} \\ \sigma_{33} \\ \sigma_{23} \\ \sigma_{13} \\ \sigma_{12} \end{bmatrix} = \underbrace{\begin{bmatrix} C_{11} & C_{12} & C_{13} & C_{14} & C_{15} & C_{16} \\ & C_{22} & C_{23} & C_{24} & C_{25} & C_{26} \\ & & C_{33} & C_{34} & C_{35} & C_{36} \\ & & & C_{44} & C_{45} & C_{46} \\ & sym. & & & C_{55} & C_{56} \\ & & & & & C_{66} \end{bmatrix}}_{\mathbf{C}} \underbrace{\begin{bmatrix} \varepsilon_{11} \\ \varepsilon_{22} \\ \varepsilon_{33} \\ 2\varepsilon_{23} \\ 2\varepsilon_{13} \\ 2\varepsilon_{12} \end{bmatrix}}_{\vec{\varepsilon}} \quad (5.26)$$

which can be written concisely as

$$\boxed{\vec{\sigma} = \mathbf{C}\vec{\varepsilon}} \quad (5.27)$$

This is the general matrix form of stress–strain or constitutive equation for anisotropic materials. The **C** matrix is symmetric. It can be computed for anisotropic composite materials [8] and for isotropic materials using its Young's modulus and Poisson's ratio (see Section 2.2.2 in textbook [5]). The C matrix can also be expressed as $[C_{\alpha\beta}]$, where $\alpha, \beta = 1, 2, \ldots, 6$. The use of Voigt notation is important for engineering applications because it enables practical means to obtain these material constants through engineering measurements. We can now produce a 6×6 stiffness matrix first, and then convert it to a $3 \times 3 \times 3 \times 3$ tensor.

We mention, but without further elaboration, that there is an alternative handcrafted way to convert between a 6×6 compliance matrix and a $3 \times 3 \times 3 \times 3$ 4th strain tensor. Interested readers may refer to Ref. [9].

5.8 Codes for conversion between a 2D matrix to fourth-order tensor*

We write the following codes for conversion between a C(6,6) matrix and a 4th tensor C(3,3,3,3). The code function C2toC4() converts 2D C(6,6) matrix (in the Voigt notation) to its 4th tensor C(3,3,3,3) form. Code function C4toC2() converts a 4th tensor C(3,3,3,3) to its 2D C(6,6) matrix.

```
 1  def C2toC4(C2):
 2      '''To convert C(6,6) matrix (the Voigt notation) to
 3         4th tensor C(3,3,3,3).
 4      '''
 5      C4 = np.zeros((3,3,3,3))                    # Initialization
 6
 7      # Pass over all C(6,6) to parts of C(3,3,3,3)
 8      C4[0,0,0,0],C4[0,0,1,1],C4[0,0,2,2] = C2[0,0],C2[0,1],C2[0,2]
 9      C4[0,0,1,2],C4[0,0,0,2],C4[0,0,0,1] = C2[0,3],C2[0,4],C2[0,5]
10
11      C4[1,1,0,0],C4[1,1,1,1],C4[1,1,2,2] = C2[1,0],C2[1,1],C2[1,2]
12      C4[1,1,1,2],C4[1,1,0,2],C4[1,1,0,1] = C2[1,3],C2[1,4],C2[1,5]
13
14      C4[2,2,0,0],C4[2,2,1,1],C4[2,2,2,2] = C2[2,0],C2[2,1],C2[2,2]
15      C4[2,2,1,2],C4[2,2,0,2],C4[2,2,0,1] = C2[2,3],C2[2,4],C2[2,5]
16
17      C4[1,2,0,0],C4[1,2,1,1],C4[1,2,2,2] = C2[3,0],C2[3,1],C2[3,2]
18      C4[1,2,1,2],C4[1,2,0,2],C4[1,2,0,1] = C2[3,3],C2[3,4],C2[3,5]
19
20      C4[0,2,0,0],C4[0,2,1,1],C4[0,2,2,2] = C2[4,0],C2[4,1],C2[4,2]
21      C4[0,2,1,2],C4[0,2,0,2],C4[0,2,0,1] = C2[4,3],C2[4,4],C2[4,5]
22
23      C4[0,1,0,0],C4[0,1,1,1],C4[0,1,2,2] = C2[5,0],C2[5,1],C2[5,2]
24      C4[0,1,1,2],C4[0,1,0,2],C4[0,1,0,1] = C2[5,3],C2[5,4],C2[5,5]
25
26      # Impose (minor) symmetric conditions
27      apply_symmetry(C4, key = "all", tol=1.e-4)
28
29      return C4
```

```
 1  def apply_symmetry(C4, key = "all", tol=1.e-2):
 2      '''Impose (minor) symmetric conditions
 3      '''
 4      if key == "all" or key == "ij":
 5          for k in range(3):
 6              for l in range(3):
 7                  for i in range(3):
 8                      for j in range(i+1,3):
 9                          if abs(C4[j,i,k,l]) <= tol:
10                              C4[j,i,k,l]=C4[i,j,k,l]
11
12      if key == "all" or key == "kl":
13          for k in range(3):
14              for l in range(k+1,3):
15                  for i in range(3):
16                      for j in range(3):
17                          if abs(C4[i,j,l,k]) <= tol:
18                              C4[i,j,l,k]=C4[i,j,k,l]
19
20      if key == "all" or key == "ijkl":
21          for k in range(3):
22              for l in range(3):
23                  for i in range(k+1,3):
24                      for j in range(l+1,3):
25                          if abs(C4[i,j,k,l]) <= tol:
26                              C4[i,j,k,l]=C4[k,l,i,j]
27      return C4
```

```
 1  def C4toC2(C4):
 2      '''To convert 4th tensor C(3,3,3,3) to C(6,6) matrix
 3         (the Voigt notation).
 4      '''
 5      C2 = np.zeros((6,6))
 6      C2[0,0],C2[0,1],C2[0,2]=C4[0,0,0,0],C4[0,0,1,1],C4[0,0,2,2]
 7      C2[0,3],C2[0,4],C2[0,5]=C4[0,0,1,2],C4[0,0,0,2],C4[0,0,0,1]
 8
 9      C2[1,0],C2[1,1],C2[1,2]=C4[1,1,0,0],C4[1,1,1,1],C4[1,1,2,2]
10      C2[1,3],C2[1,4],C2[1,5]=C4[1,1,1,2],C4[1,1,0,2],C4[1,1,0,1]
11
12      C2[2,0],C2[2,1],C2[2,2]=C4[2,2,0,0],C4[2,2,1,1],C4[2,2,2,2]
13      C2[2,3],C2[2,4],C2[2,5]=C4[2,2,1,2],C4[2,2,0,2],C4[2,2,0,1]
14
15      C2[3,0],C2[3,1],C2[3,2]=C4[1,2,0,0],C4[1,2,1,1],C4[1,2,2,2]
16      C2[3,3],C2[3,4],C2[3,5]=C4[1,2,1,2],C4[1,2,0,2],C4[1,2,0,1]
17
18      C2[4,0],C2[4,1],C2[4,2]=C4[0,2,0,0],C4[0,2,1,1],C4[0,2,2,2]
19      C2[4,3],C2[4,4],C2[4,5]=C4[0,2,1,2],C4[0,2,0,2],C4[0,2,0,1]
20
21      C2[5,0],C2[5,1],C2[5,2]=C4[0,1,0,0],C4[0,1,1,1],C4[0,1,2,2]
22      C2[5,3],C2[5,4],C2[5,5]=C4[0,1,1,2],C4[0,1,0,2],C4[0,1,0,1]
23
24      return C2
```

5.9 Monoclinic materials, 21 → 13

For special materials, the 2D Voigt matrix can be further simplified if there is some plane of symmetry in terms of material properties. Linearly elastic monoclinic solid has one plane of material symmetry, say $z = 0$. This means that the shear strains ε_{13} and ε_{23} should not result in any normal stresses. Thus, $C_{14} = C_{24} = C_{34} = 0$ and $C_{15} = C_{25} = C_{35} = 0$. In addition, the shear strains ε_{12} should not result in any shear stresses σ_{23} and σ_{23}, due to the mirror symmetry with respect to $z = 0$. This forces 8 constants to zero, and $\mathbf{C}$ has now 13 independent elasticity constants as shown below:

$$\mathbf{C} = [C_{\alpha\beta}] = \begin{bmatrix} C_{11} & C_{12} & C_{13} & 0 & 0 & C_{16} \\ & C_{22} & C_{23} & 0 & 0 & C_{26} \\ & & C_{33} & 0 & 0 & C_{36} \\ & & & C_{44} & C_{45} & 0 \\ & sym. & & & C_{55} & 0 \\ & & & & & C_{66} \end{bmatrix} \quad (5.28)$$

5.10 Orthotropic materials, 13 → 9

If the material has three mutually orthogonal planes of mirror symmetry, the number of independent elastic constants reduces to 9. The arguments on this are similar to these we made in the previous sub-section. The $\mathbf{C}$ matrix becomes

$$\mathbf{C} = [C_{\alpha\beta}] = \begin{bmatrix} C_{11} & C_{12} & C_{13} & 0 & 0 & 0 \\ & C_{22} & C_{23} & 0 & 0 & 0 \\ & & C_{33} & 0 & 0 & 0 \\ & & & C_{44} & 0 & 0 \\ & sym. & & & C_{55} & 0 \\ & & & & & C_{66} \end{bmatrix} \quad (5.29)$$

5.11 Transversely isotropic materials, 9 → 5

When the mechanical properties of the material are axial-symmetric about an axis that is normal to a plane of isotropy, say x–y plane, the independent number of elasticity constants becomes 5, and the **C** matrix can be given by

$$\mathbf{C} = [C_{\alpha\beta}] = \begin{bmatrix} C_{11} & C_{12} & C_{13} & 0 & 0 & 0 \\ & C_{11} & C_{13} & 0 & 0 & 0 \\ & & C_{33} & 0 & 0 & 0 \\ & & & C_{44} & 0 & 0 \\ & \text{sym.} & & & C_{44} & 0 \\ & & & & & \frac{1}{2}(C_{11} - C_{12}) \end{bmatrix} \quad (5.30)$$

5.12 Cubic materials, 5 → 3

Cubic materials possess three mutually orthogonal planes of mirror symmetry, and the elasticity constants are the same along these three orthogonal directions. Its independent number of elasticity constants becomes 3, and the **C** matrix can be written as

$$\mathbf{C} = [C_{\alpha\beta}] = \begin{bmatrix} C_{11} & C_{12} & C_{12} & 0 & 0 & 0 \\ & C_{11} & C_{12} & 0 & 0 & 0 \\ & & C_{11} & 0 & 0 & 0 \\ & & & C_{44} & 0 & 0 \\ & \text{sym.} & & & C_{44} & 0 \\ & & & & & C_{44} \end{bmatrix} \quad (5.31)$$

5.13 Isotropic materials, 3 → 2

The material properties of an isotropic material are direction independent. Therefore, all the elastic constants are invariant with respect to coordinate transformation. The constitutive equation for isotropic materials is thus the simplest and also most widely used. It has a number of different forms. This section provides the major ones.

5.13.1 *For 3D problems*

Due to the direction-independence, any plane would be a plane of mirror symmetry for isotropic materials. Its independent number of elasticity constants becomes 2, and the **C** matrix can be given by

$$\mathbf{C} = [C_{\alpha\beta}] = \begin{bmatrix} C_{11} & C_{12} & C_{12} & 0 & 0 & 0 \\ & C_{11} & C_{12} & 0 & 0 & 0 \\ & & C_{11} & 0 & 0 & 0 \\ & & & (C_{11}-C_{12})/2 & 0 & 0 \\ & sym. & & & (C_{11}-C_{12})/2 & 0 \\ & & & & & (C_{11}-C_{12})/2 \end{bmatrix}$$

(5.32)

where

$$\boxed{\begin{aligned} C_{11} &= \frac{E(1-\nu)}{(1-2\nu)(1+\nu)} = 2G+\lambda \\ C_{12} &= \lambda = \frac{E\nu}{(1-2\nu)(1+\nu)} = \frac{2G\nu}{1-2\nu} = \frac{G(E-2G)}{3G-E} \\ G &= \mu = \frac{E}{2(1+\nu)} \end{aligned}}$$

(5.33)

In which, E is Young's Modulus, ν is Poisson's ratio, and G the shear modulus. They are also called the engineering constants, and can be obtained from engineering experiments. Formel definition for the shear modulus G, and the derivation of relations in Eq. (5.33) will be given later, when discussing about the compliance matrix where the engineering constants are defined.

Due to the last equation in Eq. (5.33), we have only two independent elastic constants, which is any pair of these three: E, ν, and G. Due to the various formulations for solid mechanics problems, these elastic constants are also written in other forms, such as the bulk modulus K for volume deformation, P-wave modulus M for wave propagation problems. A comprehensive list of conversion formulas for elastic constants can be found at the Young modulus (https://en.wikipedia.org/wiki/Young%27s_modulus) page.

The constitutive equation for isotropic materials can be written explicitly as

$$\begin{aligned} \sigma_{11} &= \lambda e + 2G\varepsilon_{11} \\ \sigma_{22} &= \lambda e + 2G\varepsilon_{22} \\ \sigma_{33} &= \lambda e + 2G\varepsilon_{33} \\ \sigma_{23} &= 2G\varepsilon_{23} \\ \sigma_{12} &= 2G\varepsilon_{12} \\ \sigma_{13} &= 2G\varepsilon_{13} \end{aligned}$$

(5.34)

where
$$e = \varepsilon_{11} + \varepsilon_{22} + \varepsilon_{33} \tag{5.35}$$

Equation (5.34) can written in a concise indicial form:
$$\boxed{\sigma_{ij} = 2G\varepsilon_{ij} + \lambda\delta_{ij}\varepsilon_{kk}} \tag{5.36}$$

This equation contains only two constants: G and λ. Both are called Lamé's constants.

Using Eq. (5.36), the 4th elasticity tensor for isotropic materials can be given as
$$C_{ijkl} = G(\delta_{ik}\delta_{jl} + \delta_{il}\delta_{jk}) + \lambda\delta_{ij}\delta_{kl} \tag{5.37}$$

Equation (5.37) can be confirmed easily by multiplying it with ε_{kl} from the right, which shall lead to Eq. (5.36).

By setting $j = i$ in Eq. (5.36), we obtain:
$$\sigma_{kk} = (2G + 3\lambda)\varepsilon_{kk} = 3K\varepsilon_{kk} \tag{5.38}$$

where K known as the **bulk modulus**, given by
$$K = \lambda + \frac{2}{3}G = \frac{E}{3(1-2\nu)} \tag{5.39}$$

We have 3 in front of K because σ_{kk} is 3 times the hydrostatic stress. It is seen that ν cannot be 0.5. When ν approaches to 0.5, the bulk modulus approaches to infinity, implying that the material becomes incompressible. This the root of the so-called volumetric locking issue in displacement methods for solid mechanics problems [11, 12].

Substituting Eq. (5.38) into Eq. (5.36), we obtain strains in terms of stresses:
$$\varepsilon_{ij} = \frac{\lambda\delta_{ij}}{2G(2G + 3\lambda)}\sigma_{kk} + \frac{1}{2G}\sigma_{ij} \tag{5.40}$$

Equation (5.36) can also be written in matrix form:
$$\boldsymbol{\sigma} = 2G\boldsymbol{\varepsilon} + \lambda\text{tr}(\boldsymbol{\varepsilon})\mathbf{I} \tag{5.41}$$

Alternatively, $\mathbf{C}$ matrix in Eq. (5.32) for isotropic materials can be written explicitly in these engineering constants:

$$\mathbf{C} = [C_{\alpha\beta}] = \begin{bmatrix} \lambda + 2G & \lambda & \lambda & 0 & 0 & 0 \\ & \lambda + 2G & \lambda & 0 & 0 & 0 \\ & & \lambda + 2G & 0 & 0 & 0 \\ & & & G & 0 & 0 \\ & \text{sym.} & & & G & 0 \\ & & & & & G \end{bmatrix} \tag{5.42}$$

Using Eqs. (5.33) and (5.42), the stress-strain relations for isotropic materials can be written in the following explicit form, in terms of Young's modulus and Poisson's ratio.

$$\sigma_{11} = \frac{E}{(1+\nu)(1-2\nu)}[(1-\nu)\varepsilon_{11} + \nu(\varepsilon_{22} + \varepsilon_{33})]$$

$$\sigma_{22} = \frac{E}{(1+\nu)(1-2\nu)}[(1-\nu)\varepsilon_{22} + \nu(\varepsilon_{11} + \varepsilon_{33})]$$

$$\sigma_{33} = \frac{E}{(1+\nu)(1-2\nu)}[(1-\nu)\varepsilon_{33} + \nu(\varepsilon_{11} + \varepsilon_{22})]$$

$$\sigma_{23} = \frac{E}{(1+\nu)}\varepsilon_{23}; \quad \sigma_{13} = \frac{E}{(1+\nu)}\varepsilon_{13}; \quad \sigma_{12} = \frac{E}{(1+\nu)}\varepsilon_{12}$$

(5.43)

We emphasize that for isotropic materials, the elastic constants and hence their fourth-order elastic tensor are direction-independent, will not change with the coordinate rotation, and no coordinate transformation is needed for the elasticity tensor for isotropic materials. We will demonstrate this important feature in later examples.

5.13.2 *For stress and strain invariants*

We can also write the constitution equations for stress and strain invariants. Such relations are particularly useful because all their invariants are independent of coordinates. They can be obtained using the definitions of stress invariants (see Chapter 3) and strain invariants (see Chapter 4), and the constitutive equations Eq. (5.41). These are found as

$$I_1 = (2G + 3\lambda)\bar{I}_1$$
$$I_2 = \lambda(4G + 3\lambda)\bar{I}_1^2 + 4G^2\bar{I}_2 \qquad (5.44)$$
$$I_3 = \lambda^2(2G + \lambda)\bar{I}_1^3 + 4\lambda G^2\bar{I}_1\bar{I}_2 + 8G^3\bar{I}_3$$

Studying general 3D problems can often be difficult. Many problems can be simplified as 2D, in which many stress or strain components are zero. The next sections derive various constitutive equations for 2D problems.

5.13.3 *For 2D plane stress problems*

One type of 2D problem is called 2D **plane stress** problems. The nonzero stresses stay in a plane, say in the x_1-x_2 plane, we shall have $\sigma_{33} = \sigma_{23} = \sigma_{13} = 0$. Thus, the 3rd, 4th and 5th equations in Eq. (5.43) give

$$\varepsilon_{33} = \frac{-\nu}{1-\nu}(\varepsilon_{11} + \varepsilon_{22}); \quad \varepsilon_{23} = 0; \quad \varepsilon_{13} = 0 \qquad (5.45)$$

Note that the normal strain ε_{33} is nonzero. Substituting the first equation in Eqs. (5.45) to (5.43), with a little lengthy derivation, the stress-strain relations reduces to:

$$\sigma_{11} = \frac{E}{(1-\nu^2)}(\varepsilon_{11} + \nu\varepsilon_{22})$$

$$\sigma_{22} = \frac{E}{(1-\nu^2)}(\varepsilon_{22} + \nu\varepsilon_{11}) \qquad (5.46)$$

$$\sigma_{12} = \frac{E}{(1+\nu)}\varepsilon_{12}$$

This is often written in matrix form:

$$\begin{bmatrix}\sigma_{11}\\ \sigma_{22}\\ \sigma_{12}\end{bmatrix} = \underbrace{\frac{E}{(1-\nu^2)}\begin{bmatrix}1 & \nu & 0\\ \nu & 1 & 0\\ 0 & 0 & \frac{(1-\nu)}{2}\end{bmatrix}}_{C}\begin{bmatrix}\varepsilon_{11}\\ \varepsilon_{22}\\ 2\varepsilon_{12}\end{bmatrix}$$

$$= \underbrace{\begin{bmatrix}c_{11} & c_{12} & 0\\ c_{12} & c_{22} & 0\\ 0 & 0 & c_{33}\end{bmatrix}}_{C}\begin{bmatrix}\varepsilon_{11}\\ \varepsilon_{22}\\ 2\varepsilon_{12}\end{bmatrix} \qquad (5.47)$$

5.13.4 For 2D plane strain problems

Another type of 2D problem is the 2D **plane strain** problem. In this case the nonzero strains stay in a plane, say in the x_1–x_2 plane, and we shall have $\varepsilon_{13} = \varepsilon_{23} = \varepsilon_{33} = 0$. Equation (5.43) becomes

$$\sigma_{11} = \frac{E}{(1+\nu)(1-2\nu)}[(1-\nu)\varepsilon_{11} + \nu\varepsilon_{22}]$$

$$\sigma_{22} = \frac{E}{(1+\nu)(1-2\nu)}[\nu\varepsilon_{11} + (1-\nu)\varepsilon_{22}]$$

$$\sigma_{12} = \frac{E}{(1+\nu)}\varepsilon_{12} \qquad (5.48)$$

$$\sigma_{33} = \frac{\nu E}{(1+\nu)(1-2\nu)}[\varepsilon_{22} + \varepsilon_{11}]; \quad \sigma_{23} = 0; \quad \sigma_{13} = 0$$

Note in this case that σ_{33} is nonzero. The first 3 equations in Eq. (5.48) is often written in matrix form:

$$\begin{bmatrix}\sigma_{11}\\\sigma_{22}\\\sigma_{12}\end{bmatrix} = \frac{E(1-\nu)}{(1+\nu)(1-2\nu)}\begin{bmatrix}1 & \frac{\nu}{1-\nu} & 0\\\frac{\nu}{1-\nu} & 1 & 0\\0 & 0 & \frac{(1-2\nu)}{2(1-\nu)}\end{bmatrix}\begin{bmatrix}\varepsilon_{11}\\\varepsilon_{22}\\2\varepsilon_{12}\end{bmatrix}$$

$$= \underbrace{\begin{bmatrix}c_{11} & c_{12} & 0\\c_{12} & c_{22} & 0\\0 & 0 & c_{33}\end{bmatrix}}_{C}\begin{bmatrix}\varepsilon_{11}\\\varepsilon_{22}\\2\varepsilon_{12}\end{bmatrix}$$

(5.49)

Note that we can obtain Eq. (5.49) simply using the following substitutions in Eq. (5.47) (plane stress to plane strain).

$$\text{first: substitute } \nu \text{ by } \frac{\nu}{(1-\nu)}$$

$$\text{then: substitute } E \text{ by } \frac{E}{(1-\nu^2)} \quad (5.50)$$

$$\text{no change in } G$$

5.14 Mean and deviatoric stress and strain tensors

The mean stress is defined as the averaged normal stresses:

$$\sigma_m = \frac{\sigma_{kk}}{3} = \frac{\sigma_{11}+\sigma_{22}+\sigma_{33}}{3} \quad (5.51)$$

It is essentially the same as the so-called hydrostatic stress. The mean stress tensor is formed using the mean stress:

$$\sigma_{ij} = \sigma_m \delta_{ij} \quad (5.52)$$

Or in matrix form:

$$\boldsymbol{\sigma}_m = \begin{bmatrix}\sigma_m & 0 & 0\\0 & \sigma_m & 0\\0 & 0 & \sigma_m\end{bmatrix} \quad (5.53)$$

The **deviatoric stress** tensor is defined as the stress tensor minus the mean stress tensor:

$$\sigma_d = [s_{ij}] = \begin{bmatrix} \sigma_{11} - \sigma_m & \sigma_{12} & \sigma_{13} \\ \sigma_{12} & \sigma_{22} - \sigma_m & \sigma_{23} \\ \sigma_{13} & \sigma_{23} & \sigma_{33} - \sigma_m \end{bmatrix} \quad (5.54)$$

Or

$$s_{ij} = \sigma_{ij} - \frac{\sigma_{kk}}{3}\delta_{ij} \quad (5.55)$$

The **deviatoric strain** tensor is defined in the similar way as

$$e_{ij} = \varepsilon_{ij} - \frac{\varepsilon_{kk}}{3}\delta_{ij} \quad (5.56)$$

The deviatoric stress and strain tensors are related as follows:

$$s_{ij} = 2G e_{ij} \quad (5.57)$$

They are related only by the shear modulus.

The mean stress and mean strain are related as

$$\sigma_m = \underbrace{\lambda + \frac{2}{3}G}_{K}\, \varepsilon_{kk} \quad (5.58)$$

where K is the bulk modulus given in Eq. (5.39).

5.15 Compliance matrix

The **C** matrix is also called stiffness matrix of materials because it represents how stiff the materials are. Frequently, we use also compliance matrix of materials. It can be obtained by inverting the stiffness matrix, as long as the material is stable (it does not flow like a fluid, **C** is positive definite, hence invertible).

$$\mathbf{S} = [S_{\alpha\beta}] = \mathbf{C}^{-1} \quad (5.59)$$

5.15.1 *Compliance matrix for orthotropic materials*

Many composite materials used in engineering are orthotropic. We often need to obtain first the compliance matrix through measuring Young's modulus and Poisson's ratios, and then compute the stiffness matrix via inversion.

For orthotropic materials, we have

$$\begin{bmatrix} \varepsilon_{11} \\ \varepsilon_{22} \\ \varepsilon_{33} \\ 2\varepsilon_{23} \\ 2\varepsilon_{13} \\ 2\varepsilon_{12} \end{bmatrix}_{\vec{\varepsilon}} = \underbrace{\begin{bmatrix} \frac{1}{E_1} & \frac{-\nu_{21}}{E_2} & \frac{-\nu_{31}}{E_3} & 0 & 0 & 0 \\ \frac{-\nu_{12}}{E_1} & \frac{1}{E_2} & \frac{-\nu_{32}}{E_3} & 0 & 0 & 0 \\ \frac{-\nu_{13}}{E_1} & \frac{-\nu_{23}}{E_2} & \frac{1}{E_3} & 0 & 0 & 0 \\ 0 & 0 & 0 & \frac{1}{G_{23}} & 0 & 0 \\ 0 & 0 & 0 & 0 & \frac{1}{G_{13}} & 0 \\ 0 & 0 & 0 & 0 & 0 & \frac{1}{G_{12}} \end{bmatrix}}_{S} \begin{bmatrix} \sigma_{11} \\ \sigma_{22} \\ \sigma_{33} \\ \sigma_{23} \\ \sigma_{13} \\ \sigma_{12} \end{bmatrix}_{\vec{\sigma}} \quad (5.60)$$

Or

$$\boxed{\vec{\varepsilon} = \mathbf{S}\vec{\sigma}} \quad (5.61)$$

In Eq. (5.60), E_i is Young's modulus in the x_i direction, G_{ij} is the shear modulus between fibers in x_i-direction and x_j-direction defined as

$$G_{ij} := \frac{\sigma_{ij}}{\gamma_{ij}} = \frac{\sigma_{ij}}{2\varepsilon_{ij}} \quad \text{(no summation)} \quad (5.62)$$

Poisson's ratio, ν_{ij}, is defined as

$$\nu_{ij} := -\frac{\varepsilon_{jj}}{\varepsilon_{ii}} \quad \text{(no summation)} \quad (5.63)$$

Due to the symmetry, we have the following symmetric relations:

$$\frac{\nu_{12}}{E_1} = \frac{\nu_{21}}{E_2}; \quad \frac{\nu_{13}}{E_1} = \frac{\nu_{31}}{E_3}; \quad \frac{\nu_{23}}{E_2} = \frac{\nu_{32}}{E_3} \quad (5.64)$$

Due to the anisotropy, experiments discussed in Section 5.1 need to be conducted using samples in three directions. Since Young's moduli, Poisson's ratios, and shear moduli can be measured from experiments, it is often easier to obtain the compliance matrix, and then convert it to stiffness matrix.

5.15.2 *Python code for elasticity matrix for orthotropic materials (3D problems)*

We write the following code function to form a 6 × 6 compliance matrix using Young's moduli and Poisson's ratios, and then compute the 6 × 6 stiffness matrix via matrix inversion. We will use it multiple times in the later examples.

```python
def E_SnC3D(E1, E2, E3, m12, m13, m23, G23, G13, G12):
    '''Numpy code to compute the S and C matrix in Voigt notation for
    given Young's moduli and Poisson's ratios of orthotropic materials
    for 3D problems.
    '''
    S = np.zeros((6,6))                    #initialization
    
    m21, m31, m32 = m12/E1*E2, m13/E1*E3, m23/E2*E3
    
    # compute the compliance matrix S
    S[0,0], S[1,1], S[2,2] = 1./E1, 1./E2, 1./E3
    S[0,1], S[0,2], S[1,2] = -m21/E2, -m31/E3, -m32/E3
    S[3,3], S[4,4], S[5,5] = 1./G23, 1./G13, 1./G12
    S[1,0], S[2,0], S[2,1] = S[0,1], S[0,2], S[1,2]
    
    # compute C matrix
    C = np.linalg.inv(S)
    
    return C, S
```

```python
def E_SnC3Dsp(E1, E2, E3, m12, m13, m23, G23, G13, G12):
    '''Sympy code to compute the S and C matrix in Voigt notation for
    given Young's moduli and Poisson's ratios of orthotropic materials
    for 3D problems.
    '''
    S = sp.zeros(6,6)                    #initialization
    
    m21, m31, m32 = m12/E1*E2, m13/E1*E3, m23/E2*E3
    
    # compute the compliance matrix S
    S[0,0], S[1,1], S[2,2] = 1/E1, 1/E2, 1/E3
    S[0,1], S[0,2], S[1,2] = -m21/E2, -m31/E3, -m32/E3
    S[3,3], S[4,4], S[5,5] = 1/G23, 1/G13, 1/G12
    S[1,0], S[2,0], S[2,1] = S[0,1], S[0,2], S[1,2]
    
    # compute C matrix
    C = S.inv()
    
    return C, S
```

5.15.3 Plane stress problems

For **plane stress** problems, we shall have $\sigma_{33} = \sigma_{23} = \sigma_{13} = 0$. Equation (5.60) can be simplified as

$$\begin{bmatrix} \varepsilon_{11} \\ \varepsilon_{22} \\ 2\varepsilon_{12} \end{bmatrix} = \underbrace{\begin{bmatrix} \frac{1}{E_1} & -\frac{\nu_{21}}{E_2} & 0 \\ -\frac{\nu_{12}}{E_1} & \frac{1}{E_2} & 0 \\ 0 & 0 & \frac{1}{G_{12}} \end{bmatrix}}_{S} \begin{bmatrix} \sigma_{11} \\ \sigma_{22} \\ \sigma_{12} \end{bmatrix} \qquad (5.65)$$

Strain ε_{33} is given by

$$\varepsilon_{33} = -\left[\frac{\nu_{13}}{E_1}\sigma_{11} + \frac{\nu_{23}}{E_2}\sigma_{22}\right] \qquad (5.66)$$

In reverse, Eq. (5.65) becomes

$$\begin{bmatrix} \sigma_{11} \\ \sigma_{22} \\ \sigma_{12} \end{bmatrix} = \frac{1}{1-\nu_{12}\nu_{21}}\underbrace{\begin{bmatrix} E_1 & \nu_{21}E_1 & 0 \\ \nu_{12}E_2 & E_2 & 0 \\ 0 & 0 & G_{12}(1-\nu_{12}\nu_{21}) \end{bmatrix}}_{C}\begin{bmatrix} \varepsilon_{11} \\ \varepsilon_{22} \\ 2\varepsilon_{12} \end{bmatrix} \qquad (5.67)$$

For isotropic materials, simply set $E_i = E$, $\nu_{ij} = \nu$, and $G_{ij} = G$.

5.15.4 Python code for elasticity matrix for orthotropic materials (2D plane stress)

The following code function forms a 3 × 3 compliance matrix using Young's moduli, Poisson's ratios, and shear moduli. And then computes the 3 × 3 stiffness matrix via inversion. We will use it multiple times in the later examples. This code is for 2D plane stress problems.

```
1  def E_SnC2D_stress(E1, E2, v12, G12):
2      '''Numpy code to compute S and C matrix in Voigt notation for given
3         Young's moduli Ei, and Poisson's ratios v12, shear modulus G12
4         of orthotropic materials for 2D plan stress problems.
5      '''
6      S = np.zeros((3,3))    # sp: S = sp.zeros(3,3)         #initialization
7      v21 = v12/E1*E2
8
9      # compute the compliance matrix S
10     S[0,0], S[1,1] = 1/E1, 1/E2
11     S[0,1]         = -v21/E2
12     S[1,0]         = S[0,1]
13     S[2,2]         = 1/G12
14
15     # compute C matrix
16     C = np.linalg.inv(S)   # sp: S = sp.inv(S)
17
18     return C, S
```

5.15.5 Plane strain problems

For **plane strain** problems ($\varepsilon_{13} = \varepsilon_{23} = \varepsilon_{33} = 0$), the derivation of compliance matrix for orthotropic materials is quite tedious and prone to making mistakes. We thus use Sympy to help in the derivation. We define first some symbolic variables.

```
1  E1, E2, E3 = symbols("E1, E2, E3")
2  σ11, σ22 = symbols("σ11, σ22")
3  v12, v21, v13, v31, v23, v32 = symbols("v12, v21, v13, v31, v23, v32")
```

Since $\varepsilon_{33} = 0$ for plane strain problems, we can compute the stress component σ_{33}, using the 3rd equation in Eq. (5.60):

```
1  σ33 = E3/E1*v13*σ11 + E3/E2*v23*σ22
2  σ33
```

$$\frac{E_3 \nu_{23} \sigma_{22}}{E_2} + \frac{E_3 \nu_{13} \sigma_{11}}{E_1}$$

Using now the first equation in (5.60), we compute ε11:

```
1  ε11 = σ11/E1 - σ22/E2*v21 - σ33/E3*v31
2  ε11.factor(σ11).simplify()
```

$$-\frac{\nu_{21}\sigma_{22}}{E_2} - \frac{\nu_{23}\nu_{31}\sigma_{22}}{E_2} - \frac{\nu_{13}\nu_{31}\sigma_{11}}{E_1} + \frac{\sigma_{11}}{E_1}$$

Next, using the second equation in (5.60), compute ε_{22}:

```
1  ε22 = -σ11/E1*v12 + σ22/E2 - σ33/E3*v32
2  ε22.expand().factor(σ22)
```

$$-\frac{E_2\nu_{12}\sigma_{11} + E_2\nu_{13}\nu_{32}\sigma_{11} + \sigma_{22}\left(E_1\nu_{23}\nu_{32} - E_1\right)}{E_1 E_2}$$

We form the strain-stress matrix as

$$\begin{bmatrix} \varepsilon_{11} \\ \varepsilon_{22} \\ 2\varepsilon_{12} \end{bmatrix} = \underbrace{\begin{bmatrix} \frac{1-\nu_{13}\nu_{31}}{E_1} & -\frac{\nu_{12}+\nu_{13}\nu_{32}}{E_1} & 0 \\ -\frac{\nu_{12}+\nu_{13}\nu_{32}}{E_1} & \frac{1-\nu_{23}\nu_{32}}{E_2} & 0 \\ 0 & 0 & \frac{1}{G_{12}} \end{bmatrix}}_{S} \begin{bmatrix} \sigma_{11} \\ \sigma_{22} \\ \sigma_{12} \end{bmatrix} \quad (5.68)$$

Stress σ_{33} is given by

$$\sigma_{33} = \frac{E_3\nu_{23}\sigma_{22}}{E_2} + \frac{E_3\nu_{13}\sigma_{11}}{E_1} = \nu_{32}\sigma_{22} + \nu_{31}\sigma_{11} \quad (5.69)$$

Equation (5.68) can be inverted to obtain the stress-strain relation. We again use Sympy to help us get this done. First, we form the 2 by 2 submatrix in in the upper-left corner in Eq. (5.68):

```
1  S_2by2 = Matrix([[(1-v13*v31)/E1,   -(v12+v13*v32)/E1],
2                   [-(v12+v13*v32)/E1, (1-v23*v32)/E2  ]])
3  A = S_2by2.det()*E1*E1*E2    # a coefficient to simplify expression
4  A
```

$$E_1\nu_{13}\nu_{23}\nu_{31}\nu_{32} - E_1\nu_{13}\nu_{31} - E_1\nu_{23}\nu_{32} + E_1 - E_2\nu_{12}^2 - 2E_2\nu_{12}\nu_{13}\nu_{32} - E_2\nu_{13}^2\nu_{32}^2$$

```
1  C_2by2 = S_2by2.inv()*A
2  C_2by2
```

$$\begin{bmatrix} -E_1^2\nu_{23}\nu_{32} + E_1^2 & E_1 E_2\nu_{12} + E_1 E_2\nu_{13}\nu_{32} \\ E_1 E_2\nu_{12} + E_1 E_2\nu_{13}\nu_{32} & -E_1 E_2\nu_{13}\nu_{31} + E_1 E_2 \end{bmatrix}$$

In reverse, Eq. (5.68) becomes

$$\begin{bmatrix} \sigma_{11} \\ \sigma_{22} \\ \sigma_{12} \end{bmatrix} = \frac{1}{A} \underbrace{\begin{bmatrix} -E_1^2 \nu_{23} \nu_{32} + E_1^2 & E_1 E_2 \nu_{12} + E_1 E_2 \nu_{13} \nu_{32} & 0 \\ E_1 E_2 \nu_{12} + E_1 E_2 \nu_{13} \nu_{32} & -E_1 E_2 \nu_{13} \nu_{31} + E_1 E_2 & 0 \\ 0 & 0 & G_{12} A \end{bmatrix}}_{C} \begin{bmatrix} \varepsilon_{11} \\ \varepsilon_{22} \\ 2\varepsilon_{12} \end{bmatrix}$$

(5.70)

where

$$A = E_1 \nu_{13} \nu_{23} \nu_{31} \nu_{32} - E_1 \nu_{13} \nu_{31} - E_1 \nu_{23} \nu_{32} + E_1 - E_2 \nu_{12}^2$$
$$- 2 E_2 \nu_{12} \nu_{13} \nu_{32} - E_2 \nu_{13}^2 \nu_{32}^2 \quad (5.71)$$

For isotropic materials, we simply set $E_i = E$, $\nu_{ij} = \nu$, and $G_{ij} = G$. It is trivial but a little lengthy to confirm that such a setting leads to Eq. (5.49).

5.15.6 Python code for elasticity matrix for orthotropic materials (2D plane strain)

The following function forms a 3 × 3 compliance matrix using Young's moduli, Poisson's ratios, and shear modulus, and then computes the 3×3 stiffness matrix via inversion. This code is for 2D plane strain problems.

```
1  def E_SnC2D_strain(E1, E2, E3, v12, v13, v23, G12):
2      '''Compute the S and C matrix in Voigt notation for given Young's
3      moduli Ei, Poisson's ratios vij, and shear modulus G12 of
4      orthotropic materials for 2D plan strain problems.
5      '''
6      S = np.zeros((3,3))                    #initialization
7      v21 = v12/E1*E2
8      v31 = v13/E1*E3
9      v32 = v23/E2*E3
10
11     # compute the compliance matrix S
12     S[0,0] = (1. -v13*v31)/E1
13     S[0,1] =-(v12+v13*v32)/E1
14     S[1,1] = (1. -v23*v32)/E2
15     S[1,0]         = S[0,1]
16     S[2,2]         = 1./G12
17
18     # compute C matrix
19     C = np.linalg.inv(S)
20
21     return C, S
```

5.15.7 Compliance matrix for isotropic materials

For isotropic materials, we have a much simpler form.

$$\mathbf{S} = [S_{\alpha\beta}] = \begin{bmatrix} \frac{1}{E} & \frac{-\nu}{E} & \frac{-\nu}{E} & 0 & 0 & 0 \\ & \frac{1}{E} & \frac{-\nu}{E} & 0 & 0 & 0 \\ & & \frac{1}{E} & 0 & 0 & 0 \\ & & & \frac{1}{G} & 0 & 0 \\ & \text{sym.} & & & \frac{1}{G} & 0 \\ & & & & & \frac{1}{G} \end{bmatrix} \quad (5.72)$$

Using Eqs. (5.61) and (5.72), the constitutive equation in terms of strain components can be written in a familiar form of

$$\begin{aligned} \varepsilon_{11} &= \frac{1}{E}[\sigma_{11} - \nu\sigma_{22} - \nu\sigma_{33}] \\ \varepsilon_{22} &= \frac{1}{E}[\sigma_{22} - \nu\sigma_{11} - \nu\sigma_{33}] \\ \varepsilon_{33} &= \frac{1}{E}[\sigma_{33} - \nu\sigma_{11} - \nu\sigma_{22}] \\ \varepsilon_{23} &= \frac{1}{2G}\sigma_{23} = \frac{1+\nu}{E}\sigma_{23} \\ \varepsilon_{13} &= \frac{1}{2G}\sigma_{13} = \frac{1+\nu}{E}\sigma_{13} \\ \varepsilon_{12} &= \frac{1}{2G}\sigma_{12} = \frac{1+\nu}{E}\sigma_{12} \end{aligned} \quad (5.73)$$

5.15.8 Derivation of the C matrix for isotropic materials

Using Eq. (5.72), we are now ready to derive the **C** matrix for isotropic materials defined in Eq. (5.32), assuming Young's modulus E, shear modulus G, and Poisson's ratio ν are given. We shall use the code function E_SnC3Dsp(), and set $E_i = E$, $G_{ij} = G$, and $\nu_{ij} = \nu$ because of the isotropy.

```
1  E, ν, G = symbols('E, ν, G', positive=True)
2  C, S = E_SnC3Dsp(E, E, E, ν, ν, ν, G, G, G)
3  C
```

$$\begin{bmatrix} \frac{E\nu-E}{2\nu^2+\nu-1} & -\frac{E\nu}{2\nu^2+\nu-1} & -\frac{E\nu}{2\nu^2+\nu-1} & 0 & 0 & 0 \\ -\frac{E\nu}{2\nu^2+\nu-1} & \frac{E\nu-E}{2\nu^2+\nu-1} & -\frac{E\nu}{2\nu^2+\nu-1} & 0 & 0 & 0 \\ -\frac{E\nu}{2\nu^2+\nu-1} & -\frac{E\nu}{2\nu^2+\nu-1} & \frac{E\nu-E}{2\nu^2+\nu-1} & 0 & 0 & 0 \\ 0 & 0 & 0 & G & 0 & 0 \\ 0 & 0 & 0 & 0 & G & 0 \\ 0 & 0 & 0 & 0 & 0 & G \end{bmatrix}$$

This matrix gives the relations in Eq. (5.33).

5.15.9 Compliance equations for 2D problems

For 2D **plane stress** problems, the nonzero stresses stay, say in the x_1–x_2 plane, we shall have $\sigma_{33} = \sigma_{23} = \sigma_{13} = 0$. In this case, Eq. (5.73) becomes

$$\boxed{\begin{aligned} \varepsilon_{11} &= \frac{1}{E}(\sigma_{11} - \nu\sigma_{22}) \\ \varepsilon_{22} &= \frac{1}{E}(\sigma_{22} - \nu\sigma_{11}) \\ \varepsilon_{12} &= \frac{1}{2G}\sigma_{12} = \frac{1+\nu}{E}\sigma_{12} \end{aligned}} \quad (5.74)$$

The strain ε_{33} is not zero, and is given by

$$\varepsilon_{33} = -\frac{\nu}{E}[\sigma_{11} + \sigma_{22}] \quad (5.75)$$

after σ_{11} and σ_{22} are found.

For 2D **plane strain** problems, the nonzero strains stay, say in the x_1–x_2 plane, we shall have $\varepsilon_{33} = \varepsilon_{23} = \varepsilon_{13} = 0$. In this case, the constitutive equation for strain components can be written in a familiar form of

$$\boxed{\begin{aligned} \varepsilon_{11} &= \frac{1}{E}[(1-\nu^2)\sigma_{11} - \nu(1+\nu)\sigma_{22}] \\ \varepsilon_{22} &= \frac{1}{E}[(1-\nu^2)\sigma_{22} - \nu(1+\nu)\sigma_{11}] \\ \varepsilon_{12} &= \frac{1}{2G}\sigma_{12} = \frac{1+\nu}{E}\sigma_{12} \end{aligned}} \quad (5.76)$$

Stress σ_{33} is given by

$$\sigma_{33} = \nu\sigma_{11} + \nu\sigma_{22} \tag{5.77}$$

which is the same as Eq. (5.69).

We can obtain Eq. (5.76) using the substitution defined in Eqs. (5.50) to (5.74) (plane stress to plane strain). In reverse, we can obtain Eq. (5.74) using the following substitutions in Eq. (5.76) (plane strain to plane stress).

$$\text{first: substitute } \frac{E}{(1-\nu^2)} \text{ by } E$$

$$\text{then: substitute } \nu \text{ by } \frac{\nu}{(1+\nu)} \tag{5.78}$$

No change in G

5.15.10 Example: Proof of relation between G, E and ν

Consider a small thin square cell of solid that undergoes pure shearing, as shown in Fig. 5.3. Since it is thin, and no force is applied in the z-direction, all the stress components $\sigma_{zz} = \sigma_{zy} = \sigma_{zx} = 0$. It is thus a plane stress problem.

For the pure shear stress state given in Fig. 5.3, we have $\sigma_{xx} = \sigma_{yy} = 0$, $\sigma_{xy} = \tau$. Perform a coordinate transformation from x–y to X–Y with $\theta = 45°$, using the double angle formula (derived in Chapter 3):

$$\sigma_{XX} = \frac{1}{2}(\sigma_{xx} + \sigma_{yy}) + \frac{1}{2}(\sigma_{xx} - \sigma_{yy})\cos 2\theta + \sigma_{xy}\sin 2\theta = \tau$$

$$\sigma_{YY} = \frac{1}{2}(\sigma_{xx} + \sigma_{yy}) - \frac{1}{2}(\sigma_{xx} - \sigma_{yy})\cos 2\theta - \sigma_{xy}\sin 2\theta = -\tau \tag{5.79}$$

$$\sigma_{XY} = -\frac{1}{2}(\sigma_{xx} - \sigma_{yy})\sin 2\theta + \sigma_{xy}\cos 2\theta = 0$$

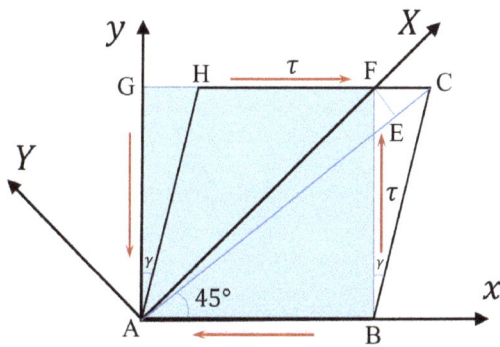

Figure 5.3. Schematic view of a small square solid subjected to pure shearing.

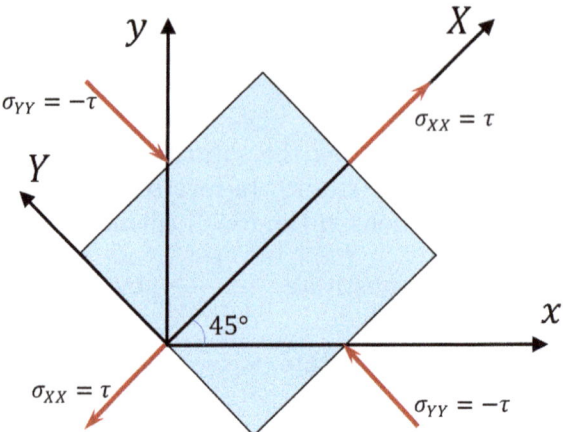

Figure 5.4. Stress state after 45-degree coordinate rotation from the pure stress state.

The transformed stress state of $\sigma_{XX} = \tau, \sigma_{YY} = -\tau, \sigma_{XY} = 0$ is shown in Fig. 5.4.

The code used to find this is given below:

```
1  σxx, σyy, σxy = symbols("σ_xx, σ_yy, σ_xy")
2  E, ν, θ, τ = symbols("E, ν, θ, τ ")
3
4  dic = {σxx:0, σyy:0, σxy:τ, θ:sp.pi/4}    # θ = 45о
5  σXX, σYY, σXY = gr.tensorT2angle(σxx, σyy, σxy, θ, py='sympy')
6
7  σXX_= σXX.subs(dic)
8  σYY_= σYY.subs(dic)
9  σXY_= σXY.subs(dic)
10 print(f"σXX={σXX_}, σYY={σYY_}, σXY={σXY_}")    # σXX=τ, σYY=-τ, σXY=0
```

σXX=τ, σYY=-τ, σXY=0

Using Eq. (5.74), the strain component in the X–Y coordinates can be found as

$$\varepsilon_{XX} = \tau \frac{\nu + 1}{E} \tag{5.80}$$

```
1  εXX = (σXX_ - ν*σYY_)/E              # use strain-stress relation
2  print(f"εXX={εXX}")
```

εXX=(ν*τ + τ)/E

Considering small strains, the length EC (see Fig. 5.3) becomes $|EC| = \varepsilon_{XX}|AE|$, where $|AE| = |AB|\sqrt{2}$, which gives

$$|EC| = \varepsilon_{XX}|AB|\sqrt{2} \tag{5.81}$$

On the other hand, $|FC| = |EC|\sqrt{2}$, and the shear strain becomes

$$\gamma = \frac{|FC|}{|AB|} = 2\varepsilon_{XX} \tag{5.82}$$

With the help of Eq. (5.80), we then found

$$G := \frac{\tau}{\gamma} = \frac{\tau}{2\varepsilon_{XX}} = \frac{E}{2(\nu+1)} \tag{5.83}$$

or

$$\boxed{G = \frac{E}{2(\nu+1)}} \tag{5.84}$$

This is the last equation given in Eq. (5.33). It is a widely used formula that relates E, G and ν.

The code to get this done is given as follows:

```
1  AB = symbols("AB")
2  AE = sp.sqrt(2)*AB
3  EC = εXX*AE
4  FC = EC*sp.sqrt(2)
5  γ  = FC/AB
6  G  = (τ/γ).simplify()
7  gr.printM(G, 'Shear modulus G=')
```

Shear modulus G = $\frac{E}{2(\nu+1)}$

Equations given in this chapter are not difficult to derive, as long as the concepts are clear. Some of the equations are quite lengthy in derivation. In this chapter, we tried to provide various forms of equations for convenience in use.

5.16 A comprehensive example: Displacement, strain and stress analysis

With relations between displacements-strain and strain-stress established, we are now ready to look at a comprehensive example created in reference to an exercise problem given in Ref. [1].

5.16.1 *Setting of the problem*

Consider 2D square solid cell with dimension of 1 cm × 1 cm defined in x–y coordinates. It is formed by four vertices: A(0, 1), B(1, 1), C(1, 0), O (0, 0). The displacement in the z-direction is zero: $w = 0$. The displacements in the x–y plane are given in the following formulas.

$$u = c_1 x + c_2 y + c_3 xy$$
$$v = c_4 x + c_5 y + c_6 xy \quad (5.85)$$

where $c_1, c_2, \ldots, c_6$ are constants. Each of the displacement components is given in a so-called **bi-linear function** in x and y because if one fixes x, the function is linear in y, and vice versa. It is linear in two directions.

Through measurements, the displacements (u, v) at points A, B, and C are found as, respectively, $(0.004, 0.002)$, $(0.002, 0.003)$, and $(0.001, 0.001)$, all in millimeters. Point O stays at the origin. %(.0, .00125), (−.00125, .00125), and (.0025, .00125),

The data for material property are given as $E = 200$ GPa, $\nu = 0.3$.

Our tasks are as follows:

1. Determine the functions for all the displacement components.
2. Compute the gradient of the displacement vector functions.
3. Compute the strain functions in the solid, and the values of the strains at point B.
4. Compute the normal strains at point B along the $\overrightarrow{OB}$ direction, and that along $\overrightarrow{AC}$ direction.
5. Compute the shear strains at point B between BA and BC.
6. Compute the principal strains and strain invariants.
7. Rotate the coordinates by 30° about z-axis, and find the displacements at point B in the new coordinates system.
8. Rotate the coordinates by 30° about z-axis, and find the strains at point B in the new coordinates system.

9. Assume the square cell is in plane strain condition, compute the stress functions in the solid, and the values of the strains at point B.
10. Compute the principal stresses, and stress invariants.
11. Rotate the coordinates by 30° about z-axis, and find the stresses at point B in the new coordinates system.

5.16.2 Solution to the problem

5.16.2.1 Analysis of the problem

For a clearer intuitive understanding of the problem, we first plot the domain of the problem: the original and the deformed square cell.

```
1   x, y = symbols("x, y ")                            # 2D coordinates
2   O = {x:0., y:0.}                                   # the origin
3   dO = np.array([ .0,     .0])
4
5   A = {x:0., y:1.}                                   # Coordinates at vertex A
6   dA = np.array([.004, .002])                        # displacement at A
7
8   B = {x:1., y:1.}                                   # Coordinates at vertex B
9   dB = np.array([.002, .003])                        # displacement at B
10
11  C = {x:1., y:0.}                                   # Coordinates at vertex C
12  dC = np.array([.001, .001])                        # displacement at C
13
14  # Create a list of the x- and y-coordinates of the vertices
15  xi = [A[x], B[x], C[x], O[x], A[x]]
16  yi = [A[y], B[y], C[y], O[y], A[y]]
17
18  plt.plot(xi, yi, marker='o')       # Plot the undeformed square cell
19
20  k = 20.                            # to magnify the displacements to be visible.
21  xp = [A[x]+dA[0]*k,B[x]+dB[0]*k,C[x]+dC[0]*k,O[x]+dO[0]*k,A[x]+dA[0]*k]
22  yp = [A[y]+dA[1]*k,B[y]+dB[1]*k,C[y]+dC[1]*k,O[y]+dO[1]*k,A[y]+dA[1]*k]
23
24  plt.plot(xp, yp, linestyle='--', marker='*')   # deformed quadrilateral
25
26  plt.text(A[x], A[y], 'A', fontsize=12, ha='right')# Label the vertices
27  plt.text(B[x], B[y], 'B', fontsize=12, ha='left')
28  plt.text(C[x], C[y], 'C', fontsize=12, ha='left')
29  plt.text(O[x], O[y], 'O', fontsize=12, ha='right')
30
31  plt.xlabel('x-axis'); plt.ylabel('y-axis')
32  plt.axis('equal')
33  plt.axis([-0.1, 1.1, -0.1, 1.1])                   # Set the axis limits
34  plt.show()
```

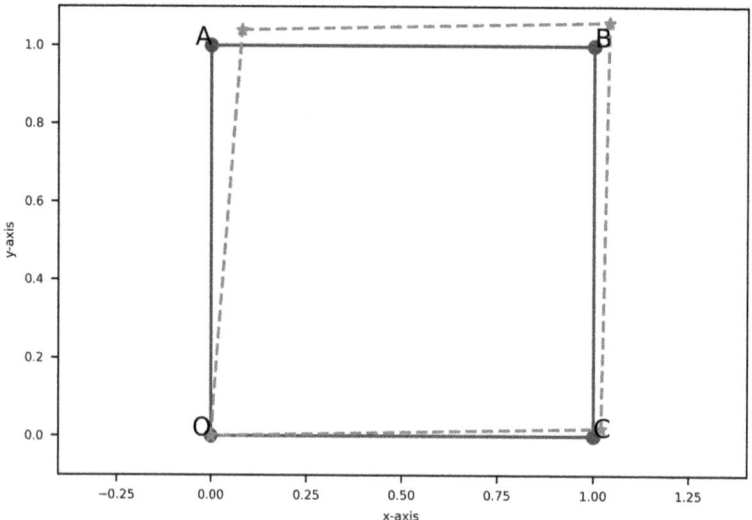

5.16.2.2 Find the displacement functions

```
1  # 1 Determine the functions for all the displacement components.
2  x, y = symbols("x, y ")
3  X = Matrix([x, y])
4
5  c1, c2, c3, c4, c5, c6 = symbols("c1, c2, c3, c4, c5, c6")
6
7  u = c1*x + c2*y + c3 *x*y
8  v = c4*x + c5*y + c6 *x*y
9
10 solution=sp.solve([u.subs(A)-dA[0],u.subs(B)-dB[0],u.subs(C)-dC[0],\
11                    v.subs(A)-dA[1],v.subs(B)-dB[1],v.subs(C)-dC[1]],\
12                    [c1, c2, c3, c4, c5, c6])
13 solution
```

$\{c_1 : 0.001,\ c_2 : 0.004,\ c_3 : -0.003,\ c_4 : 0.001,\ c_5 : 0.002,\ c_6 : 0.0\}$

```
1  # The displacement functions are found as:
2  u = u.subs(solution)
3  v = v.subs(solution)
4
5  U = Matrix([u, v])
6  gr.printM(U.T, '\nThe displacement functions are found as:')
```

The displacement functions are found as

$$\begin{bmatrix} -0.003xy + 0.001x + 0.004y & 0.001x + 0.002y \end{bmatrix}$$

5.16.2.3 Compute the gradient of the displacement

```
1  # 2: Compute the gradient of the displacement vector functions.
2
3  U3D = Matrix([u, v, 0])   # set 0 for w, to create a 3D displacements
4
5  # Find the gradient of the displacement vector functions
6  U_gradient = gr.grad_vf(U3D[:2,:2], X) # get 2D displacement gradient
7                            # or use: vector.grad_vec(U3D)[:2,:2]
8  printM(U_gradient,'Gradient of displacement vector functions',n_dgt=4)
```

Gradient of displacement vector functions

$$\begin{bmatrix} 0.001 - 0.003y & 0.001 \\ 0.004 - 0.003x & 0.002 \end{bmatrix}$$

Since the displacements are bi-linear functions in the entire cell, the gradient contains linear functions. This means that the strain field will be linear.

5.16.2.4 Find the strain tensor functions

```
1  #3 Compute the strain functions in the solid, and strain values at B.
2
3  strains = 0.5*(U_gradient +U_gradient.T)
4  printM(strains,'Strain tensor', n_dgt=4)
```

Strain tensor

$$\begin{bmatrix} 0.001 - 0.003y & 0.0025 - 0.0015x \\ 0.0025 - 0.0015x & 0.002 \end{bmatrix}$$

The strains are indeed linear functions.

```
1  εBv = strains.subs(B)              # B: coordinates at point B
2  printM(εBv, 'Strain tensor values at point B:', n_dgt=5)
```

Strain tensor values at point B:

$$\begin{bmatrix} -0.002 & 0.001 \\ 0.001 & 0.002 \end{bmatrix}$$

5.16.2.5 Compute the normal strains

```
1   # 4 Compute the normal strains at point B along OB, and AC.
2   εB_M = np.array(εBv, dtype=float)                    # convert to numpy array
3
4   # For OB, find the unit vector along OB
5   vector_B = np.array([B[x],B[y]],dtype=float)         # vector of OB
6   unit_OB = vector_B/lg.norm(vector_B)                 # Normal vector OB
7
8   NS_OB = unit_OB@εB_M@unit_OB                         # Normal strain of fiber O->B
9   print(f'Unit vector along OB = {unit_OB}')
10  print(f'Normal strain on vector OB  = {NS_OB}')
11
12  # Similar process for AC, use Eq.(2.28)
13  unit_AC = np.array([C[x]-A[x],C[y]-A[y]],dtype=float)
14  unit_AC = unit_AC/np.sqrt((C[x]-A[x])**2+(C[y]-A[y])**2)
15  print(f'Unit vector along AC = {unit_AC}')
16
17  NS_AC = unit_AC@εB_M@unit_AC
18  print(f'Normal strain on vector AC = {NS_AC:.5f}')
```

```
Unit vector along OB = [ 7.071e-01  7.071e-01]
Normal strain on vector OB  = 0.001
Unit vector along AC = [ 7.071e-01 -7.071e-01]
Normal strain on vector AC = -0.00100
```

5.16.2.6 Compute the shear strains

```
1   # 5 Compute the shear strains at point B between BA and BC.
2   # vector BA:
3   unit_BA = np.array([A[x]-B[x], A[y]-B[y]])           # use Eq.(2.28)
4   unit_BA = unit_BA/np.sqrt((A[x]-B[x])**2+(A[y]-B[y])**2)
5   print(f'The unit vector along vector BA = {unit_BA}')
6
7   NS_BA = unit_BA@εB_M@unit_BA                         # Normal strain of fiber BA
8   print(f'Normal strain along vector BA  = {NS_BA:.5f}')
9
10  # vector BC           xB-xA, yB-yA, zB-zA,
11  unit_BC = np.array([C[x]-B[x], C[y]-B[y]])           # fiber along BC
12  unit_BC = unit_BC/np.sqrt((C[x]-B[x])**2+(C[y]-B[y])**2)
13  print(f'The unit vector BC = {unit_BC}')
14
15  E_BA_BC = unit_BA@εB_M@unit_BC   # Shear strain between fiber BA and BC
16  print(f'Shear strain between BA and BC  = {E_BA_BC:.5f}')
```

```
The unit vector along vector BA = [-1.000e+00  0.000e+00]
Normal strain along vector BA  = -0.00200
The unit vector BC = [ 0.000e+00 -1.000e+00]
Shear strain between BA and BC  = 0.00100
```

5.16.2.7 Compute the principal strains and strain invariants

```
1  # 6 Compute the principal strains, and strain invariants.
2  εB_3D = np.zeros((3,3))              # create a 3D tensor with zeros
3  εB_3D[:2,:2] = εB_M                  # create a 3D tensor
4  print(εB_3D)
5  eigenValues, eigenVectors = gr.principalS(εB_3D)
6
7  I1, I2, I3 = gr.M_invariant(eigenValues)
8  print(f'Principal strains (Eigenvalues) ={eigenValues}')
9  print(f'The first  strain invariant I1={I1:.6e}')
10 print(f'The second strain invariant I2={I2:.6e}')
11 print(f'The third  strain invariant I3={I3:.6e}, discard.')
```

```
[[-2.000e-03  1.000e-03  0.000e+00]
 [ 1.000e-03  2.000e-03  0.000e+00]
 [ 0.000e+00  0.000e+00  0.000e+00]]
Principal stress/strain (Eigenvalues):
[ 2.236e-03  0.000e+00 -2.236e-03]

Principal stress/strain directions:
[[-2.298e-01  0.000e+00 -9.732e-01]
 [-9.732e-01  0.000e+00  2.298e-01]
 [ 0.000e+00  1.000e+00  0.000e+00]]

Possible angles (n1,x)=103.28252558853899∘ or 76.71747441146101∘
Principal strains (Eigenvalues) =[ 2.236e-03  0.000e+00 -2.236e-03]
The first  strain invariant I1=0.000000e+00
The second strain invariant I2=-5.000000e-06
The third  strain invariant I3=-0.000000e+00, discard.
```

One may complete this task using gr.M_eigen2D(εB_M), and gr.M_invariant2D(eigenValues).

We obtained 2 possible angles. To determine which one, we simply use:

```
1  #Tr, about = gr.transferM(103.2825, about = 'z')
2  Tr, about = gr.transferM(76.71747441146101, about = 'z')
3  εp_3D = Tr@εB_3D@Tr.T
4
5  print(f'Transformed strains: \n{εp_3D}')
```

```
Transformed strains:
[[ 2.236e-03 -6.728e-19  0.000e+00]
 [-6.246e-19 -2.236e-03  0.000e+00]
 [ 0.000e+00  0.000e+00  0.000e+00]]
```

Thus, 76.71° is the correct one (to machine accuracy). Readers may test the other one.

5.16.2.8 Displacements in the new coordinates system

```
1  #7 Rotate the coordinates by 30° about z-axis, and find the
2  # displacements at point B in the new coordinates system.
3
4  Tr, about = gr.transferM(30, about = 'z')
5  print(Tr)
6  Tu = Tr[:2,:2]@dB
7  print(f'Transformed displacements, Tu: {Tu}')
```

```
[[ 8.660e-01  5.000e-01  0.000e+00]
 [-5.000e-01  8.660e-01  0.000e+00]
 [ 0.000e+00  0.000e+00  1.000e+00]]
Transformed displacements, Tu: [ 3.232e-03  1.598e-03]
```

5.16.2.9 Strains in the new coordinates system

```
1  # 8) Rotate the coordinates by 30° about z-axis, and find the strains
2  #     at point B in the new coordinates system.
3  ɛp_3D = Tr@ɛB_3D@Tr.T
4
5  print(f'Transformed strains: \n{ɛp_3D}')
```

```
Transformed strains:
[[-1.340e-04  2.232e-03  0.000e+00]
 [ 2.232e-03  1.340e-04  0.000e+00]
 [ 0.000e+00  0.000e+00  0.000e+00]]
```

5.16.2.10 Compute the stresses

Since displacement is in the z-direction, w, is set to zero, all the strain components related to z are zero. Therefore, the problem is a **plane strain** problem. Using Eq. (5.48) or (5.49), the stress tensor is found using the following code.

```
1   # 2D plane strain problem:
2
3   E_ = 200e9; v_=0.3
4   c11 = E_*(1-v_)/(1+v_)/(1-2*v_)
5   c12 = c11*v_/(1-v_)
6   c33 = E_/(1+v_)/2          # G
7
8   C = np.array([[c11, c12,  0],
9                 [c12, c11,  0],
10                [ 0,   0,  c33]])
11
12  εv = np.array([εB_M[0,0],εB_M[1,1],2*εB_M[0,1]])    # strain vector
13  print(f"Strains  at B in vector form = {εv}")
14
15  σv = C@εv                                            # stress vector
16  print(f"Stresses at B in vector form = {σv} Pa")
17
18  σ33 = c11*v_*(εB_M[0,0] + εB_M[1,1])
19  print(f"Stress σ33 at B = {σ33:.3e} Pa")
```

```
Strains  at B in vector form = [-2.000e-03  2.000e-03  2.000e-03]
Stresses at B in vector form = [-3.077e+08  3.077e+08  1.538e+08] Pa
Stress σ33 at B = 0.000e+00 Pa
```

5.16.2.11 Compute the principal stresses, and stress invariants

```
1   σ_M = np.zeros((3,3))          # create a 3D tensor with zeros
2   σ_M[0,0] = σv[0]               # create a 3D tensor
3   σ_M[1,1] = σv[1]
4   σ_M[0,1] = σv[2]
5   σ_M[1,0] = σv[2]
6   σ_M[2,2] = σ33
7   print(σ_M)
8   eigenValues, eigenVectors = gr.M_eigen(σ_M)
9
10  I1, I2, I3 = gr.M_invariant(eigenValues)
11  print(f'Principal stress (Eigenvalues) ={eigenValues}')
12  print(f'The first  stress invariant I1={I1:.6e}')
13  print(f'The second stress invariant I2={I2:.6e}')
14  print(f'The third  stress invariant I3={I3:.6e}')
```

```
[[-3.077e+08  1.538e+08  0.000e+00]
 [ 1.538e+08  3.077e+08  0.000e+00]
 [ 0.000e+00  0.000e+00  0.000e+00]]
Principal stress (Eigenvalues) =[ 3.440e+08  0.000e+00 -3.440e+08]
The first  stress invariant I1=5.960464e-08
The second stress invariant I2=-1.183432e+17
The third  stress invariant I3=-0.000000e+00
```

5.16.2.12 Rotate the coordinates by 30° about z-axis, and find the stresses at point B in the new coordinates system

```
1  σp_M = Tr@σ_M@Tr.T                    # transformed stress tensor
2
3  print(f'Transformed stress tensor: \n{σp_M}')
```

```
Transformed stress tensor:
[[-2.061e+07  3.434e+08  0.000e+00]
 [ 3.434e+08  2.061e+07  0.000e+00]
 [ 0.000e+00  0.000e+00  0.000e+00]]
```

This example demonstrated that when the displacement functions are found, it is straightforward to obtain strains, and then stresses. Various states of stresses and strains, including the principal stresses and strains, can also be obtained with ease. Therefore, in most applications we often would like to find the displacements for given externally applied forces. This would need other important equations: equilibrium equations, which will be discussed in the next chapter.

5.17 Thermal expansion effects

5.17.1 *Thermal strains*

Structural components may work in environments where the temperature changes. Since most of the materials expand (or contract) when temperature changes, additional strains are introduced inside the materials.

If the structure component is not overconstrained, the thermal strain just results in (often small) dimension change. It will not introduce additional stress. In such cases, it is often less a problem in terms of stress induced failure.

If, however, the structure component is overconstrained (indeterminate), the material is not allowed to expand freely. The thermal strain can result in additional stress, known as **thermal stress**. Even if the thermal strain is small, the resulting stress can be significant, which can be part of the cause to failure. This section addresses such a problem for bar members that are indeterminate.

We assume the temperature change is small and hence the thermal expansion is small and linear in temperature change. It is quantified by $\alpha \Delta T$, where α is the thermal expansion coefficient of the material carrying a unit typically of $[(m/m)/°C]$, which is strain (unit 1) per degree. And ΔT is the change in the temperature that carries a unit typically $[°C]$. Therefore, $\alpha \Delta T$ gives a strain. This strain is simply added to the strain resulting from the derivatives of the displacement.

For orthotropic materials, the thermal expansion coefficient may differ in coordinate directions. The three normal strains resulting from the thermal expansions are given as follows:

$$\varepsilon_{T11} = \alpha_1 \Delta T; \quad \varepsilon_{T22} = \alpha_2 \Delta T; \quad \varepsilon_{T33} = \alpha_3 \Delta T \tag{5.86}$$

where α_i is the thermal expansion coefficient of the material in the ith coordinate direction. No shear strain will be generated from the thermal expansions.

For isotropic materials, α_i is set as α.

5.17.2 Strain-stress-temperature relation

The constitutive equation Eq. (5.60) is rewritten as

$$\underbrace{\begin{bmatrix} \varepsilon_{11} \\ \varepsilon_{22} \\ \varepsilon_{33} \\ 2\varepsilon_{23} \\ 2\varepsilon_{13} \\ 2\varepsilon_{12} \end{bmatrix}}_{\vec{\varepsilon}} = \underbrace{\begin{bmatrix} \frac{1}{E_1} & \frac{-\nu_{21}}{E_2} & \frac{-\nu_{31}}{E_3} & 0 & 0 & 0 \\ \frac{-\nu_{12}}{E_1} & \frac{1}{E_2} & \frac{-\nu_{32}}{E_3} & 0 & 0 & 0 \\ \frac{-\nu_{13}}{E_1} & \frac{-\nu_{23}}{E_2} & \frac{1}{E_3} & 0 & 0 & 0 \\ 0 & 0 & 0 & \frac{1}{G_{23}} & 0 & 0 \\ 0 & 0 & 0 & 0 & \frac{1}{G_{13}} & 0 \\ 0 & 0 & 0 & 0 & 0 & \frac{1}{G_{12}} \end{bmatrix}}_{\mathbf{S}} \underbrace{\begin{bmatrix} \sigma_{11} \\ \sigma_{22} \\ \sigma_{33} \\ \sigma_{23} \\ \sigma_{13} \\ \sigma_{12} \end{bmatrix}}_{\vec{\sigma}} + \underbrace{\begin{bmatrix} \alpha_1 \Delta T \\ \alpha_2 \Delta T \\ \alpha_3 \Delta T \\ 0 \\ 0 \\ 0 \end{bmatrix}}_{\varepsilon_T} \tag{5.87}$$

Or

$$\boxed{\vec{\varepsilon} = \mathbf{S}\vec{\sigma} + \varepsilon_T} \tag{5.88}$$

5.17.3 Stress-strain-temperature relation

Stress-strain-temperature relation can be rewritten as

$$\boxed{\vec{\sigma} = \mathbf{C}\left[\vec{\varepsilon} - \varepsilon_T\right]} \tag{5.89}$$

where $\mathbf{C} = \mathbf{S}^{-1}$.

The same concept and similar formulas of constitutive relations apply for all other cases discussed in previous section in this chapter. Hence, we will not repeat here.

5.18 Examples for elasticity tensor transformation*

5.18.1 Examples on arbitrary symmetric 6 by 6 matrix

First, we conduct tests on functions C2toC4() and C4toC2(C40) created earlier using an arbitrary symmetric 6 by 6 matrix. We create such a matrix:

```
1  # Produce a simple symmetric matrix for code testing purposes.
2  C_test = np.zeros((6,6))
3
4  for i in range(6):
5      for j in range(6):
6          C_test[i, j] = (i+1)*(j+1)
7
8  print(C_test, C_test.shape)
```

```
[[ 1.000e+00  2.000e+00  3.000e+00  4.000e+00  5.000e+00  6.000e+00]
 [ 2.000e+00  4.000e+00  6.000e+00  8.000e+00  1.000e+01  1.200e+01]
 [ 3.000e+00  6.000e+00  9.000e+00  1.200e+01  1.500e+01  1.800e+01]
 [ 4.000e+00  8.000e+00  1.200e+01  1.600e+01  2.000e+01  2.400e+01]
 [ 5.000e+00  1.000e+01  1.500e+01  2.000e+01  2.500e+01  3.000e+01]
 [ 6.000e+00  1.200e+01  1.800e+01  2.400e+01  3.000e+01  3.600e+01]] (6, 6)
```

```
1  # Test on reproducibility of conversion C(6,6) <-> C(3,3,3,3)
2  # for arbitrary symmetric matrix.
3  print(C_test,"\n")                           # Original C(6,6) matrix
4
5  C40 = C2toC4(C_test)                         # Convert C(6,6) to C(3,3,3,3)
6  C20 = C4toC2(C40)                            # Convert C(3,3,3,3) to C(6,6)
7
8  print(C20)                                   # Back-converted C(6,6) matrix
9  print(f"Is close? {np.allclose(lg.norm(C20),lg.norm(C_test),1  =1e-9)}")
```

```
[[ 1.000e+00  2.000e+00  3.000e+00  4.000e+00  5.000e+00  6.000e+00]
 [ 2.000e+00  4.000e+00  6.000e+00  8.000e+00  1.000e+01  1.200e+01]
 [ 3.000e+00  6.000e+00  9.000e+00  1.200e+01  1.500e+01  1.800e+01]
 [ 4.000e+00  8.000e+00  1.200e+01  1.600e+01  2.000e+01  2.400e+01]
 [ 5.000e+00  1.000e+01  1.500e+01  2.000e+01  2.500e+01  3.000e+01]
 [ 6.000e+00  1.200e+01  1.800e+01  2.400e+01  3.000e+01  3.600e+01]]

[[ 1.000e+00  2.000e+00  3.000e+00  4.000e+00  5.000e+00  6.000e+00]
 [ 2.000e+00  4.000e+00  6.000e+00  8.000e+00  1.000e+01  1.200e+01]
 [ 3.000e+00  6.000e+00  9.000e+00  1.200e+01  1.500e+01  1.800e+01]
 [ 4.000e+00  8.000e+00  1.200e+01  1.600e+01  2.000e+01  2.400e+01]
 [ 5.000e+00  1.000e+01  1.500e+01  2.000e+01  2.500e+01  3.000e+01]
 [ 6.000e+00  1.200e+01  1.800e+01  2.400e+01  3.000e+01  3.600e+01]]
Is close? True
```

The conversion forth and back worked fine.

Next, test C2toC4() and C4toC2() using a random transformation matrix. The testing steps are given in the comments to the code line.

```
1  # Test C2toC4() and C4toC2()
2  T, _ = gr.transferM(60., about = 'random')    # Transformation Matrix
3
4  C0 = C2toC4(C_test)                            # Convert C(6,6) to C(3,3,3,3)
5  Cp = Tensor4_transfer(T, C0)                   # Perform transformation
6
7  print(f"Is close? {np.allclose(lg.norm(C0),lg.norm(Cp),atol=1e-4)}")
```

Is close? True

The fourth-order tensor transformation did not change the norm. This is expected because of the unitary property of the transformation matrix.

```
1  C20 = C4toC2(Cp)                               # Convert C(3,3,3,3) to C(6,6)
2  print(C20)
3  print(f"Close? {np.allclose(lg.norm(C20),lg.norm(C_test),atol=1e-4)}")
```

```
[[ 1.699e-03  2.492e-01 -3.604e-03  2.809e-01  6.418e-02  1.692e-01]
 [ 2.492e-01  3.656e+01 -5.286e-01  4.120e+01  9.414e+00  2.482e+01]
 [-3.604e-03 -5.286e-01  7.644e-03 -5.958e-01 -1.361e-01 -3.590e-01]
 [ 2.809e-01  4.120e+01 -5.958e-01  4.644e+01  1.061e+01  2.798e+01]
 [ 6.418e-02  9.414e+00 -1.361e-01  1.061e+01  2.424e+00  6.392e+00]
 [ 1.692e-01  2.482e+01 -3.590e-01  2.798e+01  6.392e+00  1.686e+01]]
Close? False
```

The components in the matrix changed after the transformation, as they should be. The norm does not change.

```
1  C40 = C2toC4(C20)                              # Convert C(6,6) to C(3,3,3,3)
2  Cp = Tensor4_transfer(T.T, C40)                # Perform inverse transformation
3  C0 = C4toC2(Cp)                                #Convert C(3,3,3,3) to C(6,6) to print out
4
5  print(C0)
6  print(f"Close? {np.allclose(lg.norm(C0),lg.norm(C_test),atol=1e-4)}")
```

```
[[ 1.000e+00  2.000e+00  3.000e+00  4.000e+00  5.000e+00  6.000e+00]
 [ 2.000e+00  4.000e+00  6.000e+00  8.000e+00  1.000e+01  1.200e+01]
 [ 3.000e+00  6.000e+00  9.000e+00  1.200e+01  1.500e+01  1.800e+01]
 [ 4.000e+00  8.000e+00  1.200e+01  1.600e+01  2.000e+01  2.400e+01]
 [ 5.000e+00  1.000e+01  1.500e+01  2.000e+01  2.500e+01  3.000e+01]
 [ 6.000e+00  1.200e+01  1.800e+01  2.400e+01  3.000e+01  3.600e+01]]
Close? True
```

Using the codes above, we have confirmed the following.

- The transformed stiffness matrix will change, for a given general 6 by 6 symmetric matrix.
- The fourth-order tensor transformation does not change the norm.
- The inverse-transformed stiffness matrix recovers the original one.

5.18.2 Examples on elasticity matrices for steel

Steel can be treated as isotropic materials at macroscale. Let us compute its **C** matrix using the function E_SnC3D() defined earlier, and then compute its compliance matrix.

```
1  E_steel,v_steel=77.1e9,0.3333 # Young's modulus, GPa & Poisson's ratio
2  G_s = .5*E_steel/(1+v_steel)              # shear modulus (GPa)
3  ρ = 7600.                      # density (Kg/m^3) # not used here
4  print(f"Shear modulus = {G_s:2.4e}")
5
6  C_steel, S_steel = E_SnC3D(E_steel,E_steel,E_steel,\
7                             v_steel,v_steel,v_steel,G_s,G_s,G_s)
8  # (77.1e9,77.1e9,77.1e9,.3333,.3333,.3333,28.91e9,28.91e9,28.91e9)
9
10 S_steel_ = lg.inv(C_steel)
11
12 print(f"Matrix of elastic constants C {C_steel.shape}:\n{C_steel}")
13 print(f"Matrix of compliance constants S_{S_steel_.shape}:\n{S_steel_}")
14 print(f"Matrix of original S {S_steel.shape}:\n{S_steel}")
```

```
Shear modulus = 2.8913e+10
Matrix of elastic constants C (6, 6):
[[ 1.156e+11  5.781e+10  5.781e+10  0.000e+00  0.000e+00  0.000e+00]
 [ 5.781e+10  1.156e+11  5.781e+10  0.000e+00  0.000e+00  0.000e+00]
 [ 5.781e+10  5.781e+10  1.156e+11  0.000e+00  0.000e+00  0.000e+00]
 [ 0.000e+00  0.000e+00  0.000e+00  2.891e+10  0.000e+00  0.000e+00]
 [ 0.000e+00  0.000e+00  0.000e+00  0.000e+00  2.891e+10  0.000e+00]
 [ 0.000e+00  0.000e+00  0.000e+00  0.000e+00  0.000e+00  2.891e+10]]
Matrix of compliance constants S_(6, 6):
[[ 1.297e-11 -4.323e-12 -4.323e-12  0.000e+00  0.000e+00  0.000e+00]
 [-4.323e-12  1.297e-11 -4.323e-12  0.000e+00  0.000e+00  0.000e+00]
 [-4.323e-12 -4.323e-12  1.297e-11  0.000e+00  0.000e+00  0.000e+00]
 [ 0.000e+00  0.000e+00  0.000e+00  3.459e-11  0.000e+00  0.000e+00]
 [ 0.000e+00  0.000e+00  0.000e+00  0.000e+00  3.459e-11  0.000e+00]
 [ 0.000e+00  0.000e+00  0.000e+00  0.000e+00  0.000e+00  3.459e-11]]
Matrix of original S (6, 6):
[[ 1.297e-11 -4.323e-12 -4.323e-12  0.000e+00  0.000e+00  0.000e+00]
 [-4.323e-12  1.297e-11 -4.323e-12  0.000e+00  0.000e+00  0.000e+00]
 [-4.323e-12 -4.323e-12  1.297e-11  0.000e+00  0.000e+00  0.000e+00]
 [ 0.000e+00  0.000e+00  0.000e+00  3.459e-11  0.000e+00  0.000e+00]
 [ 0.000e+00  0.000e+00  0.000e+00  0.000e+00  3.459e-11  0.000e+00]
 [ 0.000e+00  0.000e+00  0.000e+00  0.000e+00  0.000e+00  3.459e-11]]
```

We prepare some transformation matrices for testing our codes for 4th tensor transformation.

Elasticity Tensor and Constitutive Equations

```
1  # Test on reproducibility of conversion C(6,6) <-> C(3,3,3,3)
2  # for real elastic material, using C2toC4() and C4toC2()
3
4  C4_steel = C2toC4(C_steel)        #Convert C(6,6) to C(3,3,3,3)
5  C2_steel = C4toC2(C4_steel)       #Convert C(3,3,3,3) to C(6,6)
6
7  print(C2_steel)                   #Back-converted C(6,6) matrix
8  print(np.linalg.norm(C2_steel) == np.linalg.norm(C_steel))
```

```
[[ 1.156e+11  5.781e+10  5.781e+10  0.000e+00  0.000e+00  0.000e+00]
 [ 5.781e+10  1.156e+11  5.781e+10  0.000e+00  0.000e+00  0.000e+00]
 [ 5.781e+10  5.781e+10  1.156e+11  0.000e+00  0.000e+00  0.000e+00]
 [ 0.000e+00  0.000e+00  0.000e+00  2.891e+10  0.000e+00  0.000e+00]
 [ 0.000e+00  0.000e+00  0.000e+00  0.000e+00  2.891e+10  0.000e+00]
 [ 0.000e+00  0.000e+00  0.000e+00  0.000e+00  0.000e+00  2.891e+10]]
True
```

The conversion forth and back worked correctly.

```
1  # Test C2toC4() and C4toC2() with a random transformation matrix
2  T, _ = gr.transferM(60., about = 'random' )    #transformation matrix
3
4  C4_steel = C2toC4(C_steel)                #Convert C(6,6) to C(3,3,3,3)
5  C4p_steel = Tensor4_transfer(T, C4_steel) # Perform transformation
6  print(f"Is close? "
7        f"{np.allclose(lg.norm(C4_steel),lg.norm(C4p_steel),atol=1e-4)}")
8
9  C2_steel = C4toC2(C4p_steel)              #Convert C(3,3,3,3) to C(6,6)
10 print(f"Close? "
11       f"{np.allclose(lg.norm(C2_steel),lg.norm(C_steel),atol=1e-4)}")
```

```
Is close? True
Close? True
```

The fourth-order tensor transformation did not change the norm. Let us check on forward and backward transformations.

```
1  # Test our 4th order tensor transformation, forward and backward:
2  print(C_steel,'\n')                       #Original C(6,6) matrix
3
4  C4_steel = C2toC4(C_steel)        # Convert C(6,6) to C(3,3,3,3)
5  C4p_steel= Tensor4_transfer(T,C4_steel)   # forward transformation
6  C2_steel = C4toC2(C4p_steel)      # Convert C(3,3,3,3) to C(6,6)
7
8  print(C2_steel)                           #Transformed stiffness matrix
9  print(f"Is C_steel symmetric?"
10       f" {np.allclose(C2_steel,C2_steel.T,atol=1e-04)}\n")
11
12 C4p_steel= C2toC4(C2_steel)               #Convert C(6,6) to C(3,3,3,3)
13 C4_steel = Tensor4_transfer(T.T,C4p_steel)   # back-transformation
14 C2_steel = C4toC2(C4_steel)       # Convert C(3,3,3,3) to C(6,6)
15
16 print(C2_steel)                   #Back-transformed stiffness matrix
17 print(f"Is C2 back to original? "
18       f"{np.allclose(C_steel,C2_steel,atol=1e-04)}")
```

```
[[ 1.156e+11  5.781e+10  5.781e+10  0.000e+00  0.000e+00  0.000e+00]
 [ 5.781e+10  1.156e+11  5.781e+10  0.000e+00  0.000e+00  0.000e+00]
 [ 5.781e+10  5.781e+10  1.156e+11  0.000e+00  0.000e+00  0.000e+00]
 [ 0.000e+00  0.000e+00  0.000e+00  2.891e+10  0.000e+00  0.000e+00]
 [ 0.000e+00  0.000e+00  0.000e+00  0.000e+00  2.891e+10  0.000e+00]
 [ 0.000e+00  0.000e+00  0.000e+00  0.000e+00  0.000e+00  2.891e+10]]

[[ 1.156e+11  5.781e+10  5.781e+10 -3.759e-06 -5.847e-06 -8.269e-05]
 [ 5.781e+10  1.156e+11  5.781e+10  1.931e-06 -9.491e-06 -8.362e-05]
 [ 5.781e+10  5.781e+10  1.156e+11 -1.174e-05 -3.947e-06 -3.801e-05]
 [-6.211e-06 -4.498e-06 -6.426e-06  2.891e+10 -2.286e-05 -4.059e-06]
 [-1.807e-06 -2.649e-06 -6.417e-06 -2.118e-05  2.891e+10 -2.093e-06]
 [-7.874e-05 -7.977e-05 -4.143e-05 -2.603e-06  2.734e-06  2.891e+10]]
Is C_steel symmetric? True

[[ 1.156e+11  5.781e+10  5.781e+10 -4.150e-05 -8.103e-05 -1.110e-04]
 [ 5.781e+10  1.156e+11  5.781e+10 -7.889e-05 -4.730e-05 -1.099e-04]
 [ 5.781e+10  5.781e+10  1.156e+11 -7.497e-05 -8.225e-05 -5.216e-05]
 [-4.130e-05 -7.285e-05 -7.307e-05  2.891e+10 -3.012e-05 -1.997e-05]
 [-7.656e-05 -3.787e-05 -8.583e-05 -2.465e-05  2.891e+10 -1.898e-05]
 [-1.224e-04 -1.161e-04 -5.348e-05 -2.686e-05 -2.175e-05  2.891e+10]]
Is C2 back to original? False
```

Using the codes above, we have confirmed the following.

- The transformed stiffness matrix will not change if the material is isotropic.

5.18.3 Examples on elasticity matrices for steel for 2D plane stress

```
1  E_steel,v_steel=77.1e9,0.3333   # Young's modulus,GPa & Poisson's ratio
2  G_s = .5*E_steel/(1+v_steel)
3  print(f"Shear modulus = {G_s:2.4e}")                # shear modulus (GPa)
4
5  C_steel, S_steel = E_SnC2D_stress(E_steel, E_steel, v_steel, G_s)
6  S_steel_ = lg.inv(C_steel)
7
8  print(f"Matrix of elastic constants C {C_steel.shape}:\n{C_steel}")
9  print(f"Matrix of compliance constants S_{S_steel_.shape}:\n{S_steel_}")
10 print(f"Matrix of original S {S_steel.shape}:\n{S_steel}")
```

```
Shear modulus = 2.8913e+10
Matrix of elastic constants C (3, 3):
[[ 8.674e+10  2.891e+10  0.000e+00]
 [ 2.891e+10  8.674e+10  0.000e+00]
 [ 0.000e+00  0.000e+00  2.891e+10]]
Matrix of compliance constants S_(3, 3):
[[ 1.297e-11 -4.323e-12  0.000e+00]
 [-4.323e-12  1.297e-11  0.000e+00]
 [ 0.000e+00  0.000e+00  3.459e-11]]
Matrix of original S (3, 3):
[[ 1.297e-11 -4.323e-12  0.000e+00]
 [-4.323e-12  1.297e-11  0.000e+00]
 [ 0.000e+00  0.000e+00  3.459e-11]]
```

5.18.4 *Examples on elasticity matrices for steel for 2D plane strain*

```
1  E_steel,v_steel=77.1e9,0.3333    # Young's modulus,GPa & Poisson's ratio
2  G_s = .5*E_steel/(1+v_steel)                  # shear modulus (GPa)
3  print(f"Shear modulus = {G_s:2.4e}")
4
5  C_steel, S_steel = E_SnC2D_strain(E_steel, E_steel, E_steel,\
6                                   v_steel, v_steel, v_steel, G_s)
7  S_steel_ = lg.inv(C_steel)
8
9  print(f"Matrix of elastic constants C {C_steel.shape}:\n{C_steel}")
10 print(f"Matrix of compliance constants S_{S_steel_.shape}:\n{S_steel_}")
11 print(f"Matrix of original S {S_steel.shape}:\n{S_steel}")
```

```
Shear modulus = 2.8913e+10
Matrix of elastic constants C (3, 3):
[[ 1.156e+11  5.781e+10  0.000e+00]
 [ 5.781e+10  1.156e+11  0.000e+00]
 [ 0.000e+00  0.000e+00  2.891e+10]]
Matrix of compliance constants S_(3, 3):
[[ 1.153e-11 -5.764e-12  0.000e+00]
 [-5.764e-12  1.153e-11  0.000e+00]
 [ 0.000e+00  0.000e+00  3.459e-11]]
Matrix of original S (3, 3):
[[ 1.153e-11 -5.764e-12  0.000e+00]
 [-5.764e-12  1.153e-11  0.000e+00]
 [ 0.000e+00  0.000e+00  3.459e-11]]
```

The compliance matrix obtained above can be used to double checked against Eq. (5.76):

```
1  print(f's11={(1.-v_steel**2)/E_steel}')
2  print(f's12={-v_steel*(1.+v_steel)/E_steel}')
3  print(f's33={2*(1.+v_steel)/E_steel}')
```

s11=1.1529326977950713e-11
s12=-5.763798832684825e-12
s33=3.4586251621271074e-11

5.18.5 Examples on elasticity matrices for carbon/epoxy composite**

Carbon/epoxy composite has a very strong anisotropy because of the reinforcement with carbon fibers. Let us compute the **C** matrix using the function E_SnC3D() defined above.

```
1  #Carbon-Reinforced Plastic Composite:(E1,E2,E3,m12,m13,m23,G23,G13,G12)
2  C_CRP, S_CRP = E_SnC3D(142.17e9,9.255e9,9.255e9,.334,.25,.25,\
3                         4.795e9,4.795e9,4.795e9)
4  rho = 1900.                                  # density (Kg/m^3)
5  print(f"Material stiffness  matrix:\n{C_CRP}")
6  print(f"Material compliance matrix:\n{S_CRP}")
```

```
Material stiffness  matrix:
[[ 1.443e+11  3.974e+09  3.342e+09  0.000e+00  0.000e+00  0.000e+00]
 [ 3.974e+09  9.981e+09  2.560e+09  0.000e+00  0.000e+00  0.000e+00]
 [ 3.342e+09  2.560e+09  9.949e+09  0.000e+00  0.000e+00  0.000e+00]
 [ 0.000e+00  0.000e+00  0.000e+00  4.795e+09  0.000e+00  0.000e+00]
 [ 0.000e+00  0.000e+00  0.000e+00  0.000e+00  4.795e+09  0.000e+00]
 [ 0.000e+00  0.000e+00  0.000e+00  0.000e+00  0.000e+00  4.795e+09]]
Material compliance matrix:
[[ 7.034e-12 -2.349e-12 -1.758e-12  0.000e+00  0.000e+00  0.000e+00]
 [-2.349e-12  1.080e-10 -2.701e-11  0.000e+00  0.000e+00  0.000e+00]
 [-1.758e-12 -2.701e-11  1.080e-10  0.000e+00  0.000e+00  0.000e+00]
 [ 0.000e+00  0.000e+00  0.000e+00  2.086e-10  0.000e+00  0.000e+00]
 [ 0.000e+00  0.000e+00  0.000e+00  0.000e+00  2.086e-10  0.000e+00]
 [ 0.000e+00  0.000e+00  0.000e+00  0.000e+00  0.000e+00  2.086e-10]]
```

```
1  # Test on reproducibility of conversion C(6,6) <-> C(3,3,3,3)
2  # for real elastic material
3
4  C4_CRP = C2toC4(C_CRP)                       # Convert C(6,6) to C(3,3,3,3)
5  C2_CRP = C4toC2(C4_CRP)                      # Convert C(3,3,3,3) to C(6,6)
6
7  print(C2_CRP)                                # Back-converted C(6,6) matrix
8  print(np.linalg.norm(C2_CRP) == np.linalg.norm(C_CRP))
```

```
[[ 1.443e+11  3.974e+09  3.342e+09  0.000e+00  0.000e+00  0.000e+00]
 [ 3.974e+09  9.981e+09  2.560e+09  0.000e+00  0.000e+00  0.000e+00]
 [ 3.342e+09  2.560e+09  9.949e+09  0.000e+00  0.000e+00  0.000e+00]
 [ 0.000e+00  0.000e+00  0.000e+00  4.795e+09  0.000e+00  0.000e+00]
 [ 0.000e+00  0.000e+00  0.000e+00  0.000e+00  4.795e+09  0.000e+00]
 [ 0.000e+00  0.000e+00  0.000e+00  0.000e+00  0.000e+00  4.795e+09]]
True
```

Conversion forth and back worked fine.

```
1  # Test C2toC4() and C4toC2()
2  T, _ = gr.transferM(60., about = 'random')
3
4  C4_CRP = C2toC4(C_CRP)                       #Convert C(6,6) to C(3,3,3,3)
5  C4p_CRP = Tensor4_transfer(T, C4_CRP) #4th order tensor transformation
6  C2_CRP = C4toC2(C4p_CRP)
7
8  print(C2_CRP)
9  print(f"Is C2 symmetric? {np.allclose(C2_CRP,C2_CRP.T,atol=1e-04)}\n")
10 print(f"Is close? "
11        f"{np.allclose(lg.norm(C4_CRP),lg.norm(C4p_CRP),atol=1e-4)}")
```

```
[[ 4.748e+10  8.336e+09  3.033e+10 -1.221e+10 -3.160e+10  1.387e+10]
 [ 8.336e+09  1.128e+10  7.002e+09 -2.512e+09 -5.005e+09  2.503e+09]
 [ 3.033e+10  7.002e+09  3.392e+10 -9.660e+09 -2.514e+10  1.101e+10]
 [-1.221e+10 -2.512e+09 -9.660e+09  8.846e+09  1.077e+10 -4.670e+09]
 [-3.160e+10 -5.005e+09 -2.514e+10  1.077e+10  3.127e+10 -1.191e+10]
 [ 1.387e+10  2.503e+09  1.101e+10 -4.670e+09 -1.191e+10  1.006e+10]]
Is C2 symmetric? True

Is close? True
```

The stiffness matrix is changed, but still symmetric, and the norm is unchanged.

```
1  # Test our 4th order tensor transformation
2  print(f"Original C_CRP: \n{C_CRP}\n")           #Original C(6,6) matrix
3
4  C4_CRP = Tensor4_transfer(T.T,C4p_CRP)          #Inverse transformation
5  C2_CRP = C4toC2(C4_CRP)                         #Convert C(3,3,3,3) to C(6,6)
6
7  print(f"After transformation:\n{C2_CRP}\n")     #Transformed matrix
8  print(f"Is C2 symmetric? {np.allclose(C2_CRP,C2_CRP.T,atol=1e-04)}\n")
9  print(f"Is C2 back to original? {np.allclose(C2_CRP,C_CRP,atol=1e-03)}")
```

```
Original C_CRP:
[[ 1.443e+11  3.974e+09  3.342e+09  0.000e+00  0.000e+00  0.000e+00]
 [ 3.974e+09  9.981e+09  2.560e+09  0.000e+00  0.000e+00  0.000e+00]
 [ 3.342e+09  2.560e+09  9.949e+09  0.000e+00  0.000e+00  0.000e+00]
 [ 0.000e+00  0.000e+00  0.000e+00  4.795e+09  0.000e+00  0.000e+00]
 [ 0.000e+00  0.000e+00  0.000e+00  0.000e+00  4.795e+09  0.000e+00]
 [ 0.000e+00  0.000e+00  0.000e+00  0.000e+00  0.000e+00  4.795e+09]]

After transformation:
[[ 1.443e+11  3.974e+09  3.342e+09  2.085e-06 -8.959e-06  2.150e-05]
 [ 3.974e+09  9.981e+09  2.560e+09 -6.372e-06 -4.398e-06  6.379e-06]
 [ 3.342e+09  2.560e+09  9.949e+09 -3.480e-06 -3.051e-06  6.145e-07]
 [-6.205e-07 -5.577e-06 -3.720e-06  4.795e+09  8.087e-07 -1.662e-07]
 [-1.384e-05 -4.524e-07 -1.816e-06  1.084e-07  4.795e+09 -4.319e-06]
 [ 2.431e-05  5.402e-07  2.402e-07  1.142e-06 -3.561e-07  4.795e+09]]

Is C2 symmetric? True

Is C2 back to original? True
```

We have confirmed the following:

- The transformed stiffness matrix will change if the material is anisotropic, such as the CRP composite.
- The transformation does not change the norm.
- The back-transformed stiffness matrix recovers the original one.

5.19 Remarks

1. Constitutive equation or stress-strain equations are dealt with in this chapter. The theoretically most general fourth-order tensor with 81 elastic constants is first introduced. The symmetry of the tensor is proven. The symmetry leads to 21 elastic constants for anisotropic solid materials. The fourth-order elasticity tensor can then be represented by a 6×6 matrix, which is much easier to use in engineering applications.
2. Procedure and codes are provided to reduce the fourth-order tensor form to symmetric matrix form that is more often used in engineering applications.
3. For isotropic elastic materials, the material constants become 2, which is most often encountered in engineering structures and devices. Major forms of constitutive equations and codes are thus provided.
4. The concept of stiffness matrix and compliance matrix are introduced. For stable solid materials (that do not flow), both matrices are symmetric positive definite (SPD), and hence one can be inverted to produce the other.

5. To obtain these material constants through experimental measurements, we use often the compliance matrix that contains explicitly Young's moduli, shear moduli, and Poisson's ratios.
6. Thermal effects can be taken into consideration via thermal strains in the constitutive equations.

The following chapter discusses equilibrium equations.

References

[1] A.P. Boresi, R.J. Schmidt and O.M. Sidebottom, *Advanced Mechanics of Materials*. John Wiley & Sons, New York, 1985.
[2] J.M. Gere and S.P. Timoshenko, *Mechanics of Materials*, Van Nostrand Reinhold Company, New York, 1972.
[3] S. Timoshenko and J.N. Goodier, *Theory of Elasticity*. McGraw Hill, New York, 1970. http://books.google.com/books?id=yFISAAAAIAAJ&dq=theory+of+elasticity&ei=ICiKSsr3G4jwkQSbxMyPCg.
[4] Z.L. Xu, *Elasticity*, Vols. 1&2. People's Publisher, China, 1979.
[5] G.R. Liu and S.S. Quek, *The Finite Element Method: A Practical Course*. Butterworth-Heinemann, New York, 2013.
[6] W. Michael, David Rubin and Erhard Krempl. *The Elastic Solid* [Chapter 5], 2010.
[7] Lev Davidovich Landau, Evegnii Mikhailovich Lifshitz, A.M. Kosevich et al., *Theory of Elasticity: Volume 7*, 1986.
[8] G.R. Liu and Z.C. Xi, *Elastic Waves in Anisotropic Laminates*. CRC Press, New York, 2001.
[9] T.T.C. Ting, *Anisotropic Elasticity: Theory and Applications*, 1996.
[10] A.P. Boresi, K. Chong and J.D. Lee, *Elasticity in Engineering Mechanics*. John Wiley & Sons, New York, 2010.
[11] G.R. Liu and T.T. Nguyen, *Smoothed Finite Element Methods*. Taylor and Francis Group, New York, 2010.
[12] G.R. Liu, *Mesh Free Methods: Moving Beyond the Finite Element Method*. Taylor and Francis Group, New York, 2010.

Chapter 6

Equilibrium Equations

```
1  # Place cursor in this cell, and press Ctrl+Enter to import dependences.
2  import sys                          # for accessing the computer system
3  sys.path.append('../grbin/')   # Change to the directory in your system
4
5  from commonImports import *        # Import dependences from '../grbin/'
6  import grcodes as gr               # Import the module of the author
7  importlib.reload(gr)               # When grcodes is modified, reload it
8
9  from continuum_mechanics import vector
10 init_printing(use_unicode=True)    # For latex-like quality printing
11 np.set_printoptions(precision=4,suppress=True)  # Digits in print-outs
```

6.1 Introduction

Consider a three-dimensional (3D) solid material, occupying a physical domain Ω bounded by surface S and defined in a 3D Cartesian coordinate. The solid is subjected to body forces **b** over Ω. The displacements of parts of the boundary S_u are constrained, so that the solid cannot have rigid body movements. The rest of the boundary S_t is subjected to tractions **t** that are given or zero, and $S = S_u \cup S_t$, as shown in Fig. 6.1.

In the previous chapters, we discussed where the stresses are coming from, how strains relate to displacements, and then how the stress and strain are related. We need now one more important piece: how the stress and forces are working together to satisfy the equilibrium conditions. This chapter provides this piece. The conclusion is that the divergence of the stresses at any point in solid material is compensated by the body force distributed in the solid, leading to the **equilibrium equation** in the form of partial differential equations (PDEs). It governs the behavior of the loaded solid material inside its domain and is hence known as the **governing equation**.

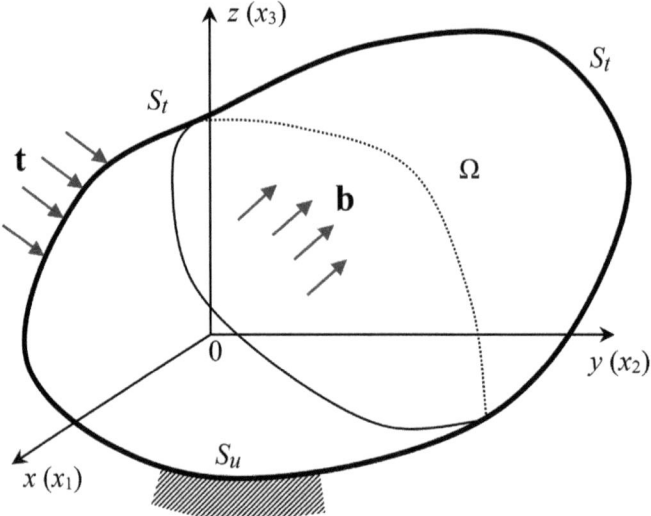

Figure 6.1. A 3D solid constrained on the displacement boundary S_u. It is loaded by body forces **b** over the domain and tractions **t** over S_t.

At the end of this chapter, the different types of boundary conditions (BCs) will also be discussed. The BCs must be satisfied when solving the governing equation for loaded solids.

6.2 Equilibrium equation in terms of stresses

Stresses in loaded solid materials will in general vary from location to location. In a coordinate system, they will be functions of coordinates. Therefore, a stress tensor becomes a tensor function. Such variations shall relate to the equilibrium state at any point within the loaded solid. We examine such variations, leading to one of the most important equations governing the behavior of the solid: the equilibrium equation.

6.2.1 3D free-body diagram

Consider a solid material subjected to external forces, resulting in stresses distributed over the solids, in a way detailed in Chapter 3. Assume that the solid media is continuous and the stresses are sufficiently smooth, and hence their functions are **differentiable** at least up to first order.

We first derive the equilibrium based on the laws of mechanics: The sum of all the forces at any point in the solid must vanish. To do this, we isolate

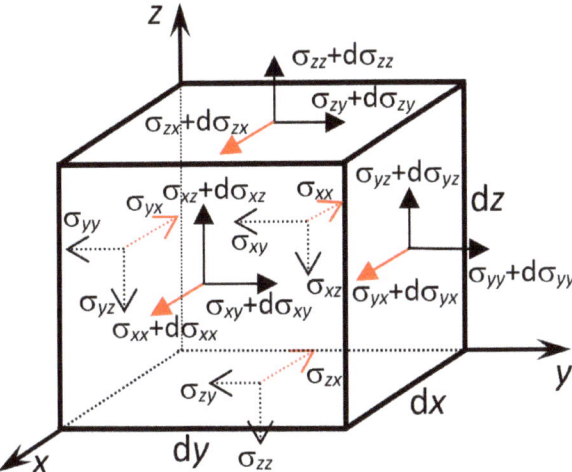

Figure 6.2. A representative infinitely small brick from a solid with dimensions of dx, dy, and dz. It serves as a typical 3D free-body diagram. All stresses are shown on the all the six surfaces of the free-body diagram, following the standard sign convention. Stresses in the x-direction are highlighted in red. Since the stresses are functions, a change in coordinates results in a change in the stresses.

a small representative cell from the solid, which is a 3D free-body diagram, as shown in Fig. 6.2. The directions of all these surfaces of the brick are in x-, y-, or z-directions.

6.2.2 The standard sign convention

The **standard sign convention** defined in this book is as follows:

1. Positive displacement components are in the positive axis direction.
2. Positive rotation or twist with respect to an axis is counterclockwise about the axis.
3. All the externally applied forces point to the positive axis direction.
4. All the externally applied moments or torques with respect to an axis are counterclockwise with respect to the axis.
5. Positive internal forces or stresses on a surface follow the direction of the surface.
6. Positive internal moments or toques on a surface are with respect to the direction of the surface.

The stress components drawn in Fig. 6.2 follow exactly this convention, which will be used in all the chapters.

6.2.3 Equilibrium equation in x-direction

Let us now consider only, as an example, all the stress components in the x-direction. The sum of the forces resulting from these stress components and the externally applied body force must vanish. We thus have

$$(\sigma_{xx} + d\sigma_{xx})dydz - \sigma_{xx}dydz$$
$$+ (\sigma_{yx} + d\sigma_{yx})dxdz - \sigma_{yx}dxdz \qquad (6.1)$$
$$+ (\sigma_{zx} + d\sigma_{zx})dxdy - \sigma_{zx}dxdy + b_x dxdydz = \rho \ddot{u}_x dxdydz$$

where b_x is the x component of the external body-force vector **b**, which may be a function of the coordinates and time t, ρ is the density of the solid material, and $\ddot{u}_x$ is the x component of the acceleration due to **b** changing with time, which is also a function of the coordinates.

Here, we used an overhead "˙" to represent the first derivative with respect to time, and thus "¨" becomes the second time derivative.

Equation (6.1) can be simplified to

$$d\sigma_{xx}dydz + d\sigma_{yx}dxdz + d\sigma_{zx}dxdy + b_x dxdydz = \rho \ddot{u}_x dxdydz \qquad (6.2)$$

Since the stress components are assumed differentiable, we have

$$d\sigma_{xx} = \frac{\partial \sigma_{xx}}{\partial x}dx; \quad d\sigma_{yx} = \frac{\partial \sigma_{yx}}{\partial y}dy; \quad d\sigma_{zx} = \frac{\partial \sigma_{zx}}{\partial z}dz \qquad (6.3)$$

Substituting Eq. (6.3) into Eq. (6.2) leads to

$$\frac{\partial \sigma_{xx}}{\partial x}dxdydz + \frac{\partial \sigma_{yx}}{\partial y}dydxdz + \frac{\partial \sigma_{zx}}{\partial z}dzdxdy + b_x dxdydz = \rho \ddot{u}_x dxdydz \qquad (6.4)$$

which can be simplified (the volume $dxdydz$ is nonzero) to

$$\frac{\partial \sigma_{xx}}{\partial x} + \frac{\partial \sigma_{yx}}{\partial y} + \frac{\partial \sigma_{zx}}{\partial z} + b_x = \rho \ddot{u}_x \qquad (6.5)$$

This is the equation obtained by enforcing the equilibrium of all these forces in the x-direction.

Note that all three stress components in Eq. (6.5) are the ones on three different surfaces but in the same x-direction. Thus, we can form the following vector:

$$\sigma_x = \begin{bmatrix} \sigma_{xx} \\ \sigma_{yx} \\ \sigma_{zx} \end{bmatrix} \qquad (6.6)$$

These stress components are on, respectively, x, y, and z surfaces and can vary with the coordinates. The sum of their derivatives $\frac{\partial \sigma_{xx}}{\partial x} + \frac{\partial \sigma_{yx}}{\partial y} + \frac{\partial \sigma_{zx}}{\partial z}$ (with respect to the coordinate variations) forms the so-called **divergence** of the stress vector. It is the spatial rate of losses of all stresses in the x-direction. It must be **compensated** by the external force (density) in the same x-direction in order to satisfy the equilibrium, which is Eq. (6.5). Note that we treat the inertial force as an externally applied force but with a negative sign.

6.2.4 Equilibrium in y- and z-directions

The same can be done for the forces in the y- and z-directions, which gives

$$\frac{\partial \sigma_{xy}}{\partial x} + \frac{\partial \sigma_{yy}}{\partial y} + \frac{\partial \sigma_{zy}}{\partial z} + b_y = \rho \ddot{u}_y$$

$$\frac{\partial \sigma_{xz}}{\partial x} + \frac{\partial \sigma_{yz}}{\partial y} + \frac{\partial \sigma_{zz}}{\partial z} + b_z = \rho \ddot{u}_z$$

(6.7)

Equations (6.5) and (6.7) are the set of three equilibrium equations of 3D solids subjected to external body forces b_i, $i = x, y, z$, each of which can be a function of the coordinates.

6.2.5 Equilibrium equation in indicial notation

In indicial notation, this set of equilibrium equations can be written simply as

$$\sigma_{ji,j} + b_i = \rho \ddot{u}_i \quad \in \Omega \tag{6.8}$$

This is the governing equation at any point in a loaded solid. Note that the contraction is on the first index of the stress tensor (standing for surfaces). The unknowns in the governing equation are these six stress components.

We assume that the application of all the forces is sufficiently slow and hence has no dynamic effects. For such a **static problem**, we simply drop the inertia term on the right-hand side. Equation (6.8) becomes

$$\sigma_{ji,j} + b_i = 0 \quad \in \Omega \tag{6.9}$$

In this book, our focus is on static problems.

6.2.6 Equilibrium equation in 2D and 1D

For 2D problems in the x–y plane, for example, the equilibrium equation can be obtained by simply removing all terms related to z, which gives

$$\frac{\partial \sigma_{xx}}{\partial x} + \frac{\partial \sigma_{yx}}{\partial y} + b_x = \rho \ddot{u}_x$$
$$\frac{\partial \sigma_{xy}}{\partial x} + \frac{\partial \sigma_{yy}}{\partial y} + b_y = \rho \ddot{u}_y \qquad (6.10)$$

Equation (6.9) still holds for 2D problems, but the range of i and j is from 1 to 2.

For 1D problems, the equilibrium equation is simply

$$\frac{\partial \sigma_{xx}}{\partial x} + b_x = \rho \ddot{u} \qquad (6.11)$$

which is the equation we derived at the beginning of Chapter 3 (when omitting the inertia term). For 1D cases, we may drop the indexes when all variables are treated as scalars.

6.2.7 Equilibrium equation via divergence of stress tensor function

In the previous section, we found that the divergence of the vector function becomes a single function of coordinates. Expanding this finding, we may expect that a divergence of a tensor function produces a vector function. We also found that a stress vector divergence is directly related to the body force. We shall expect that the stress tensor divergence relates to the body force in all directions, leading to the set of equilibrium equations, but in a pure mathematical way.

Consider a stress tensor that can be represented by a 3×3 matrix in space $\mathbb{R}^3$. We first define the stress tensor symbolically as follows:

```
1  Sxx, Sxy, Sxz = symbols("\u03C3_xx \u03C3_xy \u03C3_xz", cls=Function)
2  Syx, Syy, Syz = symbols("\u03C3_yx \u03C3_yy \u03C3_yz", cls=Function)
3  Szx, Szy, Szz = symbols("\u03C3_zx \u03C3_zy \u03C3_zz", cls=Function)
```

Since each of the stress components could be a function of the coordinates, we now define a tensor function as a matrix of functions:

```
1  x, y, z = symbols("x, y, z")
2  X = Matrix([x, y, z])
3
4  Sf = Matrix([[Sxx(x, y, z), Sxy(x, y, z), Sxz(x, y, z)],
5               [Syx(x, y, z), Syy(x, y, z), Syz(x, y, z)],
6               [Szx(x, y, z), Szy(x, y, z), Szz(x, y, z)]])
7  Sf                        # a tensor of 9 functions: stress tensor function
```

$$\begin{bmatrix} \sigma_{xx}(x,y,z) & \sigma_{xy}(x,y,z) & \sigma_{xz}(x,y,z) \\ \sigma_{yx}(x,y,z) & \sigma_{yy}(x,y,z) & \sigma_{yz}(x,y,z) \\ \sigma_{zx}(x,y,z) & \sigma_{zy}(x,y,z) & \sigma_{zz}(x,y,z) \end{bmatrix}$$

Note that a column in the stress matrix is a vector of stresses in a direction. Let us now take the divergence of the tensor function:

```
1  def div_tensorF(σ, X):
2      '''Compute the divergence of a tensor function σ.
3         input: σ, tensor function in sp.Matrix
4         return: the divergence of a tensor function
5      '''
6      n = σ.shape[0]
7      div_tf = [(sp.Add(*[sp.diff(σ[i,j], X[i]) for i in range(n)]))\
8                for j in range(n)]
9
10     return sp.Matrix(div_tf)        # divergence of a tensor function
11
```

```
1  div_Sf = div_tensorF(Sf, X)         # Or use: vector.div_tensor(Sf.T)
2  div_Sf
```

$$\begin{bmatrix} \frac{\partial}{\partial x}\sigma_{xx}(x,y,z) + \frac{\partial}{\partial y}\sigma_{yx}(x,y,z) + \frac{\partial}{\partial z}\sigma_{zx}(x,y,z) \\ \frac{\partial}{\partial x}\sigma_{xy}(x,y,z) + \frac{\partial}{\partial y}\sigma_{yy}(x,y,z) + \frac{\partial}{\partial z}\sigma_{zy}(x,y,z) \\ \frac{\partial}{\partial x}\sigma_{xz}(x,y,z) + \frac{\partial}{\partial y}\sigma_{yz}(x,y,z) + \frac{\partial}{\partial z}\sigma_{zz}(x,y,z) \end{bmatrix}$$

```
1  div_Sf.shape                        # it should be a vector
```

(3, 1)

It now becomes a vector of length 3, as expected. Each component in this vector is the divergence of the vector of stresses in one of the three directions. This divergence is the spatial rate of loss of the stresses in the direction at any point in the solid, which should be compensated by the body force applied at that point. Their sum shall vanish for the solid staying in equilibrium. This is the equilibrium equation in terms of the stress components and given body forces:

$$\begin{bmatrix} \frac{\partial}{\partial x}\sigma_{xx}(x,y,z) + \frac{\partial}{\partial y}\sigma_{yx}(x,y,z) + \frac{\partial}{\partial z}\sigma_{zx}(x,y,z) \\ \frac{\partial}{\partial x}\sigma_{xy}(x,y,z) + \frac{\partial}{\partial y}\sigma_{yy}(x,y,z) + \frac{\partial}{\partial z}\sigma_{zy}(x,y,z) \\ \frac{\partial}{\partial x}\sigma_{xz}(x,y,z) + \frac{\partial}{\partial y}\sigma_{yz}(x,y,z) + \frac{\partial}{\partial z}\sigma_{zz}(x,y,z) \end{bmatrix} + \begin{bmatrix} b_x \\ b_y \\ b_z \end{bmatrix} = \rho \begin{bmatrix} \ddot{u}_x \\ \ddot{u}_y \\ \ddot{u}_z \end{bmatrix} \quad (6.12)$$

which is the same as Eqs. (6.5) and (6.7). Note again that the inertia force is treated as a body force (with a negative sign).

> **Remark.** The divergence of the stress vector (formed by stress components in one of the x-, y-, and z-directions) at any point in the solid, together with the applied body force at that point, must vanish to ensure equilibrium. In other words, **the divergence of the stress tensor function together with the body-force vector shall vanish for a solid staying in equilibrium.**

Finally, we mention that when working with other coordinate systems, deriving the equilibrium equations using the technique given in Section 6.2 can be quite challenging and prone to error. The use of divergence of stress tensor can be much easier in one code line, which can be done in future volumes of the book series when working with other curvilinear coordinate systems for more general high-dimensional solid mechanics problems.

6.3 Equilibrium equations in displacements

6.3.1 *General formulation in 3D**

In Chapter 4, we derived a set of equations for small strains in terms of the derivatives of the displacements. In indicial notation, it is given as

$$\varepsilon_{ij} = \frac{1}{2}\left(u_{i,j} + u_{j,i}\right) \quad (6.13)$$

In Chapter 5, we introduced the constitutive equation for elastic solids that can be given in the most general form with a fourth-order elasticity tensor:

$$\sigma_{ij} = C_{ijkl}\varepsilon_{kl} \quad (6.14)$$

We shall now put all these together with Eq. (6.9) to derive the equilibrium equations in terms of displacements that have only three unknown displacement functions instead of six unknowns in Eq. (6.9). First, substituting Eqs. (6.13) into (6.14) gives

$$\sigma_{ij} = \frac{1}{2}C_{ijkl}(u_{k,l} + u_{l,k}) \qquad (6.15)$$

Since $C_{ijkl} = C_{ijlk}$ (due to the symmetry of the strain tensor discussed in Chapter 5), the foregoing equation becomes

$$\sigma_{ij} = \frac{1}{2}C_{ijkl}(u_{k,l} + u_{l,k}) = \frac{1}{2}[C_{ijkl}u_{k,l} + C_{ijkl}u_{l,k}]$$
$$= \frac{1}{2}[C_{ijkl}u_{k,l} + C_{ijlk}u_{k,l}] = \frac{1}{2}[C_{ijkl}u_{k,l} + C_{ijkl}u_{k,l}] \qquad (6.16)$$

or

$$\sigma_{ij} = C_{ijkl}u_{k,l} \qquad (6.17)$$

We obtain the stresses in terms of the displacements. Next, substituting Eq. (6.17) into Eq. (6.9), we arrive at

$$C_{ijkl}u_{k,lj} + b_i = \rho\ddot{u}_i \quad \in \Omega \qquad (6.18)$$

For static problems, Eq. (6.18) becomes

$$C_{ijkl}u_{k,lj} + b_i = 0 \quad \in \Omega \qquad (6.19)$$

This is the equilibrium equation purely in displacements. It is rather a quite complicated set of PDEs because of the fourth-order tensor. It can be reduced using the Viot notation for specific types of materials, as discussed in detail in the previous chapter.

For isotropic materials, it was found that the fourth-order tensor has the following form:

$$C_{ijkl} = G(\delta_{ik}\delta_{jl} + \delta_{il}\delta_{jk}) + \lambda\delta_{ij}\delta_{kl} \qquad (6.20)$$

Substituting Eq. (6.20) into (6.17), we obtain stresses in terms of the displacements:

$$\sigma_{ij} = G(\delta_{ik}\delta_{jl}u_{k,l} + \delta_{il}\delta_{jk}u_{k,l}) + \lambda\delta_{ij}\delta_{kl}u_{k,l}$$
$$= G(u_{i,j} + u_{j,i}) + \lambda\delta_{ij}u_{k,k} \qquad (6.21)$$

Using Eq. (6.8), we obtain the equilibrium equation in terms of the displacements:

$$Gu_{i,jj} + (G + \lambda)u_{j,ji} + b_i = \rho\ddot{u}_i \qquad (6.22)$$

If we write Eq. (6.22) explicitly, it has the form of

$$G\left(\frac{\partial^2 u}{\partial x^2} + \frac{\partial^2 u}{\partial y^2} + \frac{\partial^2 u}{\partial z^2}\right) + (G+\lambda)\left(\frac{\partial^2 u}{\partial x^2} + \frac{\partial^2 v}{\partial x \partial y} + \frac{\partial^2 w}{\partial x \partial z}\right) + b_x = \rho \ddot{u}$$

$$G\left(\frac{\partial^2 v}{\partial x^2} + \frac{\partial^2 v}{\partial y^2} + \frac{\partial^2 v}{\partial z^2}\right) + (G+\lambda)\left(\frac{\partial^2 u}{\partial x \partial y} + \frac{\partial^2 v}{\partial y^2} + \frac{\partial^2 w}{\partial y \partial z}\right) + b_y = \rho \ddot{v}$$

$$G\left(\frac{\partial^2 w}{\partial x^2} + \frac{\partial^2 w}{\partial y^2} + \frac{\partial^2 w}{\partial z^2}\right) + (G+\lambda)\left(\frac{\partial^2 u}{\partial x \partial z} + \frac{\partial^2 v}{\partial y \partial z} + \frac{\partial^2 w}{\partial z^2}\right) + b_z = \rho \ddot{w}$$

(6.23)

This is truly a set of "monster" equations. Finding an analytic solution is difficult, except for a very few drastically simplified problems.

6.3.2 Formulations of 2D problems

There is a class of problems in solid mechanics having special features which allow us to simplify it from 3D to 2D. Such 2D problems are of two major types: plane stress and plane strain.

6.3.2.1 Plane strain problems

When one of the displacement components, say w, is zero, we have a **plane strain** problem. In this case, all the nonzero strains stay within the x–y plane, which are called in-plane strains. In this case, all the variables are independent of z. A typical case is a dam built to block water flow, as shown in Fig. 6.3.

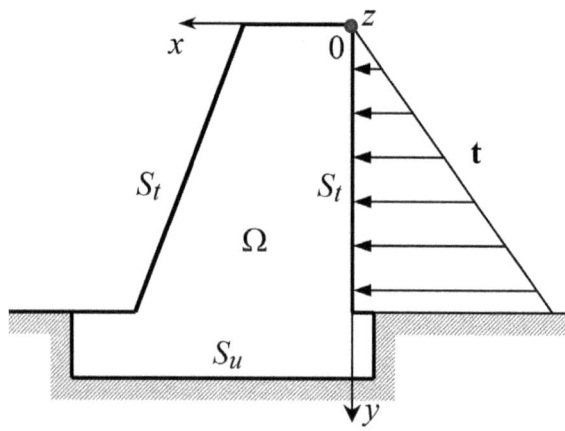

Figure 6.3. The cross-section of a water dam subjected to water pressure.

Since the dam is very long along the z-direction, it is rigid in the z-direction. In addition, the water pressure does not change in the z-direction, and w can be regarded as zero. Therefore, the problem can be simplified as 2D by considering only the 2D cross-section of the dam. Equation (6.23) can be reduced to

$$G\left(\frac{\partial^2 u}{\partial x^2} + \frac{\partial^2 u}{\partial y^2}\right) + (G+\lambda)\left(\frac{\partial^2 u}{\partial x^2} + \frac{\partial^2 v}{\partial x \partial y}\right) + b_x = \rho\ddot{u}$$
$$G\left(\frac{\partial^2 v}{\partial x^2} + \frac{\partial^2 v}{\partial y^2}\right) + (G+\lambda)\left(\frac{\partial^2 v}{\partial y^2} + \frac{\partial^2 u}{\partial x \partial y}\right) + b_y = \rho\ddot{v}$$

(6.24)

For plane strain problems, the stress in the z-direction is not necessarily zero and can be obtained from the in-plane stresses, as discussed in the previous chapter.

6.3.2.2 Plane stress problems

When the stress components in one of the coordinate directions, say in the z-direction, are zero, we have a **plane stress** problem. In such a case, all the nonzero stresses stay in the x–y plane, and all the variables also do not depend on z. A typical case is a thin plate. Its thickness in the z-direction is small compared to that in its other dimensions. In addition, all the external forces on it is only within the x–y plane. Since the two surfaces of the thin plate are free of stresses, all the stress components in the z-direction can be regarded as zero, as shown in Fig. 6.4.

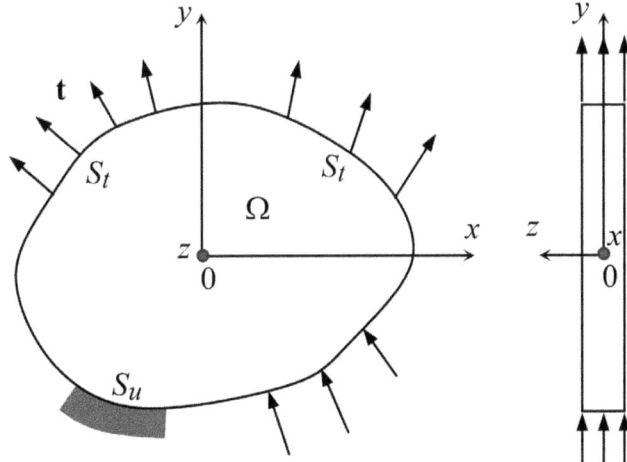

Figure 6.4. The cross-section of a thin plate subjected to loadings only within the plane of the plate.

To obtain the governing equations for 2D plane stress problems, we simply replace $\lambda = \frac{2G\nu}{1-2\nu}$ in Eq. (6.23) with $\frac{2G\nu}{1-\nu}$ because ν needs to be replaced by $\frac{\nu}{1+\nu}$ when converting a plane strain problem into a plane stress one, as discussed in the previous chapter. After such replacements followed by simplification, we obtain the equilibrium equations for 2D plane stress problems:

$$G\left(\frac{\partial^2 u}{\partial x^2} + \frac{\partial^2 u}{\partial y^2}\right) + G\frac{1+\nu}{1-\nu}\left(\frac{\partial^2 u}{\partial x^2} + \frac{\partial^2 v}{\partial x \partial y}\right) + b_x = \rho \ddot{u}$$

$$G\left(\frac{\partial^2 v}{\partial x^2} + \frac{\partial^2 v}{\partial y^2}\right) + G\frac{1+\nu}{1-\nu}\left(\frac{\partial^2 v}{\partial y^2} + \frac{\partial^2 u}{\partial x \partial y}\right) + b_y = \rho \ddot{v}$$

(6.25)

The strain in the z-direction, ε_{zz}, is not zero because the thickness of the plate can change when it is loaded in the x-y plane due to Poisson's effect. It can be calculated using these in-plane strains.

The difference between these two types of 2D problems is only in the material constants at the front of the derivative terms. Therefore, they can be dealt with in largely the same way. Plane strain problems behave stiffer than the plane stress ones because of the displacement constraints on the z-direction.

Note that even for 2D problems, seeking an analytical solution is still a challenge. Solution can be found only feasible for some simple problems with regular geometry.

6.3.3 Formulations of 1D bar problems

Consider one-dimensional (1D) problems defined in the x-coordinate. We can have three types:

1. 1D yz-constrained (both y- and z-directions are constrained),
2. 1D y-constrained (or z-constrained),
3. 1D bar (with a free circumferential surface).

6.3.3.1 1D yz-constrained problem

In this case, the displacements v and w are zero. The loading does not change in the y- and z-directions. The nonzero strain is only ε_{xx}. All the variables do not change with y and z. This is a 1D bar with all its circumferential surfaces fully constrained. In this case, Eq. (6.23) is drastically reduced by simply removing all terms with respect to y and z:

$$(2G + \lambda)\frac{\partial^2 u}{\partial x^2} + b_x = \rho \ddot{u} \qquad (6.26)$$

or

$$\frac{E(1-\nu)}{(1-2\nu)(1+\nu)}\frac{\partial^2 u}{\partial x^2} + b_x = \rho \ddot{u} \qquad (6.27)$$

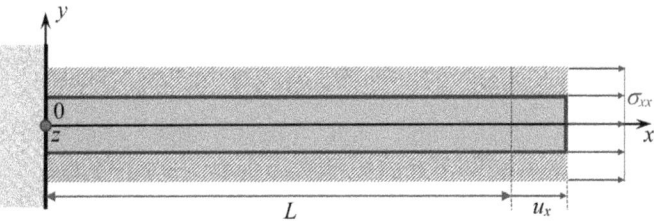

Figure 6.5. A 1D bar constrained in the y-direction. It becomes a 1D strain problem: the nonzero strains are in the x- and z-directions due to Poisson's effect.

6.3.3.2 1D y-constrained problem

If the displacement, say v, is zero, for example, if a 2D domain is infinite in the y-direction, and also the loading does not change in the y-direction, as shown in Fig. 6.5, we then have a 1D y-constrained problem. The nonzero strain is ε_{xx} with $\varepsilon_{zz} = -\nu\varepsilon_{xx}$. All the variables do not change with y.

In this case, Eq. (6.25) is further reduced to

$$G\left(1 + \frac{1+\nu}{1-\nu}\right)\frac{\partial^2 u}{\partial x^2} + b_x = \rho\ddot{u} \tag{6.28}$$

or

$$\frac{E}{1-\nu^2}\frac{\partial^2 u}{\partial x^2} + b_x = \rho\ddot{u} \tag{6.29}$$

The problem defined in Eq. (6.29) is a **1D y-constrained problem**. In this case, the stress σ_{yy} is not zero because the bar is not allowed to shrink or expand freely in the y-direction, resulting in some stress σ_{yy}. Also, $\varepsilon_{zz} = -\nu\varepsilon_{xx}$ is nonzero because the z-direction is not constrained. This problem is equivalent to the **1D z-constrained problem**, where the stress σ_{zz} is not zero because the bar is not allowed to shrink or expand freely in the z-direction, and $\varepsilon_{yy} = \nu\varepsilon_{xx}$ is nonzero because the y-direction is not constrained.

6.3.3.3 1D bar problem

If the 1D problem is a longitudinal bar with its circumferential surface free from any stress, we have a **1D stress problem**. This is the case we studied at the beginning of Chapter 3, and we derived the governing equation in terms of stress. To obtain its governing equation in displacement, we would need to, once again, do a substitution: replacing $\frac{E}{1-\nu^2}$ with E in Eq. (6.29), which gives

$$E\frac{\partial^2 u}{\partial x^2} + b_x = \rho\ddot{u} \tag{6.30}$$

Note in this case that the strains ε_{yy} and ε_{zz} are not zero because the bar is allowed to shrink or expand freely in the y- and z-directions (Poisson's effect).

Equations (6.27), (6.29), and (6.30) are 1D deferential equations (DEs) that can be solved with ease for many problems. Comparing Eqs. (6.29) and (6.30), we find that the y-constraint makes the bar stiffer by a factor of $\frac{1}{1-\nu^2}$. If $\nu = 0.3$, the factor is about 1.1, or $\sim$10% stiffer. Comparing Eqs. (6.27) and (6.30), we find that the yz-constraint makes the bar stiffer by a factor of $\frac{(1-\nu)}{(1-2\nu)(1+\nu)}$. If $\nu = 0.3$, the factor is about 1.35, or $\sim$35% stiffer. The constraint effects are only because of Poisson's ratio. Such a stiffing effect disappears only if Poisson's ratio is zero.

Since Eqs. (6.27) and (6.29) differ from Eq. (6.30) by a material constant, we only need to solve the latter because it is far more common in engineering applications. We will do so in great detail in the following chapter.

6.4 Boundary conditions

The equilibrium equations (in either stresses or displacements) govern the solid material inside the domain, as shown in Eq. (6.19). In actual engineering problems of mechanics of solid materials, there are always some kinds of conditions given at the boundary of the domain. These BCs must also be satisfied. In general, there are two major types of BCs for solids: displacement BCs and stress (or force) BCs.

6.4.1 *Displacement boundary conditions*

The displacement BCs can be expressed as follows:

$$u_i = u_{Si} \quad \text{on } S_u \in S \quad (6.31)$$

where u_{Si} denote the specified displacements u_i on the boundary surface S_u, which can be zero, but may not necessarily be.

6.4.2 *Stress boundary conditions*

The stress BCs can be expressed as follows:

$$\sigma_{ij}n_i = t_{nj} \quad \text{on } S_t \in S \quad (6.32)$$

where n_j is the jth component of the unit normal at a point on the boundary surface S_t and t_{nj} is the specified traction on the boundary S_t with the normal vector $\mathbf{n}$, which can be zero, but may not necessarily be. Equation (6.32) is essentially Eq. (3.38) used on the stress boundary.

6.5 On techniques for solving solid mechanics problems

To find a solution to a solid mechanics problem, we need to solve **a system of equations** consisting of the following:

- the **equilibrium equation** (also called the governing equation) Eq. (6.22);
- the **boundary conditions** Eqs. (6.31) and/or (6.32);
- the **compatibility equations** derived in Section 4.13.

Since the governing equation can be written in terms of stress, as in Eq. (6.9), or in terms of displacement, as in Eq. (6.19), we have largely two possible methods to solve this system of equations: the stress method and the displacement method. These methods are useful for higher-dimensional problems (2D and 3D). Each method has its own strategy, tricks, and strengths in solving different types of problems. We will discuss these methods in future volumes of this book series.

In this book, our objective is to solve problems for structural members, which can often be reduced to 1D. In these cases, the stress and displacement methods are essentially the same. We will use the displacement method in the later chapters for various types of problems because it is straightforward and easy to code to obtain systematic solutions for various types of problems, including indeterminate ones.

6.6 Remarks

1. This chapter presents a set of system equations for solid mechanics problems. Using a free-body diagram of a stressed 3D solid, we establish a set of equilibrium equations in all directions. It is a set of PDEs with the stress components as the unknowns. Thus, there are six unknowns.
2. Since the stress tensor is a function of coordinates, we also thus use the divergence of the stress tensor function together with the body-force vector to derive mathematically the same set of equilibrium equations.
3. The PDEs are then converted into ones with displacements as the unknowns. This reduces the number of unknowns to three.
4. The PDEs are reduced to 2D and 1D.
5. Both displacement and traction BCs are introduced. The BCs need to be satisfied in addition to the set of equilibrium PDEs.

Various solution methods will be introduced in future volumes on how to solve these sets of system equations for 2D and 3D solid mechanics problems.

The following chapters of this book will focus on problems in mechanics of materials, which can often be simplified into 1D problems or easily solvable

2D problems. Hence, efficient solvers can be devised for various problems involving elastic solid materials. Techniques on how to derive the simplified equations will be discussed in great detail, followed by solution procedures and Python codes, in later chapters.

References

[1] A.P. Boresi, R.J. Schmidt and O.M. Sidebottom, *Advanced Mechanics of Materials*, New York, 1985.
[2] G.R. Liu and S.S. Quek, *The Finite Element Method: A Practical Course*. Butterworth-Heinemann, New York, 2013.
[3] G.R. Liu and T.T. Nguyen, *Smoothed Finite Element Methods*. Taylor and Francis Group, New York, 2010.

Chapter 7

Solution to Axially Loaded Bars

```
1  # Place cursor in this cell, and press Ctrl+Enter to import dependences.
2  import sys                          # for accessing the computer system
3  sys.path.append('../grbin/')  # Change to the directory in your system
4
5  from commonImports import *    # Import dependences from '../grbin/'
6  import grcodes as gr           # Import the module of the author
7  importlib.reload(gr)           # When grcodes is modified, reload it
8
9  from continuum_mechanics import vector
10 init_printing(use_unicode=True)     # For Latex-like quality printing
11 np.set_printoptions(precision=4,suppress=True,
12         formatter={'float': '{:0.4e}'.format})   # Digits in print-outs
```

Bars or rods are the most widely used structural members. A bar takes axial loads that are applied along the axial direction of the bar. Solutions to it in terms of both axial displacement and stress are useful in many applications. A typical example is the widely used truss structures. Therefore, a bar is also referred as a truss member.

7.1 Governing equation: Second-order differential equation

Consider a bar subjected to a distributed body force b_f, as shown in Fig. 7.1. It is fixed at $x = 0$. At $x = l$, it may be subjected to a concentrated force F (force driven) or with a fixed given displacement (displacement driven). The problem is essentially one-dimensional (1D), implying that our concern is mainly with the unknown displacement and stress/strain varying with the coordinate x.

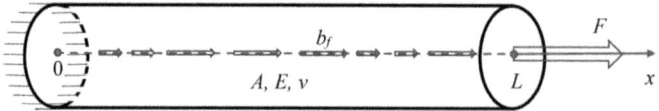

Figure 7.1. A 1D bar fixed at the left end, subjected to a distributed body force b_f [N/m] and a concentrated force F [N] at the right end.

The governing equation for 1D bar problems has been derived in Chapter 6 and is rewritten as follows in terms of displacement:

$$\frac{\partial^2 u}{\partial x^2} = \frac{-b_f}{EA} \qquad (7.1)$$

where b_f (N/m) is the body force within a 1D bar or truss member in the axial direction. For bars with uniform cross-sectional area A (m^2), it relates to the body force per volume b_x (N/m^3) as

$$b_f = b_x A \qquad (7.2)$$

Equation (7.1) is a second-order differential equation (DE). It can be solved easily by simply integrating twice. As long as the b_f function can be integrated analytically, we can obtain the exact solutions for bars with different types of boundary conditions (BCs). Since the DE is of second order, the integrations result in two unknown constants. Since the 1D bar always has two boundaries (left and right ends), which provide us the exactly needed number of BCs to determine these two constants, we have a logical path to find the exact solution effectively. Let us use the following example to demonstrate this process.

7.2 Solution procedure

First, we use an example to derive, by hand, the formulas for displacement and stress solutions, considering the simplest case where the body force is zero, so that we can have a good idea about how Eq. (7.1) can be solved. We will then write a Python code that can be used to derive formulas for more complicated problems. The code follows essentially the same process, but allows the computer to do the heavy lifting of derivation.

7.2.1 *Example: A bar subjected to a force at its right end*

Consider a bar with length L and a uniform cross-sectional area A. It is made of a material with Young's modulus E. It is fixed at the left end, where $x = 0$, and subjected to a force at its right end, denoted as F at $x = L$. Derive the formulas for displacement and stress in the bar.

Solution: Since there is no body force, Eq. (7.1) is reduced to

$$\frac{\partial^2 u}{\partial x^2} = 0 \tag{7.3}$$

The general solution of u that satisfies Eq. (7.3) can be found by integration. Integrating once gives

$$\frac{\partial u}{\partial x} = c_0 \tag{7.4}$$

where c_0 is an integral constant. Integrating again gives

$$u = c_0 x + c_1 \tag{7.5}$$

where c_1 is another integral constant. Equation (7.5) is the **general solution** for any bar without body force. It has a total of two constants. For a given specified problem defined in this example, we shall use the BCs to determine these constants.

At $x = 0$, the bar is fixed, and hence, we have the following BC:

$$u = 0 \quad \text{at } x = 0 \tag{7.6}$$

To satisfy Eq. (7.6), c_1 in Eq. (7.5) has to be zero, which leads to

$$u = c_0 x \tag{7.7}$$

At $x = L$, the bar is loaded by F, and hence, we have the following BC:

$$\sigma A = F, \quad \text{or} \quad AE\varepsilon = F, \quad \text{or} \quad AE\frac{\partial u}{\partial x} = F \quad \text{at } x = L \tag{7.8}$$

Substituting Eq. (7.7) into Eq. (7.8), we obtain

$$AEc_0 = F, \quad \text{or} \quad c_0 = \frac{F}{AE} \tag{7.9}$$

Finally, the displacement solution is found as

$$u = \frac{F}{AE} x \tag{7.10}$$

At $x = L$, we have displacement u_R as

$$\boxed{u_R = \frac{L}{AE} F} \tag{7.11}$$

where $\frac{AE}{L}$ is known as the **axial stiffness of a bar** and AE is the axial stiffness per length. In reverse, $\frac{L}{AE}$ is called the **axial flexibility of a bar**.

Equation (7.11) can also be written as

$$\boxed{F = \frac{AE}{L} u_R} \qquad (7.12)$$

The stress solution is next found as

$$\sigma = E\varepsilon = E\frac{\partial u}{\partial x} = \frac{F}{A} \qquad (7.13)$$

Clearly, this is the correct solution, as expected. This is in fact the solution we found at the beginning of Chapter 3. Here, we found it by using a standard solution procedure for DEs.

The same procedure can find solutions for bars subjected to body forces as long as the force function is integrable analytically. We shall use Sympy to do the derivation for us.

7.2.2 Python code for deriving solutions for bars

We write a Python code function to get it done in a step-by-step approach integrating Eq. (7.1), the same procedure that we did by hand in the previous example. The code is self-explanatory with comment lines:

```
1  def solver1D2(E, A, bf, l, bc0, bcl, key='disp'):
2
3      '''Solves mechanics problems for 1D bars subjected to body
4      force bf: u_xx=-bf(x)/EA, with BCs: u(0)=bc0; u(l)=bcl or fR=bcl.
5      Input: bf, body force; l, the length of the bar;
6              bcl, condition at x=l;
7              key="disp", displacement condition at a boundary;
8              otherwise a force boundary condition
9      Return: u (displacement), E*u_x (stress), u_xx (body force term)
10     Example: Consider a bar fixed at x=0, loaded with fR at x=L and q:
11             bf = q
12             u = solver1D2(E, A, bf, l, uL, fR, key='force')
13     '''
14     x, c0, c1 = sp.symbols('x, c0, c1')
15
16     # Integrate twice to obtain the general solutions in u_x and u:
17     # u_x: du/dx
18     u_x = sp.integrate(-bf/E/A,(x))+ c0   #c0: integration constant
19     u   = sp.integrate(u_x,(x))+ c1       #c1: integration constant
20
21     # Solve for the integration constants:
22     if key == "disp":
23         consts=sp.solve([u.subs(x,0)-bc0, u.subs(x,l)-bcl], [c0, c1])
24     else:
25         consts=sp.solve([u.subs(x,0)-bc0,u_x.subs(x,l)-bcl/E/A],[c0,c1])
26
```

```
27      # Substitute the constants back to the general solutions, to obtain
28      # solutions that satisfy the BCs:
29      u   =   u.subs({c0:consts[c0],c1:consts[c1]})
30      u   =   u.expand().simplify()
31      u_x = u_x.subs({c0:consts[c0],c1:consts[c1]})
32      u_x = u_x.expand().simplify()
33      u_xx = sp.diff(u_x,x)
34
35      print(f'output from solver1D2(): u, σ, u_xx')
36      return u.expand(), (E*u_x).expand(), u_xx.expand()
```

Let us use this code function to find solutions for two 1D problems: one with force boundary condition on the right end and the other with a displacement condition there. For both cases, the bar may be subjected to a given body force. We fix the bar at $x=0$ because it is needed to have our mechanics problem **well defined**, indicating that it has a physically meaningful solution with nonzero strains or stresses. By "fix", we mean that the displacement is pre-described with a fixed value that can be zero or a finite value.

7.3 Example: Constant body force and concentrated force at the right end

Consider a bar subjected to a constant body force and a concentrated force at $x=l$, all in its axial direction, as shown in Fig. 7.1. We compute the full solution: axial displacement, strain, and stress in the bar.

7.3.1 Method 1: Step-by-step approach integration

```
1  E, A, l = symbols('E, A, l', nonnegative=True)    #L: Length of the bar
2  x, q, bf, c0, c1 = symbols('x, q, b_f, c0, c1')
3
4  fL, fR, uL,uR = symbols('f_L, f_R, u_L, u_R')
5  # fL: internal force at the Left-end; fR: that at the Right-end
6  # uL: displacement at the Light-end; uR: displacement at the Right-end
7
8  bf = sp.Function('b_f')(x)                        # 1D body-force (N/m)
9  title = ["Displacement", "Stress"]
10
11 # Consider the distributed force is a constant q
12 bf = q  #0 #(2*sp.sin(x)+8)/E       one may try other force functions
13 u = solver1D2(E, A, bf, l, uL, fR, key='force')   # or: gr.solver1D2()
14
15 # check whether the DE is satisfied.
16 print(f'Is solution correct? {u[2] == -bf/E/A}')
17 u                                 # solutions: u; E*u,x; u,xx
```

output from solver1D2(): u, σ, u_xx
Is solution correct? True

$$\left(u_L + \frac{f_R x}{AE} + \frac{lqx}{AE} - \frac{qx^2}{2AE}, \ \frac{f_R}{A} + \frac{lq}{A} - \frac{qx}{A}, \ -\frac{q}{AE}\right)$$

We found a full set of formulas for solutions to the 1D bar fixed with u_L at the left end ($x = 0$), subjected to distributed force q and point force f_R at the right end ($x = l$). The formula for the displacement is summarized as follows:

$$u(x) = u_L + \frac{f_R x}{AE} + \frac{qlx}{AE} - \frac{qx^2}{2AE} \tag{7.14}$$

It is quadratic in x and has a maximum value at $x = l$:

$$u(x) = u_L + \frac{f_R l}{AE} + \frac{ql^2}{2AE} \tag{7.15}$$

When setting $q = 0$, $u_L = 0$, this is Eq. (7.11).

The formula for the stress is

$$\sigma(x) = \frac{f_R}{A} + \frac{ql}{A} - \frac{qx}{A} \tag{7.16}$$

When setting $q = 0$, this is Eq. (7.13).

Equation (7.16) is linear in x and has a maximum value at $x = 0$:

$$\sigma(x) = \frac{f_R}{A} + \frac{ql}{A} = \frac{f_R + ql}{A} \tag{7.17}$$

This is expected because the root of the bar receives the total force of $(f_R + ql)$.

Equations (7.14) and (7.16) are useful formulas for computing the solution to a bar subjected to uniform body force. The stress solution is exactly the same as we obtained in the first example, by setting $q = 0, u_L = 0$. It is also exactly the same obtained in the second example in Chapter 3 (by setting $f_R = 0$).

To obtain solutions for a bar with specified parameters, we simply substitute these parameters with fixed values. For example:

```
1  ux_cq=[u[i].subs({E:1,A:1,l:1,q:1,uL:0,fR:0}) for i in range(len(u))]
2  ux_cq                         # solution for constant body force q
```

$$\left[-\frac{x^2}{2}+x,\ 1-x,\ -1\right]$$

If the distributed body force is another type function of x, we simply change the expression for b_f in the foregoing code. For example, if the force is a sine function, we use the following code to obtain the formulas of solutions:

```
1  # Consider the distributed force is a constant q
2  bf = q*sp.sin(sp.pi*x)
3  u = solver1D2(E, A, bf, 1, uL, fR, key='force')
4  print(f'Is solution correct? {u[2] == -bf/E/A}')
5  u
```

output from solver1D2(): u, σ, u_xx
Is solution correct? True

$$\left(u_L+\frac{f_R x}{AE}-\frac{qx\cos(\pi l)}{\pi AE}+\frac{q\sin(\pi x)}{\pi^2 AE},\ \frac{f_R}{A}-\frac{q\cos(\pi l)}{\pi A}+\frac{q\cos(\pi x)}{\pi A},\ -\frac{q\sin(\pi x)}{AE}\right)$$

```
1  # For a specified set of parameters (unit values):
2  ux=[u[i].subs({E:1,A:1,l:1,q:1,uL:0,fR:0}) for i in range(len(u))]
3  ux
```

$$\left[\frac{x}{\pi}+\frac{\sin(\pi x)}{\pi^2},\ \frac{\cos(\pi x)}{\pi}+\frac{1}{\pi},\ -\sin(\pi x)\right]$$

Here, the displacement at the left end of the bar is set to zero. All the deformation and stress/strain in this case are all driven by the forces (body force and the force at the right end). This types of problem is known as a **force-driven** problem.

7.3.2 Method 2: Use the Sympy differential equation solver

One can may use the built-in sp.dsolve() to solve the differential equation directly:

```
1  E, A, l = symbols('E, A, l', nonnegative=True)
2  x, q = symbols('x, q ')
3  fR, uL, uR = symbols('f_R, u_L, u_R')
4  u = sp.Function('u')(x)                        # 1D displacement (m)
5
6  # Consider the distributed force is a constant q
7  bf = q*sp.sin(sp.pi*x)
8
9  # Define the 2nd differential equation
10 diff_eq = u.diff(x, 2) + bf/E/A
11
12 # Define the boundary conditions
13 BCs = {u.subs(x, 0): uL, u.diff(x).subs(x, l): fR/E/A}
14
15 # Solve the differential equation using sp.dsolve()
16 solution = sp.dsolve(diff_eq, u, ics=BCs)     # This is an sp.Eq object
17
18 # check whether the DE is satisfied.
19 print(f'Is solution correct? {solution.rhs.diff(x,2) == -bf/E/A}')
20 solution                                       # solution equation
```

Is solution correct? True

$$u(x) = u_L + \frac{q \sin(\pi x)}{\pi^2 AE} + \frac{x(\pi f_R - q\cos(\pi l))}{\pi AE}$$

The result obtained is the same as that obtained earlier. Note that this approach only gives the displacement.

7.3.3 Python code for plotting the solutions

Next, we write the following code to plot the results:

```
1  def plot2curveS(u, xL=0., xR=1., title="f_title"):
2
3         '''Plot the distribution of function u, the maximum values and
4         locations, as well as stationary points, and the values at the
5         stationary points, and boundaries.
6         input: u, sympy function defined in [xL, xR]
7         '''
8         x = symbols('x')
9         dx = 0.01; dxr = dx*10        # x-step
10        xi = np.arange(xL, xR+dx, dx)
11        uf = sp.lambdify((x), u[0], 'numpy')   #convert Sympy f to numpy f
12        yi = uf(xi)
```

```
13      if type(yi) != np.ndarray:              #in case, uf is a constant
14          #type(yi) == int or type(yi) == float: # or len(yi)==1:
15          xi = np.arange(xL, xR+dxr, dxr)
16          yi = float(yi)*np.ones_like(xi)
17
18      fig, ax1 = plt.subplots(figsize=(5.,1.), dpi=300)
19      fs = 8          # fontsize
20      color = 'black'
21      ax1.set_xlabel('location x', fontsize=fs)
22      ax1.set_ylabel(title[0], color=color, fontsize=fs)
23      ax1.plot(xi, yi, color=color)
24      ax1.grid(color='r',ls=':',lw=.3, which='both') # Use both tick
25      ax1.tick_params(axis='x', labelcolor=color, labelsize=fs)
26      ax1.tick_params(axis='y', labelcolor=color, labelsize=fs)
27
28      vmax = yi[yi.argmax()]
29      max_l = np.argwhere(yi == vmax)
30      ax1.plot(xi[max_l], yi[max_l], 'r*', markersize=4)
31      print(f'Maximum {title[0]} value={vmax:.3e},at x={xi[max_l][0][0]}')
32
33      uf = sp.lambdify((x), u[1], 'numpy') #convert Sympy f to numpy f
34      xi = np.arange(xL, xR+dx, dx)
35      yi2 = uf(xi)
36      if type(yi2) != np.ndarray: # or len(yi2) == 1:
37          xi = np.arange(xL, xR+dxr, dxr)
38          yi2 = float(yi2)*np.ones_like(xi)
39
40      m1, m2, m3 = np.partition(abs(yi2), 2)[0:3]
41      msl=[np.where(abs(yi2)==m1)[0][0],np.where(abs(yi2)==m2)[0][0],
42              np.where(abs(yi2)==m3)[0][0]]
43
44      vmax = yi2[yi2.argmax()]
45      max_l = np.argwhere(yi2 == vmax)
46      print(f'Maximum {title[1]} value={vmax:.3e},at x={xi[max_l][0][0]}')
47
48      if abs(xi[msl[2]]-xi[msl[1]])<2*dx:
49          if abs(yi2[msl[2]]-0.)<abs(yi2[msl[1]]-0.): msl.pop(1)
50          else: msl.pop(2)
51      if len(msl) > 2:
52          if abs(xi[msl[2]]-xi[msl[0]])<2*dx:
53              if abs(yi2[msl[2]]-0.)<abs(yi2[msl[0]]-0.): msl.pop(0)
54              else: msl.pop(2)
55      if len(msl) > 1:
56          if abs(xi[msl[1]]-xi[msl[0]])<2*dx:
57              if abs(yi2[msl[1]]-0.)<abs(yi2[msl[0]]-0.): msl.pop(0)
58              else: msl.pop(1)
59
60      ax2 = ax1.twinx() # instantiate second axes sharing the same x-axis
61
62      color = 'blue'
63      ax2.set_ylabel(title[1], color=color, fontsize=fs)
64      ax2.plot(xi, yi2, color=color)
```

```
65   ax2.plot(xi[max_1], yi2[max_1], 'r*', markersize=4)
66   ax2.plot(xi[msl], yi2[msl], 'r*', markersize=4)
67   ax1.plot(xi[msl], yi[msl], 'r*', markersize=4)
68   ax2.plot(xi[0], yi2[0], 'ro', markersize=2)
69   ax1.plot(xi[0], yi[0], 'ro', markersize=2)
70   ax2.plot(xi[-1], yi2[-1], 'ro', markersize=2)
71   ax1.plot(xi[-1], yi[-1], 'ro', markersize=2)
72   ax2.grid(color='r',ls=':',lw=.5, which='both') # Use both tick
73   ax2.tick_params(axis='x', labelcolor=color, labelsize=fs)
74   ax2.tick_params(axis='y', labelcolor=color, labelsize=fs)
75   np.set_printoptions(formatter={'float': '{: 0.3e}'.format})
76   print(f'Extreme {title[0]} values={yi[msl]},\n        at x={xi[msl]}')
77   print(f'Critical {title[1]} values={yi2[msl]},\n        at x={xi[msl]}')
78   print(f'{title[0]} values at boundary ={yi[0], yi[-1]}')
79   print(f'{title[1]} values at boundary ={yi2[0], yi2[-1]}\n')
```

We plot the solutions for the bar subjected to a constant body force $q = 1$ using the code function given above:

```
1  title = ["Displacement", "Stress"]
2  # use ux_cq, solution from method 1
3  plot2curveS(ux_cq, title=title)            # or gr.plot2curveS()
4  #plt.savefig('images/beam_cq.png', dpi=500) # save the plot to a file
5  plt.show()
```

```
Maximum Displacement value=5.000e-01,at x=1.0
Maximum Stress value=1.000e+00,at x=0.0
Extreme Displacement values=[ 5.000e-01],
       at x=[ 1.000e+00]
Critical Stress values=[ 0.000e+00],
       at x=[ 1.000e+00]
Displacement values at boundary =(0.0, 0.5)
Stress values at boundary =(1.0, 0.0)
```

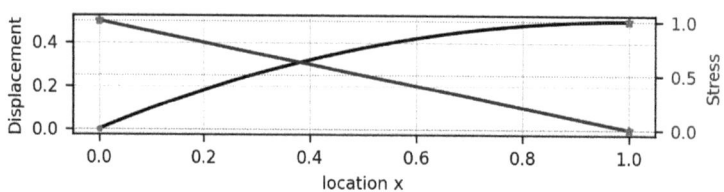

Solution to Axially Loaded Bars

It is seen that for a constant body force, the displacement is quadratic and the stress is linear along x. Let us print out specific functions and values.

```
1  gr.printM(ux_cq[0].subs(x,l), 'Displacement at the right-end',n_dgt=4)
```

Displacement at the right end: $-\dfrac{l^2}{2} + l$

```
1  gr.printM(ux_cq[1], 'The stress in the bar',n_dgt=4)
```

The stress in the bar: $1 - x$

```
1  gr.printM((ux_cq[1]/E).expand(), 'The strain in the bar',n_dgt=4)
```

The strain in the bar: $-\dfrac{x}{E} + \dfrac{1}{E}$

```
1  fL = (ux_cq[1]).subs(x,0).expand()
2  gr.printM(fL, 'Internal force at left-end:',n_dgt=4)
```

Internal force at left-end: 1

```
1  gr.printM((ux_cq[1]).subs(x,l), 'Internal force at right-end:')
```

Internal force at right-end: $1 - l$

```
1  gr.printM(-ux_cq[2], 'The distributed force along the bar:')
```

The distributed force along the bar: 1

7.4 Example: Linear body force and concentrated force at the right end

Consider a bar subjected to a linearly distributed body force and a concentrated force f_R at $x = l$, both in its axial direction. The setting of the problem is the same as that shown in Fig. 7.1, but the function for the body force is given by

$$b_f = q_0 + q\frac{x}{l} \tag{7.18}$$

where q_0 and q are given constants.

The solution can then be solved easily by simply integrating Eq. (7.1) twice by hand. Here, we use solver1D2(), which gives formulas for the displacement, stress, and the second derivative of the displacement:

```
1  # Consider the distributed force is a linear function
2  l, E, A = symbols('l, E, A')
3  q0, q, x = symbols('q0, q, x ')
4  bf = q0 + q*x/l
5
6  u = solver1D2(E, A, bf, l, uL, fR, key='force')
7  print(f'Is solution correct? {u[2] == -(bf/E/A).expand()}')
8  u
```

output from solver1D2(): u, σ, u_xx
Is solution correct? True

$$\left(u_L + \frac{f_R x}{AE} + \frac{lqx}{2AE} + \frac{lq_0 x}{AE} - \frac{q_0 x^2}{2AE} - \frac{qx^3}{6AEl}, \; \frac{f_R}{A} + \frac{lq}{2A} + \frac{lq_0}{A} - \frac{q_0 x}{A} - \frac{qx^2}{2Al}, \; -\frac{q_0}{AE} - \frac{qx}{AEl} \right)$$

This is the set of formulas for calculating the displacement in the bar in the x direction and the stress in the bar. The last expression is the term on the right-hand side of Eq. (7.1), indicating that the solution satisfies the differential equation.

```
1  # For a set of specified parameters:
2  ux=[u[i].subs({E:1,A:1,l:1,q0:.1,q:1,uL:0,fR:0}) for i in range(len(u))]
3  ux=[sp.nsimplify(ux[i]) for i in range(len(ux))]
4  ux
```

$$\left[-\frac{x^3}{6} - \frac{x^2}{20} + \frac{3x}{5}, \; -\frac{x^2}{2} - \frac{x}{10} + \frac{3}{5}, \; -x - \frac{1}{10} \right]$$

```
1  # Plot the distribution of displacement, stress and the maximum values:
2  plot2curveS(ux, title=title)
3  #plt.savefig('images/bar_cq.png', dpi=500)    # save the plot to a file
4  plt.show()
```

```
Maximum Displacement value=3.833e-01,at x=1.0
Maximum Stress value=6.000e-01,at x=0.0
Extreme Displacement values=[ 3.833e-01],
    at x=[ 1.000e+00]
Critical Stress values=[ 0.000e+00],
    at x=[ 1.000e+00]
Displacement values at boundary =(0.0, 0.3833333333333333)
Stress values at boundary =(0.6, 0.0)
```

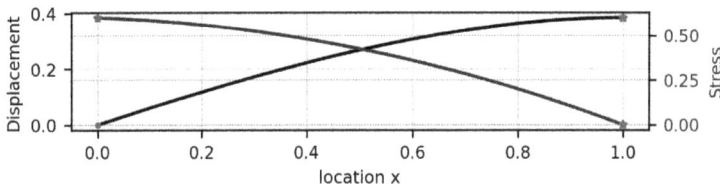

For this linear body-force case, the displacement is cubic and the stress is quadratic in x.

7.5 Example: A displacement imposed at the right end

Consider a bar with its displacement is imposed as a fixed value u_l at $x = l$. The deformation and stress/strains in this case are all driven by the imposed displacement at the boundary. This types of problem is known as **displacement-driven** problem. This is the typical case when we do a tensile test to a material specimen:

```
1  # Consider the distributed force is constant
2  bf = q                              # One may try this; q0 + q*x/l
3  u = solver1D2(E, A, bf, l, uL, uR, key='disp')
4  print(f'Is solution correct? {u[2] == -bf/E/A}')
5  u
```

```
output from solver1D2(): u, σ, u_xx
Is solution correct? True
```

$$\left(u_L - \frac{u_L x}{l} + \frac{u_R x}{l} + \frac{lqx}{2AE} - \frac{qx^2}{2AE}, -\frac{Eu_L}{l} + \frac{Eu_R}{l} + \frac{lq}{2A} - \frac{qx}{A}, -\frac{q}{AE}\right)$$

```
1  # For a set of specified parameters:
2  ux=[u[i].subs({q:0, uL:0}) for i in range(len(u))]
3  ux    # General solution for u, σ in a bar when right-end displaced by uR
```

$$\left[\frac{u_R x}{l},\ \frac{E u_R}{l},\ 0\right]$$

This gives the general solution for u and σ in a bar when right end is displaced by u_R.

If we set specified parameters with numerical values, we can plot the solution:

```
1  # For a set of specified parameters:
2  ux=[u[i].subs({E:1,A:1,l:1, q:0,uL:0,uR:1}) for i in range(len(u))]
3  ux
```

$[x,\ 1,\ 0]$

```
1  # Plot the distribution of the displacement and stress:
2  plot2curveS(ux, title=title)
3  #plt.savefig('images/bar_cq.png', dpi=500)
4  plt.show()
```

Maximum Displacement value=1.000e+00,at x=1.0
Maximum Stress value=1.000e+00,at x=0.0
Extreme Displacement values=[0.000e+00],
    at x=[0.000e+00]
Critical Stress values=[1.000e+00],
    at x=[0.000e+00]
Displacement values at boundary =(0.0, 1.0)
Stress values at boundary =(1.0, 1.0)

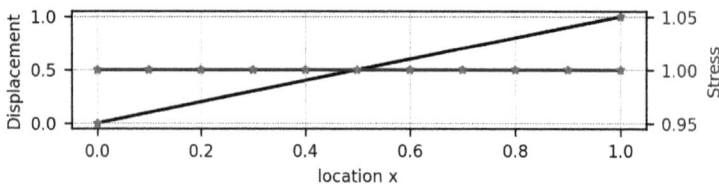

For this displacement-driven problem, the displacement is linear and the stress is constant in x:

```
1  gr.printM(u[0].subs(x,l), 'Displacement at right-end:',n_dgt=4)
```

Displacement at right end: u_R

```
1  gr.printM(u[1], 'The stress in the bar:',n_dgt=4)
```

The stress in the bar: $-\dfrac{Eu_L}{l} + \dfrac{Eu_R}{l} + \dfrac{lq}{2A} - \dfrac{qx}{A}$

```
1  gr.printM((u[1]/E).expand(), 'The strain in the bar:',n_dgt=4)
```

The strain in the bar: $-\dfrac{u_L}{l} + \dfrac{u_R}{l} + \dfrac{lq}{2AE} - \dfrac{qx}{AE}$

```
1  N0 = (u[1]*A).subs(x,0).expand()
2  gr.printM(N0, 'Internal force at left-end: f0=',n_dgt=4)
```

Internal force at left end: f0 = $-\dfrac{AEu_L}{l} + \dfrac{AEu_R}{l} + \dfrac{lq}{2}$

```
1  Nl = (u[1]*A).expand().subs(x,l)
2  gr.printM((Nl).subs(x,l), 'Internal force at the right-end:')
```

Internal force at the right end: $-\dfrac{AEu_L}{l} + \dfrac{AEu_R}{l} - \dfrac{lq}{2}$

```
1  gr.printM(-u[2]*E*A, 'The distributed force along the bar')
```

The distributed force along the bar: q

7.6 Bars subjected to discontinuous body forces

7.6.1 *Distributed forces*

In deriving the equilibrium equations, we assumed that the stresses are sufficiently smooth and their functions are **differentiable** at least at first order. When the body force is discontinuous, the stresses may not be differentiable at the force discontinuity points. Figure 7.2 shows a bar fixed at the left end, subjected to two distributed body forces q_1 and q_2 (N/m), and a concentrated force f_R at the right end.

In this case, we need to treat the bar being formed by two bars: bar-1 has length l_1 and bar-2 has length l_2. Since this problem is statically determinate, meaning that the internal forces can be obtained only by equilibrium, the forces applied at the right end of each of these two sub-bars can be obtained by simple equilibrium conditions:

For bar-2: $f_{2R} = f_R$ that is the externally applied force at the right end of the bar.

For bar-1: $f_{1R} = f_R + q_2 l_2$.

Using Eq. (7.14), the displacement solution for bar-1 is found as

$$u_1(x) = \frac{f_{1R} x}{AE} + \frac{l_1 q_1 x}{AE} - \frac{q_1 x^2}{2AE} \qquad (7.19)$$

where we used boundary condition $u_L = 0$.

At $x = l_1$, we have

$$\begin{aligned}
u_1(l_1) &= \frac{f_{1R} l_1}{AE} + \frac{q_1 l_1^2}{2AE} \\
&= \frac{l_1}{AE}\left(f_{1R} + \frac{q_1 l_1}{2}\right) \qquad (7.20) \\
&= \frac{l_1}{AE}\left(f_R + q_2 l_2 + \frac{q_1 l_1}{2}\right)
\end{aligned}$$

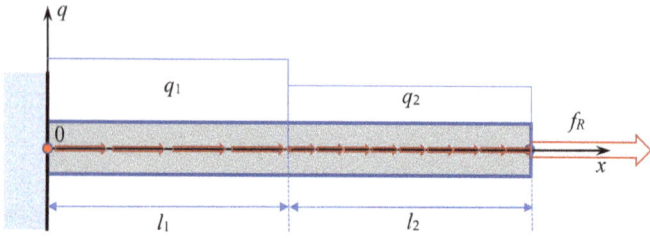

Figure 7.2. A 1D bar fixed at the left end, subjected to two distributed body-force, and a concentrated force at the right end.

The displacement solution for bar-2 is found as

$$u_2(x) = u_1(l_1) + \frac{f_{2R}x}{AE} + \frac{q_2 l_2 x}{AE} - \frac{q_2 x^2}{2AE}$$

$$= u_1(l_1) + \frac{f_R x}{AE} + \frac{q_2 l_2 x}{AE} - \frac{q_2 x^2}{2AE} \quad (7.21)$$

Here, we used the boundary condition for bar-2 at its left end: $u_L = u_1(l_1)$. Using Eq. (7.16), the stress solution for bar-1 is found as

$$\sigma_1(x) = \frac{f_R + q_2 l_2}{A} + \frac{q_1 l_1}{A} - \frac{q_1 x}{A} \quad (7.22)$$

Here, we used $f_{1R} = f_R + q_2 l_2$. The stress solution for bar-2 is found as

$$\sigma_2(x) = \frac{f_R}{A} + \frac{q_2 l_2}{A} - \frac{q_2 x}{A} \quad (7.23)$$

Clearly, if $q_1 = q_2 = 0$, we have the solution for the whole bar subjected to f_R.

7.6.2 *Point force*

For point forces, the discontinuities are at the points where the forces are applied. Figure 7.3 shows a bar fixed at the left end. It is subjected to two point forces: f_I at an interior point and f_R at the right end.

In this case, we also need to treat the bar as two bars: bar-1 has length l_1 and bar-2 has length l_2. This problem is also statically determinate. The forces applied at the right-end of each of these two sub-bars can be obtained by simple equilibrium conditions:

For bar-2: $f_{2R} = f_R$ that is the externally applied force at the right-end of the bar.

For bar-1: $f_{1R} = f_R + f_I$. Here, we need to look at the left-hand side of the point where f_I is applied.

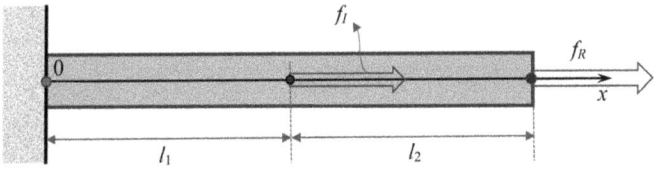

Figure 7.3. A 1D bar fixed at the left end, subjected to two point forces: one at an interior point, and one at the right end.

Using Eq. (7.14), the displacement solution for bar-1 is found as

$$u_1(x) = \frac{f_{1R}x}{AE} = \frac{(f_R + f_I)x}{AE} \qquad (7.24)$$

where we used boundary condition $u_L = 0$. At $x = l_1$, we have

$$u_1(l_1) = \frac{f_{1R}l_1}{AE} = \frac{(f_R + f_I)l_1}{AE} \qquad (7.25)$$

The displacement solution for bar-2 is found as

$$u_2(x) = \frac{(f_R + f_I)l_1}{AE} + \frac{f_R x}{AE} \qquad (7.26)$$

Here, we used the boundary condition for bar-2 at its left end: $u_L = u_1(l_1)$.

Using Eq. (7.16), the stress solution for bar-1 is found as

$$\sigma_1(x) = \frac{f_R + f_I}{A} \qquad (7.27)$$

Here, we used, $f_{1R} = f_R + f_I$. The stress solution for bar-2 is found as

$$\sigma_2(x) = \frac{f_R}{A} \qquad (7.28)$$

7.7 Statically indeterminate bars: Reaction force

If the problem is statically indeterminate, implying the forces applied at the right end of each of the bars cannot be obtained only by equilibrium conditions, extra work is needed. Figure 7.4 shows a bar fixed at both ends, subjected to two distributed body forces q_1 and q_2 (N/m).

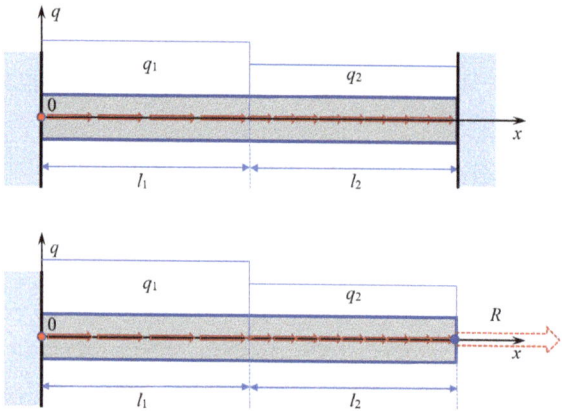

Figure 7.4. Top: a 1D bar fixed at both ends, subjected to two distributed body-force. Bottom: the constraint at the right end is removed, and replaced by an unknown reaction force R.

For this problem, we still need to treat the bar as two bars: bar-1 has length l_1 and bar-2 has length l_2 because of the body force discontinuity. This problem is statically indeterminate, and hence, the forces applied at the right end of each of these two sub-bars cannot be obtained by equilibrium conditions alone. Therefore, we remove the constraint at the right end, and replace it with an unknown force called reaction force R. After this is done, the displacements at the right end of the bar-2 can be found using the same procedure in the previous two examples. Using Eq. (7.21), and replacing f_R by R, we obtain

$$u_2(x) = \frac{l_1}{2AE}(2R + 2q_2 l_2 + q_1 l_1) + \frac{Rx}{AE} + \frac{q_2 l_2 x}{AE} - \frac{q_2 x^2}{2AE} \qquad (7.29)$$

At $x = l_2$, $u_2 = 0$, we then found

$$R = \frac{-1}{2(l_1 + l_2)}(q_1 l_1^2 + q_2 l_2^2 + 2q_2 l_1 l_2) \qquad (7.30)$$

It is clear that the reaction force is negative for positive body forces. Consider a special case of $q_1 = q_2 = q$, Eq. (7.30) becomes

$$R = \frac{-q}{2(l_1 + l_2)}(l_1^2 + l_2^2 + 2l_1 l_2) = \frac{-q}{2(l_1 + l_2)}(l_1 + l_2)^2 = \frac{-q(l_1 + l_2)}{2} \qquad (7.31)$$

Finally, substituting Eq. (7.30), for (7.19), (7.21) we obtain the displacement solution. Substituting Eq. (7.30), for (7.22), (7.23) we obtain the stress solution.

The above technique works well for simple problems with a small number of overconstraints. For large structural system, the overconstraints can be very large in number. More effective approaches such as the finite element method [1] should be used.

7.8 Thermal stress

7.8.1 *Thermal displacement*

Structural components may work in an environment where the temperature changes. Due to the thermal expansion of materials and the overconstrained (indeterminate) nature of the bar, the material is not allowed to expand freely. This can result in additional stress, known as **thermal stress**. The general constitutive equation is given in Chapter 5. For 1D bars, we have

$$\varepsilon_{Txx} = \alpha \Delta T \tag{7.32}$$

where ε_{Txx} is the thermal strain. The displacement corresponding to the temperature change can be found from the following differential equations:

$$\frac{\partial u_T}{\partial x} = \varepsilon_{Txx} = \alpha \Delta T \tag{7.33}$$

Integrating the foregoing equation gives

$$u_T = \alpha \Delta T x + c \tag{7.34}$$

where c is the integral constant. Assuming $u_T = 0$ at $x = 0$, we found $c = 0$, and we have

$$u_T = \alpha \Delta T x \tag{7.35}$$

This is the displacement at x, when the bar is fixed at $x = 0$ and subjected to a temperature change ΔT.

7.8.2 Thermal forces

Such a thermal displacement can be simply added into the displacement expression obtained earlier. For the example of fixed–fixed bar subjected to distributed body forces discussed in the previous section, we shall use the same trick of replacing the constraint on the left end by an unknown reaction force R. Using Eq. (7.29), we have the combined displacement in bar-2 as

$$u_2(x) = \frac{l_1}{2AE}(2R + 2q_2l_2 + q_1l_1) + \underbrace{\alpha \Delta T l_1}_{u_{T1}} + \frac{Rx}{AE} + \frac{q_2l_2x}{AE} - \frac{q_2x^2}{2AE} + \underbrace{\alpha \Delta T x}_{u_{T2}}$$

$$\tag{7.36}$$

where the last term is the extra displacement caused by the temperature change.

At $x = l_2$, $u_2 = 0$, we then found

$$R = \frac{-1}{2(l_1 + l_2)}(q_1 l_1^2 + q_2 l_2^2 + 2q_2 l_1 l_2) - AE(\alpha \Delta T) \tag{7.37}$$

The last term is the thermal forces resulting from the temperature change. Even if all the distributed forces are zero, there will still be a stress built up in the bar. For positive temperature change, the reaction force is negative as expected. The formula for R is found using the following code:

```
1  E, A, l1, l2, x = symbols('E, A, l_1, l_2, x')
2  R, ΔT, q1, q2, α, x = symbols('R, ΔT, q1, q2, α, x', nonnegative=True)
3
4  eqn = l1*(2*R+2*q2*l2+q1*l1)/(2*A*E) + α*ΔT*l1 + \
5        l2*(2*R        +q2*l2)/(2*A*E) + α*ΔT*l2
6  sln = sp.solve(eqn, R)
7  sln
```

$$\left[\frac{-AEl_1 \Delta T \alpha - AEl_2 \Delta T \alpha - \frac{l_1^2 q_1}{2} - l_1 l_2 q_2 - \frac{l_2^2 q_2}{2}}{l_1 + l_2} \right]$$

After the reaction force is found, the displacements in these two bars should be obtained using Eqs. (7.19)–(7.21). The combined displacement formula for bar-1 becomes

$$u_1(x) = \frac{(R + q_2 l_2)x}{AE} + \frac{l_1 q_1 x}{AE} - \frac{q_1 x^2}{2AE} + \alpha \Delta T x \qquad (7.38)$$

At $x = l_1$, we have

$$u_1(l_1) = \frac{(R + q_2 l_2)l_1}{AE} + \frac{q_1 l_1^2}{2AE} = \frac{l_1}{2AE}(2f_R + q_1 l_1) + \alpha \Delta T l_1 \qquad (7.39)$$

The combined displacement formula for bar-2 is

$$u_2(x) = u_1(l_1) + \frac{Rx}{AE} + \frac{q_2 l_2 x}{AE} - \frac{q_2 x^2}{2AE} + \alpha \Delta T x \qquad (7.40)$$

7.8.3 Computing thermal stresses

7.8.3.1 Approach 1: Use reaction and external forces

The thermal stress can be obtained by treating the reaction force R as the externally applied force in addition to other externally applied forces. In this approach, the formulas for stress remain unchanged. For stress in bar-2, for example, the formula becomes

$$\sigma_{xx} = \frac{R}{A} + \frac{q_2 l_2}{A} - \frac{q_2 x}{A} \qquad (7.41)$$

The thermal effects to the stress solution are introduced thorough the reaction forces resulting from the temperature change. Again, the thermal stress should be be computed from all forces, including the mechanical forces and the thermal forces.

7.8.3.2 Approach 2: Use the combined displacement

Alternatively, we can use the combined displacement and the strain–displacement relation to compute the strain. Then, compute the stress using the constitutive equation. In this case, we need to use the constitutive equation given in Chapter 5:

$$\sigma_{xx} = E\left[\frac{du_{\text{combined}}}{dx} - \frac{du_T}{dx}\right] = E\left[\frac{du_{\text{combined}}}{dx} - \alpha\Delta T\right] \quad (7.42)$$

This is because the displacement gradient caused by temperature change should not result in stresses and hence needs to be removed. Using the foregoing equation for bar-2, for example, we shall get the same formula as given in Eq. (7.41).

7.9 Example: Stress in an overconstrained bar loaded mechanically and thermally

Consider a bar with length 1.0 subjected to $q_1 = 1.0$ over $0 < x < 0.5$ and $q_2 = 0.0$ over $0.5 < x < 1.0$. Let $A = 1.0$ and $E = 1.0$. Compute the reaction forces, the displacement and stress distributed over the entire bar. The bar may be loaded mechanically and thermally.

Solution: We first write a code function to calculate the displacement and stress at x in an arbitrary 1D bar with length of $l_1 + l_2$, subjected to 2 distributed body-forces q_1 and q_2, respectively, over l_1 and l_2, for any given force at the right end. This code also considers the thermal expansion, which is set as zero as default.

```
def twobars_bf(x, A, E, l1, l2, q1, q2, fR, aT=0, barId=1):
    '''Calculate the displacement and stress at x in a 1D bar subjected
    to 2 distributed body-force q1 & q2 over l1 and l2.
    fR: force applied at the right-end of the bar
    A and E: area and Youngs modulus of the bar
    aT: α*ΔT (thermal expansion coefficient * change in temperature)
    barId: if 1 (default), calculate the results over l1
           if 2, calculate the results over l2
    Return: displacement u, and stress σ
    '''
    if barId == 1:
        u = ((fR+q2*l2)*x+l1*q1*x-q1*x**2/2)/(A*E) + aT*x
        σ = (fR+l2*q2 + q1*l1-q1*x)/A
    else:
        u1R = l1*((fR+q2*l2) + l1*q1/2)/(A*E) + aT*l1
        u = u1R + (fR*x+l2*q2*x-q2*x**2/2)/(A*E) + aT*x
        σ = (fR + q2*l2-q2*x)/A
    return u, σ
```

Solution to Axially Loaded Bars

Set the values for the variables involved in this problem, assuming there is no temperature change.

```
1   q1 = 1.0; q2 = 0.0
2   l1 = 0.5; l2 = 0.5
3   E  = 1.0; A  = 1.0
4   α  = 1.0e-4                          # thermal expansion coefficient
5   dT = 0.0                             # temperature change
6   αT = α*dT
7   RT = A*E*αT                          # Reaction force by temperature change
8
9   # Compute the reaction force at the right-end of the bar
10  fR = -(q1*l1**2 + q2*l2**2+2*q2*l1*l2)/(2*l1 + 2*l2)-RT    # fR=-0.125
11  print(f"Reaction force at the right-end = {fR}")
12
13  dx = 0.002
14  x1 = np.arange(0.0, l1+dx, dx)       # coordinates in l1
15  x2 = np.arange(0.0, l2+dx, dx)       # coordinates in l2
16
17  # Compute the displacement and stress:
18  u1, σ1 = twobars_bf(x1, A, E, l1, l2, q1, q2, fR, aT=αT, barId=1)
19  u2, σ2 = twobars_bf(x2, A, E, l1, l2, q1, q2, fR, aT=αT, barId=2)
20
21  x  = np.concatenate((x1, l1+x2))     # put the results in a single array
22  ux = np.concatenate((u1, u2))
23  sx = np.concatenate((σ1, σ2))
24
25  ux_max = np.max(abs(ux))             # find the absolute maximum value
26  sx_max = np.max(abs(sx))
27  print(f"The absolute maximum displacement: {ux_max}")
28  print(f"The absolute maximum stress:       {sx_max}")
29
30  step = max(int(len(x)/5), 1)
31  print(f"Location x:\n {x[::step]}")           # print out some results
32  print(f"Displacement ux:\n {ux[::step]}")
33  print(f"Stress σ:\n {sx[::step]}")
34  print(f"Normal force N:\n {A*sx[::step]}")
```

```
Reaction force at the right end = −0.125
The absolute maximum displacement: 0.07031200000000001
The absolute maximum stress:       0.375
Location x:
 [ 0.000e+00  2.000e-01  4.000e-01  5.980e-01  7.980e-01
   9.980e-01]
Displacement ux:
 [ 0.000e+00  5.500e-02  7.000e-02  5.025e-02  2.525e-02
   2.500e-04]
```

Stress σ:
[3.750e-01 1.750e-01 -2.500e-02 -1.250e-01 -1.250e-01
 -1.250e-01]
Normal force N:
[3.750e-01 1.750e-01 -2.500e-02 -1.250e-01 -1.250e-01
 -1.250e-01]

We now plot the distribution of the solutions. In order to plot both displacement and stress together in the same figure, we divide the solutions by its corresponding absolute maximum value. Therefore, the unit of the value on these curves are their corresponding maximum values.

```
# Plot the distribution curves:

x_data = [x, x]                         # put x coordinate values to a list
y_data = [ux/max(ux_max,1.e-9), sx/max(sx_max,1.e-9)]   # put to a list
labels=['Displacement', 'Stress']       # define the label
gr.plot_fs(x_data, y_data, labels, 'x',    'u(x), $σ$(x)',   'temp')
                                       # x-label, y-label,              file-name
```

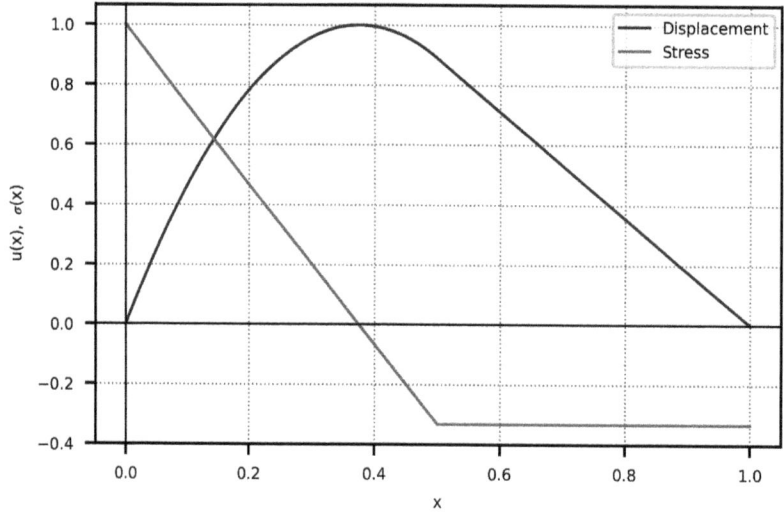

Note in the foregoing examples (and some examples in the future), we used unit values for the parameters involved in the problem. Users may use the actual values for a given practical problem. Any unit system can be used, but all the units for the same problem must be consistent.

The code given above can be used for many other problems. For example, if we set $l_1 = 1.0$ and $l_2 = 0$ and keep others unchanged, we obtain the reaction forces, displacement and stress for a fixed–fixed bar subjected to a uniform body force. Readers may play with it.

It is a good exercise for readers to write a code to compute a similar case, considering a fixed–fixed bar with a point force in the interior of the bar, as shown in Fig. 7.3.

7.10 Example: Thermal stress in an overconstrained bar

Consider a bar with length 1.0 subjected to a temperature change of $\Delta T = 1.0$. Let $A = 1.0$ and $E = 1.0$. Compute the reaction forces, the displacement and stress distributed over the entire bar.

Solution: We use the same code function twobars_bf() written earlier, but with the following setting, so that we see the stress resulting purely by a temperature change:

```
1   q1 = 0.0; q2 = 0        # no body force, consider only the thermal effects
2   l1 = 1.0; l2 = 0.0
3   E  = 1.0; A  = 1.0
4   α  = 1.0; dT = 1.0                   # with temperature change dT: ΔT
5   aT = α*dT                             # thermal expansion per length
6   RT = A*E*aT                           # Reaction force by temperature change
7
8   # Compute the reaction force at the right-end of the bar
9   fR = -(q1*l1**2 + q2*l2**2+2*q2*l1*l2)/(2*l1 + 2*l2) - RT
10  print(f"Reaction force at the right-end = {fR}")
11
12  dx = 0.02
13  x1  = np.arange(0.0, l1+dx, dx)        # coordinates in l1
14  x2  = np.arange(0.0, l2+dx, dx)        # coordinates in l2
15
16  # Compute the displacement and stress:
17  u1, σ1 = twobars_bf(x1, A, E, l1, l2, q1, q2, fR, aT, barId=1)
18  u2, σ2 = twobars_bf(x2, A, E, l1, l2, q1, q2, fR, aT, barId=2)
19
20  x  = np.concatenate((x1, l1+x2))    # put the results in a single array
21  ux = np.concatenate((u1, u2))
22  sx = np.concatenate((σ1, σ2))
23
24  ux_max = np.max(abs(ux))            # find the absolute maximum value
25  sx_max = np.max(abs(sx))
26  print(f"The absolute maximum displacement: {ux_max}")
27  print(f"The absolute maximum stress:       {sx_max}")
28
29  step = max(int(len(x)/5), 1)
30  print(f"Location x:\n {x[::step]}")          # print out some results
31  print(f"Displacement ux:\n {ux[::step]}")
32  print(f"Stress σ:\n {sx[::step]}")
```

```
Reaction force at the right end = -1.0
The absolute maximum displacement: 0.0
The absolute maximum stress:         1.0
Location x:
[ 0.000e+00  2.000e-01  4.000e-01  6.000e-01  8.000e-01
  1.000e+00]
Displacement ux:
[ 0.000e+00  0.000e+00  0.000e+00  0.000e+00  0.000e+00
  0.000e+00]
Stress σ:
[-1.000e+00 -1.000e+00 -1.000e+00 -1.000e+00 -1.000e+00
 -1.000e+00]
```

```
1  # Plot the distribution curves:
2
3  x_data = [x, x]                        # put x coordinate values to a list
4  y_data = [ux/max(ux_max,1.e-9), sx/max(sx_max,1.e-9)]   # put to a list
5  labels=['Displacement', 'Stress']                  # define the label
6  gr.plot_fs(x_data, y_data, labels, 'x',   'u(x), $σ$(x)',   'temp')
7                                         # x-label, y-label,      file-name
```

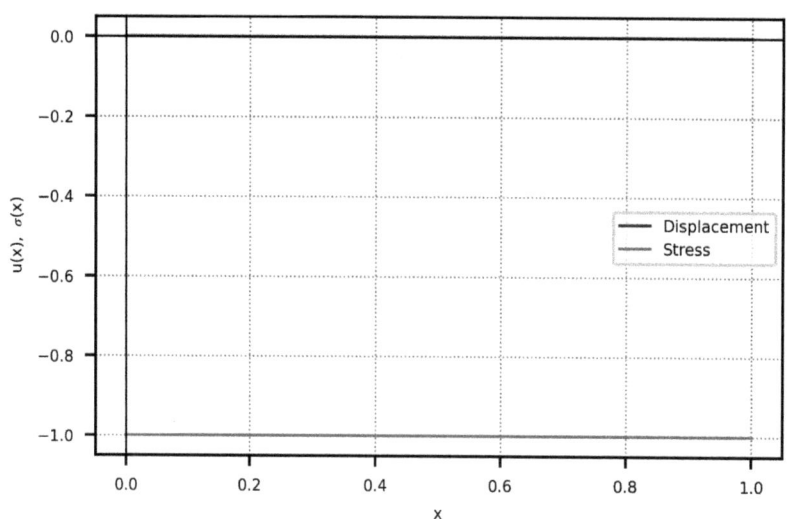

It is seen that the displacement is zero in the entire bar because it is fixed at both ends. The stress is -1 which is induced by the temperature change. This stress is purely caused by the temperature change because there are no external forces applied on the bar.

It is a good exercise for readers to write a code to compute a similar case, considering a fixed–fixed bar with a distributed force or point force in the interior of the bar, as shown in Figs. 7.3 and 7.4, but also with a temperature change.

7.11 Example: A composite bar subjected to both mechanical and thermal loads

Consider a bar of length L consisting of three thin bars of homogeneous isotropic materials, as shown in Fig. 7.5. The bars are all fixed on a wall at the left ends and are rigidly connected at the right ends. The bars on the top and bottom (bar-1) are identical with Young's modulus E_1, thermal expansion coefficient of α_1, and each with area A_1. The bar in the middle (bar-2) is with Young's modulus E_2, area A_2, and thermal expansion coefficient of α_2.

The data are given as follows: $E_1 = 200$ GPa, $E_2 = 70$ GPa, $A_1 = 5$ mm^2, $A_2 = 8$ mm^2 and $L = 100$ mm, $\alpha_1 = 11.0E-6/°C$, $\alpha_2 = 22.0E-6/°C$, $\Delta T = 300°C$, and $F = 50$ N.

Considering only the displacement in the x-direction, complete the following tasks:

1. Consider only external force F applied at the right end of the bar, derive the formulas for computing the normal stresses and the internal normal forces in each of the three bars as well as their elongations.
2. Consider only temperature change ΔT over the entire bars, derive the formulas for computing the normal stresses and the internal normal forces in each of the three bars as well as their elongations.
3. Using the data given, compute numerical values for normal stresses, internal normal forces in each of the three bars, and the elongation of the bars,

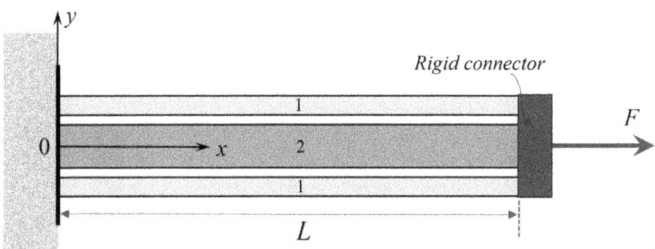

Figure 7.5. A bar consisting of three thin bars of homogeneous isotropic material connected by a rigid block.

when the composite bar is subjected to both external force and temperature changes.
4. Compute the principal stresses and the maximum stress and their directions in bar-1.

7.11.1 Solution to (1)

Consider only external force F applied at the right end of the bar and derive the formulas for computing the normal stresses and the internal normal forces in each of the three bars as well as their elongations.

Since the problem is statically indeterminate, using equilibrium conditions alone is not enough to find the solution. We thus release the rigid connector, and replace its interaction with the three bars by three unknown forces, F_1, F_2, and F_1, as shown in Fig. 7.6.

For the free body diagram of the rigid block, we shall have the following **equilibrium equation**:

$$-2F_1 - F_2 + F = 0 \qquad (7.43)$$

From the diagram for the three bars, we have

$$u_{1L} = \frac{F_1 L}{A_1 E_1}$$

$$u_{2L} = \frac{F_2 L}{A_2 E_2} \qquad (7.44)$$

Since these three bars are connected together by the rigid block, we shall have

$$u_{1L} = u_{2L} \qquad (7.45)$$

This is known as the so-called **continuity equation**, implying in this case that the structure is still in one piece and is not broken. These three bars are

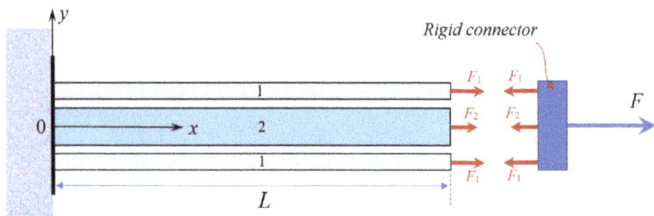

Figure 7.6. Free body diagram of the bar consisting of three thin bars and the reactions with the rigid block.

always attached to the rigid connector without gaps and overlaps. Equations (7.44) and (7.45) give

$$F_1 = \frac{A_1 E_1}{A_2 E_2} F_2 \qquad (7.46)$$

Substituting Eq. (7.46) into (7.43), we obtain

$$F_2 = \frac{A_2 E_2}{2 A_1 E_1 + A_2 E_2} F \qquad (7.47)$$

Substituting Eq. (7.47) into (7.46) gives

$$F_1 = \frac{A_1 E_1}{2 A_1 E_1 + A_2 E_2} F \qquad (7.48)$$

After these two forces are found, the displacement at $x = L$ is found as

$$u_{1L} = u_{2L} = \frac{FL}{2 A_1 E_1 + A_2 E_2} \qquad (7.49)$$

This is the elongation of these three bars.

The strains in these three bars are

$$\varepsilon_{bar-1} = \varepsilon_{bar-2} = \frac{F}{2 A_1 E_1 + A_2 E_2} \qquad (7.50)$$

The stresses in these three bars are found as

$$\sigma_{bar-1} = \frac{E_1 F}{2 A_1 E_1 + A_2 E_2}$$

$$\sigma_{bar-2} = \frac{E_2 F}{2 A_1 E_1 + A_2 E_2} \qquad (7.51)$$

This process is simple enough to work out by hand, as shown. Simple codes can be written to obtain the same results for checking to ensure no error is made:

```
1  E1, A1, E2, A2, L = symbols('E_1, A_1, E_2, A_2, L', positive=True)
2  σbar1F, σbar2F, F, F1F, F2F = symbols('σ_bar1F, σ_bar2F, F, F_1F, F_2F')
3  u1L, u2L, u1LF, u2LF = symbols('u_1L, u_2L, u_1LF, u_2LF')
4
5  u1L = F1F*L/(A1*E1)           # displacementof bar-1 at x=L by force F
6  u2L = F2F*L/(A2*E2)           # displacementof bar-2 at x=L by force F
7
8  slnFF = sp.solve([-2*F1F-F2F+F, u1L-u2L], [F1F,F2F])
9  slnFF            # solution of internal forces in bars by external Force F
```

$$\left\{ F_{1F} : \frac{A_1 E_1 F}{2 A_1 E_1 + A_2 E_2},\ F_{2F} : \frac{A_2 E_2 F}{2 A_1 E_1 + A_2 E_2} \right\}$$

```
1  slnσF = {σbar1F: slnFF[F1F]/A1, σbar2F:slnFF[F2F]/A2}
2  slnσF                  # solution of stresses in bars by external Force F
```

$$\left\{ \sigma_{bar1F} : \frac{E_1 F}{2A_1 E_1 + A_2 E_2},\ \sigma_{bar2F} : \frac{E_2 F}{2A_1 E_1 + A_2 E_2} \right\}$$

```
1  slnUF = {u1LF: slnFF[F1F]*L/(A1*E1), u2LF:slnFF[F2F]*L/(A2*E2)}
2  slnUF  # solution of displacements of bars at x=L, i.e. elongation by F
```

$$\left\{ u_{1LF} : \frac{FL}{2A_1 E_1 + A_2 E_2},\ u_{2LF} : \frac{FL}{2A_1 E_1 + A_2 E_2} \right\}$$

7.11.2 Solution to (2)

Consider only temperature change ΔT over the entire bars and derive the formulas for computing the normal stresses and the internal normal forces in each of the three bars, as well as their elongations.

When the bars experiences temperature change ΔT, the displacements in each of these bars shall have an additional term. Equation (7.44) becomes

$$u_{1L} = \frac{F_1 L}{A_1 E_1} + \alpha_1 \Delta T L$$

$$u_{2L} = \frac{F_2 L}{A_2 E_2} + \alpha_2 \Delta T L \quad (7.52)$$

Using Eqs. (7.52) and (7.45) gives

$$F_1 = \frac{F_2 A_1 E_1}{A_2 E_2} + A_1 E_1 (\alpha_2 - \alpha_1) \Delta T \quad (7.53)$$

Since $F = 0$, Eq. (7.43) gives

$$-2F_1 - F_2 = 0 \quad (7.54)$$

Substituting Eq. (7.54) into (7.53) gives

$$F_1 A_2 E_2 = -2F_1 A_1 E_1 + A_1 E_1 A_2 E_2 (\alpha_2 - \alpha_1) \Delta T \quad (7.55)$$

$$F_1 = \frac{A_1 E_1 A_2 E_2 (\alpha_2 - \alpha_1) \Delta T}{2F_1 A_1 + A_2 E_2} \quad (7.56)$$

Substituting it into Eq. (7.54), we find

$$F_2 = -\frac{2A_1 E_1 A_2 E_2 (\alpha_2 - \alpha_1) \Delta T}{2F_1 A_1 + A_2 E_2} \quad (7.57)$$

The displacements are then found using Eq. (7.52):

$$u_{1L} = \frac{A_2 E_2 (\alpha_2 - \alpha_1) \Delta T L}{2E_1 A_1 + A_2 E_2} + \alpha_1 \Delta T L$$

$$u_{2L} = -\frac{2A_1 E_1 (\alpha_2 - \alpha_1) \Delta T L}{2E_1 A_1 + A_2 E_2} + \alpha_2 \Delta T L \quad (7.58)$$

These are the elongations. The stresses are found as

$$\sigma_{bar-1} = \frac{F_1}{A_1} = \frac{E_1 A_2 E_2 (\alpha_2 - \alpha_1) \Delta T}{2E_1 A_1 + A_2 E_2}$$

$$\sigma_{bar-2} = \frac{F_2}{A_2} = -\frac{2A_1 E_1 E_2 (\alpha_2 - \alpha_1) \Delta T}{2E_1 A_1 + A_2 E_2} \quad (7.59)$$

The codes to get the same formulas are as follows:

```
1  α1, α2, T, F1T, F2T = symbols('α_1, α_2, T, F_1T, F_2T')  # T denotes ΔT
2  u1LT, u2LT, σbar1T, σbar2T = symbols('u_1LT, u_2LT, σ_bar1T, σ_bar2T')
3
4  # Due to the rigid connector, T shall also results in forces: F1T & F2T
5  u1L = F1T*L/(A1*E1)+α1*T*L       # displacement of bar-1 at x=L by T
6  u2L = F2T*L/(A2*E2)+α2*T*L       # displacement of bar-2 at x=L by T
7
8  slnFT = sp.solve([-2*F1T-F2T, u1L-u2L], [F1T,F2T])    # set F = 0
9  slnFT[F1T] = (slnFT[F1T]).simplify()
10 slnFT[F2T] = (slnFT[F2T]).simplify()
11
12 slnFT   # solution of internal forces in bars by T
```

$$\left\{ F_{1T} : \frac{A_1 A_2 E_1 E_2 T (-\alpha_1 + \alpha_2)}{2A_1 E_1 + A_2 E_2}, \quad F_{2T} : \frac{2A_1 A_2 E_1 E_2 T (\alpha_1 - \alpha_2)}{2A_1 E_1 + A_2 E_2} \right\}$$

It is clear that the forces result from the difference of the thermal expansion coefficients of these two bars. If $\alpha_1 = \alpha_2$, there will not be forces (and hence thermal stresses) resulting from the temperature change:

```
1  slnσT = {σbar1T:(slnFT[F1T]/A1).simplify(), \
2           σbar2T:(slnFT[F2T]/A2).simplify()}
3  slnσT                            # solution of stresses in bars by T
```

$$\left\{\sigma_{bar1T} : \frac{A_2 E_1 E_2 T \left(-\alpha_1 + \alpha_2\right)}{2A_1 E_1 + A_2 E_2}, \ \sigma_{bar2T} : \frac{2A_1 E_1 E_2 T \left(\alpha_1 - \alpha_2\right)}{2A_1 E_1 + A_2 E_2}\right\}$$

```
1  slnUT = { u1LT: (slnFT[F1T]*L/(A1*E1)).simplify()+α1*T*L, \
2             u2LT: (slnFT[F2T]*L/(A2*E2)).simplify()+α2*T*L}
3  print(slnUT[u1LT].simplify() == slnUT[u2LT].simplify())
4  slnUT  # solution of displacements of bars at x=L, i.e. elongation by T
```

True

$$\left\{u_{1LT} : \frac{A_2 E_2 L T \left(-\alpha_1 + \alpha_2\right)}{2A_1 E_1 + A_2 E_2} + LT\alpha_1, \ u_{2LT} : \frac{2A_1 E_1 L T \left(\alpha_1 - \alpha_2\right)}{2A_1 E_1 + A_2 E_2} + LT\alpha_2\right\}$$

It is clear that even if $\alpha_1 = \alpha_2$, there will still be displacements in these bars because of the thermal expansion.

The elongation of the bars must be the same, we double check using the following code:

```
1  print("Same u_L?:", slnUT[u1LT].simplify()==slnUT[u2LT].simplify())
```

Same u_L?: True

7.11.3 Solution to (3)

Using the data given, compute numerical values for normal stresses, internal normal forces in each of the three bars, and the elongation of the bars when the bar is subjected to both external force and temperature changes.

We can find the formulas for effects from both force F and temperature change T by simply superpositioning. Let us do this via coding:

```
1  F1, F2 = symbols('F1, F2')              # total forces by both F and T
2  Fs = {F1:(slnFF[F1F]+slnFT[F1T]).simplify(), \
3        F2:(slnFF[F2F]+slnFT[F2T]).simplify()}
4  Fs                                      # solution of forces in bars by both F and T
```

$$\left\{F_1 : \frac{A_1 E_1 \left(-A_2 E_2 T \left(\alpha_1 - \alpha_2\right) + F\right)}{2A_1 E_1 + A_2 E_2}, \ F_2 : \frac{A_2 E_2 \cdot \left(2A_1 E_1 T \left(\alpha_1 - \alpha_2\right) + F\right)}{2A_1 E_1 + A_2 E_2}\right\}$$

```
1  σbar1, σbar2 = symbols('σ_bar1, σ_bar2')    # total stresses by F and T
2  σbars = {σbar1:(slnσF[σbar1F]+slnσT[σbar1T]).simplify(), \
3           σbar2:(slnσF[σbar1F]+slnσT[σbar2T]).simplify()}
4  σbars                                       # solution of stresses in bars by both F and T
```

$$\left\{\sigma_{bar1}: \frac{E_1\left(-A_2 E_2 T\left(\alpha_1-\alpha_2\right)+F\right)}{2A_1 E_1+A_2 E_2},\ \sigma_{bar2}: \frac{E_1 \cdot\left(2A_1 E_2 T\left(\alpha_1-\alpha_2\right)+F\right)}{2A_1 E_1+A_2 E_2}\right\}$$

```
1  u1L, u2L = symbols('u_1L, u_2L ')  # total displacements at x=L by F&T
2  UsL = {u1L:(slnUF[u1LF]+slnUT[u1LT]).simplify(), \
3         u2L:(slnUF[u2LF]+slnUT[u2LT]).simplify()}
4  UsL                # solution of displacements of bars at x=L by both F and T
```

$$\left\{u_{1L}: \frac{L\left(2A_1 E_1 T\alpha_1+A_2 E_2 T\alpha_2+F\right)}{2A_1 E_1+A_2 E_2},\ u_{2L}: \frac{L\left(2A_1 E_1 T\alpha_1+A_2 E_2 T\alpha_2+F\right)}{2A_1 E_1+A_2 E_2}\right\}$$

```
1  print("Same u_L?:", UsL[u1L].simplify()==UsL[u1L].simplify())
```

Same u_L?: True

For given data, we can compute the numerical results with ease:

```
1  # Assign values to all parameters. The unites are in Pa, m^2, N, C
2
3  data = {E1:2e11, E2:7e10, A1:5/1000/1000, A2:8/1000/1000, \
4          L:100/1000, F:50, T:300, α1:11e-6, α2:22e-6}
5
6  Fs_val = np.array([Fs[F1].subs(data), Fs[F2].subs(data)], dtype=float)
7  print("Normal Force in bars  =",Fs_val,"(N)")
8
9  σbars_val = np.array([σbars[σbar1].subs(data), \
10                        σbars[σbar2].subs(data)], dtype=float)
11 print(f"Normal stress in bars = {σbars_val} (Pa)")
12
13 UsL_val = np.array([UsL[u1L].subs(data), \
14                     UsL[u1L].subs(data)], dtype=float)
15 print(f"Displacement of bars at x=L: {UsL_val} (m)")
```

Normal Force in bars = [7.414e+02 -1.433e+03] (N)
Normal stress in bars = [1.483e+08 -1.766e+08] (Pa)
Displacement of bars at x=L: [4.041e-04 4.041e-04] (m)

It is clear that the equilibrium of all the forces and the continuity of the displacements are satisfied. Using the same code, readers may compute the results by setting $F=0$ or $T=0$ to examine the numerical results to see if it makes sense.

7.11.4 Solution to (4)

Compute the principal stresses and the maximum stress and their directions in bar-1.

Since the stress in the bars is only in the x-direction, the shear stress is zero in the x–y coordinate. Therefore, the largest principal stress is the normal stress we found earlier. The other principal stresses are all zero:

```
1  PrincipalStress_1 = np.array([σbars_val[0], 0., 0.])
2  print('Principal Stress in bar-1 = ', PrincipalStress_1,'(Pa)')
3  PrincipalStress_2 = np.array([σbars_val[1], 0., 0.])
4  print('Principal Stress in bar-2 = ', PrincipalStress_2,'(Pa)')
5  print('The direction is all in the axial direction of the bars')
```

```
Principal Stress in bar-1 =  [ 1.483e+08  0.000e+00  0.000e+00] (Pa)
Principal Stress in bar-2 =  [-1.766e+08  0.000e+00  0.000e+00] (Pa)
The direction is all in the axial direction of the bars
```

7.12 Remarks

This chapter solves a class of problems of mechanics of materials: 1D bars subjected to types of mechanical and thermal loadings. The bar is assumed long and the loading is axial. The full set of solutions are obtained, including displacements, stresses and strains. A number of useful Python codes are provided and examined using various examples.

The following chapter solves another class of problem of structure members: beams subjected to bending.

Reference

[1] G.R. Liu and S.S. Quek, *The Finite Element Method: A Practical Course.* Butterworth-Heinemann, New York, 2013.

Chapter 8

Solution to Bended Beams

```
1  # Place cursor in this cell, and press Ctrl+Enter to import dependences.
2  import sys                         # for accessing the computer system
3  sys.path.append('../grbin/')   # Change to the directory in your system
4
5  from commonImports import *        # Import dependences from '../grbin/'
6  import grcodes as gr               # Import the module of the author
7  importlib.reload(gr)               # When grcodes is modified, reload it
8
9  from continuum_mechanics import vector
10 from continuum_mechanics.solids import sym_grad, strain_stress
11 from matplotlib import colors
12 from collections import OrderedDict
13 init_printing(use_unicode=True)    # For Latex-like quality printing
14 #np.set_printoptions(precision=4,suppress=True)  # Digits in print-outs
15 np.set_printoptions(formatter={'float': '{: 0.4e}'.format})
```

Beams may be the most widely used structure members. When one sees a structure, most likely there are some beam members in it. A beam takes transverse loads and moments, resulting in deflections, rotations, and internal shear force and moments. This chapter develops equations that govern the behavior of beams, methods and formulas and codes leading to solutions to loaded beams in terms of deflection, rotation, and internal shear force and moments.

8.1 Governing equation, fourth-order differential equation

8.1.1 *Thin beam theory*

Consider a thin beam member with its axial axis long the x-direction, as shown in Fig. 8.1. Its cross-section dimensions are much smaller than its length. It is similar to a 1D bar studied in the previous chapter in terms of slender geometry. The difference is that a beam is subjected to forces in the transverse y-direction, or moments with respect to z axis, resulting in bending deformation called **deflection** and internal shear force and moments.

The primary variable for the beam is the deflection, the displacement in the y-direction v of its neutral axis, where the normal stress σ_{xx} is zero. The deflection is primary because all other variables such as cross-section rotation, normal stress, shear forces, and moments can be obtained from it. Since the deflection v changes only along the x axis, it is essentially a one-dimensional (1D) problem, after a proper simplification process.

Consider transverse loads are only in the y-direction. The defection will not occur in the z-direction. Therefore, it can be treated as at most a 2D problem, for which there are two unknown displacement components: u and v. To further simplify the problem to 1D, we must find a way to relate one of the displacements, say u, to v. This needs to be done in a way that captures the major mechanics behavior of the beam under bending. The so-called **Euler–Bernoulli beam theory** captures well such behavior of thin beams, and hence most widely used. This theory makes two major assumptions:

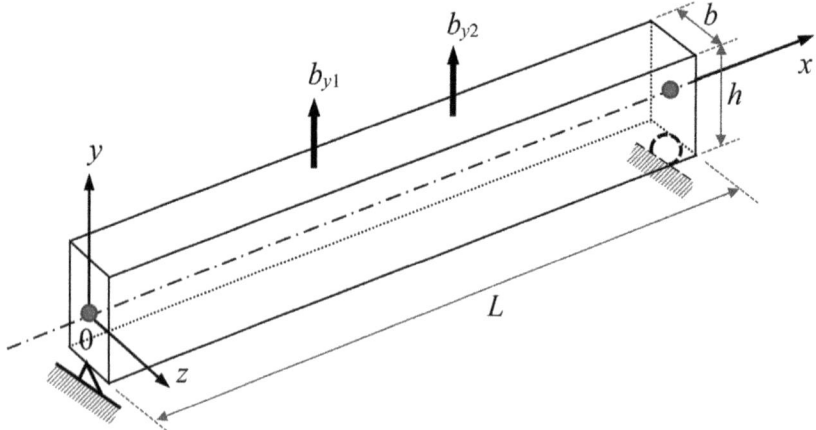

Figure 8.1. A 1D beam member. It is subjected to forces in the transverse y-direction, resulting in bending deformation.

1. The cross-section of the beam remains straight during bending.
2. The cross-section of the beam (perpendicular to its axis) remains perpendicular to its axis during bending.

Figure 8.2 is a schematic diagram of a deflected beam.

8.1.2 *Normal stress and the resulted moment*

These two assumptions mean that the shear strain ε_{xy} is negligible. Otherwise the cross-section and the neutral axis will not be perpendicular, when the beam is bent. This leads to

$$\varepsilon_{xy} = \frac{1}{2}\left(\frac{\partial u}{\partial y} + \frac{\partial v}{\partial x}\right) = 0 \tag{8.1}$$

where $\frac{\partial v}{\partial x}$ is the **slope** of the deflection. Due to the second assumption, the rotation of the cross-section θ is the same as the slope of the deflection (see Fig. 8.2). We thus have

$$\theta = \frac{\partial v}{\partial x} \tag{8.2}$$

Equation (8.1) becomes

$$\frac{\partial u}{\partial y} = -\theta \tag{8.3}$$

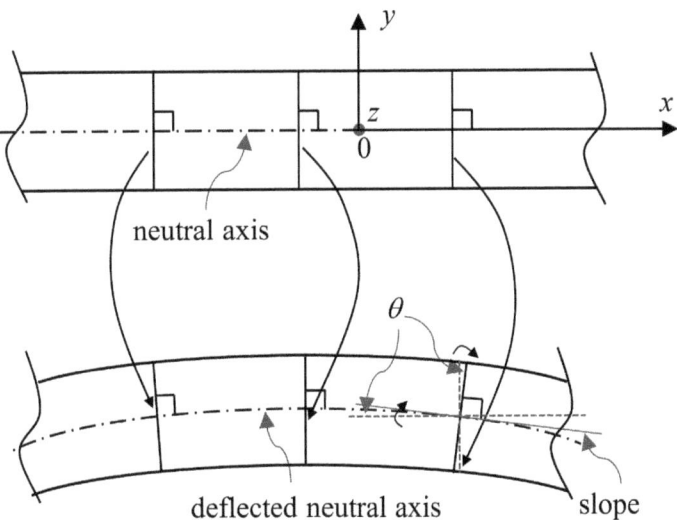

Figure 8.2. A thin beam under bending. This neutral axis that representing the beam has a deflection, while the cross-section of the beam has a rotation. The slope of the deflection and the rotation angle is the same due to the Euler–Bernoulli assumptions.

Integration of Eq. (8.3) with respect to y gives

$$u = -y\theta + c \tag{8.4}$$

where c is an integral constant. Since the neutral axis ($y = 0$) does not deform in the x direction, $u = 0$. This condition leads to $c = 0$. Equation (8.4) becomes

$$u = -y\theta = -y\frac{\partial v}{\partial x} \tag{8.5}$$

The displacement u is written in the derivative of deflection v. Thus, two unknown displacements become only one, which is deflection v. We need to write other variables also in terms of only v.

Using Eq. (8.5), the normal strain can now be evaluated using

$$\varepsilon_{xx} = \frac{\partial u}{\partial x} = -y\frac{\partial^2 v}{\partial x^2} \tag{8.6}$$

The normal strain ε_{xx} is written in terms of the second derivative of deflection v. It is linear in y, implying that the normal strain distribution along y is a straight line.

Note that there is only one independent variable x involved for beams under bending. Therefore, the partial derivative can be replaced by total derivative.

For this 1D beam, the normal stress relates the normal strain by Hooke's law, and hence can be obtained using

$$\sigma_{xx} = E\varepsilon_{xx} = -yE\frac{d^2v}{dx^2} \tag{8.7}$$

This means the normal stress changes also linearly with y, as shown in Fig. 8.3.

The linearly distributed normal stress results in moment M_z on the cross-section, which can be evaluated using,

$$M_z = -\int_A y\sigma_{xx} dA = E \underbrace{\left(\int_A y^2 dA\right)}_{I_z} \frac{d^2v}{dx^2} = EI_z\frac{d^2v}{dx^2} \tag{8.8}$$

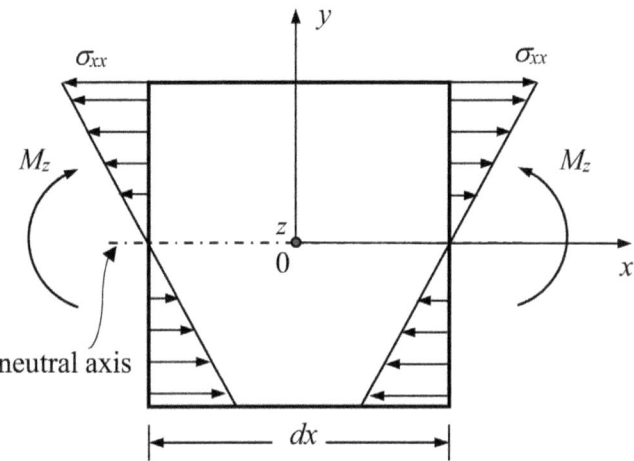

Figure 8.3. Normal stress (positive) distribution along the cross-section of beams, and the resultant moment (positive), following the standard sign convention.

or

$$M_z = EI_z \frac{d^2 v}{dx^2} \quad (8.9)$$

Note that on a positive plane, positive moment must be counterclockwise. The minus sign is because the positive normal stress results in negative moment. Equation (8.9) shows that the moment is related to the deflection in the second derivative, in which I_z is the **second moment of area** of the cross-section of the beam about z axis. It is calculated using

$$I_z = \int_A y^2 \, dA \quad (8.10)$$

This is the general equation for computing I_z of the cross-section of a beam. Its value dependents on shape and dimensions of cross-section over which the integration is performed. Note that because y is squared, the height of the cross-section influences significantly the I_z value.

Using Eqs. (8.9) and (8.7), the normal stress relates to the moment as

$$\sigma_{xx} = -y \frac{M_z}{I_z} \quad (8.11)$$

This is a useful equation for computing normal stress on the cross-section of the beam for given moment. We see that:

1. The normal stress is proportional to the moment on the cross-section, implying that larger moment results in large normal stress.

2. The normal stress is linear in y, and the largest stress is registered at y_{max} or $-y_{max}$ on the cross-section.
3. Increasing I_z reduces the stress. Since I_z relates to the shape of the cross-section, it is thus important to design a good shape with large I_z value.

Since the moment relates to the deflection v in Eq. (8.9). The normal stress is also related to displacement v.

8.1.3 *On second moment of area of beam cross-section*

The normal stress is inversely proportional to I_z of the cross-section, implying that a larger I_z reduces the normal stress. In other words, we would like to have a beam with large I_z.

For beams with rectangular cross-section with height h and base b, as shown in Fig. 8.4 (see also Fig. 8.1), I_z can be computed easily using Eq. (8.10):

$$I_z = \int_A y^2 dA = \int_{-h/2}^{h/2} y^2 b\, dy = 2\int_0^{h/2} y^2 b\, dy = 2b\frac{1}{3}y^3\Big|_0^{h/2} = \frac{bh^3}{12} \quad (8.12)$$

To make I_z larger, we can increase b and/or h. Since h has a cubic power, increasing h is far more effective than increasing b for given same amount of material. This means that we would like to make the beam as tall as possible, which is what the engineering designer is doing. However, there is a limit to

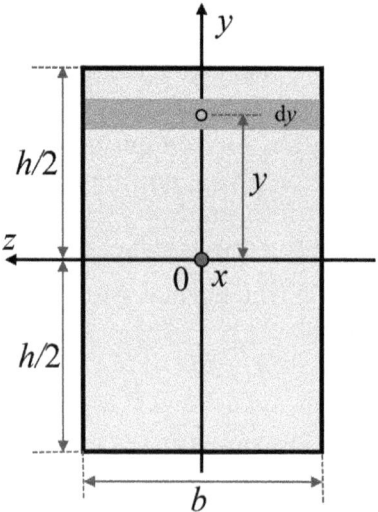

Figure 8.4. Rectangular cross-section of a beam.

it because beam with too large h/b ratio will have stability issues, which will be discussed in Chapter 12.

A better alternative is to design the beam cross-section with a better shape. The objective is to move the material as farther as possible away from the neutral axis of the beam. A typical such design is the I-beam. This kind of study relates to the property of an area, which will be discussed in more detail in Chapter 10.

8.1.4 Equilibrium equation for beams

We now need to establish the equilibrium equation of beams subjected to transverse loading. This is done using the usual trick of free body diagram, which is created by isolating a small cell from the beam with loads on. Figure 8.5 shows such a cell.

Here, ρ is the density of the material. The external body-force is assumed to be b_y that can change along x. It carries a unit of $[N/m]$. When the force is dynamic or the beam is subjected to dynamic excitation, there can also be inertia force, which depends on the mass of the free-body and its acceleration, as shown in Fig. 8.5. On the cross-section, there are a possible moment and a shear force caused by body-force in the transversal direction. Both moments and shear forces are internal because we see them only when the cross-section is exposed. They both may vary along x, and thus have increments corresponding to the coordinate increment dx.

First, all the moments about z axis must be in equilibrium. This leads to

$$dM_z + V dx - \bigl(b_y(x) - \rho A\ddot{v}\bigr) dx \frac{1}{2} dx = 0 \qquad (8.13)$$

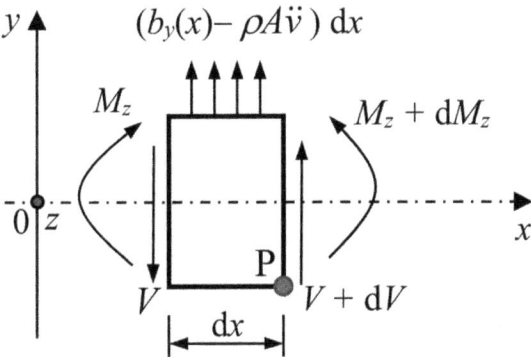

Figure 8.5. A small free-body diagram from a beam subjected to loading, including possible inertia force. All the forces (moments) shown are all positive by our sign convention.

Since the third term is second order in the small dx, it can be neglected. We obtain

$$\boxed{\frac{dM_z}{dx} = -V} \qquad (8.14)$$

This means that the shear force on the cross-section is the negated derivative of the moment.

Next, all these forces in the y direction must be in equilibrium. This gives

$$dV + \left(b_y(x) - \rho A\ddot{v}\right)dx = 0 \qquad (8.15)$$

or

$$\frac{dV}{dx} = -b_y(x) + \rho A\ddot{v} \qquad (8.16)$$

Using Eq. (8.9), we found the shear force in terms of deflection:

$$\boxed{V = -EI_z \frac{d^3v}{dx^3}} \qquad (8.17)$$

The shear force relates to the deflection in its third derivative. Substituting Eqs. (8.17) to (8.16), we finally arrived at the dynamic equilibrium equation for beams:

$$EI_z \frac{d^4v}{dx^4} + \rho A\ddot{v} = b_y \qquad (8.18)$$

For static cases, we have

$$\boxed{\frac{d^4v}{dx^4} = \frac{b_y}{EI_z}} \qquad (8.19)$$

where EI_z is often called bending stiffness (per length) of the beam. Equation (8.19) is the well-known **beam equation** based on the Euler–Bernoulli theory, and is the **governing equation** for thin beams with only one unknown function v. It is quite similar to Eq. (7.1), but it is a fourth-order differential equation (DE). It can be solved through integration as long as the body force b_y is integrable. The integration will result in four integral constants, which need to be determined using boundary conditions (BCs). The process is essentially the same as we did in the previous chapter for second-order DEs, but a little more complicated because of the higher-order differential.

In this section, we put subscript z for I and M to indicate the bending is with respect to the z-axis. In the following sections, we may drop z to lighten the subscripts.

8.2 Boundary conditions for beams

8.2.1 *Possible BCs for beams*

Based on the thin beam theory, at a cross-section (a point on its neutral axis to be more precise) has two displacement degrees of freedom (DoFs): deflection v and rotation θ. Their corresponding force variables are the shear force V and moment M. Thus, there are a total of four possible variables: v, θ, V, and M, at any cross-section. There are always two paired complementary relations. The first pair is v with V, implying that if we know v at a cross-section, we do not in general know V, and if we know V we do not generally know v. The same applies to the pair of θ with M. Thus, we can only pre-scribe two variables at a cross-section.

Now, any given beam will have two boundaries: left- and right-ends. At each of these two boundary points, there are two possible BCs. The possible BCs for a boundary point can be given in Fig. 8.6. As seen, we cannot pre-scribe any complementary pair of variables at a boundary point.

Since there will always be two boundary points for a given beam, we always have four BCs. This is just the right number we need to solve the fourth-order DE given in Eq. (8.19).

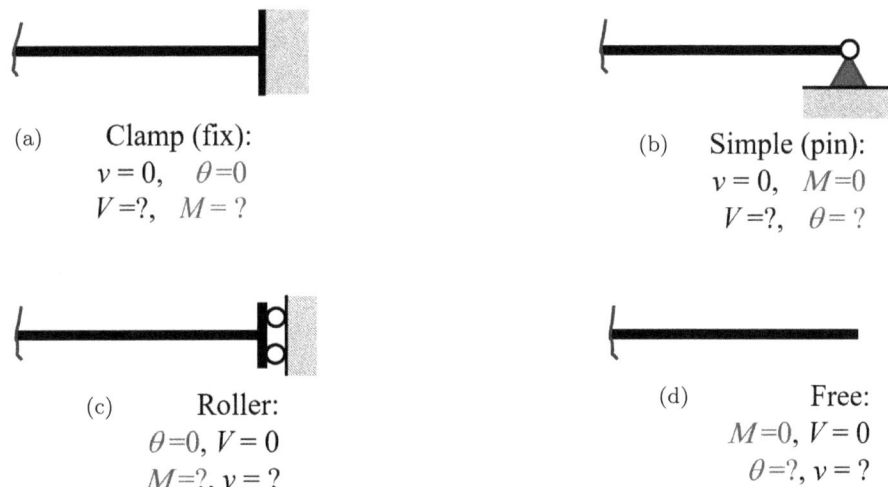

Figure 8.6. Possible boundary conditions for a beam based on the thin beam theory. In many cases, these two variables are given as zero, but in general these can also be pre-specified values.

8.2.2 Admissible BCs for beams

It is important to note that in a set of four BCs, there must be at least two conditions on these two displacement DoFs. This is because we must constrain the beam for rigid body movements, or we will not have a unique solution. For example, we cannot have both ends of the beam have condition (d) shown in Fig. 8.6. Since the beam can move freely without any force being generated. If condition (d) is given for one end, the other end must be condition (a), which is the often-encountered cantilever beam. Also, (d) cannot go with (b) or (c) because such cases have only one displacement variable constrained. The other rigid motion is still possible.

On the other hand, we can have overprescribed with displacement DoFs because our method is displacement based. It works naturally well for indeterminate beams. For example, it is perfectly fine for beams with both ends clamped that is known as c-c condition. Simple-clamp (s-c) and roller-clamp (r-c) are also fine, implying that unique solutions can be found.

8.3 Example: A cantilever beam subjected to a force at its right-end

Considering a simplest case of cantilever beam fixed at the left-end and subjected to a concentrated force F [N] at the right-end, without body force, as shown in Fig. 8.7. The beam has length L and a uniform cross-section with its second moment of area given as I.

We derive, by hand, the formulas for solutions to a beam as functions of x, so that we can have a good idea on how Eq. (8.19) can be solved. We will then write a Python code that can be used to derive formulas for more

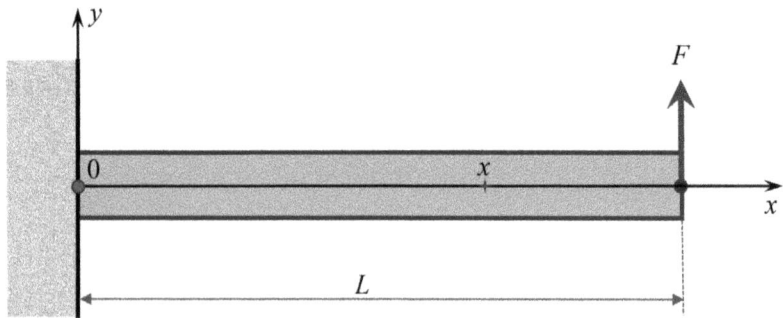

Figure 8.7. A cantilever beam fixed at the left-end, subjected to a concentrated force at the right-end.

complicated problems. The code follows essentially the same process, but allows the computer to do the heavy lifting of derivation.

Solution: Since there is no body force, Eq. (8.19) is reduced to

$$\frac{d^4v}{dx^4} = 0 \tag{8.20}$$

The general solution of the deflection v that satisfies Eq. (8.20) can be found by integration four times, which gives

$$v = \frac{1}{6}c_3 x^3 + \frac{1}{2}c_2 x^2 + c_1 x + c_0 \tag{8.21}$$

where $c_i, i = 0, 1, 2, 3$ are integral constants. Equation (8.21) is the **general solution** for any beam without body force. It has a total of four constants because the DE is of fourth order. For a given specified problem defined in this example, we shall use the BCs to determine these constants.

At $x = 0$, the beam is fixed and hence we have the following BCs:

$$\begin{cases} v = 0 \\ \theta = 0 \end{cases} \text{at } x = 0 \tag{8.22}$$

To satisfy Eq. (8.22), $c_0 = 0$ and $c_1 = 0$. Equation (8.21) is reduced to

$$v = \frac{1}{6}c_3 x^3 + \frac{1}{2}c_2 x^2 \tag{8.23}$$

At $x = L$, we have the following BCs:

$$\begin{cases} V = F \\ M = 0 \end{cases} \text{at } x = L \tag{8.24}$$

Using Eqs. (8.17) and (8.9), we have

$$\begin{cases} -EI c_3 = F \\ EI(c_3 L + c_2) = 0 \end{cases} \text{at } x = L \tag{8.25}$$

which gives

$$\begin{cases} c_3 = -\dfrac{F}{EI} \\ c_2 = \dfrac{FL}{EI} \end{cases} \text{at } x = L \tag{8.26}$$

Substituting the foregoing equation to Eq. (8.23), we obtain the **solution for the deflection** as follows:

$$v = -\frac{F}{6EI}x^3 + \frac{FL}{2EI}x^2 \tag{8.27}$$

At the right-end of the beam, $x = L$, the deflection is found as

$$v_L = \frac{L^3}{3EI} F \quad (8.28)$$

which is the largest deflection on the cantilever beam subject to point force at its free end.

Once the deflection is found, we can easily find all other solutions (typical for a displacement method). The **solution for the rotation** is obtained using Eq. (8.2):

$$\theta = \frac{\partial v}{\partial x} = -\frac{F}{2EI} x^2 + \frac{FL}{EI} x \quad (8.29)$$

At the right-end of the beam, $x = L$, the rotation is found as

$$\theta_L = \frac{L^2}{2EI} F \quad (8.30)$$

The **moment** is found using Eq. (8.9):

$$M = EI \frac{d^2 v}{dx^2} = F(L - x) \quad (8.31)$$

It changes linearly along the beam. The **shear force** is found using Eqs. (8.17) and (8.27):

$$V = -EI \frac{d^3 v}{dx^3} = F \quad (8.32)$$

It is constant along the beam, which can be easily confirm by equilibrium.

The correctness of these solutions can be confirmed using the general solver: gr.solver1D4() to be developed in the next section.

The **normal stress** is found using Eq. (8.11):

$$\sigma_{xx} = -y \frac{F(L - x)}{I} \quad (8.33)$$

The same procedure can find solutions for beams subjected to body forces, as long as the force function is integrable analytically. We shall use Sympy to do the derivation for us.

8.4 A Python beam solver

Our Python beam solver, solver1D4(), is developed based on the following considerations:

1. Since Eq. (8.19) is a fourth-order DE, and we need to integrate it four times using Sympy. This results in 4 integral constants. For beams of length l, we have two boundaries at $x = 0$ (left-end) and $x = l$ (right-end). At each boundary, we have two of the four possible variables: deflection v, rotation θ, shear force V and moment M. Therefore, we shall have just enough 4 equations to determine these integral constants. After these 4 constants are determined, we can have analytic formulas for v, $\frac{dv}{dx}$, $\frac{d^2v}{dx^2}$, $\frac{d^3v}{dx^3}$, and $\frac{d^4v}{dx^4}$. These formulas give the full solution set in a Python list.
2. The solver1D4() accepts the following combinations of BCs: c-c, s-c, c-s, f-c, and c-f. In which c stands for clamp, s for simple, and f for free. These set of BCs are all admissible.
3. As discussed before, because the mechanics problem is **complimentary**, the displacement variables are naturally paired with the force variables. For example, for the s-c BC, v and M are specified at $x = 0$, and v and θ are specified at $x = l$. This gives 4 equations for determining these 4 integral constants. Their corresponding counterparts V and θ at $x = 0$, and V and M at $x = l$ can then be computed using the output formulas from solver1D4().
4. The computation is done symbolically. Hence, we can set all the variables specified on the BCs as symbolic variables, and the derived general formulas are in closed-form. Such a set of formulas can then be used to compute many problems of beams with complicated loading, with multiple spans, and even frame structures. If some or all of these variables are zero, we can simply set so using sp.subs(). Therefore, it is flexible and useful.
5. The shear force and moment used in solver1D4() are **internal**. When replacing those by the applied external force/moment at the boundary, the **standard sign convention** applies: on the right-boundary (positive boundary), the directions of the internal force/moment are the same as these of the external point force and moment applied there; On the left-boundary (negative boundary), the directions the internal force/moment are opposite to these of the externally applied point force/moment. This is because positive externally applied force always follows the direction of the coordinates, and positive externally applied moment always follows the right-hand rule. Internal forces/moments, however, follow the direction of the surface the forces/moments.

8.4.1 *Python code of beam solver*

The following code uses a step-by-step approach to perform the integration four times using Sympy, and then uses BCs of symbolic variables to determine the integral constants. The code is self-explanatory with comments given.

```python
 1  def solver1D4(E, I, by, l, v0, θ0, vl, θl, V0, M0, Vl, Ml, key='c-c'):
 2
 3      '''Solves the Beam Equation for integrable distributed body force:
 4              u,x4=-by(x)/EI, with various boundary conditions (BCs):
 5              s-s, c-c, s-c, c-s, f-c, c-f.
 6      Input: EI: bending stiffness factor; by, body force; l, the length
 7              of the beam; v0, θ0, V0, M0, deflection, rotation,
 8              (internal) shear force, (internal) moment at x=0;
 9              vl, θl, Vl, Ml, are those at x=l.
10      Return: u, v_x, v_x2, v_x3, v_x4    up to 4th derivatives of v
11      '''
12      x = sp.symbols('x')
13      c0, c1, c2, c3 = sp.symbols('c0, c1, c2, c3')  #integration constant
14      EI = E*I                    # I is the Iz in our chosen coordinates.
15
16      # Integrate 4 times:
17      v_x3= sp.integrate(by/EI,(x, 0, x))+ c0    #ci: integration constant
18      v_x2= sp.integrate( v_x3,(x, 0, x))+ c1
19      v_x = sp.integrate( v_x2,(x, 0, x))+ c2
20      v   = sp.integrate(  v_x,(x, 0, x))+ c3
21
22      # Solve for the 4 integration constants:
23      if   key == "s-s":
24          cs=sp.solve([v.subs(x,0)-v0, v_x2.subs(x,0)-M0/EI,
25                      v.subs(x,l)-vl, v_x2.subs(x,l)-Ml/EI],[c0,c1,c2,c3])
26      elif key == "c-c":
27          cs=sp.solve([v.subs(x,0)-v0, v_x.subs(x,0)-θ0,
28                      v.subs(x,l)-vl, v_x.subs(x,l)-θl], [c0,c1,c2,c3])
29      elif key == "s-c":
30          cs=sp.solve([v.subs(x,0)-v0, v_x2.subs(x,0)-M0/EI,
31                      v.subs(x,l)-vl, v_x.subs(x,l)-θl], [c0,c1,c2,c3])
32          #print('solver1D4:',cs[c0],cs[c1],cs[c2],cs[c3])
33      elif key == "c-s":
34          cs=sp.solve([v.subs(x,0)-v0, v_x.subs(x,0)-θ0,
35                      v.subs(x,l)-vl, v_x2.subs(x,l)-Ml/EI],[c0,c1,c2,c3])
36      elif key == "c-f":
37          cs=sp.solve([v.subs(x,0)-v0, v_x.subs(x,0)-θ0,
38              v_x3.subs(x,l)+Vl/EI, v_x2.subs(x,l)-Ml/EI], [c0,c1,c2,c3])
39      elif key == "f-c":
40          cs=sp.solve([v_x3.subs(x,0)+V0/EI, v_x2.subs(x,0)-M0/EI,
41                      v.subs(x,l)-vl, v_x.subs(x,l)-θl], [c0,c1,c2,c3])
42      else:
43          print("Please specify boundary condition type.")
44          sys.exit()
45
46      # Substitute the constants back to the integral solutions
47      v   =   v.subs({c0:cs[c0],c1:cs[c1],c2:cs[c2],c3:cs[c3]})
48      v   =   v.expand().simplify().expand()
49
50      v_x = v_x.subs({c0:cs[c0],c1:cs[c1],c2:cs[c2],c3:cs[c3]})
51      v_x = v_x.expand().simplify().expand()
52
53      v_x2=v_x2.subs({c0:cs[c0],c1:cs[c1],c2:cs[c2],c3:cs[c3]})
```

```
54      v_x2=v_x2.expand().simplify().expand()
55
56      v_x3=v_x3.subs({c0:cs[c0],c1:cs[c1],c2:cs[c2],c3:cs[c3]})
57      v_x3=v_x3.expand().simplify().expand()
58
59      v_x4 = sp.diff(v_x3,x).expand()
60
61      print("Outputs form solver1D4(): v, θ, M, V, qy")
62      return v,v_x,(EI*v_x2).expand(),(-EI*v_x3).expand(),(v_x4).expand()
```

Let us use this function to find solutions for some bending problems. We define first necessary symbolic variables.

```
1  E, I, EI, A, l, ξ = symbols('E, I, EI, A, l, ξ ', nonnegative=True)
2  x, q = symbols('x, q ')             # geometry & force
3  c0, c1, c2, c3 = symbols('c0, c1, c2, c3')         #integration constant
4
5  # for displacement boundary conditions (DBCs):
6  v0, θ0, vl, θl = symbols('v0, θ0, v_l, θ_l')       # DBCs
7
8  # for force boundary conditions (FBCs):
9  V0, M0, Vl, Ml = symbols('V0, M0, V_l, M_l')       # FBCs
10
11 by = sp.Function('b_y')(x)                         # body force function
```

8.4.2 Example: Clamped-clamped beams

Consider beams of length l with clamped-clamped (c-c) displacement boundary conditions (DBCs) given in symbolic variables, as shown in Fig. 8.8. The cross-section has a second moment of area I. It is made of material with Young's modulus E. It is subjected to a distributed force $q(x)$, with the x-axis starting at the left-end of the beam and being leftwards. Derive formulas for solutions, including deflection, rotation, moment, and shear force in the beam.

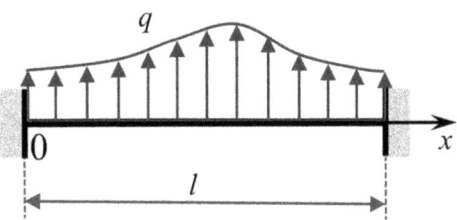

Figure 8.8. A beam with clamped-clamped (c-c) displacement boundary conditions (DBCs). It is subjected to a distributed force $q(x)$.

We consider first uniformly distributed force.

```
1  # The title strings are used later for plots.
2  title = [["Deflection", "Rotation"], ["Moment", "Shear force"]]
3
4  # Consider constant distributed force
5  by = q                    # [N/m], one may try other force functions
6
7  v = solver1D4(E, I, by, l, v0, θ0, vl, θl, V0, M0, Vl, Ml,key='c-c')
8  print(f'Is solution correct? {by/E/I==v[4]}; The solution v(x) is:')
9  v[0]                      # solution of general displacement function
```

Outputs form solver1D4(): v, θ, M, V, qy
Is solution correct? True; The solution v(x) is

$$v_0 + x\theta_0 - \frac{2x^2\theta_0}{l} - \frac{x^2\theta_l}{l} - \frac{3v_0 x^2}{l^2} + \frac{3v_l x^2}{l^2} + \frac{x^3\theta_0}{l^2} + \frac{x^3\theta_l}{l^2} + \frac{2v_0 x^3}{l^3} - \frac{2v_l x^3}{l^3}$$
$$+ \frac{l^2 q x^2}{24EI} - \frac{l q x^3}{12EI} + \frac{q x^4}{24EI}$$

This expression is a little lengthy with all the symbolic variables. It is a general solution and hence is useful to many beam problems. In the solution expression, v_0, θ_0 are the deflection and rotation at the left-end, and v_l, θ_l are the deflection and rotation at the right-end of the beam. This allows users to replace these variables at the boundary with known values. For example, if these variables are all zero, we can simply set so, as in the following code.

```
1  # Set all the pre-scribed displacements to zero:
2
3  vx=[v[i].subs({v0:0,θ0:0,vl:0,θl:0}) for i in range(len(v))]
4  vx                        # Outputs are: v, θ, Mx, Vx, qy
```

$$\left[\frac{l^2 q x^2}{24EI} - \frac{l q x^3}{12EI} + \frac{q x^4}{24EI}, \frac{l^2 q x}{12EI} - \frac{l q x^2}{4EI} + \frac{q x^3}{6EI},\right.$$
$$\left.\frac{l^2 q}{12} - \frac{l q x}{2} + \frac{q x^2}{2}, \frac{l q}{2} - q x, \frac{q}{EI}\right]$$

This is a set of solutions that may be found in textbooks or other open literature. A better presentation of these solutions is in a so-called natural coordinate system. We can obtain these using the following code.

```
1  # Convert to the natural coordinate: ξ = x/l (dimensionless)
2  vξ = [vx[i].subs({x:ξ*l}) for i in range(len(vx))]
3  vξ
```

$$\left[\frac{l^4q\xi^4}{24EI} - \frac{l^4q\xi^3}{12EI} + \frac{l^4q\xi^2}{24EI}, \frac{l^3q\xi^3}{6EI} - \frac{l^3q\xi^2}{4EI} + \frac{l^3q\xi}{12EI},\right.$$

$$\left.\frac{l^2q\xi^2}{2} - \frac{l^2q\xi}{2} + \frac{l^2q}{12}, -lq\xi + \frac{lq}{2}, \frac{q}{EI}\right]$$

The above results can be written in the following easy-to-manage forms.

$$v(\xi) = \underbrace{\frac{l^4q}{EI}}_{\text{displacement unit}} \underbrace{\left[\frac{\xi^4}{24} - \frac{\xi^3}{12} + \frac{\xi^2}{24}\right]}_{\bar{v}(\xi)} = \frac{l^4q}{EI}\bar{v}(\xi)$$

$$\theta(\xi) = \underbrace{\frac{l^3q}{EI}}_{\text{rotation unit}} \underbrace{\left[\frac{\xi^3}{6} - \frac{\xi^2}{4} + \frac{\xi}{12}\right]}_{\bar{\theta}(\xi)} = \frac{l^3q}{EI}\bar{\theta}(\xi) \quad (8.34)$$

$$M(\xi) = \underbrace{\frac{l^2q}{EI}}_{\text{moment unit}} \underbrace{\left[\frac{\xi^2}{2} - \frac{\xi}{2} + \frac{1}{12}\right]}_{\bar{M}(\xi)} = \frac{l^2q}{EI}\bar{M}(\xi)$$

$$V(\xi) = \underbrace{ql}_{\text{force unit}} \underbrace{\left[-\xi + \frac{1}{2}\right]}_{\bar{V}(\xi)} = ql\bar{V}(\xi)$$

where $\xi = x/l$ is the so-called natural coordinate, which is normalized coordinate by the length of the beam l. Thus, ξ is dimensionless and varies in [0 1]. Each variable has its own unit. For example, the unit for the displacement is $\frac{l^4q}{EI}$ and it will be naturally with a length unit (m). The bar on top of a variable stands for dimensionless function depending only on $\xi \in [0\ 1]$, and hence convenient to analyze, especially in symbolic computations. For example, we need only to deal with the dimensionless function of variables to find the locations of the maximum or minimum values. In such an analysis we can set all the variable units to 1, without affecting the generality. We will practice this in the following analysis. After the solutions are found using the dimensionless functions, one can re-install physical solutions simply by multiplying the corresponding units, using Eq. (8.34).

8.4.3 *Use of the Sympy differential equation solver*

Alternatively, one my use the built-in sp.dsolve() to solve the differential equation directly.

```
1   E, I, EI, l = symbols('E, I, EI, l', nonnegative=True)
2   x, q, M0 = symbols('x, q, M0 ')
3   v0, vl, θ0, θl = symbols('v0, v_l, θ0, θ_l')
4   vv = sp.Function('vv')(x)                  # deflection of the beam, v (m)
5
6   # Consider the distributed force is a constant q
7   by = q #sp.sin(sp.pi*x)
8
9   # Define the 2nd differential equation
10  diff_eq = vv.diff(x, 4) - by/EI
11
12  # Define the boundary conditions, consider clamped-clamped, "c-c":
13  BCs={vv.subs(x,0): v0, vv.diff(x).subs(x,0): θ0,
14       vv.subs(x,l): vl, vv.diff(x).subs(x,l): θl}
15
16  #simple-simple supports, "s-s"
17  #BCs={vv.subs(x,0): v0, vv.diff(x,2).subs(x,0): M0/EI,
18  #    vv.subs(x,L): vL, vv.diff(x,2).subs(x,L): ML/EI}
19
20  # Solve the differential equation using sp.dsolve()
21  solution = sp.dsolve(diff_eq, vv, ics=BCs)   # This is an sp.Eq object
22
23  # check whether the DE is satisfied.
24  print(f'Is solution correct? {solution.rhs.diff(x,4) == by/EI}')
25  solution.rhs.expand()                                         # solution
```

Is solution correct? True

$$v_0 + x\theta_0 - \frac{2x^2\theta_0}{l} - \frac{x^2\theta_l}{l} - \frac{3v_0 x^2}{l^2} + \frac{3v_l x^2}{l^2} + \frac{x^3\theta_0}{l^2} + \frac{x^3\theta_l}{l^2} + \frac{2v_0 x^3}{l^3}$$
$$- \frac{2v_l x^3}{l^3} + \frac{l^2 q x^2}{24EI} - \frac{l q x^3}{12EI} + \frac{q x^4}{24EI}$$

The solution obtained is the same as that using method 1. Since our beam solver given in method 1 has all the different BCs built in, it is much easier to use. We will use solver1D4() in the following examples. For conveniences, we write a few more utility codes.

8.5 Python code for finding the maximum and minimum values

We write a code to find the maximum and minimum values of a given function.

```
1  def roundS(expr, n_d):         # to be used in maxminS()
2      '''To limit the number of digits to keep in a variable.
3      Usage: roundS(expr, n_d), where n_d: number of digits to keep.
4      '''
5      return expr.xreplace({n.evalf():round(n,n_d)
6                            for n in expr.atoms(sp.Number)})
```

```
1  def maxminS(f, title="value"):
2      '''Find maximum location and values of a symbolic function f.
3      '''
4      ndg = 8                      # number of digits
5      df = sp.diff(f, x)
6      ddf = sp.diff(df, x)
7      df0_points = sp.solve(df, x)      #find stationary points
8      df0s = [roundS(point.evalf(),ndg) for point in df0_points]
9      print(f"Stationary points for {title}: {df0s}")
10
11     for point in df0s:
12         fv = roundS((f.subs(x,point)).evalf(), ndg)
13         ddfv = ddf.subs({x:point})
14         if ddfv < 0:
15             print(f"At x={point}, local maximum {title}={fv}")
16         elif ddfv > 0:
17             print(f"At x={point}, local minimum {title}={fv}")
18         else:
19             print(f"At x={point}, max or min {title}={fv}")
20
21     df0s.append(0)           # Add in points on the boundaries
22     df0s.append(1)
23     fs_df0 = [f.subs(x, point).evalf() for point in df0s]
24     f_max = roundS(max(fs_df0), ndg)
25     x_max=[pnt for pnt in df0s if ma.isclose(f.subs(x,pnt),max(fs_df0))]
26
27     print(f"\nAt x={x_max}, Max {title}={f_max}\n")
```

Since the location of the extreme values, maximum and/or minimum values, depends only on the function part of the solution. We can set all the parameters to unity and then find these extreme values and their corresponding locations.

```
1  # Set all factors to unit to the solution obtained using beam solver.
2  dic = {E:1,I:1,l:1,q:1,v0:0,θ0:0,vl:0,θl:0}
3  vx=[v[i].subs(dic) for i in range(len(v))]
4  vx                  # Outputs are: v, θ, Mx, Vx, qy
```

$$\left[\frac{x^4}{24} - \frac{x^3}{12} + \frac{x^2}{24},\ \frac{x^3}{6} - \frac{x^2}{4} + \frac{x}{12},\ \frac{x^2}{2} - \frac{x}{2} + \frac{1}{12},\ \frac{1}{2} - x,\ 1\right]$$

```
1  maxminS(vx[0], title="deflection")
2  maxminS(vx[1], title="rotation")
```

```
Stationary points for deflection: [0, 0.50000000, 1.00000000000000]
At x=0, local minimum deflection=0
At x=0.50000000, local maximum deflection=0.00260417
At x=1.00000000000000, local minimum deflection=0

At x=[0.50000000], Max deflection=0.00260417

Stationary points for rotation: [0.21132487, 0.78867513]
At x=0.21132487, local maximum rotation=0.00801875
At x=0.78867513, local minimum rotation=-0.00801875

At x=[0.21132487], Max rotation=0.00801875
```

```
1  maxminS(vx[2], title="moment")
2  maxminS(vx[3], title="shear force")
```

```
Stationary points for moment: [0.50000000]
At x=0.50000000, local minimum moment=-0.04166667

At x=[0, 1], Max moment=0.08333333

Stationary points for shear force: []

At x=[0], Max shear force=0.50000000
```

8.6 Python code for plotting the solution marked with extreme values

The most effective way to analyze the solution obtained is to plot it in graphs. The following code plots the distribution along the beam for all four solutions: deflection, rotation, moment and shear force.

```
1  # plot the distribution of the solutions:
2
3  for i in range(len(title)):
4      gr.plot2curveS(vx[2*i:2*(i+1)], 0., 1., title=title[i])
5
6  #fig.tight_layout()
7  #plt.savefig('images/beam_cq.png', dpi=500)    # save the plot to file
8  plt.show()
```

```
Maximum Deflection value=2.604e-03, at x=0.5
Maximum Rotation value=8.018e-03, at x=0.21
Extreme Deflection values=[ 0.000e+00  2.604e-03  0.000e+00],
    at x=[ 0.000e+00  5.000e-01  1.000e+00]
Critical Rotation values=[ 0.000e+00 -6.939e-18 -1.388e-17],
    at x=[ 0.000e+00  5.000e-01  1.000e+00]
Deflection values at boundary =(0.0, 0.0)
Rotation values at boundary =(0.0, -1.3877787807814457e-17)

Maximum Moment value=8.333e-02, at x=0.0
Maximum Shear force value=5.000e-01, at x=0.0
Extreme Moment values=[-4.167e-02],
    at x=[ 5.000e-01]
Critical Shear force values=[ 0.000e+00],
    at x=[ 5.000e-01]
Moment values at boundary =(0.08333333333333333, 0.08333333333333333)
Shear force values at boundary =(0.5, -0.5)
```

8.7 Solutions at specific locations

We can obtain solutions at different specific locations. This is useful not only for obtaining the desired solutions, but also for building up equations for solving more complicated problems. The codes are as follows:

```
1  printM(v[0].subs(x,1), 'Deflection at x=1',n_dgt=4)
```

Deflection at x=1: v_l

```
1  printM(v[0].subs(x,1/2), 'Deflection at x=1/2:',n_dgt=4)
```

Deflection at x=1/2: $\dfrac{l\theta_0}{8} - \dfrac{l\theta_l}{8} + \dfrac{v_0}{2} + \dfrac{v_l}{2} + \dfrac{l^4 q}{384 EI}$

```
1  printM(v[0].subs({x:l/2,v0:0, θ0:0, vl:0, θl:0}),
2             'Deflection at x=l/2 when all DBCs are set to zero:',n_dgt=4)
```

Deflection at x=l/2 when all DBCs are set to zero: $\dfrac{l^4 q}{384 EI}$

This is the well-known formula for computing the largest deflection in a c-c beam subject to uniformly distributed force.

```
1  printM((v[2].subs(x,0)).expand(), 'Moment at x=0:')
```

Moment at x=0: $-\dfrac{4EI\theta_0}{l} - \dfrac{2EI\theta_l}{l} - \dfrac{6EIv_0}{l^2} + \dfrac{6EIv_l}{l^2} + \dfrac{l^2 q}{12}$

```
1  printM((v[2].subs(x,l)).expand(), 'Moment at x=l:')
```

Moment at x=l: $\dfrac{2EI\theta_0}{l} + \dfrac{4EI\theta_l}{l} + \dfrac{6EIv_0}{l^2} - \dfrac{6EIv_l}{l^2} + \dfrac{l^2 q}{12}$

```
1  printM((v[2].subs(x,l/2)).expand(), 'Moment at x=l/2:')
```

Moment at x=l/2: $-\dfrac{EI\theta_0}{l} + \dfrac{EI\theta_l}{l} - \dfrac{l^2 q}{24}$

```
1  printM((v[3].subs(x,l)).expand(), 'Shear force at x=l:')
```

Shear force at x=l: $-\dfrac{6EI\theta_0}{l^2} - \dfrac{6EI\theta_l}{l^2} - \dfrac{12EIv_0}{l^3} + \dfrac{12EIv_l}{l^3} - \dfrac{lq}{2}$

```
1  printM((v[3].subs({x:l,v0:0, θ0:0, vl:0, θl:0})).expand(),
2             'Shear force at x=l when all fixed:')
```

Shear force at x=l when all fixed: $-\dfrac{lq}{2}$

```
1  printM(v[4], 'Body force on the beam:')
```

Body force on the beam: $\dfrac{q}{EI}$

This recovers the right-hand side of the fourth-order DE for beams.

8.8 Example: Simple-simple supported beams

Consider beams of length l with cross-section having a second moment of area I. It is made of material with Young's modulus E. Its boundaries are simple-simple supports (s-s), as shown in Fig. 8.9.

Solution to Bended Beams

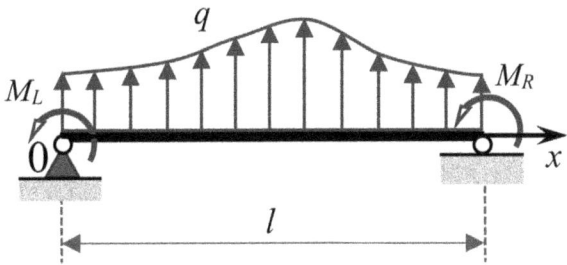

Figure 8.9. A beam with simple-simple (s-s) displacement boundary conditions (DBCs). It is subjected to a distributed force $q(x)$ over the beam and moment M_L at its left- and M_R at its right-end.

Solution: Use our solver: solver1D4() to obtain the general solutions for the beam, considering uniformly distributed force.

```
1  v = solver1D4(E, I, by, l, v0, θ0, vl, θl, V0, M0, Vl, Ml, key='s-s')
2
3  print(f'Is solution correct? {by/E/I==v[4]};   The solution u(x) is:')
4  v[0]                       # solution of general displacement function
```

Outputs form solver1D4(): v, θ, M, V, qy
Is solution correct? True;   The solution u(x) is

$$v_0 - \frac{v_0 x}{l} + \frac{v_l x}{l} - \frac{M_0 l x}{3EI} + \frac{M_0 x^2}{2EI} - \frac{M_0 x^3}{6EIl} - \frac{M_l l x}{6EI} + \frac{M_l x^3}{6EIl} + \frac{l^3 q x}{24EI} - \frac{l q x^3}{12EI} + \frac{q x^4}{24EI}$$

In the solution expression, v_0, M_0 are the deflection and (internal) moment at the left-end, and v_l, M_l are those at the right-end of the beam. This allows users to replace these variables at the boundary with known values. For example, if $v_0 = 0$, $M_0 = -M_L$, and $v_l = 0$, $M_l = M_R$, where M_L and M_R are the externally applied moments at the beam ends, as shown in Fig. 8.9, we can simply set so, as in the following code.

```
1  ML, MR = symbols('M_L, M_R')
2  dic = {v0:0, vl:0, Ml:MR, M0:-ML}
3          # at negative boundary ⇑, sign reversal
4  vM=[v[i].subs(dic) for i in range(len(v))]
5  vM[0]                       # Outputs v
```

$$-\frac{M_L l x}{3EI} - \frac{M_L x^2}{2EI} + \frac{M_L x^3}{6EIl} - \frac{M_R l x}{6EI} + \frac{M_R x^3}{6EIl} + \frac{l^3 q x}{24EI} - \frac{l q x^3}{12EI} + \frac{q x^4}{24EI}$$

Note that setting $M_0 = -M_L$ with a minus sign is due to our standard sign convention. At the left-end, the boundary is in the negative x-direction, and hence the internal moment is clockwise. However, the external moment M_L is counterclockwise (positive), as shown in Fig. 8.9. Therefore, we shall set $M_0 = -M_L$. Let us we check the solution of moment distributed along the beam:

```
1  gr.printM(vM[2], "Moment solution in the beam as a function of x:")
```

Moment solution in the beam as a function of x:

$$-M_L + \frac{M_L x}{l} + \frac{M_R x}{l} - \frac{lqx}{2} + \frac{qx^2}{2}$$

It is clear that at $x = 0$, we obtain the correct result of $M_0 = -M_L$. We can confirm also the moment solution at $x = l$, by setting $q = 0$, $M_l = 0$:

```
1  gr.printM(vM[2].subs({x:l, ML:0, q:0}), "Moment solution at x=l:")
```

Moment solution at x=l: M_R

We obtain again the correct result. Let us now check the solution for the shear force.

```
1  v[3].subs(di)c              # solution of general shear force function
```

$$-\frac{M_L}{l} - \frac{M_R}{l} + \frac{lq}{2} - qx$$

If $q = 0$, the applied moment M_R (or M_L) results in constant shear force of $V = M_R/l$ (or M_L/l) along the beam. If $M_R = M_L = 0$, the shear force is a linear function of x, at $x = 0$, $V = lq/2$, and at $x = l$, $V = -lq/2$. All the results are correct.

Let us now check the solution for deflection.

```
1  # Check the solution at the boundary:
2  printM(v[0].subs(x,l), 'Deflection at x=l:',n_dgt=4)
```

Deflection at x=l: v_l

We obtain the correct solution (which were set as the boundary condition). The distribution of the deflection over the beam by only q is found as follows.

```
1  gr.printM(v[0].subs({v0:0, vl:0,M0:0, Ml:0}), "Deflection v(x):")
```

Deflection v(x): $\dfrac{l^3 qx}{24EI} - \dfrac{lqx^3}{12EI} + \dfrac{qx^4}{24EI}$

This is the well-known formula for computing the deflection function in an s-s beam subject to uniformly distributed force. It is a fourth-order polynomial.

```
1  # get the solution at a location within the beam:
2  printM(v[0].subs(dic).subs(x,l/2), 'Deflection at x=l/2:',n_dgt=4)
```

Deflection at x=l/2: $\dfrac{M_L l^2}{16EI} - \dfrac{M_R l^2}{16EI} + \dfrac{5l^4 q}{384EI}$

```
1  # get the deflection at the middle span by only distributed force q:
2  printM(v[0].subs(dic).subs({ML:0, MR:0,x:l/2}), 'Deflection at x=l/2:')
```

Deflection at x=l/2: $\dfrac{5l^4 q}{384EI}$

This is the well-known formula for computing the largest deflection in an s-s beam subject to uniformly distributed force. It is five times that of the c-c beam.

```
1  # get the rotation at beam's left-end by only distributed force q:
2  printM(v[1].subs(dic).subs({ML:0, MR:0,x:0}), 'Rotation at x=s:')
```

Rotation at x=s: $\dfrac{l^3 q}{24EI}$

This is the well-known formula for computing the largest rotation in an s-s beam subject to uniformly distributed force, which is registered at the left-end (or at the right-end, but with a negative sign).

```
1  # Check whether the governing equation is satisfied:
2  gr.printM(v[4], 'Body force on the beam:')
```

Body force on the beam: $\dfrac{q}{EI}$

```
1  # Get the distribution of the solutions, by setting variables values:
2  vx=[v[i].subs({E:1,I:1,l:1,q:1,v0:0,vl:0,Ml:0,M0:0})
3             for i in range(len(v))]
4  vx
```

$\left[\dfrac{x^4}{24} - \dfrac{x^3}{12} + \dfrac{x}{24},\ \dfrac{x^3}{6} - \dfrac{x^2}{4} + \dfrac{1}{24},\ \dfrac{x^2}{2} - \dfrac{x}{2},\ \dfrac{1}{2} - x,\ 1\right]$

```
1  # Plot the distributions of solutions.
2  for i in range(len(title)):
3      gr.plot2curveS(vx[2*i:2*(i+1)], 0., 1., title=title[i])
4
5  plt.show()
```

```
Maximum Deflection value=1.302e-02, at x=0.5
Maximum Rotation value=4.167e-02, at x=0.0
Extreme Deflection values=[ 1.302e-02],
    at x=[ 5.000e-01]
Critical Rotation values=[-6.939e-18],
    at x=[ 5.000e-01]
Deflection values at boundary =(0.0, 0.0)
Rotation values at boundary =(0.041666666666666664, -0.04166666666666668)

Maximum Moment value=0.000e+00, at x=0.0
Maximum Shear force value=5.000e-01, at x=0.0
Extreme Moment values=[-1.250e-01],
    at x=[ 5.000e-01]
Critical Shear force values=[ 0.000e+00],
    at x=[ 5.000e-01]
Moment values at boundary =(0.0, 0.0)
Shear force values at boundary =(0.5, -0.5)
```

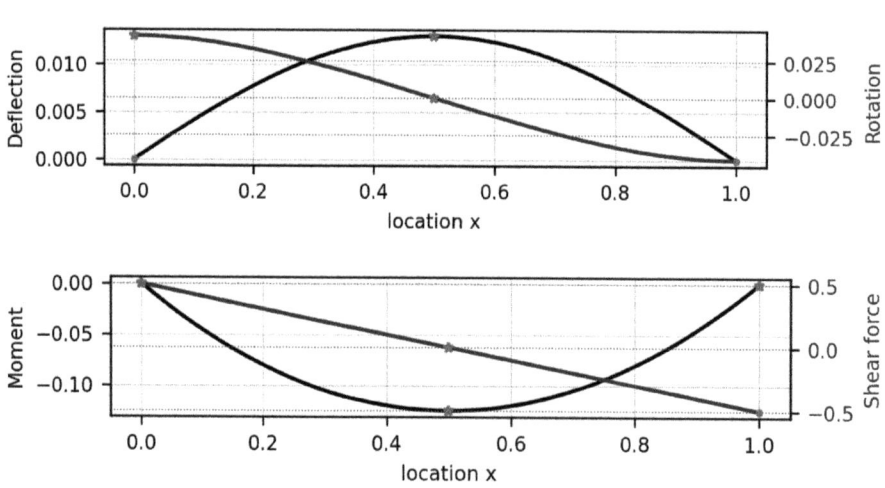

Let us get the formulas for rotations of the s-s beam.

```
1  v=solver1D4(E, I, by, 1, v0, θ0, v1, θ1, V0, M0, V1, M1, key = 's-s')
2  print(f'Is solution correct? {by/E/I==v[4]}; The solution u(x) is: ')
3  v[1]                    # solution of general function for rotation
```

```
Outputs form solver1D4(): v, θ, M, V, qy
Is solution correct? True; The solution u(x) is
```

$$-\frac{v_0}{l} + \frac{v_l}{l} - \frac{M_0 l}{3EI} + \frac{M_0 x}{EI} - \frac{M_0 x^2}{2EIl} - \frac{M_l l}{6EI} + \frac{M_l x^2}{2EIl} + \frac{l^3 q}{24EI} - \frac{lqx^2}{4EI} + \frac{qx^3}{6EI}$$

```
1  printM(v[1].subs(dic).subs(x,0), 'Rotation at x=0:',n_dgt=4)
```

Rotation at x=0: $\dfrac{M_L l}{3EI} - \dfrac{M_R l}{6EI} + \dfrac{l^3 q}{24EI}$

We found correct solutions of rotation at $x = 0$, which are resulting from the external loads.

```
1  printM(v[1].subs(dic).subs(x,1), 'Rotation at x=1:',n_dgt=4)
```

Rotation at x=1: $-\dfrac{M_L l}{6EI} + \dfrac{M_R l}{3EI} - \dfrac{l^3 q}{24EI}$

```
1  # Deflection solutions when setting all the fixed DBCs to 0:
2  v_ss = v[0].subs({v0:0, θ0:0, vl:0, θl:0})
3
4  printM(v_ss, 'Deflection function when all DBCs=0:',n_dgt=4)
```

Deflection function when all DBCs=0:

$$-\frac{M_0 lx}{3EI} + \frac{M_0 x^2}{2EI} - \frac{M_0 x^3}{6EIl} - \frac{M_l lx}{6EI} + \frac{M_l x^3}{6EIl} + \frac{l^3 qx}{24EI} - \frac{lqx^3}{12EI} + \frac{qx^4}{24EI}$$

```
1  # Deflection when setting all the fixed DBCs to 0, and q=0, M0=0
2  v_ss_Ml = v[0].subs({v0:0, θ0:0, vl:0, θl:0, q:0, M0:0}).subs(dic)
3
4  printM(v_ss_Ml, 'Deflection function, only Ml is applied at x=1:')
```

Deflection function, only Ml is applied at x=1: $-\dfrac{M_R lx}{6EI} + \dfrac{M_R x^3}{6EIl}$

This is the formula for computing the deflection of an s-s beam when a concentrated moment is applied at the right-end. As seen, solver1D4() can produce all kinds of solutions for beams. Readers may play with it further.

8.9 Example: Simple-clamped beams

Consider beams of length l with cross-section having a second moment of area I. It is made of material with Young's modulus E. It is simple-clamped (s-c) at its two ends, as shown in Fig. 8.10.

Solution: Use our solver: solver1D4() to obtain the general solutions for the beam, considering uniformly distribute force q.

```
1  ML, q = symbols('M_L, q')
2  by = q
3  v=solver1D4(E, I, by, l, v0, θ0, vl, θl, V0, M0, Vl, Ml, key= 's-c')
4  print(f'Is solution correct? {by/E/I==v[4]}; The solution u(x) is:')
5  dic = {v0:0,vl:0,θl:0,M0:-ML}        # for setting boundary conditions
6  # at negative boundary ↑, sign reversal
7
8  v[0].subs(dic)                  # solution for deflection as function of x
```

Outputs form solver1D4(): v, θ, M, V, qy
Is solution correct? True; The solution u(x) is

$$\frac{M_L l x}{4EI} - \frac{M_L x^2}{2EI} + \frac{M_L x^3}{4EIl} + \frac{l^3 q x}{48EI} - \frac{l q x^3}{16EI} + \frac{q x^4}{24EI}$$

Similar to the previous example, one can find various solutions for the beam by setting desired values to these variables. For example, solutions at $x = l$ caused by the body force q can be found as follows:

```
1  M_l = [v[i].subs(dic).subs({x:0, ML:0}) for i in range(len(v))]
2  M_l                       # solutions for v, θ, M, V, qy at x=0
```

$$\left[0, \ \frac{l^3 q}{48EI}, \ 0, \ \frac{3lq}{8}, \ \frac{q}{EI}\right]$$

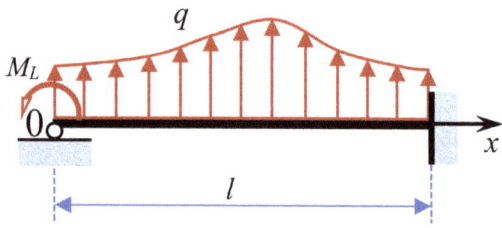

Figure 8.10. A beam with simple-clamped (s-c) displacement boundary conditions (DBCs). It is subjected to a distributed force $q(x)$ and moment M_L at the left-end.

This time we plot all the solutions in graphs.

```
1  dic1 = {E:1,I:1,l:1,q:1,ML:0}           # set values for variables
2  vx=[v[i].subs(dic).subs(dic1) for i in range(len(v))]
3  vx
```

$$\left[\frac{x^4}{24}-\frac{x^3}{16}+\frac{x}{48},\ \frac{x^3}{6}-\frac{3x^2}{16}+\frac{1}{48},\ \frac{x^2}{2}-\frac{3x}{8},\ \frac{3}{8}-x,\ 1\right]$$

```
1  # Plot all these curves for  v, θ, M, V:
2  for i in range(len(title)):
3      gr.plot2curveS(vx[2*i:2*(i+1)], 0., 1., title=title[i])
4
5  plt.show()
```

Maximum Deflection value=5.416e-03, at x=0.42
Maximum Rotation value=2.083e-02, at x=0.0
Extreme Deflection values=[-3.469e-18 5.416e-03],
    at x=[1.000e+00 4.200e-01]
Critical Rotation values=[-1.041e-17 1.063e-04],
    at x=[1.000e+00 4.200e-01]
Deflection values at boundary =(0.0, -3.469446951953614e-18)
Rotation values at boundary =(0.0208333333333333332, -1.0408340855860843e-17)

Maximum Moment value=1.250e-01, at x=1.0
Maximum Shear force value=3.750e-01, at x=0.0
Extreme Moment values=[-7.030e-02],
    at x=[3.700e-01]
Critical Shear force values=[5.000e-03],
    at x=[3.700e-01]
Moment values at boundary =(0.0, 0.125)
Shear force values at boundary =(0.375, -0.625)

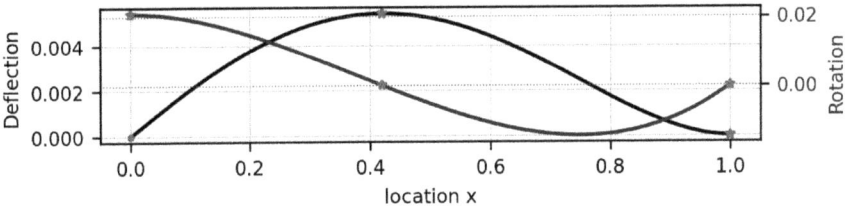

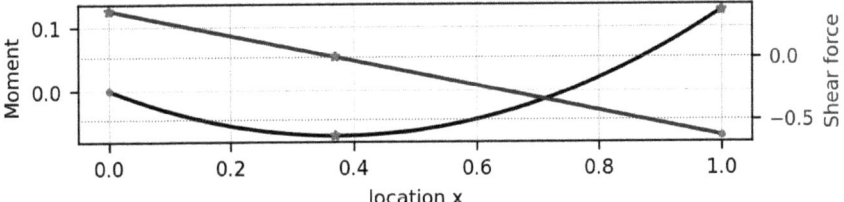

```
1  dicML = {v0:0, θ0:0, vl:0, θl:0, q:0, M0:-ML}
2  printM(v[0].subs(dicML),
3          'Deflection function by moment ML, when all DBCs=0:',n_dgt=4)
```

Deflection function by moment ML, when all DBCs=0: $\dfrac{M_L l x}{4EI} - \dfrac{M_L x^2}{2EI} + \dfrac{M_L x^3}{4EIl}$

```
1  printM(v[1].subs(dicML),
2          'Rotation distribution by moment ML, when all DBCs=0:',n_dgt=4)
```

Rotation distribution by moment ML, when all DBCs=0:

$$\dfrac{M_L l}{4EI} - \dfrac{M_L x}{EI} + \dfrac{3 M_L x^2}{4EIl}$$

```
1  printM((v[2].subs(dicML)).expand(),
2          'Moment distribution by moment ML, when all DBCs=0:')
```

Moment distribution by moment ML, when all DBCs=0: $-M_L + \dfrac{3 M_L x}{2l}$

```
1  printM((v[3].subs(dicML)).expand(),
2          'Shear force distribution by moment ML, when all DBCs=0:')
```

Shear force distribution by moment ML, when all DBCs=0: $-\dfrac{3 M_L}{2l}$

The shear force is constant along the beam because the body force is set to zero.

8.10 Example: Free-clamp supported beams

Consider beams with cross-section having a second moment of area I. It is made of material with Young's modulus E. It is free-clamp supported (f-c), as shown in Fig. 8.11.

A full set of solutions is found using solver1D4(). First, let us find the general solution.

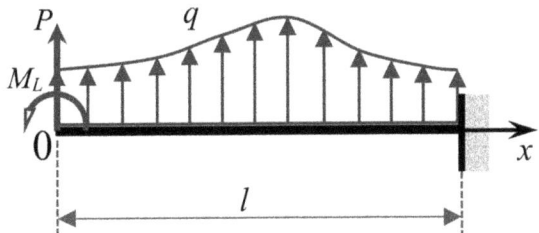

Figure 8.11. A cantilever beam with free-clamp (s-c) displacement boundary conditions (DBCs). It is subjected to a distributed force $q(x)$, a concentrated force P and a concentrated moment M at the left-end.

```
1  q, M0, v0, v1, θ0, θ1 = symbols('q, M0, v0, v1, θ0, θ1')
2  # Consider the distributed force is a constant q
3  by = q
4  v = solver1D4(E, I, by, l, v0, θ0, v1, θ1, V0, M0, V1, M1, key='f-c')
5  print(f'Is solution correct? {by/E/I==v[4]}; The solution u(x) is:')
6  v[0]
```

Outputs form solver1D4(): v, θ, M, V, qy
Is solution correct? True; The solution u(x) is

$$-l\theta l + vl + x\theta l + \frac{M_0 l^2}{2EI} - \frac{M_0 lx}{EI} + \frac{M_0 x^2}{2EI} - \frac{V_0 l^3}{3EI} + \frac{V_0 l^2 x}{2EI} - \frac{V_0 x^3}{6EI}$$
$$+ \frac{l^4 q}{8EI} - \frac{l^3 qx}{6EI} + \frac{qx^4}{24EI}$$

Let us consider $q = 0, M = 0$, and only P is on.

```
1  # Cantilever beam under tip load P (positive upward) at the left-tip:
2  P, M = symbols('P, M')
3  dicP = {q:0, v1:0, θ1:0, M0:0, V0:-P}
4          # at negative boundary ↑, sign reversal
5  vx= [v[i].subs(dicP) for i in range(len(v))]
6  vx                                    # solution in [v, θ, M, V, qy]
```

$$\left[\frac{Pl^3}{3EI} - \frac{Pl^2 x}{2EI} + \frac{Px^3}{6EI}, -\frac{Pl^2}{2EI} + \frac{Px^2}{2EI}, Px, -P, 0\right]$$

It is seen that the shear force in the beam is a constant of $-P$ along the beam. The moment in the beam varies linearly, Px, along the beam.

The deflection distribution is third-order polynomial, and at $x = 0$ it can be found as follows:

```
1  vx[0].subs(x,0)
```

$$\frac{Pl^3}{3EI}$$

This is the well-known formula for computing the largest deflection in a cantilever beam subject to a point force at its free end, which is registered at the free-end.

```
1  # Cantilever beam under external M (positive counterclockwise) at x=0:
2  ML = symbols('M_L')
3  vx= [v[i].subs({q:0, vl:0, θl:0, M0:-ML, V0:0}) for i in range(len(v))]
4  vx           # at negative boundary ↑, sign reversal
```

$$\left[-\frac{M_L l^2}{2EI} + \frac{M_L l x}{EI} - \frac{M_L x^2}{2EI},\; \frac{M_L l}{EI} - \frac{M_L x}{EI},\; -M_L,\; 0,\; 0\right]$$

It is seen in this case that the moment in the beam is a constant of $-M_L$ along the beam. The rotation at $x = 0$ is found as follows:

```
1  vx[1].subs(x,0)
```

$$\frac{M_L l}{EI}$$

We have the widely used rotation-moment relation for cantilever beam. We write it down for later reference.

$$\theta_L = \frac{l}{EI} M_L \quad \text{or} \quad M_L = \frac{EI}{l}\theta_L \qquad (8.35)$$

where $\frac{EI}{l}$ is called **bending stiffness** for beams with length l. It is equivalent to the spring constant k.

Let us plot the solutions in graphs for $q = 0$ and $M_L = 1$.

```
1  vx1 = [vx[i].subs({E:1,I:1,l:1,q:1,ML:1}) for i in range(len(v))]
2  vx1
```

$$\left[-\frac{x^2}{2} + x - \frac{1}{2},\; 1 - x,\; -1,\; 0,\; 0\right]$$

```
1  title = [["Deflection", "Rotation"], ["Moment", "Shear force"]]
2  for i in range(len(title)):
3      gr.plot2curveS(vx1[2*i:2*(i+1)], 0., 1., title=title[i])
4
5  plt.show()
```

```
Maximum Deflection value=0.000e+00, at x=1.0
Maximum Rotation value=1.000e+00, at x=0.0
Extreme Deflection values=[ 0.000e+00],
    at x=[ 1.000e+00]
Critical Rotation values=[ 0.000e+00],
    at x=[ 1.000e+00]
Deflection values at boundary =(-0.5, 0.0)
Rotation values at boundary =(1.0, 0.0)

Maximum Moment value=-1.000e+00, at x=0.0
Maximum Shear force value=0.000e+00, at x=0.0
Extreme Moment values=[-1.000e+00],
    at x=[ 0.000e+00]
Critical Shear force values=[ 0.000e+00],
    at x=[ 0.000e+00]
Moment values at boundary =(-1.0, -1.0)
Shear force values at boundary =(0.0, 0.0)
```

It is seen that the moment solution is constant (-1 because the positive external moment is on the negative boundary of the beam), and shear force is zero. This means that the bean is under **pure bending** without any shear force been generated, as expected for an f-c beam loaded with a moment at the free-end.

Next, let us plot the case of $q = 0$, $M = 0$, and only P is on the f-c beam.

```
1  # Cantilever beam under external P (positive upwards) at x=0:
2  P = symbols('P')
3  vx= [v[i].subs({q:0, vl:0, θl:0, M0:0, V0:-P}) for i in range(len(v))]
4  vx           #  at negative boundary ⇑, sign reversal
```

$$\left[\frac{Pl^3}{3EI} - \frac{Pl^2 x}{2EI} + \frac{Px^3}{6EI}, -\frac{Pl^2}{2EI} + \frac{Px^2}{2EI}, Px, -P, 0\right]$$

```
1  vx1 = [vx[i].subs({E:1,I:1,l:1,q:1,P:1}) for i in range(len(v))]
2  vx1
```

$$\left[\frac{x^3}{6} - \frac{x}{2} + \frac{1}{3}, \frac{x^2}{2} - \frac{1}{2}, x, -1, 0\right]$$

```
1  title = [["Deflection", "Rotation"], ["Moment", "Shear force"]]
2  for i in range(len(title)):
3      gr.plot2curveS(vx1[2*i:2*(i+1)], 0., 1., title=title[i])
4
5  plt.show()
```

Maximum Deflection value=3.333e-01, at x=0.0
Maximum Rotation value=0.000e+00, at x=1.0
Extreme Deflection values=[-5.551e-17],
    at x=[1.000e+00]
Critical Rotation values=[0.000e+00],
    at x=[1.000e+00]
Deflection values at boundary =(0.3333333333333333, -5.551115123125783e-17)
Rotation values at boundary =(-0.5, 0.0)

Maximum Moment value=1.000e+00, at x=1.0
Maximum Shear force value=-1.000e+00, at x=0.0
Extreme Moment values=[0.000e+00],
    at x=[0.000e+00]
Critical Shear force values=[-1.000e+00],
    at x=[0.000e+00]
Moment values at boundary =(0.0, 1.0)
Shear force values at boundary =(-1.0, -1.0)

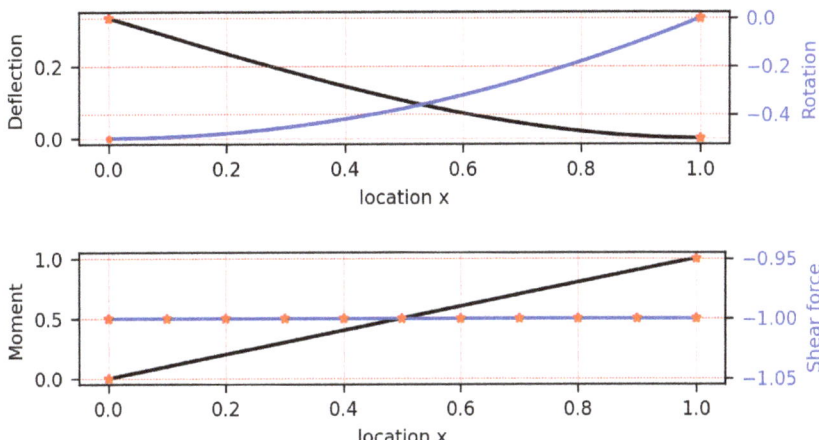

This time, the shear force solution is constant (−1 because the positive external force is on the negative boundary of the beam), and moment increases from linearly from 0. This is the case we studied by-hand in Section 8.3. The difference is the orientation of the beam is reversed.

8.11 Example: Clamped-clamped beam subjected to a point force at the middle span

Consider a beam with length l with c-c BC, as shown in Fig. 8.12(a). The beam has the same Young's modulus E, and its cross-section has a second moment of area I. It is subjected to a point force at the middle span A ($x = l/2$), with x-axis starting at the left-end of the beam and being leftwards. Derive formulas for solutions, including deflection, rotation, moment, and shear force in the beam.

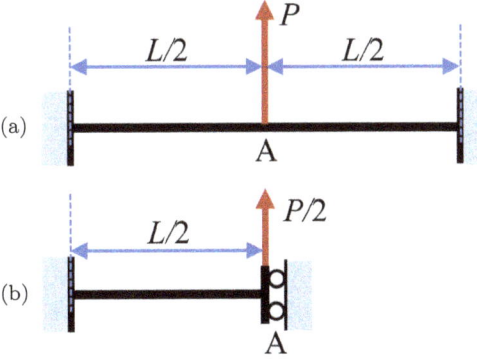

Figure 8.12. (a) A beam with clamped-clamped (c-c) displacement boundary conditions (DBCs). It is subjected to a concentrated force P at the middle span. (b) Consider only half of the beam by making use of the symmetry.

Solution: This problem looks simple, but it needs some considerations. First because force P is a point force, the solution may not be continuous and needs to be given piecewise. Since the problem is symmetric with respect to $x = l/2$, the rotation at A must be zero. Thus, we can consider only a half of the beam, say the left half, with also the force being halved. Now, we have a beam with length of $l/2$ and clamp-roller (c-r) BCs, as shown in Fig. 8.12b. The general solution for the c-r beam can be found using our beam solver, solver1D4(), with c-c BC and then set the rotation at A to zero. The process and codes are as follows:

```
1  title = [["Deflection", "Rotation"], ["Moment", "Shear force"]]
2
3  by = 0                                          # Not distributed force
4                           # the length is halved
5  v = solver1D4(E, I, by, l/2, v0, θ0, v1, θ1, V0, M0, V1, M1,key='c-c')
6  print(f'Is solution correct? {by/E/I==v[4]}; The solution u(x) is:')
7  v[0]                     # solution in a general displacement function
```

Outputs form solver1D4(): v, θ, M, V, qy
Is solution correct? True; The solution u(x) is

$$v_0 + x\theta_0 - \frac{4x^2\theta_0}{l} - \frac{2x^2\theta l}{l} - \frac{12v_0x^2}{l^2} + \frac{12vlx^2}{l^2} + \frac{4x^3\theta_0}{l^2} + \frac{4x^3\theta l}{l^2} + \frac{16v_0x^3}{l^3} - \frac{16vlx^3}{l^3}$$

```
1  # Set DBCs at the left-end, and rotation at A to zero:
2
3  vx=[v[i].subs({v0:0,θ0:0,θ1:0}) for i in range(len(v))]
4  vx
```

$$\left[\frac{12vlx^2}{l^2} - \frac{16vlx^3}{l^3}, \frac{24vlx}{l^2} - \frac{48vlx^2}{l^3}, \frac{24EIvl}{l^2} - \frac{96EIvlx}{l^3}, \frac{96EIvl}{l^3}, 0\right]$$

This set of solution has an unknown v_l. At point A, we can find the shear force using the solution obtained:

```
1  fy2=vx[3].subs(x,1/2).expand()        # at point A
2  gr.printM(fy2, 'Shear force at point A, fy2:',n_dgt=4)
```

Shear force at point A, fy2: $\dfrac{96EIvl}{l^3}$

The shear force at point A must be $P/2$, we then have

```
1  P = symbols("P")
2  sln = sp.solve(fy2 - P/2, vl)
3  gr.printM(sln[0], 'deflection at A, v_l = ',n_dgt=4)
```

deflection at A, v_l = $\dfrac{Pl^3}{192EI}$

This is the well-known formula for computing the largest deflection in a c-c beam subject to a point force at the middle span.

Finally, we put v_l back to the solution set:

```
1  # substitute v_l to the solution obtained earlier:
2
3  sln_final =[vx[i].subs({vl:sln[0]}) for i in range(len(v))]
4  sln_final                  # solution set in [v, θ, M, V, qy]
```

$$\left[\dfrac{Plx^2}{16EI} - \dfrac{Px^3}{12EI},\ \dfrac{Plx}{8EI} - \dfrac{Px^2}{4EI},\ \dfrac{Pl}{8} - \dfrac{Px}{2},\ \dfrac{P}{2},\ 0\right]$$

This is the final full solution set for a c-c beam subject to a point force at the middle span, which can be found in textbooks and the open literature.

8.12 Plot piecewise solutions

For the foregoing problem, the solution has two pieces: one for left-span and one for the right-span of the beam. We write the following code to plot the solutions.

```
1  # Set the parameters to unit values:
2
3  sln_x =[sln_final[i].subs({P:1,l:1,E:1,I:1}) for i in range(len(v))]
4  sln_x
```

$$\left[-\frac{x^3}{12}+\frac{x^2}{16},\ -\frac{x^2}{4}+\frac{x}{8},\ \frac{1}{8}-\frac{x}{2},\ \frac{1}{2},\ 0\right]$$

```
1   # set parameters: #P = 1.0; E = 1.0; I = 1.0
2   l1 = 0.5; l2 = 0.5
3
4   dx = 0.02
5   x1 = np.arange(0.0, l1, dx)          # coordinates in L1, left span
6   x2 = np.arange(0.0, l2, dx)          # coordinates in L2, right span
7   X2 = -x2 + l1    # coord. transfer, right-span is opposite to left-span
8   signs=[1, -1, 1, -1]  # symmetry of solutions: v is sym, θ is anti-sym
9
10  # convert Sympy solution func. to numpy arrays.
11  n = len(sln_x)-1
12  x = symbols("x")
13  sln = [sp.lambdify(x, sln_x[i], 'numpy') for i in range(n)]
14  sx1 = [           sln[i](x1)  for i in range(n)]
15  sx2 = [signs[i]*sln[i](X2) for i in range(n)]
16
17  # The shear force function could be a constant. Need to generate array
18  if type(sx1[-1]) != np.ndarray:       # in case, func. is a constant
19      sx1[-1] = float(sx1[-1])*np.ones_like(x1)
20      sx2[-1] = float(sx2[-1])*np.ones_like(X2)
21
22  X  = np.concatenate((x1, l1+x2))   # put the results in a single array
23  sx = [np.concatenate((sx1[i], sx2[i])) for i in range(n)]
24
25  sx_max = [np.max(abs(sx[i])) for i in range(n)]  # absolute max values
26  print(f"The absolute maximum solutions: {sx_max}")
```

The absolute maximum solutions: [0.005208333333333334, 0.0156, 0.125, 0.5]

```
1   # Plot the distribution curves:
2   x_data = [X for i in range(n)]               # put x values to a list
3   y_data = [sx[i]/sx_max[i] for i in range(n)] # func. values to a list
4
5   labels=["Deflection", "Rotation", "Moment", "Shear force"]
6   gr.plot_fs(x_data, y_data, labels, 'x',      'Solutions',  'temp')
7                                        # x-label, y-label,   file-name
```

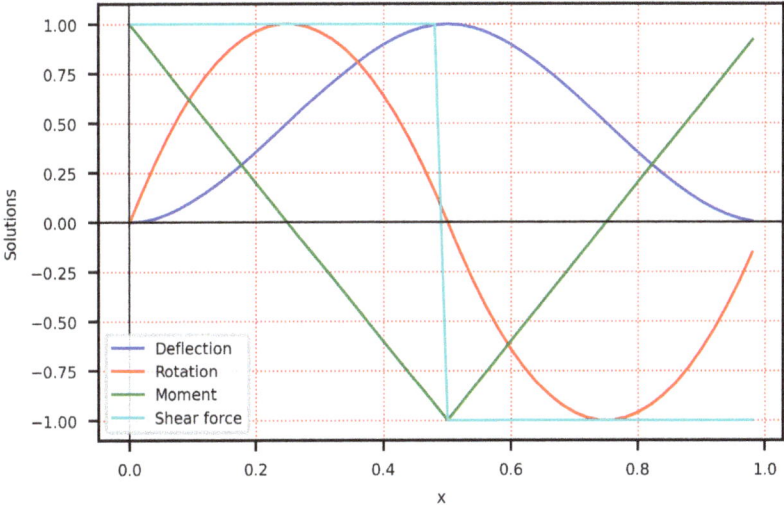

The discontinuity of the solutions especially for the moment and shear force is clearly shown.

Readers may try to find the solution to the same problem, but replacing P with M at the same location.

8.13 Example: Simply supported beam with two spans

Consider a beam with two spans. Span-1 has a length l_1, and span-2 has a length l_2. All the members has the same Young's modulus E, and its cross-section has a second moment of area I. The beam is simple-simple-clamp supported (s-s-c), as shown in Fig. 8.13. It is subjected to a distributed force q over span-2.

1. Derive formulas for solutions, including deflection, rotation, moment, and shear force in the beam.
2. Plot the distribution of all these solutions.

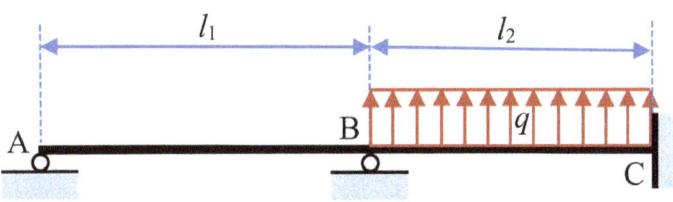

Figure 8.13. A beam with two spans. It is simple-simple-clamp supported.

Solution: Since the beam has two spans, the solution will be piecewise. Let the x-axis start at point A for span-1 and at point B for span-2. Both are positive rightwards.

For each of the spans, we can find the general solution using our beam solver solver1D4(). The process and codes are as follows:

```
1  # Define variables:
2  title = [["Deflection", "Rotation"], ["Moment", "Shear force"]]
3  E, I, l1, l2, q = symbols("E, I, l1, l2, q")
4  x = symbols("x")
5
6  # for displacement boundary conditions (DBCs):
7  v10, θ10, v11, θ11 = symbols('v10, θ10, v_11, θ_11')  # DBCs for span-1
8  v20, θ20, v21, θ21 = symbols('v20, θ20, v_21, θ_21')  # DBCs for span-2
9
10 # for force boundary conditions (FBCs):
11 V10, M10, V11, M11 = symbols('V10, M10, V_11, M_11')  # FBCs for span-1
12 V20, M20, V21, M21 = symbols('V20, M20, V_21, M_21')  # FBCs for span-2
```

8.13.1 For span-1

Let us seek the general solution for span-1.

```
1  # Distributed force is zero for span-1
2  by = 0                                                    # BC is s-s
3  v1 = solver1D4(E,I,by,l1, v10,θ10,v11,θ11, V10, M10, V11, M11,key='s-s')
4  print(f'Is solution correct? {by/E/I==v1[4]}; The solution v(x) is:')
5  v1[0]                    # solution of general displacement function for span-1
```

Outputs form solver1D4(): v, θ, M, V, qy
Is solution correct? True; The solution v(x) is

$$v_{10} - \frac{v_{10}x}{l_1} + \frac{v_{11}x}{l_1} - \frac{M_{10}l_1 x}{3EI} + \frac{M_{10}x^2}{2EI} - \frac{M_{10}x^3}{6EIl_1} - \frac{M_{11}l_1 x}{6EI} + \frac{M_{11}x^3}{6EIl_1}$$

```
1  # Set known BCs for span-1:
2
3  v1x=[v1[i].subs({v10:0,M10:0,v11:0}) for i in range(len(v1))]
4  v1x                                  # solution of v, θ, Mx, Vx, qy
```

$$\left[-\frac{M_{1l}l_1 x}{6EI} + \frac{M_{1l}x^3}{6EIl_1}, \ -\frac{M_{1l}l_1}{6EI} + \frac{M_{1l}x^2}{2EIl_1}, \ \frac{M_{1l}x}{l_1}, \ -\frac{M_{1l}}{l_1}, \ 0\right]$$

This set of solutions has an unknown M_{1l}. At point B, we can find the rotation using the general solution:

```
1  θ1B = v1x[1].subs(x,l1).expand()     # at point B of span-1
2  gr.printM(θ1B, 'rotation at point B of span-1, θ1B:',n_dgt=4)
```

rotation at point B of span-1, θ1B: $\dfrac{M_{1l}l_1}{3EI}$

8.13.2 For span-2

Next, let us seek the general solution for span-2.

```
1  # General solution for span-2: (constant q)
2                                               # BC is simple-clamp
3  v2 = solver1D4(E,I,q,l2, v20,θ20,v21, θ21, V20,M20, V21,M21,key='s-c')
4  print(f'Is solution correct? {by/E/I==v2[4]}; The solution u(x) is: ')
5  v2[0]         # solution of general displacement function for span-2
```

Outputs form solver1D4(): v, θ, M, V, qy
Is solution correct? False; The solution u(x) is

$$v_{20} - \frac{x\theta_{2l}}{2} - \frac{3v_{20}x}{2l_2} + \frac{3v_{2l}x}{2l_2} + \frac{x^3\theta_{2l}}{2l_2^2} + \frac{v_{20}x^3}{2l_2^3} - \frac{v_{2l}x^3}{2l_2^3} - \frac{M_{20}l_2 x}{4EI} + \frac{M_{20}x^2}{2EI}$$
$$- \frac{M_{20}x^3}{4EIl_2} + \frac{l_2^3 qx}{48EI} - \frac{l_2 qx^3}{16EI} + \frac{qx^4}{24EI}$$

```
1  # Set known BCs for span-2:
2
3  v2x=[v2[i].subs({v20:0,v21:0,θ21:0}) for i in range(len(v2))]
4  v2x[1]                                    # solution of θ
```

$$-\frac{M_{20}l_2}{4EI} + \frac{M_{20}x}{EI} - \frac{3M_{20}x^2}{4EIl_2} + \frac{l_2^3 q}{48EI} - \frac{3l_2 qx^2}{16EI} + \frac{qx^3}{6EI}$$

This set of solutions has an unknown M_{20}. At point B of span-2, we can find the rotation using the general solution:

```
1  θ2B = v2x[1].subs(x,0).expand()        # at point B of span-2
2  gr.printM(θ2B, 'rotation at point B of span-2, θ2B:',n_dgt=4)
```

rotation at point B of span-2, θ2B: $-\dfrac{M_{20} l_2}{4EI} + \dfrac{l_2^3 q}{48EI}$

8.13.3 *Continuity and equilibrium conditions*

At point B of the beam, we must have the following conditions:

$$\begin{aligned} \theta_{1B} &= \theta_{2B} \quad \text{continuity in rotation} \\ -M_{1B} + M_{2B} &= 0 \quad \text{equilibrium in moment} \end{aligned} \tag{8.36}$$

The free-body diagram is shown in Fig. 8.14. We now use these conditions to solve for unknowns M_{1B} and M_{2B}, using the following code.

```
1  sln = sp.solve([θ1B - θ2B, -M1l + M20], [M1l, M20])
2  sln                          # solution for these unknowns
```

$$\left\{ M_{20} : \frac{l_2^3 q}{16 l_1 + 12 l_2},\ M_{1l} : \frac{l_2^3 q}{16 l_1 + 12 l_2} \right\}$$

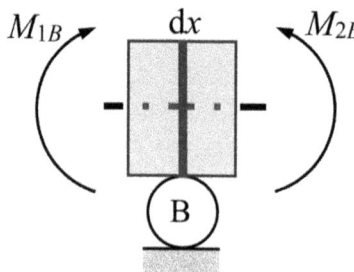

Figure 8.14. Free-body diagram of an infinitely small beam segment on support B with length dx.

8.13.4 Final solutions

Substituting these unknown moments back to the general solutions, gives the final solutions.

```
1  # Final solution in span-1:
2  vs1x = [v1x[i].subs({M11:sln[M11]}) for i in range(len(v1x))]
3  vs1x[0]  # solution of deflection v as function of x for span-1.
```

$$-\frac{l_1 l_2^3 qx}{6EI\,(16l_1 + 12l_2)} + \frac{l_2^3 qx^3}{6EIl_1 \cdot (16l_1 + 12l_2)}$$

```
1  vs1x[1]                              # solution of rotation θ
```

$$-\frac{l_1 l_2^3 q}{6EI\,(16l_1 + 12l_2)} + \frac{l_2^3 qx^2}{2EIl_1 \cdot (16l_1 + 12l_2)}$$

```
1  vs1x[2:]                             # solution of moment M and shear force V
```

$$\left[\frac{l_2^3 qx}{l_1 \cdot (16l_1 + 12l_2)},\; -\frac{l_2^3 q}{l_1 \cdot (16l_1 + 12l_2)},\; 0\right]$$

```
1  # Final solution in span-2:
2  vs2x=[v2x[i].subs({M20:sln[M20]}) for i in range(len(v2x))]
3  vs2x[0]                              # solution v(x)
```

$$-\frac{l_2^4 qx}{4EI\,(16l_1 + 12l_2)} + \frac{l_2^3 qx^2}{2EI\,(16l_1 + 12l_2)} + \frac{l_2^3 qx}{48EI} - \frac{l_2^2 qx^3}{4EI\,(16l_1 + 12l_2)}$$
$$- \frac{l_2 qx^3}{16EI} + \frac{qx^4}{24EI}$$

```
1  vs2x[1]                                          # solution of θ(x)
```

$$-\frac{l_2^4 q}{4EI\left(16l_1+12l_2\right)} + \frac{l_2^3 qx}{EI\left(16l_1+12l_2\right)} + \frac{l_2^3 q}{48EI} - \frac{3l_2^2 qx^2}{4EI\left(16l_1+12l_2\right)}$$
$$-\frac{3l_2 qx^2}{16EI} + \frac{qx^3}{6EI}$$

```
1  vs2x[2:]                     # solution of moment M(x) and shear force V(x)
```

$$\left[\frac{l_2^3 q}{16l_1+12l_2} - \frac{3l_2^2 qx}{2\cdot(16l_1+12l_2)} - \frac{3l_2 qx}{8} + \frac{qx^2}{2},\right.$$
$$\left.\frac{3l_2^2 q}{2\cdot(16l_1+12l_2)} + \frac{3l_2 q}{8} - qx, \frac{q}{EI}\right]$$

This is the final solution to our problem, which may not be found in textbooks and the open literature. When setting $l_1 = l_2$, one can obtain simplified solution that can be found in textbooks and the open literature. Readers may do so to check the solutions obtained.

8.14 Plot piecewise solutions

For the foregoing problem, the solution has two pieces: one for left-span and one for the right-span of the beam. We write the following code to plot the solutions together.

```
1  # Set the parameters to unit values:
2
3  sln1x=[vs1x[i].subs({q:1,l1:1,l2:1,E:1,I:1}) for i in range(len(vs1x))]
4  sln1x
```

$$\left[\frac{x^3}{168} - \frac{x}{168}, \frac{x^2}{56} - \frac{1}{168}, \frac{x}{28}, -\frac{1}{28}, 0\right]$$

```
1  # Set the parameters to unit values:
2
3  sln2x=[vs2x[i].subs({q:1,l1:1,l2:1,E:1,I:1}) for i in range(len(vs2x))]
4  sln2x
```

$$\left[\frac{x^4}{24} - \frac{x^3}{14} + \frac{x^2}{56} + \frac{x}{84},\ \frac{x^3}{6} - \frac{3x^2}{14} + \frac{x}{28} + \frac{1}{84},\ \frac{x^2}{2} - \frac{3x}{7} + \frac{1}{28},\ \frac{3}{7} - x,\ 1\right]$$

```
1   np.set_printoptions(formatter={'float': '{: 0.4e}'.format})
2   # set parameters: #q = 1.0; E = 1.0; I = 1.0
3   l1 = 1.0; l2 = 1.0
4
5   dx = 0.002
6   x1 = np.arange(0.0, l1, dx)        # coordinates in l1, left span
7   x2 = np.arange(0.0, l2, dx)        # coordinates in l2, right span
8
9   # convert Sympy solution func. to numpy arrays.
10  n1 = len(sln1x)-1
11  n2 = len(sln1x)-1
12  x = symbols("x")
13  sln1 = [sp.lambdify(x, sln1x[i], 'numpy') for i in range(n1)]
14  sln2 = [sp.lambdify(x, sln2x[i], 'numpy') for i in range(n2)]
15  sx1 = [sln1[i](x1)  for i in range(n1)]
16  sx2 = [sln2[i](x2)  for i in range(n2)]
17
18  # The shear force function could be a constant. Need to generate array
19  if type(sx1[-1]) != np.ndarray:      # in case, func. is a constant
20      sx1[-1] = float(sx1[-1])*np.ones_like(x1)
21
22  if type(sx2[-1]) != np.ndarray:      # in case, func. is a constant
23      sx2[-1] = float(sx2[-1])*np.ones_like(x2)
24
25  X = np.concatenate((x1, l1+x2))    # put the results in a single array
26  sx = [np.concatenate((sx1[i], sx2[i])) for i in range(n1)]
27
28  sx_max = [np.max(abs(sx[i])) for i in range(n1)]   # absolute max values
29  print(f"Absolute maximum values: {np.array(sx_max)}")
```

Absolute maximum values: [4.1733e-03 1.3507e-02 1.0600e-01 5.6943e-01]

```
1  # Plot the distribution curves:
2  x_data = [X for i in range(n1)]          # put x coordinates to a list
3  y_data = [sx[i]/sx_max[i] for i in range(n1)] # func. values to a list
4
5  labels=["Deflection", "Rotation", "Moment", "Shear force"]
6  gr.plot_fs(x_data, y_data, labels, 'x',       'Solutions',   'temp')
7                                      # x-label, y-label,    file-name
```

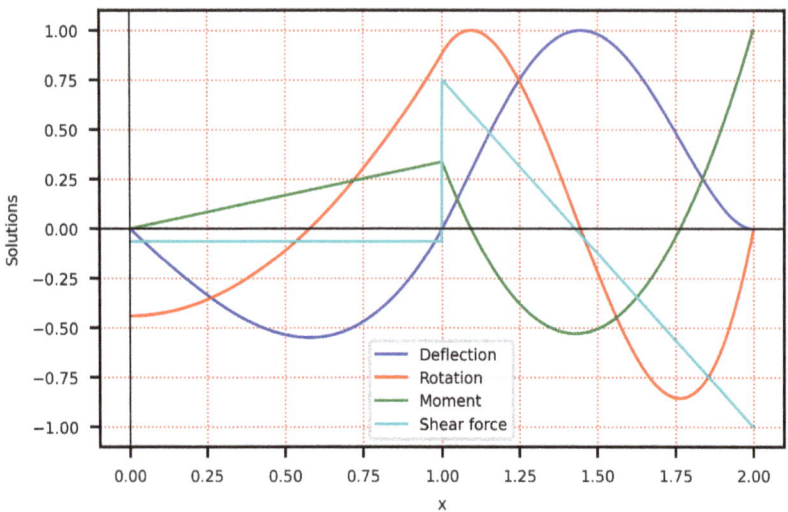

The discontinuity of the solutions especially for the moment and shear force is clearly shown.

8.15 Example: A symmetric frame subjected to a uniformly distributed load*

Consider a simple frame consisting of three members with uniform cross-section. It is symmetric, as shown in Fig. 8.15(a). Its pillar has a length l_1, and top-beam has a length $2l_2$. All the members have the same Young's modulus E, and its cross-section area is A and second moment of area I. The frame is clamped on the ground. It is subjected to a uniformly distributed force q on the top-beam of the frame.

1. Derive formulas for solutions, including deflection, rotation, moment, and shear force in the frame.
2. Plot the distribution of all these solutions, using unit for all its parameters.

Solution: Since the frame is symmetric, we can consider only half of it, as shown in in Fig. 8.15(b). The support for point C becomes a roller, implying that it constrains the horizontal movement, but allows it to move vertically at that point. The frame now has two members: member-1 is the pillar, and member-2 is the (halved) top-beam. We can now use similar strategy

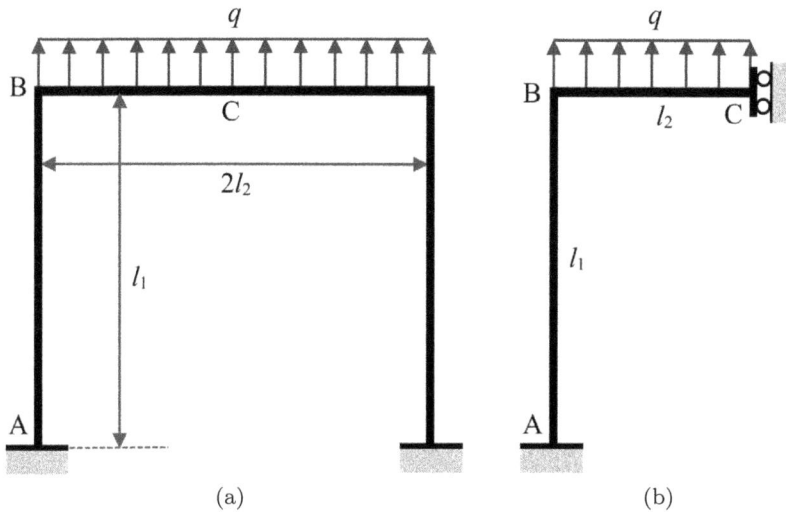

Figure 8.15. (a) A symmetric frame fixed on the grounds. It is subjected to a uniformly distributed force on the top. (b) Only half a frame needs to be analyzed, making use of the symmetry.

as for the previous example, by obtaining the general solutions for these two members separately, and then use the continuity conditions at point B and equilibrium conditions to determine these unknowns.

Let the x-axis start at point A for member-1 (mem-1) with upwards being positive, and x-axis starts at point B for member-2 (mem-2) with rightwards positive.

For each of the members, we can find the general solution using our beam solver solver1D4(). The process and codes are as follows:

```
1  # Define variables:
2  title = [["Deflection", "Rotation"], ["Moment", "Shear force"]]
3  E, I, A, l1, l2, q = symbols("E, I, A, l1, l2, q")
4  x = symbols("x")
5
6  # for displacement boundary conditions (DBCs):
7  v10, θ10, v11, θ11 = symbols('v10, θ10, v_11, θ_11')  # DBCs for mem-1
8  v20, θ20, v21, θ21 = symbols('v20, θ20, v_21, θ_21')  # DBCs for mem-2
9
10 # for force boundary conditions (FBCs):
11 V10, M10, V11, M11 = symbols('V10, M10, V_11, M_11')  # FBCs for mem-1
12 V20, M20, V21, M21 = symbols('V20, M20, V_21, M_21')  # FBCs for mem-2
```

8.15.1 For member-1

Let us seek the general solution for member-1.

```
1  # Distributed force is zero for member-1
2  by = 0                                    # BC is clamp-clamp
3  v1 = solver1D4(E,I,by,l1, v10,θ10,v11,θ11, V10,M10, V11,M11,key='c-c')
4  print(f'Is solution correct? {by/E/I==v1[4]}; The solution v(x) is: ')
5  v1[0]                                     # solution of general displacement function
```

Outputs form solver1D4(): v, θ, M, V, qy
Is solution correct? True; The solution v(x) is

$$v_{10} + x\theta_{10} - \frac{2x^2\theta_{10}}{l_1} - \frac{x^2\theta_{1l}}{l_1} - \frac{3v_{10}x^2}{l_1^2} + \frac{3v_{1l}x^2}{l_1^2} + \frac{x^3\theta_{10}}{l_1^2} + \frac{x^3\theta_{1l}}{l_1^2} + \frac{2v_{10}x^3}{l_1^3} - \frac{2v_{1l}x^3}{l_1^3}$$

```
1  # Set known BCs for member-1:
2
3  v1x=[v1[i].subs({v10:0,θ10:0}) for i in range(len(v1))]
4  v1x[:2]                                   # solution of v, θ
```

$$\left[-\frac{x^2\theta_{1l}}{l_1} + \frac{3v_{1l}x^2}{l_1^2} + \frac{x^3\theta_{1l}}{l_1^2} - \frac{2v_{1l}x^3}{l_1^3},\ -\frac{2x\theta_{1l}}{l_1} + \frac{6v_{1l}x}{l_1^2} + \frac{3x^2\theta_{1l}}{l_1^2} - \frac{6v_{1l}x^2}{l_1^3} \right]$$

```
1  v1x[2:]                                   # solution of Mx, Vx, qy
```

$$\left[-\frac{2EI\theta_{1l}}{l_1} + \frac{6EIv_{1l}}{l_1^2} + \frac{6EIx\theta_{1l}}{l_1^2} - \frac{12EIv_{1l}x}{l_1^3},\ -\frac{6EI\theta_{1l}}{l_1^2} + \frac{12EIv_{1l}}{l_1^3},\ 0 \right]$$

This set of solutions has 2 unknowns θ_{1l} and v_{1l}. At point B, we can find the rotation using the general solution:

```
1  M1B = v1x[2].subs(x,l1).expand()          # at point B of mem-1
2  gr.printM(M1B, 'Moment at point B of member-1, M1B:',n_dgt=4)
```

Moment at point B of member-1, M1B: $\dfrac{4EI\theta_{1l}}{l_1} - \dfrac{6EIv_{1l}}{l_1^2}$

We can also find the shear force at point B of member-1.

```
1  V1B = v1x[3].subs(x,l1).expand()        # at point B of mem-1
2  gr.printM(V1B, 'Shear force at point B of member-1, V1B:',n_dgt=4)
```

Shear force at point B of member-1, V1B: $-\dfrac{6EI\theta_{1l}}{l_1^2} + \dfrac{12EIv_{1l}}{l_1^3}$

8.15.2 For member-2

Next, let us seek the general solution for member-2.

```
1  # General solution for member-2: (constant q)
2                                          # BC is clamp-clamp
3  v2 = solver1D4(E,I,q,l2, v20,θ20,v2l,θ2l, V20, M20, V2l, M2l,key='c-c')
4  print(f'Is solution correct? {by/E/I==v2[4]}; The solution u(x) is:')
5  v2[0]                                   # solution of general displacement function v(x)
```

Outputs form solver1D4(): v, θ, M, V, qy
Is solution correct? False; The solution u(x) is

$$v_{20} + x\theta_{20} - \frac{2x^2\theta_{20}}{l_2} - \frac{x^2\theta_{2l}}{l_2} - \frac{3v_{20}x^2}{l_2^2} + \frac{3v_{2l}x^2}{l_2^2} + \frac{x^3\theta_{20}}{l_2^2} + \frac{x^3\theta_{2l}}{l_2^2} + \frac{2v_{20}x^3}{l_2^3}$$

$$- \frac{2v_{2l}x^3}{l_2^3} + \frac{l_2^2 qx^2}{24EI} - \frac{l_2 qx^3}{12EI} + \frac{qx^4}{24EI}$$

```
1  # Set known BCs for member-2:
2
3  v2x=[v2[i].subs({θ2l:0}) for i in range(len(v2))]
4  v2x[1]                                  # solution of θ(x)
```

$$\theta_{20} - \frac{4x\theta_{20}}{l_2} - \frac{6v_{20}x}{l_2^2} + \frac{6v_{2l}x}{l_2^2} + \frac{3x^2\theta_{20}}{l_2^2} + \frac{6v_{20}x^2}{l_2^3} - \frac{6v_{2l}x^2}{l_2^3}$$

$$+ \frac{l_2^2 qx}{12EI} - \frac{l_2 qx^2}{4EI} + \frac{qx^3}{6EI}$$

This set of solutions has 3 unknowns θ_{20}, v_{20}, v_{2l}. At point B of member-2, we can find the moment using the general solution:

```
1  M2B = v2x[2].subs(x,0).expand()        # at point B of mem-2
2  gr.printM(M2B, 'Moment at point B of top-beam, M2B:',n_dgt=4)
```

Moment at point B of top-beam, M2B

$$-\frac{4EI\theta_{20}}{l_2} - \frac{6EIv_{20}}{l_2^2} + \frac{6EIv_{2l}}{l_2^2} + \frac{l_2^2 q}{12}$$

The shear force at point B of member-2 is found as follows.

```
1  V2B = v2x[3].subs(x,0).expand()        # at point B of mem-2
2  gr.printM(V2B, 'Shear force at point B of member-2, V2B:',n_dgt=4)
```

Shear force at point B of member-2, V2B

$$-\frac{6EI\theta_{20}}{l_2^2} - \frac{12EIv_{20}}{l_2^3} + \frac{12EIv_{2l}}{l_2^3} + \frac{l_2 q}{2}$$

Also, the shear force at point C of member-2 is found as follows.

```
1  V2l = v2x[3].subs(x,l2).expand()       # at point B of mem-2
2  gr.printM(V2l, 'Shear force at point C of member-2, V2l:',n_dgt=4)
```

Shear force at point C of member-2, V2l

$$-\frac{6EI\theta_{20}}{l_2^2} - \frac{12EIv_{20}}{l_2^3} + \frac{12EIv_{2l}}{l_2^3} - \frac{l_2 q}{2}$$

8.15.3 *Continuity and equilibrium conditions at point B*

In the solutions for the frame, we have a total of 5 unknowns: θ_{1l} and v_{1l}, θ_{20}, v_{20}, and v_{2l}. We need 5 conditions to determine those.

At point B of the frame, we have the following conditions:

$$\theta_{1B} = \theta_{2B} \quad \text{continuity}$$

$$-M_{1B} + M_{2B} = 0 \quad \text{equilibrium in moment}$$

$$\frac{-V_{1B}l_2}{EA} = v_{1l} \quad \text{elongation of member-2} = \text{deflection of member-1}$$

$$\frac{-V_{2B}l_1}{EA} = -v_{20} \quad \text{elongation of member-1} = \text{deflection of member-2}$$

$$V_{2B} = ql_2 \quad \text{shear force} = \text{sum of load on member-2} \quad (8.37)$$

These are the 5 conditions we need to solve for these 5 unknowns $\theta_{1l}, v_{1l}, \theta_{20}, v_{20}$ and v_{2l}. The following code sets these 5 equations and then finds these unknowns.

```
1  eq1 = θ11-θ20
2  eq2 = -M1B+M2B
3  eq3 = -V1B*l2/(E*A)-v1l
4  eq4 = -V2B*l1/(E*A)+v20
5  eq5 = V2B-q*l2
6  sln = sp.solve([eq1, eq2, eq3, eq4, eq5], [θ11, v1l, θ20, v20, v2l])
7  sln[θ11]                                   # solution for unknown θ1L
```

$$\frac{Al_1^4 l_2^3 q + 12Il_1 l_2^4 q}{3AEIl_1^4 + 12AEIl_1^3 l_2 + 36EI^2 l_1 l_2 + 36EI^2 l_2^2}$$

```
1  sln[v1l]                                   # solution for unknown v1L
```

$$\frac{2l_1^2 l_2^4 q}{AEl_1^4 + 4AEl_1^3 l_2 + 12EIl_1 l_2 + 12EIl_2^2}$$

```
1  sln[θ20]                                   # solution for unknown θ20
```

$$\frac{Al_1^4 l_2^3 q + 12Il_1 l_2^4 q}{3AEIl_1^4 + 12AEIl_1^3 l_2 + 36EI^2 l_1 l_2 + 36EI^2 l_2^2}$$

```
1  sln[v20]                                   # solution for unknown v20
```

$$\frac{l_1 l_2 q}{AE}$$

```
1  sln[v2l]                                   # solution for unknown v2l
```

$$\frac{5A^2 l_1^4 l_2^4 q + 4A^2 l_1^3 l_2^5 q + 24AIl_1^5 l_2 q + 96AIl_1^4 l_2^2 q + 60AIl_1 l_2^5 q + 12AIl_2^6 q + 288I^2 l_1^2 l_2^2 q + 288I^2 l_1 l_2^3 q}{24A^2 EIl_1^4 + 96A^2 EIl_1^3 l_2 + 288AEI^2 l_1 l_2 + 288AEI^2 l_2^2}$$

8.15.4 Final solutions

Substituting these unknowns found back to the general solution for the frame members, gives the final solutions.

```
1  # Final solution in mem-1:
2  vs1x=[v1x[i].subs({θ11:sln[θ11],v11:sln[v11]}) for i in range(len(v1x))
3  vs1x[0].simplify()                     # solution of deflection v(x)
```

$$\frac{l_2^3 q x^2 \left(-Al_1^3 + Al_1^2 x + 6Il_2\right)}{3EI\left(Al_1^4 + 4Al_1^3 l_2 + 12Il_1 l_2 + 12Il_2^2\right)}$$

```
1  vs1x[1].simplify()                     # solution of rotation θ(x)
```

$$\frac{l_2^3 q x \left(-2Al_1^3 + 3Al_1^2 x + 12Il_2\right)}{3EI\left(Al_1^4 + 4Al_1^3 l_2 + 12Il_1 l_2 + 12Il_2^2\right)}$$

```
1  vs1x[2].simplify()                     # solution of moment M(x)
```

$$\frac{2l_2^3 q \left(-Al_1^3 + 3Al_1^2 x + 6Il_2\right)}{3\left(Al_1^4 + 4Al_1^3 l_2 + 12Il_1 l_2 + 12Il_2^2\right)}$$

```
1  vs1x[3].simplify()                     # solution of shear force V(x)
```

$$-\frac{2Al_1^2 l_2^3 q}{Al_1^4 + 4Al_1^3 l_2 + 12Il_1 l_2 + 12Il_2^2}$$

```
1  vs1x[4]                                # solution of transverse load by
```

0

This is zero as expected. Let us find the solutions for member-2. Since the formulas are very long. We set $l_2 = l_1$, so that these can be shortened for displaying.

```
1  # Final solution in mem-2:
2  dic = {θ20:sln[θ20],v20:sln[v20],v21:sln[v21]}
3  vs2x=[v2x[i].subs(dic) for i in range(len(v2x))]
4  vs2x[0].subs({l2:l1}).simplify().collect(x**4).collect(x**3)\
5                   .collect(x**2).collect(x)    # solution of v(x)
```

Solution to Bended Beams

$$\frac{q\left(120AIl_1^4 + 576I^2l_1^2 + x^4 \cdot \left(5A^2l_1^2 + 24AI\right) + x^3\left(-20A^2l_1^3 - 96AIl_1\right)\right.}{\left. + x^2 \cdot \left(16A^2l_1^4 + 48AIl_1^2\right) + x\left(8A^2l_1^5 + 96AIl_1^3\right)\right)}}{24AEI\left(5Al_1^2 + 24I\right)}$$

```
1  vs2x[1].subs({l2:l1}).simplify().collect(x**3)\
2                   .collect(x**2).collect(x)        # solution of θ(x)
```

$$\frac{q\left(2Al_1^5 + 24Il_1^3 + x^3 \cdot \left(5Al_1^2 + 24I\right) + x^2\left(-15Al_1^3 - 72Il_1\right) + x\left(8Al_1^4 + 24Il_1^2\right)\right)}{6EI\left(5Al_1^2 + 24I\right)}$$

```
1  vs2x[2].subs({l2:l1}).simplify().collect(x**2).collect(x)  # moment M(x)
```

$$\frac{q\left(8Al_1^4 + 24Il_1^2 + x^2 \cdot \left(15Al_1^2 + 72I\right) + x\left(-30Al_1^3 - 144Il_1\right)\right)}{6 \cdot \left(5Al_1^2 + 24I\right)}$$

```
1  vs2x[3].subs({l2:l1}).simplify()                  # shear force V(x)
```

$$q\left(l_1 - x\right)$$

The normal forces in these members are found as

```
1  N_mem1 = vs2x[3].subs({x:0})                      # normal forces in member-1
2  N_mem1.simplify()
```

$$l_2 q$$

```
1  N_mem2 = vs1x[3].subs({x:l1})                     # normal forces in member-2
2  N_mem2.simplify()
```

$$-\frac{2Al_1^2 l_2^3 q}{Al_1^4 + 4Al_1^3 l_2 + 12Il_1 l_2 + 12Il_2^2}$$

This is the final solution to our frame structure problem, which may not be found in textbooks and the open literature.

8.16 Plot piecewise solutions

For the foregoing problem, the solution has two pieces: one for pillar and one for the top-beam. We write the following code to plot the solutions.

```
1  # Set the parameters to unit values:
2  dic = {q:1,l1:1,l2:1,E:1,I:1,A:1}        # set all parameters to unit
3  sln1x=[vs1x[i].subs(dic) for i in range(len(vs1x))]
4  sln1x
```

$$\left[\frac{x^3}{87} + \frac{5x^2}{87}, \frac{x^2}{29} + \frac{10x}{87}, \frac{2x}{29} + \frac{10}{87}, -\frac{2}{29}, 0\right]$$

```
1  # Set the parameters to unit values:
2
3  sln2x=[vs2x[i].subs(dic) for i in range(len(vs2x))]
4  sln2x
```

$$\left[\frac{x^4}{24} - \frac{x^3}{6} + \frac{8x^2}{87} + \frac{13x}{87} + 1, \frac{x^3}{6} - \frac{x^2}{2} + \frac{16x}{87} + \frac{13}{87}, \frac{x^2}{2} - x + \frac{16}{87}, 1-x, 1\right]$$

```
1  np.set_printoptions(formatter={'float': '{: 0.4e}'.format})
2  # set parameters: #q = 1.0; E = 1.0; I = 1.0
3  l1 = 1.0; l2 = 1.0
4
5  dx = 0.002
6  x1 = np.arange(0.0, l1, dx)              # coordinates in l1, pillar
7  x2 = np.arange(0.0, l2, dx)              # coordinates in l2, top-beam
8
9  # convert Sympy solution func. to numpy arrays.
10 n1 = len(sln1x)-1
11 n2 = len(sln1x)-1
12 x = symbols("x")
13 sln1 = [sp.lambdify(x, sln1x[i], 'numpy') for i in range(n1)]
14 sln2 = [sp.lambdify(x, sln2x[i], 'numpy') for i in range(n2)]
15 sx1 = [sln1[i](x1)  for i in range(n1)]
16 sx2 = [sln2[i](x2)  for i in range(n2)]
17
18 # The shear force function could be a constant. Need to generate array
19 if type(sx1[-1]) != np.ndarray:           # in case, func. is a constant
20     sx1[-1] = float(sx1[-1])*np.ones_like(x1)
21
22 if type(sx2[-1]) != np.ndarray:           # in case, func. is a constant
23     sx2[-1] = float(sx2[-1])*np.ones_like(x2)
24
25 X  = np.concatenate((x1, l1+x2))    # put the results in a single array
26 sx = [np.concatenate((sx1[i], sx2[i])) for i in range(n1)]
27
28 sx_max = [np.max(abs(sx[i])) for i in range(n1)]  # absolute max values
29 print(f"Absolute maximum values: {np.array(sx_max)}")
```

Absolute maximum values: [1.1164e+00 1.6755e-01 3.1609e-01 1.0000e+00]

```
1   # Plot the distribution curves:
2   X_data = [X for i in range(n1)]              # put x coordinates to a list
3   y_data = [sx[i]/sx_max[i] for i in range(n1)] # func. values to a list
4   labels=["Deflection", "Rotation", "Moment", "Shear force"]
5   gr.plot_fs(X_data, y_data, labels, 'x',      'Solutions',   'temp')
6                                                # x-label,  y-label,   file-name
```

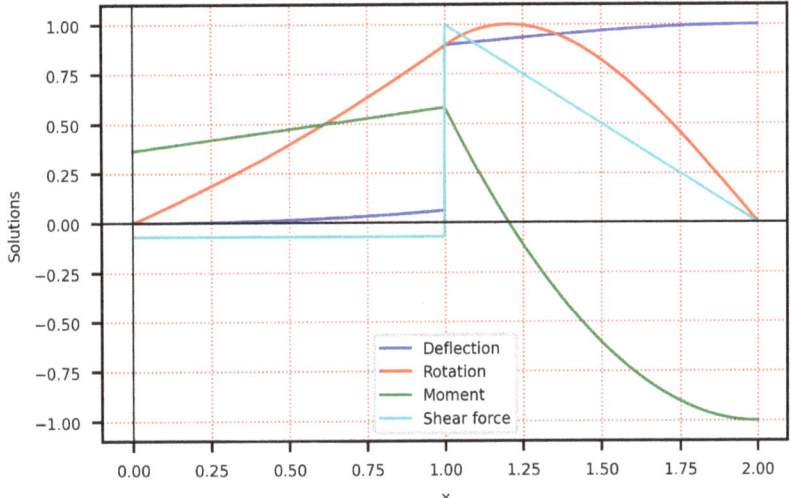

As shown, we can have endless examples done using our powerful solver1D4() for thin beams. The best approach for readers to learn may be to spend some time to play with the code: first to repeat the examples given above, and then try other examples. The following section uses solver1D4() for analysis of thermal stress in beams.

8.17 Thermal stress*

8.17.1 *Setting of the problem*

For beams that are overconstrained (indeterminate), a temperature gradient across the beam thickness can also result in additional thermal stresses, when the material is not allowed to expand freely. The general constitutive equation is given in Eq. (5.87). For thin beams, a temperature gradient relates to the curvature (derivative of the rotation of the cross-section) of the beam. The relationship can be derived as follows.

Consider a free-body with length dx and height h from a thin beam, as shown in Fig. 8.16.

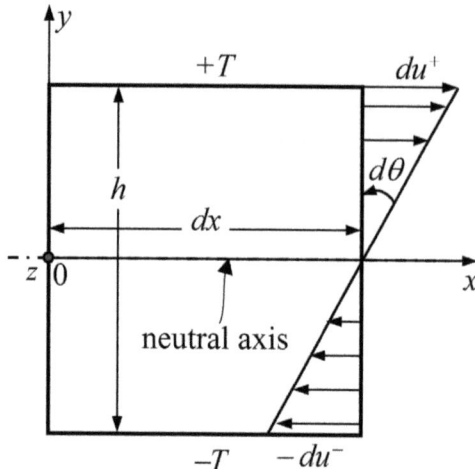

Figure 8.16. Free-body diagram for a thin beam experiencing a temperature change across its thickness.

8.17.2 Governing equation for deflection caused by the temperature gradient

The temperature change on the top surface of the beam is assumed T and that on the bottom surface is $-T$, and the temperature varies linearly across the thickness of the beam. Consider small displacements resulting from the temperature difference, the change in rotation θ_T of the cross-section is evaluated as follows:

$$d\theta_T = -\frac{du^+ dx - du^- dx}{h} = -\frac{\alpha T - (-\alpha T)}{h} dx = -\alpha \underbrace{\frac{2T}{h}}_{T_g} dx \qquad (8.38)$$

where du^+ and du^- are, respectively, the incremental displacements on the top and bottom surfaces of the beam caused the temperature change there, α is the thermal expansion coefficient of the material of the beam, and T_g is the temperature gradient. The negative sign on the front of the right-hand side is because positive gradient results in negative rotation. Equation (8.38) can be rewritten as

$$\frac{d\theta_T}{dx} = -\alpha T_g \qquad (8.39)$$

where $\frac{d\theta_T}{dx}$ is the curvature of the beam.

Using Eqs. (8.2) and (8.39), we obtain:

$$\frac{\partial^2 v_T}{\partial x^2} = -\alpha T_g \tag{8.40}$$

where v_T is the deflection resulting from the temperature gradient.

8.17.3 Solution of deflection resulting from temperature gradient

Assuming T_g is constant along x-axis, the foregoing equation can be integrated twice, which gives

$$v_T = -\frac{1}{2}\alpha T_g x^2 + c_0 x + c_1 \tag{8.41}$$

Case-I: Assuming the beam is clamped at $x = 0$, we have $v_T = 0$ and $\frac{dv_T}{dx} = 0$. This leads to both $c_0 = 0$ and $c_1 = 0$. Hence, the deflection resulting from the temperature gradient is found as

$$v_T = -\frac{\alpha T_g}{2} x^2 \tag{8.42}$$

Case-II: Assuming the beam is clamped at $x = l$, we have $v_T = 0$ and $\frac{dv_T}{dx} = 0$. This leads to $c_0 = \alpha T_g l$ and $c_1 = -\frac{\alpha T_g l^2}{2}$. Hence, the deflection resulting from the temperature gradient is found as

$$v_T = -\frac{\alpha T_g}{2}(l-x)^2 \tag{8.43}$$

The code to find this expression is given as follows:

```
1  x, l = symbols("x, l")
2  vT, Tg, α, c0, c1 = symbols("v_T, T_g, α, c0, c1 ")
3  vT = -α*Tg/2*x**2 + c0*x+c1
4  sln = sp.solve([vT.subs(x,l), vT.diff(x).subs(x,l)], [c0, c1])
5  sln
```

$$\left\{ c_0 : T_g l \alpha, \; c_1 : -\frac{T_g l^2 \alpha}{2} \right\}$$

```
1  vT1 = vT.subs({c0:sln[c0], c1:sln[c1]})
2  vT1.factor(Tg)
```

$$\frac{T_g \alpha \left(-l^2 + 2lx - x^2 \right)}{2}$$

8.17.4 Solutions of indeterminate beams in temperature gradient

Since we have already the beam solver, and the just derived formulas for the deflection caused by a temperature gradient, finding the solutions of deflection, rotation, moment and shear force in beams is straightforward. The procedure is as follows:

1. Find solution to beams for mechanical effects using the solver1D4(), as we did earlier, which gives the mechanical deflection v for the beam, in analytic formulas.
2. Add the thermal deflection v_T to the mechanical deflection v.
3. Enforce BCs to the combined deflection to determine the relevant unknowns, and to obtain the solution for the final deflection.
4. After the final deflection is obtain, use Eq. (8.2) to compute the rotation.
5. The moment can be computed using Eqs. (8.9) and (8.40):

$$M_z = EI_z \left[\frac{d^2 v_{\text{combined}}}{dx^2} - \frac{\partial^2 v_T}{\partial x^2} \right] = EI_z \left[\frac{d^2 v_{\text{combined}}}{dx^2} + \alpha T_g \right] \quad (8.44)$$

This is because of the constitutive equation change due to the temperature gradient. In other words, we need to remove the thermal curvature to compute the moment. This treatment is similar to the equation derived in the previous chapter for 1D bars.
6. Use Eq. (8.14) to compute the shear force.

Let us look at some examples.

8.17.5 Example: Simple-clamped beams subjected to temperature gradient

Consider beams of length l with cross-section having a second moment of area I. It is made of material with Young's modulus E. It is subjected to temperature gradient T_g across its thickness. Its boundaries are simple-clamped (s-c).

1. Derive formulas for solutions, including deflection, rotation, moment, and shear force in the beam.
2. Plot the distribution of all these solutions, using unit for all its parameters.

We write the following code to complete the task. First, define all necessary symbolic variables.

Solution to Bended Beams

```
1  E, I, A, l = symbols('E, I, A, l ', nonnegative=True)
2  x, q = symbols('x, q ')      # geometry & force
3
4  # variables for displacement boundary conditions (DBCs):
5  v0, θ0, vl, θl = symbols('v0, θ0, v_l, θ_l')
6
7  # variables for force boundary conditions (FBCs):
8  V0, M0, Vl, Ml = symbols('V0, M0, V_l, M_l')
9
10 by = sp.Function('b_y')(x)                    # for force function
```

8.17.5.1 Find the general displacement solution for an s-c beam

```
1  by = 0                                        # no mechanical loading
2  vM = solver1D4(E, I, by, l,v0,θ0,vl,θl,V0,M0,Vl,Ml,key='s-c')
3  print(f'Is solution correct? {by/E/I==vM[4]}; The solution u(x) is:')
4  vM[0]                                          # mechanical deflection vM(x)
```

Outputs form solver1D4(): v, θ, M, V, qy
Is solution correct? True; The solution u(x) is

$$v_0 - \frac{x\theta_l}{2} - \frac{3v_0 x}{2l} + \frac{3v_l x}{2l} + \frac{x^3 \theta_l}{2l^2} + \frac{v_0 x^3}{2l^3} - \frac{v_l x^3}{2l^3} - \frac{M_0 l x}{4EI} + \frac{M_0 x^2}{2EI} - \frac{M_0 x^3}{4EIl}$$

This is the general displacement solution for an s-c beam for possible internal mechanical forces as variables on the boundaries.

8.17.5.2 Add in the displacement caused by temperature gradient

```
1  vT, Tg, α = symbols("v_T, T_g, α")          # Tg: temperature gradient
2  vT = - α*Tg/2*(l-x)**2                       # deflection caused by temperature gradient
3  v  = vT + vM[0]                              # total displacement
4  v
```

$$-\frac{T_g \alpha (l-x)^2}{2} + v_0 - \frac{x\theta_l}{2} - \frac{3v_0 x}{2l} + \frac{3v_l x}{2l} + \frac{x^3 \theta_l}{2l^2} + \frac{v_0 x^3}{2l^3} - \frac{v_l x^3}{2l^3}$$
$$- \frac{M_0 l x}{4EI} + \frac{M_0 x^2}{2EI} - \frac{M_0 x^3}{4EIl}$$

8.17.5.3 Find the combined displacement using BCs

For our s-c beam, the clamped BCs at the right-end is zero. We use the BCs to simplify the general solution.

```
1  v = v.subs({v1:0,θ1:0,M0:0})       # set known values for BC variables
2  v                                   # simplified total displacement
```

$$-\frac{T_g\alpha(l-x)^2}{2} + v_0 - \frac{3v_0 x}{2l} + \frac{v_0 x^3}{2l^3}$$

The combined displacement at the left-end must be zero. We thus use this condition to find variable v_0.

```
1  sln = sp.solve(v.subs(x,0), v0)    # solve for displacement at left-end
2  gr.printM(sln[0], 'v0 is found as:')
```

v0 is found as: $\dfrac{T_g l^2 \alpha}{2}$

```
1  v_T = v.subs({v0:sln[0]})          # put the solution for v0 back
2  v_T  # final solution for the deflection caused the temperature change
```

$$\frac{T_g l^2 \alpha}{2} - \frac{3 T_g l x \alpha}{4} - \frac{T_g \alpha (l-x)^2}{2} + \frac{T_g x^3 \alpha}{4l}$$

This is the final solution for the deflection caused by the temperature gradient. It can then be used to find the other solutions: rotation, moment, and shear force. We put this set of solutions in a list:

```
1  S = [v_T, v_T.diff(x,1), E*I*(v_T.diff(x,2)+α*Tg), -E*I*v_T.diff(x,3)]
2  S = [si.simplify() for si in S]
3  S                                   # formulas for the set of solutions
```

$$\left[\frac{T_g x \alpha (l^2 - 2lx + x^2)}{4l},\ \frac{T_g \alpha (l(l-4x) + 3x^2)}{4l},\ \frac{3EIT_g x \alpha}{2l},\ -\frac{3EIT_g \alpha}{2l} \right]$$

These are all functions of x, except the shear force that is constant.

8.17.5.4 Plot the distribution of solutions

Finally, we plot the solutions using gr.plot2curveS() that finds also some of the extreme values (maximum, minimum, and values at the boundaries.)

To generate the plots for viewing the variation of the solutions, we set all the parameters to unit.

Solution to Bended Beams

```
1  Sx=[S[i].subs({E:1,I:1,l:1,q:1,Tg:1,a:1}) for i in range(len(S))]
2  Sx                                # This is a set functions of only x
```

$$\left[\frac{x\left(x^2 - 2x + 1\right)}{4},\ \frac{3x^2}{4} - x + \frac{1}{4},\ \frac{3x}{2},\ -\frac{3}{2}\right]$$

```
1  title = [["Deflection", "Rotation"], ["Moment", "Shear force"]]
2  for i in range(len(title)):
3      gr.plot2curveS(Sx[2*i:2*(i+1)], 0., 1., title=title[i])
4
5  plt.show()
```

Maximum Deflection value=3.703e-02, at x=0.33
Maximum Rotation value=2.500e-01, at x=0.0
Extreme Deflection values=[0.000e+00 3.703e-02],
    at x=[1.000e+00 3.300e-01]
Critical Rotation values=[0.000e+00 1.675e-03],
    at x=[1.000e+00 3.300e-01]
Deflection values at boundary =(0.0, 0.0)
Rotation values at boundary =(0.25, 0.0)

Maximum Moment value=1.500e+00, at x=1.0
Maximum Shear force value=-1.500e+00, at x=0.0
Extreme Moment values=[0.000e+00],
    at x=[0.000e+00]
Critical Shear force values=[-1.500e+00],
    at x=[0.000e+00]
Moment values at boundary =(0.0, 1.5)
Shear force values at boundary =(-1.5, -1.5)

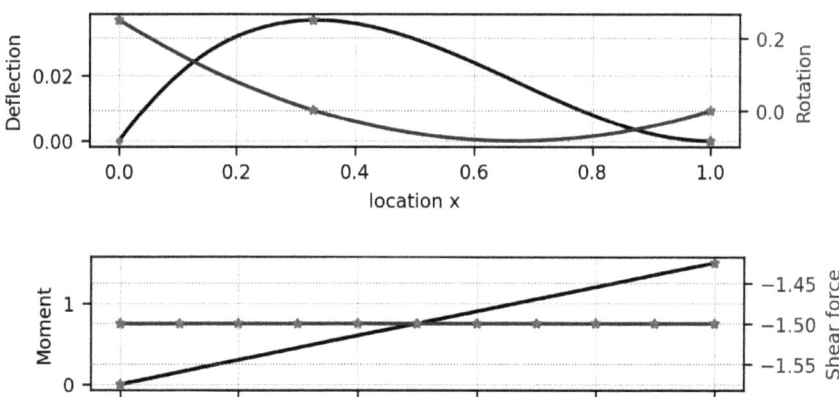

8.17.6 Example: Beams simple-simple supported subjected to temperature gradient

It is seen that although there is no mechanical loading on the beam, the beam deforms with moments and shear force built up along the beam. This is purely due to the temperature gradient.

Consider the same problem as the previous one, but change s-c to s-s support.

8.17.6.1 Find the general displacement solution for an s-s beam

```
1  by = 0                                                              # no loading
2  vM = solver1D4(E, I, by, l, v0, θ0, vl, θl, V0, M0, Vl, Ml, key='s-s')
3  print(f'Is solution correct? {by/E/I==vM[4]}; The solution u(x) is:')
4  vM[0]
```

Outputs form solver1D4(): v, θ, M, V, qy
Is solution correct? True; The solution u(x) is

$$v_0 - \frac{v_0 x}{l} + \frac{v_l x}{l} - \frac{M_0 l x}{3EI} + \frac{M_0 x^2}{2EI} - \frac{M_0 x^3}{6EIl} - \frac{M_l l x}{6EI} + \frac{M_l x^3}{6EIl}$$

This is the general displacement solution for an s-s beam with possible mechanical forces in the beam as variables on the boundaries.

8.17.6.2 Add in the displacement caused by temperature gradient

```
1  vT, Tg, α = symbols("v_T, T_g, α")     # Tg: temperature gradient
2  vT = - α*Tg/2*(l-x)**2                 # disp. caused by temperature gradient
3  v  = vT + vM[0]                        # total displacement
4  v
```

$$-\frac{T_g \alpha \left(l - x\right)^2}{2} + v_0 - \frac{v_0 x}{l} + \frac{v_l x}{l} - \frac{M_0 l x}{3EI} + \frac{M_0 x^2}{2EI} - \frac{M_0 x^3}{6EIl} - \frac{M_l l x}{6EI} + \frac{M_l x^3}{6EIl}$$

8.17.6.3 Find the combined displacement

For s-s beams, we can leave v_0 as the only unknown and set the rest zero.

```
1  v = v.subs({vl:0,θl:0,M0:0,Ml:0})  # set known values for BC variables
2  v                                   # simplified total displacement
```

$$-\frac{T_g \alpha \left(l - x\right)^2}{2} + v_0 - \frac{v_0 x}{l}$$

The combined displacement at the left-end must be zero. We thus use this condition to find variable v_0.

```
1  sln = sp.solve(v.subs(x,0), v0)   # solve for displacement at left-end
2  sln
```

$$\left[\frac{T_g l^2 \alpha}{2}\right]$$

```
1  v_T = v.subs({v0:sln[0]})          # put the solution for v0 back
2  v_T
```

$$\frac{T_g l^2 \alpha}{2} - \frac{T_g l x \alpha}{2} - \frac{T_g \alpha (l-x)^2}{2}$$

This is the final solution for the displacement. It can then be used to find the other solutions: rotation, moment, and shear force. We put this set of solutions in a list:

```
1  S = [v_T, v_T.diff(x,1), E*I*(v_T.diff(x,2)+a*Tg), -E*I*v_T.diff(x,3)]
2  S = [si.simplify() for si in S]
3  S                                   # formulas for the set of solutions
```

$$\left[\frac{T_g x \alpha (l-x)}{2}, \frac{T_g \alpha (l-2x)}{2}, 0, 0\right]$$

It is seen that although there is no mechanical loading on the beam, the beam deforms with rotation. This is purely due to the temperature gradient. The moments and shear force are zero because of the s-s support. The temperature gradient causes deflection, but no stress is built up. The s-s beam is determinate structure. It does not constraint the thermal rotation and hence will not create thermal moments. This is the major difference from the s-c beam that is indeterminate.

8.17.6.4 Plot the distribution of solutions

Finally, we plot the solutions using gr.plot2curveS() that finds also some of the extreme values (maximum, minimum, and values at the boundaries.)

To generate the plots for viewing the variation of the solutions, we set all the parameters to unit.

```
1  Sx=[S[i].subs({E:1,I:1,l:1,q:1,Tg:1,α:1}) for i in range(len(S))]
2  Sx                              # This is a set functions of only x
```

$$\left[\frac{x(1-x)}{2},\ \frac{1}{2}-x,\ 0,\ 0\right]$$

```
1  title = [["Deflection", "Rotation"], ["Moment", "Shear force"]]
2  for i in range(len(title)):
3      gr.plot2curveS(Sx[2*i:2*(i+1)], 0., 1., title=title[i])
4
5  plt.show()
```

```
Maximum Deflection value=1.250e-01, at x=0.5
Maximum Rotation value=5.000e-01, at x=0.0
Extreme Deflection values=[ 1.250e-01],
    at x=[ 5.000e-01]
Critical Rotation values=[ 0.000e+00],
    at x=[ 5.000e-01]
Deflection values at boundary =(0.0, 0.0)
Rotation values at boundary =(0.5, -0.5)

Maximum Moment value=0.000e+00, at x=0.0
Maximum Shear force value=0.000e+00, at x=0.0
Extreme Moment values=[ 0.000e+00],
    at x=[ 0.000e+00]
Critical Shear force values=[ 0.000e+00],
    at x=[ 0.000e+00]
Moment values at boundary =(0.0, 0.0)
Shear force values at boundary =(0.0, 0.0)
```

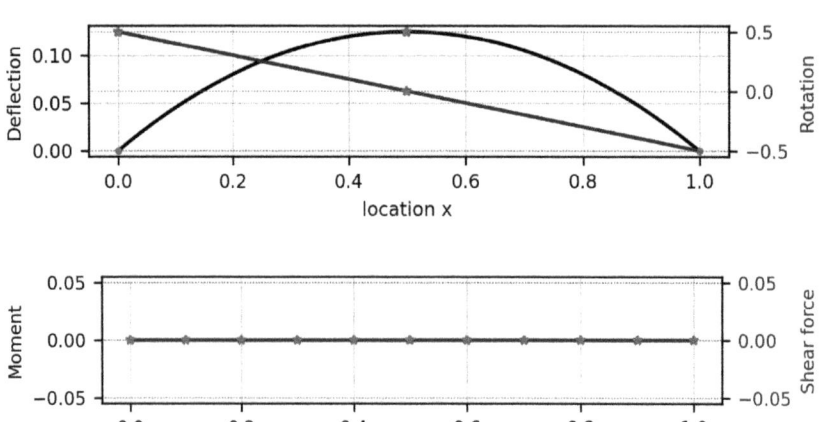

As shown there is no thermal moment and force in the beam, because it is determinant. The thermal effects are only on the deformations (deflection and rotation).

8.18 Remarks

The beam solver solver1D4() developed in this chapter is a powerful tool for thin beams. It is quite easy to use and applicable to a wide range of problems, including frames and thermal effect analysis.

In the existing textbooks and literature, there is a quite common approach by drawing shear force and moment diagrams, and then use Eq. (8.17) and/or (8.9) to obtain the deflection via integration. This may work for statically determinate beams because it may be easier for drawing the shear force and moment diagrams. However, use of beam solver solver1D4() is much more straightforward and effortless. All the diagrams, deflection, rotation, moment and shear force can all obtained in one run, with much fewer manual operations.

For large frame structures or frame with truss members, it is much easier to use a more systematic approach based on the concept of stiffness matrix assembly, as we often do in the finite element method (FEM) [1]. Using FEM, the number of members in a structure can be millions and even billions because no manual operations for setting various types of condition are needed.

The next chapter deals with another important structural member: bars subjected to torsion.

Reference

[1] G.R. Liu and S.S. Quek, *The Finite Element Method: A Practical Course.* Butterworth-Heinemann, New York, 2013.

Chapter 9

Torsion of Bars

```
1  # Place cursor in this cell, and press Ctrl+Enter to import dependences.
2  import sys                          # for accessing the computer system
3  sys.path.append('../grbin/')   # Change to the directory in your system
4
5  from commonImports import *        # Import dependences from '../grbin/'
6  import grcodes as gr                # Import the module of the author
7  importlib.reload(gr)                # When grcodes is modified, reload it
8
9  from continuum_mechanics import vector
10 init_printing(use_unicode=True)     # For latex-like quality printing
11 #np.set_printoptions(precision=4,suppress=True) # Digits in print-outs
12 np.set_printoptions(formatter={'float': '{: 0.4e}'.format})
```

Structure members or components are frequently used in power transmission or similar purposes. Figure 9.1 shows a rotor of a modern steam turbine by Siemens Pressebild (http://www.siemens.com) as a typical example. The shaft used in it is one of the most critical components in such a rotational machine undergoing torsional deformation.

Most shafts are essentially made of bars with circular cross-section, which are quite effective in many ways and are relatively easier to deal with in mechanics aspects. However, there are also bar members with other cross-sectional shapes, as shown in Fig. 9.2. These types of bars require more intensive stress analysis for proper design.

This chapter discusses torsional bar members with various cross-sectional shapes. The objective is to find the relationship between the **twist angle**

Figure 9.1. A modern steam turbine, where shafts are used for power transmission. The shaft undergoes torsional deformation.
Source: Image from Siemens Pressebild available at Wikimedia Commons under CC BY-SA 3.0 license.

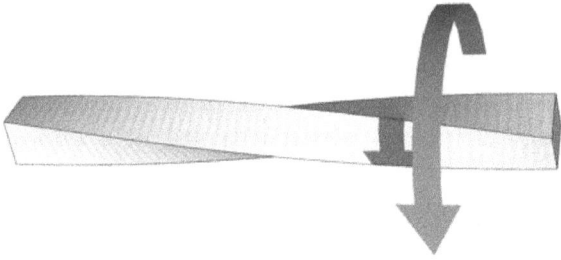

Figure 9.2. A bar with square cross-section subjected to a torque, resulting in complicated deformations in the cross-section.
Source: Image from Orion-8 at Wikimedia Commons under the CC BY 3.0 license.

(an important measure of deformation) and the applied **torque** (a typical force application) and **stresses** (critical for failure analysis) on the cross-section of the members. Since our focus is purely on torsional deformation, the members or components should have a **slender** geometry, like a bar or rod. Thus, we will call all these types of torsional members and components a **bar** because it captures the major geometric feature, allowing proper assumptions to be made. The cross-section of the bar can be arbitrary as long as its cross-sectional dimensions are sufficiently smaller than its length. The torques are applied about the axial axis of the bar.

The **major challenge** in dealing with torsional bars is its three-dimensional (3D) nature: the torque is applied about is axial axis (z-axis). The stresses, on the other hand, occur on its cross-sectional plane (x–y plane). The **art** of the technique lies in converting the 3D problem into a 1D or at most a 2D one, so that solutions can be found analytically. Detailed formulations will be provided in this chapter on how this is done.

If the bar is bulky, it needs to be treated in principle as a true 3D problem. The most effective techniques will be numerical, including the finite element method (FEM) [1], smoothed FEM [2], and even the meshfree methods [3]. All these techniques require intensive formulation and substantial coding. They are thus beyond the topic of this book.

This chapter is written in reference to the textbooks in Refs. [4–8]. Our study begins with the simplest and most fundamental case: bars with a circular cross-section, which are the most widely used in engineering.

9.1 Bars with circular cross-section

9.1.1 *Problem description*

Consider a solid cylindrical bar with a circular cross-section. It has a cross-sectional area A and length L. It is defined with the Cartesian coordinates (x, y, z). Let z be the axial coordinate, and it is perpendicular to the cross-section. It is subjected to torque T about the z-axis. An originally straight-line AB on the lateral surface of the bar becomes a helical curve AB* due to the twist deformation of the bar, as shown in Fig. 9.3. Our task is to find out how the twist angle relates to the applied torque and the resulting stresses in the bar.

9.1.2 *Assumptions*

Clearly, any cross-section of the bar undergoes shearing, which can result in shear stresses. The shearing deformation accumulates to a twist angle that may change along the z-axis. To formulate this problem properly, we assume the following:

1. The plane of the cross-section of the bar remains perpendicular to the z-axis.

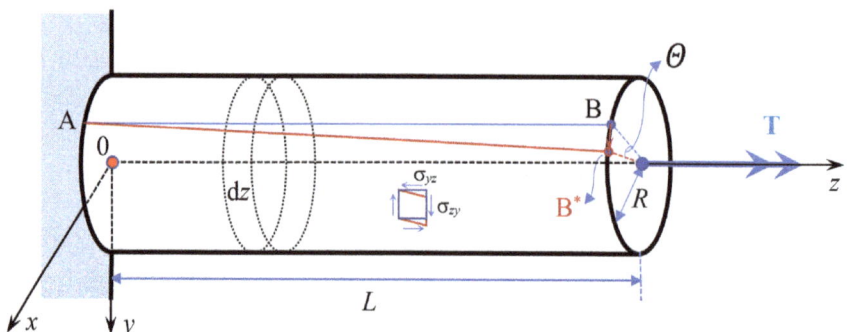

Figure 9.3. A bar with circular cross-section subjected to a torque, undergoing a twist deformation.

2. The rotation is small, and the cross-section rotates as a rigid plane about the z-axis.
3. The rotation Θ of a given section, relative to the base plane $z = 0$, depends linearly on the distance z from the base plane.

Based on these assumptions, we have

$$\Theta = \vartheta_t z \tag{9.1}$$

where ϑ_t is the twist angle per unit length and is a constant (does not change with x, y, or z). It is also sometimes called torsional curvature (the rate of angular change with respect to z).

These assumptions practically fix the dependence of the problem on z. The problem becomes 2D with displacements on the cross-section of the bar.

9.1.3 Displacements on the cross-section

Because the cross-section undergoes a rotation about the z-axis and the rotation is small, the displacements on the cross-section can be shown in Fig. 9.4.

Because the cross-section rotates rigidly, these displacement components can be obtained using simple geometry relations. They are written as

$$u = -y\Theta; \quad v = x\Theta; \quad w = 0 \tag{9.2}$$

or

$$u = -y\vartheta_t z; \quad v = x\vartheta_t z; \quad w = 0 \tag{9.3}$$

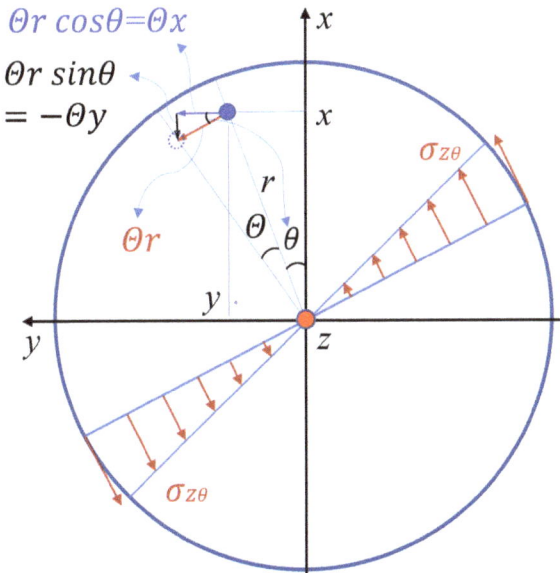

Figure 9.4. Displacement components on the cross-section of a circular bar subjected to a torque.

It appears that the displacements depend on x, y, and z: a 3D problem. It is controlled by one parameter: ϑ_t or Θ.

9.1.4 Shear strains and stresses on the cross-section

Using the small strain–displacement relations for a 3D solid given in Chapter 4 and since ϑ_t is independent of x, y, and z, we can find the strains on the bar readily:

$$\varepsilon_{xx} = \frac{\partial u}{\partial x} = 0; \quad \varepsilon_{yy} = \frac{\partial v}{\partial y} = 0; \quad \varepsilon_{zz} = \frac{\partial w}{\partial z} = 0; \quad \varepsilon_{xy} = \frac{1}{2}\left(\frac{\partial u}{\partial y} + \frac{\partial v}{\partial x}\right) = 0 \tag{9.4}$$

and

$$2\varepsilon_{zx} = \gamma_{zx} = \frac{\partial u}{\partial z} + \frac{\partial w}{\partial x} = -y\vartheta_t; \quad 2\varepsilon_{zy} = \gamma_{zy} = \frac{\partial w}{\partial y} + \frac{\partial v}{\partial z} = x\vartheta_t \tag{9.5}$$

It is clear that there are only two nonzero strain components. All these can also be obtained using the following codes. First, define the displacement vector:

```
1  x, y, z = symbols("x, y, z")              # define symbolic coordinates
2  X = Matrix([x, y, z])
3  ϑt = symbols("ϑ_t")                       # twist angle per length
4
5  U = Matrix([-y*ϑt*z, x*ϑt*z, 0]) # Assumed displacement vector [u,v,w]
6  U
```

$$\begin{bmatrix} -yz\vartheta_t \\ xz\vartheta_t \\ 0 \end{bmatrix}$$

```
1  # Compute the strains (see Chapter 4):
2  U_g = gr.grad_vf(U, X)                    # displacement gradient
3  ε = (U_g + U_g.T)/2                       # strain tensor, symmetric
4  ε
```

$$\begin{bmatrix} 0 & 0 & -\frac{y\vartheta_t}{2} \\ 0 & 0 & \frac{x\vartheta_t}{2} \\ -\frac{y\vartheta_t}{2} & \frac{x\vartheta_t}{2} & 0 \end{bmatrix}$$

The code finds again that there are only two nonzero strains, and they are both shear strains.

For isotropic materials, there can only be two shear stresses on the cross-section of the bar, which relate to shear strains by only the shear modulus G (see Chapter 5). These are

$$\sigma_{zx} = G\gamma_{zx} = -yG\vartheta_t; \quad \sigma_{zy} = G\gamma_{zy} = xG\vartheta_t \tag{9.6}$$

where the shear modulus G is a material constant. Note that these stresses are independent of z, implying that these stresses are uniform along the axial axis of the bar.

With this simple stress state, we can perform some interesting analysis on it using the following code. The stress tensor at any point in a torsion bar is given as follows:

```
1  G = symbols("G")
2  σ_tensor = G*(2*ε)                        # γ_ij = 2ε_ij
3  σ_tensor
```

$$\begin{bmatrix} 0 & 0 & -Gy\vartheta_t \\ 0 & 0 & Gx\vartheta_t \\ -Gy\vartheta_t & Gx\vartheta_t & 0 \end{bmatrix}$$

9.1.4.1 Divergence of the stress states in a circular bar

This set of stresses satisfy the equilibrium equation for 3D solids given in Chapter 6 (considering a static problem with no body force). This is because their divergences (see Chapter 3) are zeros:

```
1  div_σ_tensor = gr.div_tensorF(σ_tensor, X)   # a vector of 3 functions
2  div_σ_tensor                                 # Or use: vector.div_tensor(σ_tensor.T)
```

$$\begin{bmatrix} 0 \\ 0 \\ 0 \end{bmatrix}$$

Clearly, the stress tensor of torsion is a divergence-free tensor. Therefore, the equilibrium equations are satisfied for the free body forces in the bar.

9.1.4.2 Curl of the stress vector field in a circular bar

The stresses on the z-plane in a circular bar under torsion can form a vector function with three components: $\sigma_{zx}, \sigma_{zy}, \sigma_{zz}$. Let us take a look at the curl (see Chapter 2) of this vector field formed by these torsional stresses.

```
1  # Define a vector field as symbolic functions of (x, y, z)
2  σz = symbols("σ_z")
3  σz = σ_tensor[2,:]                           # get the vector on z-surface
4  σz
```

$$\begin{bmatrix} -Gy\vartheta_t & Gx\vartheta_t & 0 \end{bmatrix}$$

```
1  curl_F = gr.curl_vf(σz, X)
2  curl_F                                       # Curl of the torsional stress vector field
```

$$\begin{bmatrix} 0 \\ 0 \\ 2G\vartheta_t \end{bmatrix}$$

This means that:

- The rates of rotation of the torsional stress vector field around the x- and y-axis are both zero at any point on the cross-section. The field has zero circulation around these two axes.
- It has a positive constant circulation at any point on the cross-section around the z-axis. The shear stresses are in the x–y plane, and the torque (curl) is with respect to the z-axis. This is a typical feature of a circular bar under pure torsion.

Also, the sum of these two shear stress components is in the circumferential direction on the cross-section of the bar, which is the red arrow in Fig. 9.4. This can be confirmed by viewing the stress state in the polar coordinates (r, θ).

9.1.5 Distribution of shear stress on the cross-section

Again, because the cross-section undergoes a rigid body rotation about the z-axis and the rotation is small, the displacements on the cross-section in the polar coordinate (r, θ) are

$$u_r = 0; \quad u_\theta = r\vartheta_t z; \quad w = 0 \qquad (9.7)$$

The only nonzero displacement is u_θ, which is in the circumferential θ-direction (the red arrow $r\Theta$ in Fig. 9.4). This again means that the problem is 1D. Reader may derive Eq. (9.7) via a coordinate transformation using Eq. (9.3).

The stress in the polar coordinates can be obtained using the stress tensor transformation rule:

$$\boldsymbol{\sigma'} = \mathbf{T}\boldsymbol{\sigma}\mathbf{T}^\top \qquad (9.8)$$

where $\boldsymbol{\sigma'}$ is the stress tensor in the polar coordinate system, $\boldsymbol{\sigma}$ is the stress tensor in the Cartesian coordinates with components given in Eq. (9.6), and $\mathbf{T}$ is the transformation matrix defined in Section 2.10.

Let us use the following Sympy code to get it done.

```
1  # Compute the stress tensor on the cross-section of the bar:
2
3  ϑ, θ  = "\u03D1", "\u03B8"              # Define symbol in hex code
4  G, ϑt = symbols("G, ϑ_t")                # Define symbolic variables
5  r, θ  = symbols("r, θ")                  # for polar coordinates
6
7  # write the stresses in terms of r and θ:
8  σzx = -r*sp.sin(θ)*G*ϑt                                  # y=rsinθ
9  σzy =  r*sp.cos(θ)*G*ϑt                                  # x=rcosθ
10
11 S = Matrix([[ 0 ,  0 , σzx],    # stress state for torsion w.r.t z-axis
12             [ 0 ,  0 , σzy],
13             [σzx, σzy,  0 ]])
14 S
```

$$\begin{bmatrix} 0 & 0 & -Gr\vartheta_t \sin(\theta) \\ 0 & 0 & Gr\vartheta_t \cos(\theta) \\ -Gr\vartheta_t \sin(\theta) & Gr\vartheta_t \cos(\theta) & 0 \end{bmatrix}$$

Readers are not to be confused about the symbols θ and ϑ_t. The former is a coordinate of the polar system, and the latter is the twist angle per unit length.

Perform coordinate transformation for the stresses.

```
1  # coordinate transformation for the stresses:
2  T = gr.transferMts(θ)   # transformation matrix for rotation θ w.r.t. z
3  Sp=T@S@T.T              # S: stress tensor; @: the shorthand of np.matmul
4  gr.simplifyM0(Sp)                   # simply the terms in the output
```

$$\begin{bmatrix} 0 & 0 & 0 \\ 0 & 0 & Gr\vartheta_t \\ 0 & Gr\vartheta_t & 0 \end{bmatrix}$$

We found that the stress components are all zero everywhere in the polar coordinates:

$$\sigma_{rr} = \sigma_{r\theta} = \sigma_{rz} = \sigma_{\theta\theta} = \sigma_{zz} = 0 \tag{9.9}$$

The only nonzero component is

$$\sigma_{z\theta} = rG\vartheta_t \tag{9.10}$$

which changes linearly in r and is in the circumferential direction, as shown in Fig. 9.4. The largest shear stress is registered by the surface of the bar. Because there is only one shear stress on the cross-section of a bar, the stress is often denoted as τ, when dealing with torsion:

$$\tau = rG\vartheta_t \tag{9.11}$$

The strain should also have only one component: $2\varepsilon_{z\theta} = r\vartheta_t$, also changing linearly in r.

Note that, in the coordinate transformation, we used an arbitrary angle θ. However, the transformed results do not depend on θ at all. That means the result obtained is applicable to any θ location on the cross-section. This is the consequence of the rigid cross-section rotation assumption. The problem is essentially 1D.

9.1.6 *Boundary condition on the lateral surface of the bar*

Because the only nonzero stress is in the circumferential direction, it will not result in any stress on the lateral surface of the bar. Therefore, the zero-stress boundary condition on the lateral surface of the bar is also satisfied.

A rigorous proof is given by the code right above Eq. (9.9) which gives for any θ: $\sigma_{rr} = \sigma_{r\theta} = \sigma_{rz} = 0$. These are all the possible stress components on the lateral surface of a circular bar, and they all are zero regardless of θ.

Note also that $\sigma_{rr} = \sigma_{r\theta} = \sigma_{rz} = 0$ is true for any r. This means that any cylinder in the circular bar rotates together without an inter-surface stress. In other words, if we put cylinders together to form a circular bar, the assembly behaves the same as one solid circular bar in terms of pure torsion, if they all have the same twist angle.

Note that in usual applications, the shear stress distribution on the cross-section at the two ends of the bar may not be linearly distributed. In those cases, the solution is still applicable far away from the ends, based on the so-called Saint-Venant principle. This implies that our formulation applicable in general for slender bars for locations away from the two ends.

9.1.7 Equilibrium on the cross-section, twist-torque relation

We found the stresses and know their distribution on the cross-section. The amplitude of the stresses must depend on the level of the torque applied. For example, if the torque is zero, the stresses must be zero. We need now to establish the relationship between the stresses and the applied torque. For this, naturally, we must invoke equilibrium conditions.

Let us now examine the equilibrium status on the cross-section of the bar subjected to a torque T. The sum of all moments with respect to the z-axis should be zero. Thus, the torque T must result from these two shear stresses.

With the help of Fig. 9.5, we have

$$\sum M_z = 0 : T = \int_A (x\sigma_{zy} - y\sigma_{zx})dA = \int_A (x^2 G\vartheta_t + y^2 G\vartheta_t)dA$$

$$= \int_A G\vartheta_t \underbrace{(x^2 + y^2)}_{r^2} dA = G\vartheta_t \underbrace{\int_A r^2 dA}_{J} = G\vartheta_t J \qquad (9.12)$$

where J is the **polar second moment of area**. For circular cross-section, we have

$$J = \int_A r^2 dA = \int_A r^2 \underbrace{r d\theta \, dr}_{dA} = 2\pi \int_0^R r^3 dr = \frac{\pi R^4}{2} \qquad (9.13)$$

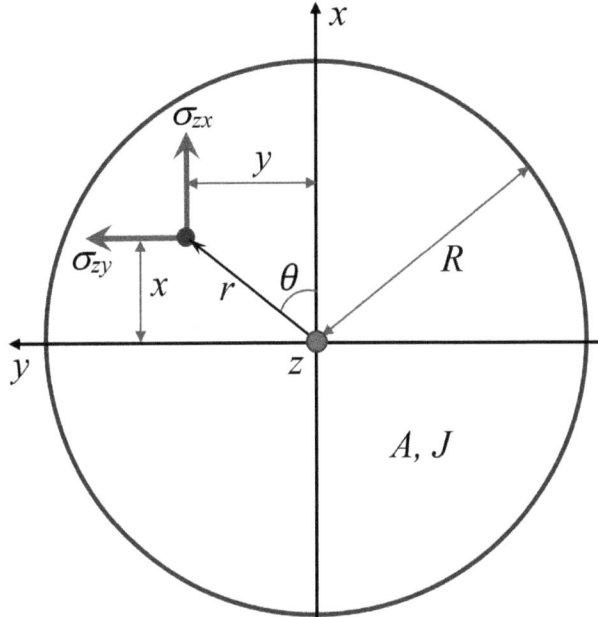

Figure 9.5. Shear stresses on the cross-section of a circular bar result in a torque.

9.1.8 *On polar second moment of area*

The polar second moment of area J is a property of the cross-section of a bar, and it is the most important one for all bars subject to torsion. The twist angle and the shear stress all depend on it. For circular cross-section, we have a very nice Eq. (9.13) for computing it. It is derived using the polar coordinate system, as shown above. The word "second" stands for r^2 in the integral in Eq. (9.13).

For other shapes of cross-section, the formulas can be very complicated, and often the polar coordinate system is not usable in the derivation. We call such a J **equivalent** polar second moment of area. We need to use the word "polar" because the second moment of area in Cartesian coordinate is a different one. A better unified name might be **torsional second moment of area** for bars with any type of cross-section.

Our strategy for dealing with bars with other types of torsional bars is to find the equivalent polar second moment of area of the cross-section of the bar. Once this is done, the formulas for computing the twist angle and the shear stress will be the same.

9.1.9 Twist-angle and torque relations

Equation (9.12) gives the twist-angle per-unit-length, ϑ_t:

$$\vartheta_t = \frac{T}{GJ} \tag{9.14}$$

If the bar is fixed at $z = 0$, the (total) twist-angle at an arbitrary location z becomes

$$\Theta = \frac{T}{GJ} z \tag{9.15}$$

The total twist angle for a bar with length L becomes

$$\boxed{\Theta = \frac{L}{GJ} T} \quad \text{for 1D bars subject to torque } T \tag{9.16}$$

where $\frac{GJ}{L}$ is the **torsional stiffness of the bar**, and GJ the **torsional stiffness per length**.

Comparing the solution for a 1D bar subjected to axial force F, which is given in Chapter 7:

$$\boxed{u_R = \frac{L}{EA} F} \quad \text{for 1D bars subject to axial force } F \tag{9.17}$$

The formula is exactly the same if we replace

$$u_R \text{ by } \Theta; \quad E \text{ by } G; \quad A \text{ by } J \tag{9.18}$$

Equation (9.16) can also be written as

$$\boxed{T = \frac{GJ}{L} \Theta} \tag{9.19}$$

In this case, $\frac{L}{GJ}$ is the **torsional flexibility of the bar**.

Using Eq. (9.11), we have the shear-stress torque relation:

$$\tau = rG\vartheta_t = \frac{Tr}{J} \tag{9.20}$$

The maximum shear stress is registered at the furthest point from the z-axis, $r = R$ (by the lateral surface of the bar):

$$\tau_{\max} = \frac{TR}{J} \tag{9.21}$$

All these formulas are useful in designing torsional bars. These formulas are also the base formulas for many other types of bars. The difference will be on the calculation of J.

9.1.10 Continuous formulation

Consider a small cell dz in the bar, as shown in Fig. 9.3. The twist angle for it becomes

$$d\Theta = \frac{T(z)}{GJ}dz; \quad \text{or} \quad \frac{d\Theta}{dz} = \frac{T(z)}{GJ} \qquad (9.22)$$

This is the differential equation (DE) for twist-torque relation, which allows the torque to be a function of z, implying that one can apply torque in a distributed manner. In this case, the equilibrium $\sum M_z = 0$ is enforced locally at any location z in the integral manner over the cross-section of the bar. Equation (9.22) is a simple 1D DE of first order and can be solved with ease.

In general, even J and/or G can also be a function of z, but the bar must be "long" to justify being 1D.

9.1.11 Bars with hollow cylindrical cross-section

In practice, bars designed for taking torsional loads are often with a hollow cylindrical cross-section, as shown in Fig. 9.6. This is because the material near the axial axis does not take much stress, as shown in Fig. 9.4, and can be removed.

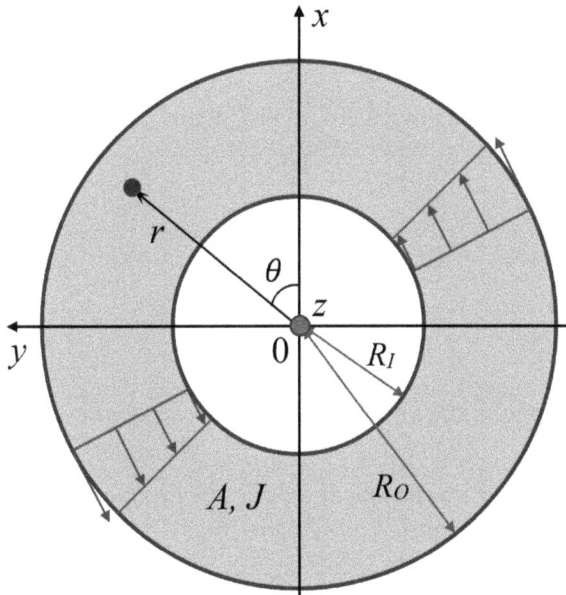

Figure 9.6. Bar with hollow cylindrical cross-section.

In such a case, all the formulations derived so far still apply. The only change is the calculation of the polar second moment of area J. It is calculated using

$$J = \int_A r^2 dA = 2\pi \int_{R_I}^{R_O} r^3 dr = \frac{\pi(R_O{}^4 - R_I{}^4)}{2} \tag{9.23}$$

where R_I is the inner radius, and R_O is the outer radius of the cylinder. The maximum shear stress is registered at the furthest point from the z-axis, $r = R_O$ (the outer radius):

$$\tau_{max} = \frac{TR_O}{J} \tag{9.24}$$

The cross-section areas is computed using

$$A = \pi(R_O{}^2 - R_I{}^2) \tag{9.25}$$

9.1.12 Bars with thin-wall cross-sections

If the hollow cylinder is very thin with thickness t, we have $R_I \approx R_O$, and $R_O - R_I = t$. Equation (9.23) gives

$$J = \frac{\pi(R_O{}^4 - R_I{}^4)}{2} = \frac{\pi(R_O{}^2 + R_I{}^2)(R_O + R_I)(R_O - R_I)}{2} \approx 2\pi R_m^3 t \tag{9.26}$$

where $R_m = \frac{R_O+R_I}{2}$ is the mean radius of the cylinder.

Equation (9.26) can be reformulated as

$$J \approx 2\pi t R_m^3 = \frac{4t(\pi R_m^2)^2}{2\pi R_m} = \frac{4t A_m^2}{l_m} \tag{9.27}$$

where A_m is the area of the circle with mean radius R_m, and is computed using

$$A_m = \pi R_m^2 \tag{9.28}$$

and l_m is the length of the mean perimeter of the circle with R_m.

It is found that the following equation works for thin-wall cylinders with arbitrary cross-section shape, as long as the cross-section is enclosed. We thus have the general formula:

$$\boxed{J = \frac{4t A_m^2}{l_m}} \tag{9.29}$$

where A_m is the area formed by the middle line of the cross-section, l_m is the length of the mean perimeter of the cross-section, and t is the thickness of the wall. This is because cross-sections regardless of the shape all have these three geometric features: A_m, l_m, and t. The J value should be proportional to t, inversely proportional to l_m, and A_m^2 is needed for dimension match, similar to the thin wall circular cylinder. Detailed rigorous (but lengthy) derivation leading to Eq. (9.29) can be found in the following section and in Sections 9.2 and 9.6.

Note that when the cross-section is not closed (have a cut), the J value is drastically reduced (see examples in Section 9.6).

9.1.13 Example: A shaft with hollow circular cross-section subject to a torque

Consider a bar made of Grade-304 stainless steel with shear modulus $G = 88$ GPa. The cross-section is hollow circular with inner radius of $R_O = 30$ mm and outer radius of $R_O = 50$ mm. It is fixed at $z = 0$, and subjected to a torque of $T = 1000$ Nm at $z = L$, as shown in Fig. 9.7.

Calculate:

1. the maximum shear stress on the cross-section of the shaft;
2. the twist angle per unit length;
3. the stress components on the cross-section and by the lateral surface in the Cartesian coordinates on the x-axis;
4. the stress components on the cross-section and by the lateral surface in the Cartesian coordinates at $\theta = 45°$ from the x-axis;
5. without calculation, give the stress components on the cross-section and by the lateral surface in the Cartesian coordinates on the y-axis;
6. compute the principal stresses and maximum shear stress in the shaft.

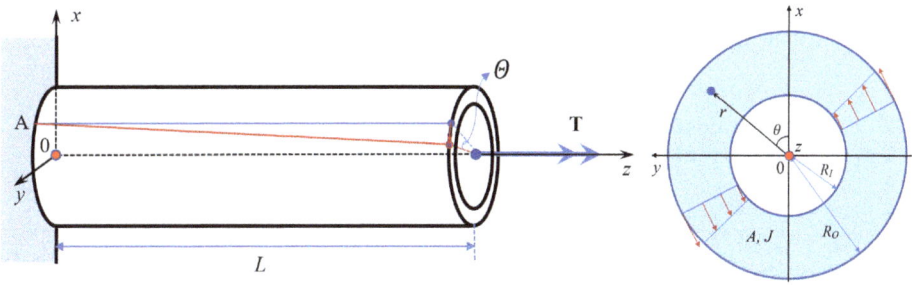

Figure 9.7. Stress on a hollow cylindrical cross-section of a uniform bar.

Solution to 1: The maximum shear stress in the shaft.

```
1  RI, RO = symbols("R_I, R_O")            # Define symbolic variables
2  T = 1000                                 # Nm, torque
3  RI = 0.03; RO = 0.05                     # m,  radii
4  J = np.pi * (RO**4 - RI**4)/2
5  print(f"The polar second moment of area J = {J} (m^4)")
6
7  τ_RO = T*RO/J
8  print(f"The maximum τ on the cross-section = {τ_RO*1.0e-6} (MPa)")
```

```
The polar second moment of area J = 8.545132017764239e-06 (m^4)
The maximum τ on the cross-section = 5.851284672496151 (MPa)
```

Solution to 2: The twist angle per unit length.

```
1  G    = 88.0e9  # Pa
2  ϑ_t = T/(G*J)
3
4  print(f"Twist-angle/length={ϑ_t:.4e} rad/m or {np.rad2deg(ϑ_t):.4e}o/m")
```

```
Twist-angle/length=1.3298e-03 rad/m or 7.6194e-02o/m
```

Solution to 3: The stress components on the cross-section and by lateral surface in the Cartesian coordinates on the x-axis.

```
1  θ    = 0.0                              # On the x-axis, θ=0
2  x    = RO*np.cos(np.deg2rad(θ))
3  y    = RO*np.sin(np.deg2rad(θ))
4
5  σzx =-ϑ_t*G*y
6  σzy = ϑ_t*G*x
7
8  print(f"On x-axis: σzx={σzx*1.0e-6:.4e}; σzy={σzy*1.0e-6:.4e} (MPa)")
```

```
On x-axis: σzx=-0.0000e+00; σzy=5.8513e+00 (MPa)
```

Solution to 4: The stress components on the cross-section and by lateral surface in the Cartesian coordinates at $\theta = 45°$ from the x-axis.

```
1  θ    = 45.0
2  x    = RO*np.cos(np.deg2rad(θ))
3  y    = RO*np.sin(np.deg2rad(θ))
4
5  σzx =-ϑ_t*G*y
6  σzy = ϑ_t*G*x
7
8  print(f"at 45o: σzx = {σzx*1.0e-6:.4e}; σzy = {σzy*1.0e-6:.4e} (MPa)")
```

```
at 45o: σzx = -4.1375e+00; σzy = 4.1375e+00 (MPa)
```

Solution to 5: On the x-axis, $\theta = 90$ degrees. The solution is found as follows using the same code for solution 3.

On y-axis: $\sigma_{zx} = -5.8513\text{e} + 00;\quad \sigma_{zy} = 0.0000\text{e} + 00$ (MPa)

Solution to 6: Compute the principal stresses and maximum shear stress.

```
1  θ    = 90.0                           # use the stress components on suface-y
2  x = RO*np.cos(np.deg2rad(θ)); y = RO*np.sin(np.deg2rad(θ))
3
4  σzx =-ϑ_t*G*y; σzy = ϑ_t*G*x
5  print(f"at 90o: σzx = {σzx:.4e}; σzy = {σzy:.4e} (Pa)")
6
7  σ_np = np.array([[ 0 ,  0 , σzx],     # stress tensor, torsion wrt z-axis
8                   [ 0 ,  0 , σzy],
9                   [σzx, σzy,  0 ]], dtype = float)
10
11 eigs, eigvs = gr.principalS(σ_np)
12 print(f"τ_max = {(eigs[0]-eigs[-1])/2:.4e}(Pa)")
```

```
at 90o: σzx = -5.8513e+06; σzy = 3.5829e-10 (Pa)
Principal stress/strain (Eigenvalues):
[ 5.8513e+06  1.0074e-74 -5.8513e+06]

Principal stress/strain directions:
[[ 7.0711e-01   6.1232e-17   7.0711e-01]
 [-4.3298e-17   1.0000e+00  -4.3298e-17]
 [-7.0711e-01  -1.6652e-48   7.0711e-01]]

Possible angles (n1,x)=45.0000000000001o or 135.0o
τ_max = 5.8513e+06(Pa)
```

```
1  # Check the angle for the principal axis.
2  T, _ = gr.transferM(45., about = 'y')        # rotation axis, angle
3  T@σ_np@T.T
```

```
array([[ 5.8513e+06, -2.5335e-10,  1.5231e-10],
       [-2.5335e-10,  0.0000e+00,  2.5335e-10],
       [-1.5231e-10,  2.5335e-10, -5.8513e+06]])
```

9.1.14 Example: A shaft with two segments subject to two torques

Consider a shaft consisting of two uniform bars perfectly connected together, as shown in Fig. 9.8. It is fixed at point A. Both bars have solid circular cross-section. It is loaded at point B with torque T_1, and at point C with T_2. The diameters of these two bars are denoted as d_1 and d_2, and their length are denoted as L_1 and L_2, respectively, for bar-1 and bar-2. Both bars are made of Grade-304 stainless steel with shear modulus $G = 88$ GPa.

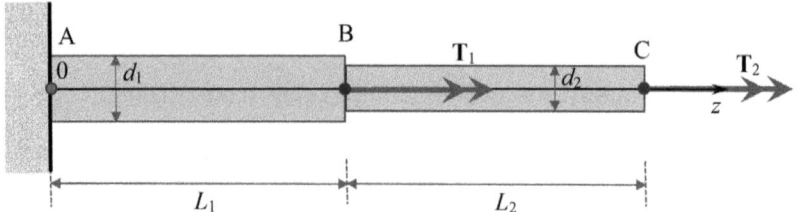

Figure 9.8. A shaft consisting of two uniform bars perfectly connected together.

1. Derive the formula for computing the twist angle at C.
2. Derive the formula for computing the maximum shear stress and its location for bar-1.
3. Determine the ratio of these two diameters d_1/d_2, so that the maximum shear stress these two bars are the same.
4. Let $T_1 = 100$ Nm, $T_2 = 50$ Nm, $L_1 = 0.5$ m and $L_2 = 1.0$ m. Compute items 1, 2, and 3.

Solution to 1: First, the 2nd polar moment of area for these two bars are:

$$J_1 = \frac{\pi(d_1/2)^4}{2} = \frac{\pi d_1^4}{32}; \quad J_2 = \frac{\pi d_2^4}{32} \tag{9.30}$$

The twist angle at C in reference to B is

$$\Theta_{BC} = \frac{T_2 L_2}{G J_2} = \frac{32 T_2 L_2}{\pi G d_2^4} \tag{9.31}$$

In bar-1, the torque is $(T_1 + T_2)$, and hence the twist angle at B in reference to A is

$$\Theta_{AB} = \frac{32(T_1 + T_2)L_1}{\pi G d_1^4} \tag{9.32}$$

The total twist angle at C becomes

$$\Theta_C = \Theta_{AB} + \Theta_{BC} = \frac{32(T_1 + T_2)L_1}{\pi G d_1^4} + \frac{32 T_2 L_2}{\pi G d_2^4}$$
$$= \frac{32}{\pi G}\left(\frac{(T_1+T_2)L_1}{d_1^4} + \frac{T_2 L_2}{d_2^4}\right) \tag{9.33}$$

Solution to 2: Compute the maximum shear stress and its location for bar-1.

For bar-1, we have $R_1 = \frac{d_1}{2}$. Thus,

$$\tau_1^{max} = \frac{(T_1 + T_2)d_1}{2 J_1} = \frac{16(T_1 + T_2)}{\pi d_1^3} \tag{9.34}$$

Solution to 3: Determine the ratio of these two diameters d_1/d_2:

For bar-2, we have $R_2 = \frac{d_2}{2}$. Hence,

$$\tau_2^{\max} = \frac{T_2 d_2}{2 J_2} = \frac{16(T_2)}{\pi d_2^3} \tag{9.35}$$

The ratio d_1/d_2 is found when $\tau_1^{\max} = \tau_2^{\max}$:

$$\frac{d_1}{d_2} = \sqrt[3]{\frac{T_1 + T_2}{T_2}} \tag{9.36}$$

Solution to 4: The numerical calculation is done using the following code.

$T_1 = 100$ Nm, $T_2 = 50$ Nm, $d_1 = 0.03$ m, $d_2 = 0.02$ m, $L_1 = 0.5$ m and $L_2 = 1.0$ m.

```
1  T1 = 100 ; T2 = 50              # Nm
2  d1 = 0.03; d2 = 0.02             # m
3  L1 = 0.5 ; L2 = 1.0              # m
4  G  = 88.0e9                      # Pa
5
6  # Question 1:
7  deg = 180./np.pi
8  T12 = T1+T2
9  θC  = 32./(np.pi*G)*(T12*L1/d1**4 + T2*L2/d2**4)
10 print(f"Q1: The twist angle at C, θC= {θC:.4e} rad/m or {θC*deg:.4e}°")
11
12 # Question 2:
13 τ1max = 16.*G*T12/(np.pi*d1**3)
14 print(f"Q2: The max shear stress in bar-1 = {τ1max*1.0e-6:.4e} MPa\n"
15      f"    on the surface of bar-1 within its entire length")
16
17 # Question 3:
18 Rd12 = (T12/T2)**(1/3)
19 print(f"Q3: The ratio of two diameters d1/d2 = {Rd12:.4e} \n")
```

Q1: The twist angle at C, θC= 4.6889e-02 rad/m or 2.6865e+00°
Q2: The max shear stress in bar-1 = 2.4899e+12 MPa
    on the surface of bar-1 within its entire length
Q3: The ratio of two diameters d1/d2 = 1.4422e+00

We can check the codes using EXAMPLE 6.2 given in Ref. [4], using the following code.

```
1  L1 = 1.0 ; L2 = 1.27              # m
2  G  = 77.0e9                       # Pa
3  T1 = 113 ; T2 = 113               # Nm
4  d1 = 0.0224; d2 = 0.0177          # m
5  print(f"The ratio of two diameters d1/d2 = {d1/d2:.4e}")
6
7  # Question 1:
8  deg = 180./np.pi
9  T12 = T1+T2
10 θC = 32./(np.pi*G)*(T12*L1/d1**4 + T2*L2/d2**4)
11 print(f"Q1: The twist angle at C, θC= {θC:.4e} rad/m or {θC*deg:.4e}o")
12
13 # Question 2:
14 τ1max = 16.*T12/(np.pi*d1**3)
15 print(f"Q2: The max shear stress in bar-1 = {τ1max*1.0e-6:.4e} MPa\n"
16       f"    on the surface of bar-1 within its entire length")
17
18 # Question 3:
19 Rd12 = (T12/T2)**(1/3)
20 print(f"Q3: The ratio of two diameters d1/d2 = {Rd12:.4e} \n")
```

```
The ratio of two diameters d1/d2 = 1.2655e+00
Q1: The twist angle at C, θC= 3.1217e-01 rad/m or 1.7886e+01o
Q2: The max shear stress in bar-1 = 1.0241e+02 MPa
    on the surface of bar-1 within its entire length
Q3: The ratio of two diameters d1/d2 = 1.2599e+00
```

9.2 Bars with enclosed thin-wall cross-sections

Consider a long bar with a uniform cross-section that is a hollow with an enclosed thin-wall, as shown in Fig. 9.9. The axis of the bar is denoted as z that is perpendicular to the plane of the cross-section, and passing through its centroid O. The thickness t of the wall needs not be constant and can change at different locations, denoted by a curve coordinate s, and local coordinate r orthogonal to s. Thus, r, s, and z axes form a local orthogonal coordinate system with the right-hand rule.

The shape of the cross-section can be arbitrary but closed. The bar is subjected to a torque $\mathbf{T}$ with respect to the z-axis, resulting in a twist Θ of the cross-section by at $z = L$.

Before our analysis, we make the following assumptions:

1. The wall thickness is very small compared to the other dimensions of the cross-section.
2. The shear stress direction is along the middle line of the thin-wall boundary.

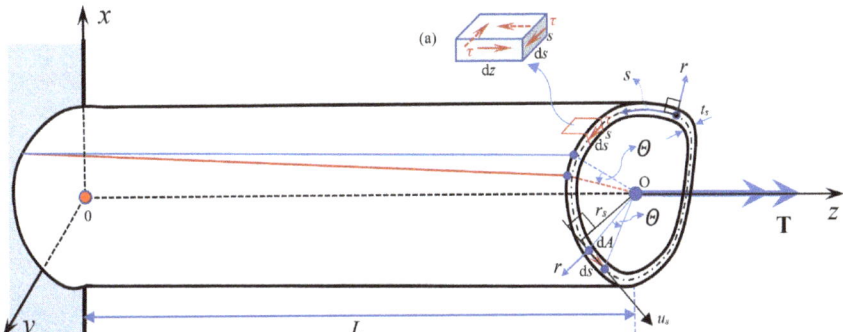

Figure 9.9. A long bar with a uniform cross-section that is a hollow with an enclosed thin-wall. It is fixed at $z = 0$ and subjected to a torque **T** with respect to the z-axis of the bar, and the entire cross-section at the right-end undergoes a rigid rotation Θ.

3. The cross-section rotates rigidly with respect to the axis of the bar under the action of a torque, so that the displacements in the x–y plane behave like a rigid disk. The displacement in the z-direction, w, is not constrained.
4. The bar is loaded only with a torque with respect to its axis.

Based on these assumptions, the stresses on the cross-section (the z-plane) become

$$\sigma_{zr} = 0, \quad \text{thin-wall, and circumferential surface is free of stresses} \quad (9.37)$$
$$\sigma_{zz} = 0, \quad \text{the bar is not loaded in the axial direction}$$

The only nonzero stress on the cross-section is σ_{zs}. For convenience, it is denoted as τ. Its direction is in the tangent direction of the middle line of the thin-wall, and its positive direction follows the standard right-hand rule for rotation with respect to the z-axis. This creates a flow of shear stress along the cross-section of the bar. Because the wall is thin, the shear stress τ is uniform across the thickness (or τ is treated as the averaged shear stress) at any fixed location s.

9.2.1 Concept of shear flow

The **shear flow** is quite an important concept when dealing with shear stresses in thin-walled structural members, and will be used multiple times in this book. It is usually denoted as q. To introduce this concept, let us cut an infinitely small free-body diagram from the thin-walled bar at an arbitrary location s, shown in Fig. 9.9(a). We can look at the equilibrium of all the forces in the z-direction:

$$\sum F_z = 0 : \tau_{s+ds} t_{s+ds} dz - \tau_s t_s dz = 0 \quad (9.38)$$

which is

$$\tau_{s+ds} t_{s+ds} = \tau_s t_s \quad (9.39)$$

This means that the product $\tau_s t_s$ does not change when s changes. It is constant along s. This conserved quantity is called **shear flow**, defined as

$$q = \tau_s t_s \quad (9.40)$$

The shear stress τ_s and the wall thickness t_s may change with s, but q is constant for a given bar with thin-walled cross-section subjected to torque **T**!

9.2.2 Torque and shear-flow relation

Because the shear stress is generated by the torque, we have (see, Fig. 9.9):

$$T = \oint \underbrace{\tau_s t_s}_{q:\text{constant}} r_s ds = 2q \oint \underbrace{\underbrace{\frac{1}{2} r_s ds}_{dA}}_{A_m} = 2q A_m \quad (9.41)$$

where A_m is the area enclosed by the middle line of the wall on the cross-section, and r is "radius" at point s which changes with s (which is different from the circular cross-section). It is the shortest distance from point O in the r-axis direction. Clearly, this formula holds only if the cross-section is closed, allowing the closed loop integration. We obtain now a very simple but important equation, known as the Bredt-Batho formula:

$$\boxed{T = 2q A_m} \quad \text{or} \quad \boxed{q = \frac{T}{2 A_m}} \quad (9.42)$$

This means that the torque is twice the shear flow multiplied by the mean cross-section area. With Eq. (9.42), the shear stress, as a function of t_s and hence of s, can be easily obtained for a given torque:

$$\boxed{\tau_s = \frac{T}{2 A_m t_s}} \quad (9.43)$$

Note here that t_s is in general a function of s. It is clear that **the maximum shear stress is registered where t_s is smallest**.

9.2.3 Twist-angle and torque relation

The final piece of formulation is the relation between the twist-angle and the torque. This clearly leads to the stiffness of the bar under torsion. Naturally, it must involve the shear strain and stress relation.

Based on the strain–displacement relations similar to Eq. (9.5), and the fact that z-axis and s-axis are always orthogonal, we have

$$\gamma_{zr} = 0 \quad \text{because} \quad \sigma_{zr} = 0$$
$$\gamma_{zs} = 2\varepsilon_{zs} = \left(\frac{\partial w}{\partial s} + \frac{\partial u_s}{\partial z}\right) \tag{9.44}$$

Because the cross-section rotates rigidly, the displacement u_s relates the twist angle Θ as (see Fig. 9.9)

$$u_s = r_s \Theta \tag{9.45}$$

Note that r_s is in general a function of s, but independent of z. Substituting Eq. (9.45) into Eq. (9.44), and using the constitutive equation for shear strain and stress, we obtain:

$$\gamma_{zs} = \frac{\tau_s}{G} = \frac{q}{Gt_s} = \left(\frac{\partial w}{\partial s} + \frac{\partial [r_s \Theta]}{\partial z}\right) = \left(\frac{\partial w}{\partial s} + r_s \frac{\partial \Theta}{\partial z}\right) \tag{9.46}$$

which gives

$$\frac{q}{Gt_s} = \left(\frac{\partial w}{\partial s} + r_s \frac{\partial \Theta}{\partial z}\right) \tag{9.47}$$

Integration of the foregoing equation along s over the entire cross-section of the bar gives

$$\oint \frac{q}{Gt_s} ds = \oint \frac{\partial w}{\partial s} ds + \oint r_s \frac{\partial \Theta}{\partial z} ds \tag{9.48}$$

Now, because the cross-section rotates rigidly, Θ does not depend on s, we have

$$\oint \frac{q}{Gt_s} ds = \underbrace{w\Big|_{s=0}^{s=l_m}}_{=0} + \frac{\partial \Theta}{\partial z} \underbrace{\oint r_s ds}_{2A_m} \tag{9.49}$$

where l_m is the perimeter length of the middle line of the cross-section and A_m is the area enclosed by the middle line of the cross-section. Because the cross-section is enclosed, and s comes back to its original location after looping around, $w\big|_{s=0}^{s=l_m} = 0$. Note also that q is independent of s, Eq. (9.49) becomes

$$\boxed{\frac{\partial \Theta}{\partial z} = \frac{q}{2A_m G} \oint \frac{1}{t_s} ds} \tag{9.50}$$

where $\frac{\partial \Theta}{\partial z}$ is the twist angle per unit length ϑ_t.

Using Eqs. (9.42) and (9.50) can also be written as

$$\frac{\partial \Theta}{\partial z} = \frac{T}{4A_m^2 G} \oint \frac{1}{t_s} ds = \frac{T}{GJ} \qquad (9.51)$$

where J is the **equivalent polar moment of area** of the cross-section, defined as

$$J = \frac{4A_m^2}{\oint \frac{1}{t_s} ds} \qquad (9.52)$$

If the thickness t_s is piecewise constant: $t_1, t_2, \ldots, t_n$ with the corresponding lengths: $l_1, l_2, \ldots, l_n$, we shall have

$$\oint \frac{1}{t_s} ds = \sum_i^n \frac{l_i}{t_i} \qquad (9.53)$$

If the bar has a uniform wall thickness t, the foregoing equation is simplified to

$$J = \frac{4 t A_m^2}{l_m} \qquad (9.54)$$

where l_m is the perimeter of the middle line of the thin wall cross-section. This is Eq. (9.29) we obtained earlier by simple induction or guessing.

If the cross-section is circular with radius R (that is constant), we have $l_m = 2\pi R_m$, and $A_m = \pi R_m^2$, we have Eq. (9.26). Readers may find the formula for J for a rectangular thin-wall cross-section.

If Eq. (9.1) holds: $\Theta = \vartheta_t z$, Eq. (9.51) leads to Eq. (9.14) and also Eq. (9.16).

9.2.4 Example: Shear stress in a thin wall bar with box cross-section

Consider a long bar of length L with box cross-section with dimensions given in Fig. 9.10. It is fixed at $z = 0$, and subjected to a torque of T at $z = L$. The sides of the box are considered as thin walls made of elastic material with shear modulus G.

1. Derive the formulas for computing its twist angle, shear flow, and shear stress in the left-wall on the cross-section.
2. Given data: $G = 25.0$ GPa, $L = 0.5$ m, $a = 58$ mm, $b = 38$ mm, $t_T = 2$ mm, $t_B = 4$ mm, $t_L = 3$ mm, $t_R = 4$ mm, $T = 5$ kNm. Compute the numerical values for the twist angle, shear flow, and the maximum shear stress.

Torsion of Bars

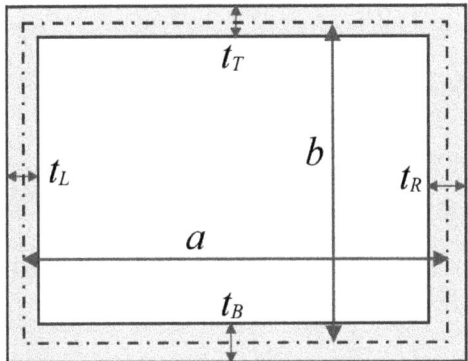

Figure 9.10. A long bar with a box cross-section of thin-walls subjected to a torque **T** with respect to its axial axis.

Solution: We write the following code to complete the task.

```
1  # 1) Derive formulas using Sympy:
2
3  t, tT, tB, tL, tR = sp.symbols("t, t_T, t_B, t_L, t_R")   # dimensions
4  a, b = sp.symbols("a, b ")
5  L, G, T, ϑt = sp.symbols("L, G, T, ϑ_t")
6
7  # compute the area enclosed by the middle lines:
8  l = 2*a + 2*b             # perimeter of the middle line of the thin wall
9  A = a*b                                                   # area
10 inverse_t_ds= a/tT + a/tB + b/tL + b/tR                   # ∫(1/t)ds
11 J   = 4*A**2/inverse_t_ds            # equivalent polar moment of area
12 gr.printM(J,"Formula for J:")
```

Formula for J: $\dfrac{4a^2 b^2}{\dfrac{a}{t_T} + \dfrac{a}{t_B} + \dfrac{b}{t_R} + \dfrac{b}{t_L}}$

```
1  ϑt = (T/(G*J)).simplify()         # twist angle per unit length
2  θ  = ϑt*L                          # total twist angle
3  gr.printM(θ,"Twist angle by T:")
```

Twist angle by T: $\dfrac{LT\left(\dfrac{a}{t_T} + \dfrac{a}{t_B} + \dfrac{b}{t_R} + \dfrac{b}{t_L}\right)}{4Ga^2 b^2}$

```
1  dic = {tT:t, tB:t, tL:t, tR:t}        # if all thickness are the same t
2  θt = θ.subs(dic).factor(t)
3  gr.printM(θt,"Twist angle for uniform t:")
```

Twist angle for uniform t: $\dfrac{LT(a+b)}{2Ga^2b^2t}$

```
1  Jt = J.subs(dic).factor(t)            # equivalent polar moment of area
2  gr.printM(Jt,"J for uniform t:")
```

J for uniform t: $\dfrac{2a^2b^2t}{a+b}$

```
1  q = T/(2*A)
2  gr.printM(q,"Shear flow by T:")
```

Shear flow by T: $\dfrac{T}{2ab}$

```
1  τ_L = q/tL
2  gr.printM(τ_L,"Shear stress on the left wall:")
```

Shear stress on the left wall: $\dfrac{T}{2abt_L}$

```
1   # 2) numerical calculation:
2
3   G_ = 25.0e9              # Pa
4   L_ = 0.5                 # m
5   a_ = 0.058; b_=0.038     # m
6
7   tT_= 0.002; tB_=0.004; tL_=0.003; tR_=0.004   # mm,
8
9   T_ = 5.0e3               # kNm.
10  data = {T:T_, G:G_, L:L_, a:a_, b:b_, tT:tT_, tB:tB_, tL:tL_, tR:tR_}
11
12  θ_data = θ.subs(data)
13  gr.printM(θ_data, "θ value (rad) for the given data:")
```

θ value (rad) for the given data: 0.338

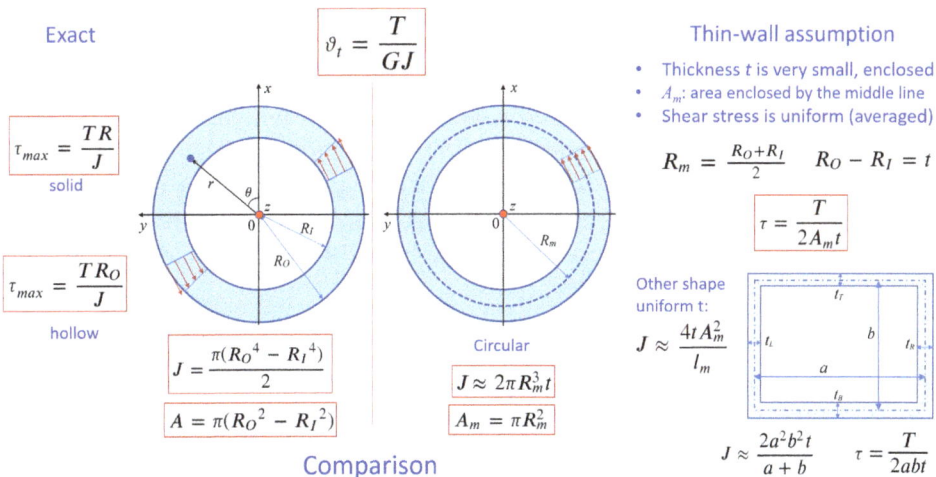

Figure 9.11. Comparison of major formulas for exact and thin-wall assumptions for torsional bars.

```
1  gr.printM(q.subs(data),"shear flow value (m·Pa) for the given data:")
```

shear flow value (m·Pa) for the given data: $1.134 \cdot 10^6$

```
1  τ_max = q.subs(data)/tT_
2  gr.printM(τ_max,"Maximum shear stress (Pa) in the top wall (thinest):")
```

Maximum shear stress (Pa) in the top wall (thinest): $5.671 \cdot 10^8$

9.2.5 A comparison of exact and thin-wall formulas

Figure 9.11 gives a comparison of major formulas for exact and thin-wall assumption for bars subjected to torques.

9.3 Bars with non-circular cross-sections*

9.3.1 Setting of the problem

Consider a torsion bar member with non-circular cross-section subjected to a torque T. It is defined in coordinates (x, y, z) with z-axis being the axial axis of the bar. The cross-section of the bar is assumed uniform along z-axis.

Because of the non-circular shape, the cross-section may not remain plane and can warp, implying that the displacement w may vary with x and y. The rotation of the bar changes along with z-axis. Therefore, the problem is three dimensional. The method used to perform analysis is the so-called Saint-Venant's semi-inverse method [4]. It approximates the displacements in such a way that it leads to a stress state $(\sigma_{zx}, \sigma_{zy})$ on the cross-section, under torque T, aiming to convert the problem to a two-dimensional one. Our formulation and notation follow largely the textbook [4]. Codes are developed for some of the derivations.

9.3.2 Assumed displacements

Consider a point B at (x, y, z) on a cross-section in an undeformed bar. When it is deformed by subjecting T, the cross-section rotates by Θ about z-axis with respect to the origin, and B is displaced by displacements (u, v, w) and moves to B*, as shown in Fig. 9.12.

The semi-inverse method assumes:

1. The rotation is the principal cause of displacements (u, v), which decouples u and v from w.
2. The twist angle changes linearly with z: $\Theta = \vartheta_t z$, where ϑ_t is the angle of twist per unit length and is a constant (does not change with (x, y, z)).

Based on these assumptions, the displacements can be written as

$$u = -y\vartheta_t z; \quad v = x\vartheta_t z; \quad w = \vartheta_t \psi(x, y) \tag{9.55}$$

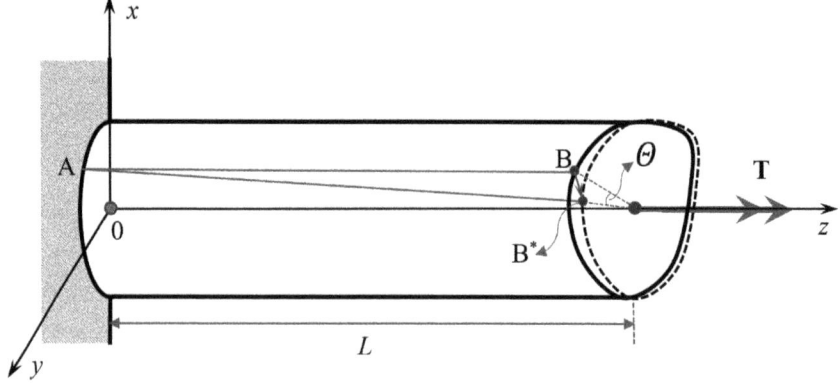

Figure 9.12. A bar with non-circular cross-section subjected to a torque. Warping may occur on the cross-section.

where ψ is the warping function representing the off-cross-section deformation on the bar cross-section. Note that in the circular cross-sectional bars, ψ is assumed to be zero. In non-circular bars, it may not be zero and we may need to find the warping function.

We have two unknowns ϑ_t and ψ that depend on (x, y). We have converted the 3D problem to 2D.

9.3.3 Shear strains in non-circular bars

Using the small strain–displacement relations given in Chapter 4, and ϑ_t is a constant, we can find the strains in the bar:

$$\varepsilon_{xx} = \frac{\partial u}{\partial x} = 0; \quad \varepsilon_{yy} = \frac{\partial v}{\partial y} = 0; \quad \varepsilon_{zz} = \frac{\partial w}{\partial z} = 0$$

$$\varepsilon_{xy} = \frac{1}{2}\left(\frac{\partial u}{\partial y} + \frac{\partial v}{\partial x}\right) = 0$$

$$2\varepsilon_{zx} = \gamma_{zx} = \frac{\partial u}{\partial z} + \frac{\partial w}{\partial x} = \vartheta_t\left(\frac{\partial \psi}{\partial x} - y\right)$$

$$2\varepsilon_{zy} = \gamma_{zy} = \frac{\partial w}{\partial y} + \frac{\partial v}{\partial z} = \vartheta_t\left(\frac{\partial \psi}{\partial y} + x\right)$$

(9.56)

The codes to obtain these equations are as follows:

```
1  x, y, z = symbols("x, y, z")              # define symbolic coordinates
2  ψ = sp.Function("ψ")(x, y)
3  X = Matrix([x, y, z])
4  ϑt = symbols("ϑ_t")                       # twist angle per length
5
6  U = Matrix([-y*ϑt*z,x*ϑt*z,ψ*ϑt])  # Assumed displacement vector[u,v,w]
7  U
```

$$\begin{bmatrix} -yz\vartheta_t \\ xz\vartheta_t \\ \vartheta_t\psi(x, y) \end{bmatrix}$$

```
1  # Compute the strains (see Chapter 4):
2  U_g = gr.grad_vf(U, X)                    # displacement gradient
3  ε = (U_g + U_g.T)/2                       # strain tensor, symmetric
4  2*ε                                       # change to γ
```

$$\begin{bmatrix} 0 & 0 & -y\vartheta_t + \vartheta_t\frac{\partial}{\partial x}\psi(x, y) \\ 0 & 0 & x\vartheta_t + \vartheta_t\frac{\partial}{\partial y}\psi(x, y) \\ -y\vartheta_t + \vartheta_t\frac{\partial}{\partial x}\psi(x, y) & x\vartheta_t + \vartheta_t\frac{\partial}{\partial y}\psi(x, y) & 0 \end{bmatrix}$$

It is clear that the only nonzero strains are:

$$\gamma_{zx} = \vartheta_t \left(\frac{\partial \psi}{\partial x} - y \right)$$
$$\gamma_{zy} = \vartheta_t \left(\frac{\partial \psi}{\partial y} + x \right) \quad (9.57)$$

It is easy to confirm that the compatibility conditions given in Chapter 4 are satisfied. Note that if ψ is zero, Eq. (9.57) becomes Eq. (9.5).

Differentiate γ_{zx} with respect to y, and γ_{zy} with respect to x, and then subtracting the results of these two, the warping function ψ is eliminated, leading to

$$\frac{\partial \gamma_{zx}}{\partial y} - \frac{\partial \gamma_{zy}}{\partial x} = 2\vartheta_t \quad (9.58)$$

This means that if γ_{zx} and γ_{zy} satisfy the foregoing equation, any function ψ is admissible. Thus, we need not worry about ψ (if we do not want to find it out). Instead, the strains resulting from the stresses to-be-found must satisfy Eq. (9.58).

9.3.4 Equilibrium equations at any point in torsion bar and stress function

Because only γ_{zx} and γ_{zy} are nonzero, the nonzero stresses are σ_{zx} and σ_{zy} (for isotropic materials). The equilibrium equations for 3D solids given in Chapter 6 are reduced to (without body force):

$$\frac{\partial \sigma_{zx}}{\partial z} = 0; \quad \frac{\partial \sigma_{zy}}{\partial z} = 0$$
$$\frac{\partial \sigma_{zy}}{\partial y} + \frac{\partial \sigma_{zx}}{\partial x} = 0 \quad (9.59)$$

The first two equations mean that σ_{zx} and σ_{zy} are independent of z, and they shall satisfy the last equation in Eq. (9.59). Such stresses can be written in terms of the so-called Prandtl **stress function** $\phi(x, y)$ as

$$\sigma_{zx} = \frac{\partial \phi(x, y)}{\partial y}; \quad \sigma_{zy} = -\frac{\partial \phi(x, y)}{\partial x} \quad (9.60)$$

Function ϕ is in fact a stream function of the field of shear stress vector $[\sigma_{zx} \ \sigma_{zy}]^T$ on a surface-z. It is easy to confirm that the last equation in Eq. (9.59) is satisfied, by substituting Eq. (9.60) into it.

The problem becomes finding the stress function $\phi(x,y)$, subjected to proper boundary conditions and overall equilibrium on the cross-section of the bar.

9.3.5 Boundary conditions for the stress function

Because the lateral surface s of a torsion bar is free of stresses, the component of the shear stress τ at any point on the cross-section by the surface S in its surface normal direction $\mathbf{n}$ must be zero.

Consider an infinitely small curve segment ds for a point P on the boundary of the cross-section shown Fig. 9.13. The corresponding infinitely small line segments along the Cartesian coordinates are denoted as dx and dy, which cut off ds on the boundary. Based on the geometry of the infinitely small triangular cell on the boundary, we shall have on boundary S:

$$\sin\theta = -\frac{dx}{ds}; \quad \cos\theta = \frac{dy}{ds} \qquad (9.61)$$

The minus sign is due to the increase in ds (counter-clockwise), resulting in a decrease in dx.

In order to obtain the shear stresses on the cross-section plane z by boundary surface S, we can simply perform coordinate transformation. Because these shear stresses are all on the same cross-sectional surface-z,

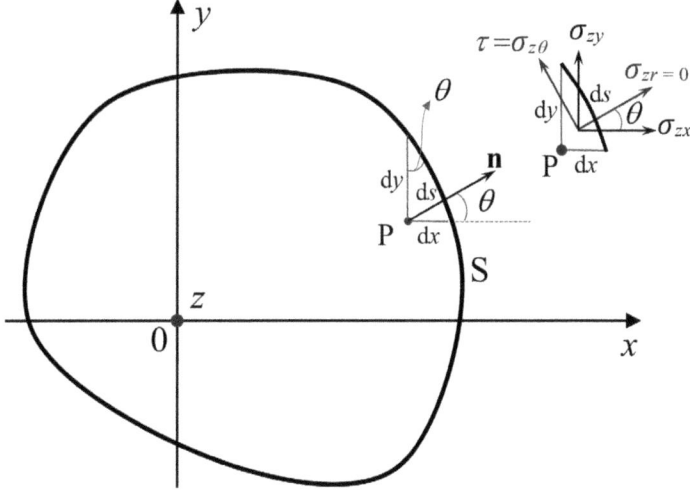

Figure 9.13. Stress components on the cross-section by the lateral surface S of a bar. They can be expressed in both the Cartesian and polar coordinates, and the coordinate transformation rule for vectors apply.

and the coordinate transformation is with respect to z-axis. This means that $(\sigma_{zx}, \sigma_{zy})$ and $(\sigma_{zr}, \sigma_{z\theta})$ carry the same first index z. The coordinate transformation rules for vectors (only one index is transferred) apply, which gives

$$\begin{bmatrix} \sigma_{zr} \\ \sigma_{z\theta} \end{bmatrix} = \underbrace{\begin{bmatrix} \cos\theta & \sin\theta \\ -\sin\theta & \cos\theta \end{bmatrix}}_{\mathbf{T}} \begin{bmatrix} \sigma_{zx} \\ \sigma_{zy} \end{bmatrix} = \begin{bmatrix} \sigma_{zx}\cos\theta + \sigma_{zy}\sin\theta \\ -\sigma_{zx}\sin\theta + \sigma_{zy}\cos\theta \end{bmatrix} \qquad (9.62)$$

Here, we used transformation matrix $\mathbf{T}$ for 2D, its formation is discussed in Section 2.10. Note that this transformation angle is between the x-axis and the normal vector $\mathbf{n}$ at a point on the boundary surface S.

We can also just follow the standard tensor transformation rule for stress tensor, given in Eq. (9.8). The code is as simple as follows:

```
1  # Define the stress tensor on the cross-section of the bar:
2
3  σzx, σzy, θ = symbols("σ_zx, σ_zy, θ")    # Define symbolic variables
4
5  S = Matrix([[ 0,   0, σzx],               # stress state for torsion w.r.t z-axis
6              [ 0,   0, σzy],
7              [σzx, σzy, 0]])
8
9  # Create sympy transformation matrix for rotation θ, w.r.t. z:
10 T = gr.transferMts(θ)
11
12 # Perform the second order tensor transformation for the stresses:
13 Sp=T@S@T.T
14 gr.simplifyM0(Sp)                         # simply the terms in the output
```

$$\begin{bmatrix} 0 & 0 & \sigma_{zx}\cos(\theta) + \sigma_{zy}\sin(\theta) \\ 0 & 0 & -\sigma_{zx}\sin(\theta) + \sigma_{zy}\cos(\theta) \\ \sigma_{zx}\cos(\theta) + \sigma_{zy}\sin(\theta) & -\sigma_{zx}\sin(\theta) + \sigma_{zy}\cos(\theta) & 0 \end{bmatrix}$$

Compared the result with that in Eq. (9.62), these two shear stresses $(\sigma_{zr}, \sigma_{z\theta})$ in the (r, θ) coordinates are exactly the same.

The first equation in Eq. (9.62) gives

$$\sigma_{zr} = \sigma_{zx}\cos\theta + \sigma_{zy}\sin\theta \qquad (9.63)$$

Because there is no shear stresses on the lateral surface s of the bar, we must have $\sigma_{zr} = 0$. Substituting Eq. (9.61) into Eq. (9.63) leads to

$$\sigma_{zx}\frac{dy}{ds} - \sigma_{zy}\frac{dx}{ds} = 0 \qquad (9.64)$$

Next, substituting Eq. (9.60) into Eq. (9.64), we obtain:

$$\frac{\partial \phi}{\partial y}\frac{dy}{ds} + \frac{\partial \phi}{\partial x}\frac{dx}{ds} = \boxed{\frac{d\phi}{ds} = 0} \tag{9.65}$$

Note the change from partial derivatives of ϕ to total derivative. This means that ϕ must be a **constant** on the boundary of the cross-section of the bar. Further because the stresses are given by partial derivatives of ϕ, they φ can be set to zero on the boundary without affecting the stress solution. Therefore, the boundary condition for ϕ is found as

$\boxed{\phi = 0}$ at any point on the boundary of the cross-section of a solid bar
(9.66)

9.3.6 Shear stress on the cross-section by the boundary

By the boundary because $\sigma_{zr} = 0$, the only nonzero stress components in the polar coordinates is $\sigma_{z\theta}$, and is often denoted as τ. It is given by the 2nd equation in Eq. (9.62):

$$\tau = \sigma_{z\theta} = -\sigma_{zx}\sin\theta + \sigma_{zy}\cos\theta \tag{9.67}$$

In other words, σ_{zx} and σ_{zy} are the two components of τ, and we have

$$\begin{aligned}\sigma_{zx} &= -\tau\sin\theta \\ \sigma_{zy} &= \tau\cos\theta\end{aligned} \tag{9.68}$$

When the foregoing equation is substituted to Eq. (9.67), it justifies our argument. Readers may observe the foregoing equation together with Fig. 9.13.

Next, substituting Eq. (9.60) into Eq. (9.67), we obtain

$$\tau = -\frac{\partial \phi}{\partial y}\sin\theta - \frac{\partial \phi}{\partial x}\cos\theta = -\frac{\partial \phi}{\partial r} \tag{9.69}$$

In obtaining Eq. (9.69), we used coordinate transformation again, but for the gradient vector of ϕ.

Equation (9.69) means that τ is the negative slope of the stress function ϕ with respect to r (in the normal direction of bar's lateral surface S with normal **n**). This finding is important when using the so-called membrane analogy for analyzing bars with complicated cross-section, which will be discussed later. Most importantly, it tells us that

the maximum absolute shear stress along the lateral surface is at location where normal slope of ϕ is the largest.

9.3.7 Equilibrium conditions for the stress function

We found the expressions for the stresses on the cross-section (in terms of the stress function). The amplitude of the stresses must depend on the level of the torque applied. Therefore, we must examine the equilibrium conditions for the stresses obtained using the stress function using Eq. (9.60), in relation to the torque on the cross-section. These conditions are

$$\sum F_x = \int_A \sigma_{zx} dx dy = \int_A \frac{\partial \phi}{\partial y} dx dy = 0$$

$$\sum F_y = \int_A \sigma_{zy} dx dy = -\int_A \frac{\partial \phi}{\partial x} dx dy = 0 \qquad (9.70)$$

$$\sum M_z = \int_A (x\sigma_{zy} - y\sigma_{zx}) dx dy = -\int_A \left(x\frac{\partial \phi}{\partial x} + y\frac{\partial \phi}{\partial y} \right) dx dy = T$$

where T is the torque on the cross-section. We need to find the conditions on stress function ϕ so that the foregoing equilibrium conditions are satisfied.

First, we evaluate the 2nd equation in Eq. (9.70). Consider the strip dy across the cross-section along x-direction, as shown in Fig. 9.14.

Because ϕ does not vary in the y-direction for this strip, we have

$$-\int_A \frac{\partial \phi}{\partial x} dx dy = -dy \int_A \frac{d\phi}{dx} dx = -dy \int_{\phi(x_A)}^{\phi(x_B)} d\phi$$

$$= -dy[\phi(x_B) - \phi(x_A)] = 0 \qquad (9.71)$$

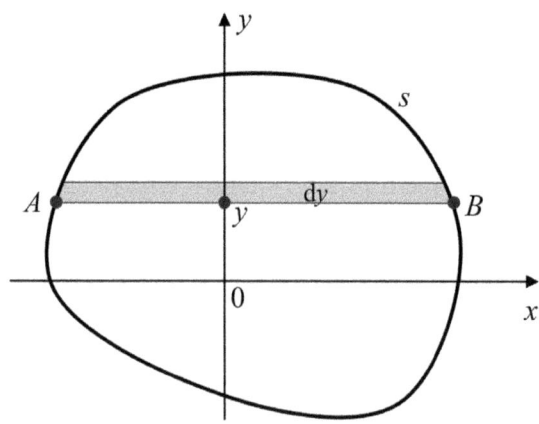

Figure 9.14. Integration along x-direction over dy strip.

This means that the sum of all the forces in the y direction vanishes. The same can be done for the 1st equation in Eq. (9.70), by considering a dx strip. Readers may get this done on a piece of paper. The first two equations in Eq. (9.70) are both satisfied. This makes sense because all the shear stresses on the cross-section can only be circling around, and the total divergence must be zero [9].

Consider now the 3rd equation in Eq. (9.70).

$$\sum M_z = -\int_A \left(x\frac{\partial \phi}{\partial x} + y\frac{\partial \phi}{\partial y} \right) dxdy$$

$$= -\int_A [x \ y] \begin{bmatrix} \frac{\partial \phi}{\partial x} \\ \frac{\partial \phi}{\partial y} \end{bmatrix} dxdy = -\int_A [x \ y] \cdot \nabla \phi \, dxdy$$

$$= -\int_s \underbrace{\phi}_{=0, \text{ on } s} [x \ y] \cdot \mathbf{n} \, dxdy + \int_A \phi \nabla \cdot \begin{bmatrix} x \\ y \end{bmatrix} dxdy$$

$$= 2\int_A \phi \, dxdy = T \tag{9.72}$$

In the 2nd to last line we used the Gauss's formula [9]. We finally found that the integral of the stress function over the area of the cross-section should equal to half of the torque on the cross-section. In other words, the volume under $\phi(x, y)$ is half of the torque T:

$$\boxed{2\int_A \phi \, dxdy = T} \tag{9.73}$$

This is an important condition for ϕ because it relates directly to the torque applied.

9.3.8 Linear elastic solution

Assume the bar is made of a linear elastic isotropic material. The stresses relate to the strains by Hooke's law:

$$\sigma_{zx} = \frac{\partial \phi}{\partial y} = G\gamma_{zx}$$

$$\sigma_{zy} = -\frac{\partial \phi}{\partial x} = G\gamma_{zy} \tag{9.74}$$

or

$$\gamma_{zx} = \frac{1}{G}\frac{\partial \phi}{\partial y}$$
$$\gamma_{zy} = -\frac{1}{G}\frac{\partial \phi}{\partial x}$$
(9.75)

Substituting Eq. (9.75) to Eq. (9.58), we obtain:

$$\boxed{\frac{\partial^2 \phi}{\partial x^2} + \frac{\partial^2 \phi}{\partial y^2} = -2G\vartheta_t} \qquad (9.76)$$

This is a Poisson's equation or a nonhomogeneous Laplace equation. If the twist angle ϑ_t is given and ϕ satisfies BC Eq. (9.66), then Eq. (9.76) uniquely determines $\phi(x, y)$. The stresses can be found via Eq. (9.74), and the torque by Eq. (9.73). We thus have a logical and workable path to the solution.

9.3.9 Techniques for constructing the stress function

Because ϕ must vanish on the boundary of the cross-section, we can make use of the equation that describes the boundary of the cross-section to construct ϕ. Assuming the boundary of the cross-section is given by $f(x, y) = 0$, ϕ can be written as

$$\phi(x, y) = Cf(x, y) \qquad (9.77)$$

where C is a constant. If $f(x, y)$ satisfies

$$\frac{\partial^2 f}{\partial x^2} + \frac{\partial^2 f}{\partial y^2} = \text{constant} \qquad (9.78)$$

we can then adjust C to have Eq. (9.76) be satisfied.

9.3.10 Example: Bars with elliptical cross-section

Consider a bar with elliptical cross-section, as shown in Fig. 9.15.

Stress function ϕ can be written as

$$\phi = C\left(\frac{x^2}{a^2} + \frac{y^2}{b^2} - 1\right) \qquad (9.79)$$

Substituting it to Eq. (9.76), we found

$$C = -\frac{a^2 b^2 G \vartheta_t}{a^2 + b^2} \qquad (9.80)$$

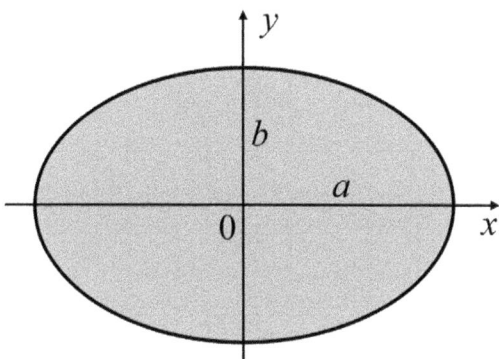

Figure 9.15. Elliptical cross-section for a torsional bar.

The shear stresses can be found via Eq. (9.74):

$$\sigma_{zx} = \frac{\partial \phi}{\partial y} = \frac{2Cy}{b^2} = -\frac{2a^2 G\vartheta_t y}{a^2 + b^2}$$

$$\sigma_{zy} = -\frac{\partial \phi}{\partial x} = -\frac{2Cx}{a^2} = \frac{2b^2 G\vartheta_t x}{a^2 + b^2}$$

(9.81)

The maximum shear stress is found as

$$\tau_{\max} = \sigma_{zy}\bigg|_{y=b} = \frac{2a^2 b G\vartheta_t}{a^2 + b^2} \qquad (9.82)$$

Torque T is obtained by substituting Eq. (9.79) into Eq. (9.73):

$$T = 2\int_A \phi dA = \frac{2C}{a^2}\int_A x^2 dA + \frac{2C}{b^2}\int_A y^2 dA - 2C\int_A dA$$

$$= \frac{2C}{a^2} I_{yy} + \frac{2C}{b^2} I_{xx} - 2CA \qquad (9.83)$$

where A is the area, I_{xx}, I_{yy} are the 2nd moments of area with respect to x- and y-axis of the cross-section of the bar. For an ellipse, the formulas are as follows:

$$I_{xx} = \frac{\pi ab^3}{4}; \quad I_{yy} = \frac{\pi ba^3}{4}; \quad A = \pi ab \qquad (9.84)$$

Using these formulas given above, we have

$$T = -\pi Cab \qquad (9.85)$$

The torque may be expressed either in terms of τ_{max} or ϑ_t by means of Eqs. (9.80), (9.82) and (9.85).

$$\tau_{max} = \frac{2T}{\pi ab^2} \tag{9.86}$$

$$\vartheta_t = \frac{T(a^2+b^2)}{\pi G b^3 a^3} = \frac{T}{GJ} \tag{9.87}$$

where GJ is the torsional stiffness per unit length, with J given by

$$\boxed{J = \frac{\pi b^3 a^3}{a^2+b^2}} \tag{9.88}$$

This is the equivalent polar moment of area of ellipses. When $a = b$, Eq. (9.88) is reduced to Eq. (9.13) for the circular cross-section.

9.3.11 Comparison of torsional stiffness between elliptic and circular bars

Let us conduct an analysis on torsional stiffness of elliptic and circular solid bars made of the same material. Assuming the radius of the circular cross-section is R, we shall set $R^2 = ab$ to ensure their areas are the same. The stiffness ratio of these two bars becomes.

$$\alpha_{stiffness} = \left.\frac{J_{ellipse}}{J_{circle}}\right|_{R^2=ab} = \left.\frac{\frac{\pi b^3 a^3}{a^2+b^2}}{\frac{\pi R^4}{2}}\right|_{R^2=ab} = \frac{\frac{b^3 a^3}{a^2+b^2}}{\frac{(ab)^2}{2}} = \frac{2ab}{a^2+b^2} \leq 1 \tag{9.89}$$

In the last step, we used inequality $a^2 + b^2 \geq 2ab$ (because $(a-b)^2 \geq 0$). This means that the stiffness of an elliptic bar is always smaller that the circular bar with the same area. They are equal only when $a = b$.

9.3.12 Example: Bars with equilateral triangle cross-section

Consider a bar with equilateral triangle cross-section, as shown in Fig. 9.16. It is subjected to torque T.

Stress function ϕ can be written as

$$\phi = \frac{G\vartheta_t}{2h}\left(x - \sqrt{3}y - \frac{2h}{3}\right)\left(x - \sqrt{3}y + \frac{2h}{3}\right)\left(x + \frac{h}{3}\right) \tag{9.90}$$

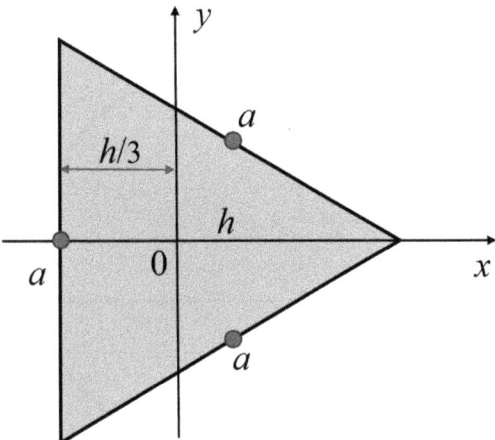

Figure 9.16. Equilateral triangle cross-section for a torsional bar.

Following the same process for the ellipse bar, the maximum shear stress is found at the center of the edges as

$$\tau_{max} = \frac{15\sqrt{3}T}{2h^3} = \frac{20T}{a^3} \qquad (9.91)$$

The twist angle per unit length is found as

$$\vartheta_t = \frac{15\sqrt{3}T}{Gh^4} = \frac{T}{GJ} \qquad (9.92)$$

where GJ is the torsional stiffness per unit length for the equilateral triangle cross-section, with J being the equivalent polar 2nd moment of area given by

$$J = \frac{h^4}{15\sqrt{3}} = \frac{\sqrt{3}a^4}{80} \qquad (9.93)$$

For cross-sections of more complex shape, analytical solutions are difficult to obtain. Approximate solutions may be obtained by Prandtl's membrane analogy for some of those.

9.4 Prandtl's elastic membrane analogy for torsion*

9.4.1 *General formulation*

Considering a membrane subjected to pressure p, resulting in a uniform membrane tension S, and undergoes a small deflection, as shown in Fig. 9.17. It is fixed on its boundary s. Our formulation and notation follow largely the textbook [4]. Codes are developed for computations and result plots.

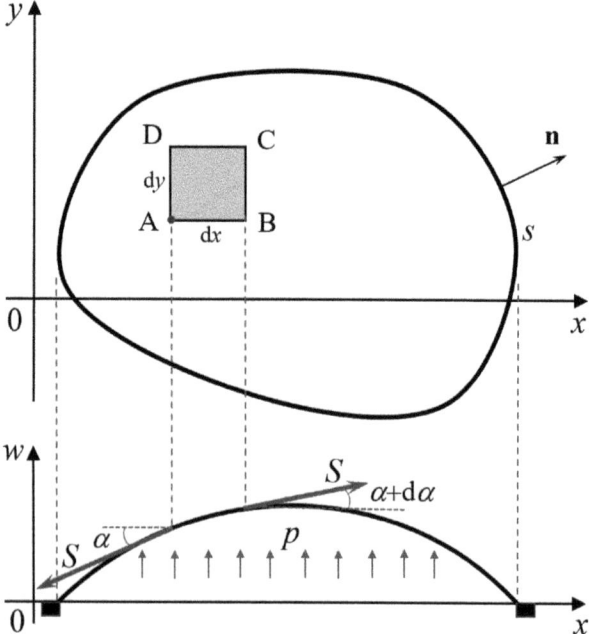

Figure 9.17. A membrane subjected to pressure and undergoes a small deflection. An infinitely small cell at a point is taken to examine the equilibrium state. Top: top-down view. Bottom: side section view.

An infinitely small cell of $dxdy$ at a point is taken to examine the equilibrium state. Assume the membrane undergoes a tension stress S and its deflection is denoted as w.

On edge A-D, the force in the membrane is

$$-Sdy \sin \alpha \approx -Sdy \tan \alpha = -Sdy \frac{\partial w}{\partial x} \qquad (9.94)$$

On edge B-C, the force is

$$Sdy \tan\left(\alpha + \frac{\partial \alpha}{\partial x} dx\right) = Sdy \frac{\partial}{\partial x}\left(w + \frac{\partial w}{\partial x} dx\right) \qquad (9.95)$$

Similarly, on edge A-B, the force in the membrane is

$$-Sdx \frac{\partial w}{\partial y} \qquad (9.96)$$

On edge D-C, it is

$$Sdx \frac{\partial}{\partial y}\left(w + \frac{\partial w}{\partial y} dy\right) \qquad (9.97)$$

The sum of all these four forces with the pressure should be zero, and dividing $dxdy$ for all terms, we obtain the equilibrium equation in terms of w:

$$\boxed{\frac{\partial^2 w}{\partial x^2} + \frac{\partial^2 w}{\partial y^2} = -\frac{p}{S}} \qquad (9.98)$$

This equation has the same form as Eq. (9.76) for the stress function. They both are Laplace equation, but with different constants on the right-hand side. Therefore, the membrane deflection and the stress function differ by a constant factor c:

$$w = c\phi \qquad (9.99)$$

where c is found by equaling terms on the right-hand side of Eqs. (9.98) and (9.76), which is

$$c = \frac{p}{2G\vartheta_t S} \qquad (9.100)$$

Finding the stress function for torsion bar becomes solving for the deflection of a pressured membrane with a shape same as the cross-section of the bar.

Examples are given in the following sections.

9.4.2 A bar with narrow rectangular cross-section*

Consider a bar with a narrow rectangular cross-section subjected to a torque **T**, as shown in Fig. 9.18. The torque will create a force couple on the cross-section that results in shear stress distributed there. Because of its narrow shape, it is expected that the shear stress will form two major flows: one near the top surface and another near the bottom surface, with mutually opposite direction. We shall find a way to quantify it.

Using the membrane analogy, we consider a membrane with a narrow rectangular shape and subjected to pressure, as shown in Fig. 9.19.

Based on the slender geometry of the deflected membrane, we can assume membrane deflection is independent of x, and varies in a parabolic way with

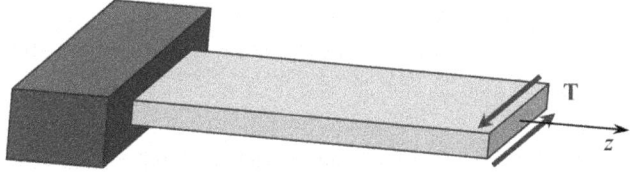

Figure 9.18. A bar with narrow rectangular cross-section subjected to a torque.

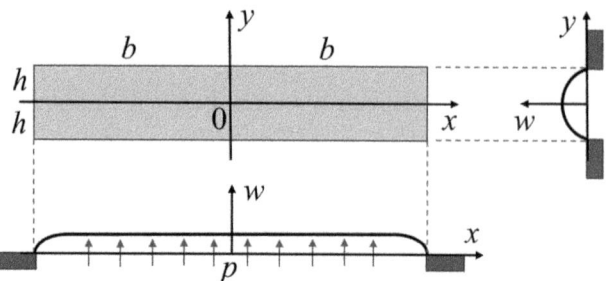

Figure 9.19. Narrow rectangular cross-section of a bar subjected to a pressure and undergoes a small deflection. It has a width of $2b$ and height of $2h$.

respect to y.

$$w = w_0 \left(1 - \frac{y^2}{h^2}\right) \quad (9.101)$$

where w_0 is the maximum deflection at $y = 0$. This deflection satisfies the boundary condition on $y = \pm h$. Except for the small regions near $x = \pm b$. Hence, this is a good approximation. The approximation error reduces with decreasing h/b. The Laplace of the assumed w becomes

$$\frac{\partial^2 w}{\partial y^2} = -\frac{w_0}{h^2} \quad (9.102)$$

It is now a 1D problem. The membrane analogy leads to $w_0/h^2 = 2cG\vartheta_t$. The stress function is found as

$$\phi = G\vartheta_t h^2 \left(1 - \frac{y^2}{h^2}\right) \quad (9.103)$$

The shear stresses can be found via Eq. (9.74):

$$\sigma_{zx} = \frac{\partial \phi}{\partial y} = -2G\vartheta_t y$$
$$\sigma_{zy} = -\frac{\partial \phi}{\partial x} = 0 \quad (9.104)$$

It is clear that σ_{zx} is linearly in y and anti-symmetric with respect to $y = 0$.

An actual distribution of the shear stress on a narrow rectangular cross-section is give in Fig. 9.20. This is produced using our code to be introduced later.

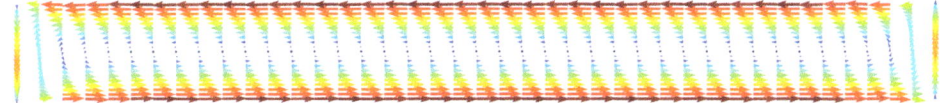

Figure 9.20. An actual case of shear stress on narrow rectangular cross-section of a bar subjected to torsion ($b = 10h$).

As shown, the shear stress flows near the top and bottom surfaces form a force couple giving the torque on the cross-section of the bar, as shown in Fig. 9.18. The resulting two forces are located about $h/3$ away from the top and bottom surfaces.

The maximum shear stress is found at $y = \pm h$:

$$\tau_{\max} = 2G\vartheta_t h \tag{9.105}$$

Torque T is obtained using Eqs. (9.73) and (9.103):

$$T = 2\int_A \phi dA = 2\int_{-b}^{b}\int_{-h}^{h} \phi dxdy = G\vartheta_t \underbrace{\tfrac{1}{3}(2b)(2h)^3}_{J} = GJ\vartheta_t \tag{9.106}$$

where J is the equivalent polar 2nd moment of area given by

$$\boxed{J = \frac{1}{3}(2b)(2h)^3} \tag{9.107}$$

Note that Eq. (9.107) can be used for any cross-section that has a narrow channel shape, and it can be curved. This is because of the membrane analogy, the deformed membrane will have the same volume regardless of whether it is curved or not, as long as it is a narrow channel. Therefore, Eq. (9.107) is very useful to calculate the stiffness per length GJ for such bars.

Using Eqs. (9.105) and (9.106), the maximum shear stress can now be written in terms of T:

$$\tau_{\max} = \frac{3T}{(2b)(2h)^2} = \frac{2Th}{J} \tag{9.108}$$

Finally, the twist angle per length can be written for given T:

$$\vartheta_t = \frac{3T}{G(2b)(2h)^3} = \frac{T}{GJ} \tag{9.109}$$

We mention that the boundary conditions are not satisfied on $x = \pm b$. The error is expected to be small if h/b is small.

9.4.3 Bars with cross-sections consisting of narrow rectangles*

Many torsion bars are with composite sections consisting of a number of narrow rectangles. The most typical one is the I-beams, as shown in Fig. 9.21. This type of bars is analyzed easily by making use of the results from the previous section.

Assuming the cross-section of a bar is formed with n narrow channels, its J value can be obtained simply by

$$J = c\frac{1}{3}\sum_{i=1}^{n}(2b_i)(2h_i)^3 \qquad (9.110)$$

This summation is justified by, again, the membrane analogy because the total volume of the deflection (hence the stress function) for a composite section with narrow channels are just simple addition of the individual channels. Each channel acts alone having a stress distribution like that shown in Fig. 9.20, forming an individual torque. The sum of the torques of all the channels of the cross-section gives the total torque T.

In Eq. (9.110), we introduced a correction coefficient c. It can be used for proper adjustment when a rectangle is not narrow. If $b_i > 10h_i$ for a rectangular part, then $c \approx 1$. If $b_i < 10h_i$, c may be reduced to, say, $c \approx 0.9$.

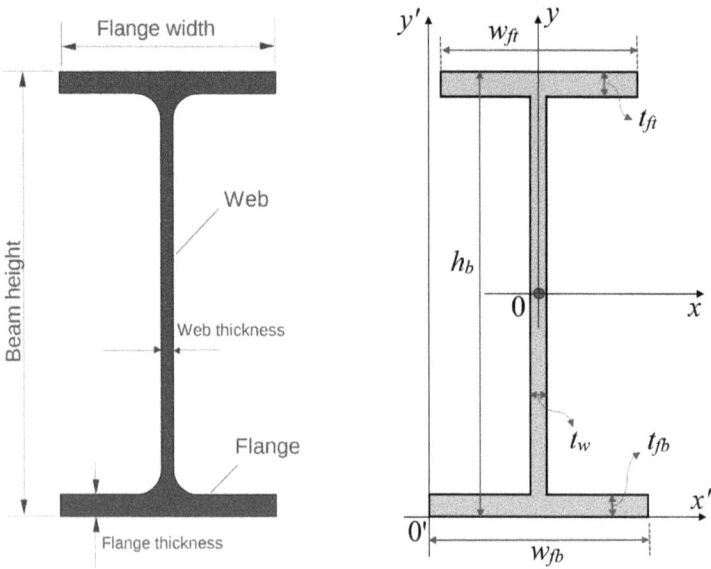

Figure 9.21. A typical I-beam with three narrow rectangles. The image on the left is from Poutrelle métallique at wikipedia commons under the CC BY 3.0 license.

It is a rough estimation. As will be seen in the next section with computer codes, we can use more precise formula.

Note also that when computing the maximum shear stress in a composite section, we shall use

$$\tau_{max} = \frac{2Th_{max}}{J} \qquad (9.111)$$

where h_{max} is $\max_i(h_i)$. This is because a wider channel has larger membrane deflection and hence larger ϕ values.

9.5 Bars with rectangular cross-section*

9.5.1 *Strategy for solution*

Consider torsion bars with the rectangular section of width $2h$ and depth $2b$, as shown in Fig. 9.22. This types of bars require more intensive analysis for proper design against torsion. Our formulation and notation follow largely the textbook [4]. Codes are developed for computations and result plots.

We need to find the stress function ϕ, and then solve the torsion problem for stresses and twist angle. Because of the symmetry of the cross-section and the Laplace operator is also symmetric, ϕ must be an even function of both x and y. The problem can be mathematically defined by

$$\begin{aligned} \nabla^2 \phi &= -2G\vartheta_t \quad \text{PDE in the domain of the cross-section} \\ \phi &= 0 \quad \text{on the boundary of the cross-section} \end{aligned} \qquad (9.112)$$

One effective approach to solve this problem is to use the method of separation of variables. To do so, we first make the Laplace PDE homogeneous

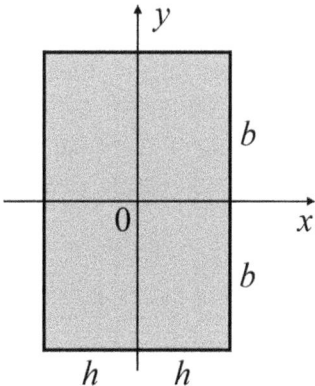

Figure 9.22. A typical rectangular cross-section for bars subjected to torsion.

(with zero right-hand side). This is done by writing ϕ in the following form, assuming $h \leq b$:

$$\phi = G\vartheta_t(h^2 - x^2) + w(x, y) \tag{9.113}$$

where w is an unknown function to be determined. It is easy to check that the first term in the foregoing equation satisfies the PDE in Eq. (9.112). It is thus a particular solution to Eq. (9.112). Now, substituting Eq. (9.113) into Eq. (9.112) gives

$$\nabla^2 w = 0 \quad \text{governing PDE}$$

$$w = \begin{cases} 0 & \text{for } x = \pm h \\ G\vartheta_t(h^2 - x^2) & \text{for } y = \pm b \end{cases} \tag{9.114}$$

We need next to solve the foregoing equation for $w(x, y)$, using the method of separation of variables. Note if assuming $h > b$, we would use $G\vartheta_t(b^2 - y^2)$ instead as the particular solution to Eq. (9.112). This is just for a better convergence in the solution to be obtained later.

9.5.2 Method of separation of variables

We write the unknown function as the product of two unknown functions $f(x)$ and $g(x)$:

$$w(x, y) = f(x)g(y) \tag{9.115}$$

Substitute Eq. (9.115) into Eq. (9.114), we obtain

$$\nabla^2 w = g(y)f''(x) + f(x)g''(y) = 0 \tag{9.116}$$

which can be rewritten as

$$\frac{f''(x)}{f(x)} = -\frac{g''(y)}{g(y)} = -\lambda^2 \tag{9.117}$$

The first term is a function of only x and the second is a function of only y. Thus, λ can only be a (unknown) constant independent of both x and y. Equation (9.117) gives two separate equations:

$$\begin{aligned} f''(x) + \lambda^2 f(x) &= 0 \\ g''(y) - \lambda^2 g(y) &= 0 \end{aligned} \tag{9.118}$$

Each of these equations is a 1D DE. The solutions to them are, respectively,

$$\begin{aligned} f(x) &= A \cos \lambda x + B \sin \lambda x \\ g(x) &= C \cosh \lambda y + D \sinh \lambda y \end{aligned} \tag{9.119}$$

The solutions contain four constants. Because the original stress function ϕ must be an even function, both $f(x)$ and $g(x)$ must be even. Thus, $B = D = 0$. Substitute Eq. (9.119) into Eq. (9.115), we obtain

$$w(x, y) = A \cos \lambda x \cosh \lambda y \qquad (9.120)$$

where A (absorbed C) is a to-be-determined constant, in addition to constant λ. This is done using boundary conditions.

To satisfy the boundary condition at $x = \pm h$, we set

$$\lambda = \frac{n\pi}{2h}, \quad n = 1, 3, 5, \ldots \qquad (9.121)$$

Equation (9.120) becomes

$$w(x, y) = \sum_{n=1,3,5,\ldots} A_n \cos \frac{n\pi x}{2h} \cosh \frac{n\pi y}{2h} \qquad (9.122)$$

Using the boundary condition at $y = \pm b$, we have

$$\sum_{n=1,3,5,\ldots} A_n \cos \frac{n\pi x}{2h} \cosh \frac{n\pi b}{2h} = G\vartheta_t(h^2 - x^2) \qquad (9.123)$$

To solve the foregoing equation for A_n, we multiply $\cos \frac{n\pi x}{2h}$ on both sides and integrate along x from $-h$ to h, and use formulas for cosine function, which gives

$$A_n \cosh \frac{n\pi b}{2h} = \frac{G\vartheta_t}{h} \int_{-h}^{h} (x^2 - h^2) \cos \frac{n\pi x}{2h} dx = \frac{-32 G \vartheta_t h^2 (-1)^{(n-1)/2}}{n^3 \pi^3} \qquad (9.124)$$

These above-integrals can be done using the following code.

```
1  x, h = symbols("x, h")
2  n = symbols('n', integer=True, odd=True)
3
4  # The integral = on the left-hand-side
5  integrate(((sp.cos(n*sp.pi*x/(2*h))**2)), (x, -h, h))
```

$$\begin{cases} h & \text{for } \frac{\pi n}{h} \neq 0 \\ 2h & \text{otherwise} \end{cases}$$

```
1  # The integral = on the right-hand-side
2  integrate((x**2-h**2)*sp.cos(n*sp.pi*x/(2*h)), (x, -h,h))
```

$$-\frac{32(-1)^{\frac{n}{2}-\frac{1}{2}} h^3}{\pi^3 n^3}$$

Finally, A_n is found as

$$A_n = \frac{-32G\vartheta_t h^2(-1)^{(n-1)/2}}{n^3\pi^3 \cosh\frac{n\pi b}{2h}}, \quad n = \text{odd} \qquad (9.125)$$

9.5.3 Final solution of the stress function

The stress function becomes

$$\phi = G\vartheta_t(h^2 - x^2) - \frac{32G\vartheta_t h^2}{\pi^3}\sum_{n=1,3,5,\ldots}\frac{(-1)^{(n-1)/2}}{n^3 \cosh\frac{n\pi b}{2h}} \cos\frac{n\pi x}{2h}\cosh\frac{n\pi y}{2h} \qquad (9.126)$$

Note that in the denominator of the coefficient of the series above, there is n^3 term. Also, the cosh function has the following Taylor expansion:

$$\cosh x = 1 + \frac{x^2}{2!} + \frac{x^4}{4!} + \cdots \qquad (9.127)$$

Therefore, the coefficients in Eq. (9.126) goes to zero quite fast when b/h is large. When b/h goes to infinity, Eq. (9.126) reduces to the one for narrow rectangles discussed in the previous section.

9.5.4 Shear stress solution

Using Eq. (9.60), the shear stresses can be found as follows:

$$\sigma_{zx} = \frac{\partial\phi}{\partial y} = -\frac{16G\vartheta_t h}{\pi^2}\sum_{n=1,3,5,\ldots}\frac{(-1)^{(n-1)/2}}{n^2 \cosh\frac{n\pi b}{2h}} \cos\frac{n\pi x}{2h}\sinh\frac{n\pi y}{2h} \qquad (9.128)$$

$$\sigma_{zy} = -\frac{\partial\phi}{\partial x} = 2G\vartheta_t x - \frac{16G\vartheta_t h}{\pi^2}\sum_{n=1,3,5,\ldots}\frac{(-1)^{(n-1)/2}}{n^2 \cosh\frac{n\pi b}{2h}} \sin\frac{n\pi x}{2h}\cosh\frac{n\pi y}{2h} \qquad (9.129)$$

9.5.5 Torque twist-angle solution and equivalent polar 2nd moment of area

The torque is found using Eq. (9.73) as

$$T = 2\int_{-b}^{b}\int_{-h}^{h}\phi\,dxdy = GJ\vartheta_t \qquad (9.130)$$

where

$$J = (2b)(2h)^3 \underbrace{\left[\frac{1}{3} - \frac{192\,h}{3\pi^5\,b}\sum_{n=1,3,5,\ldots}\frac{1}{n^5}\tanh\left(\frac{n\pi\,b}{2\,h}\right)\right]}_{k_1} = (2b)(2h)^3 k_1$$

(9.131)

where k_1 is a coefficient that depends only on b/h. When $b/h \to \infty$, $k_1 \to \frac{1}{3}$ because the summation term in the foregoing equation approaches to zero.

9.5.6 The maximum shear stress

The maximum slope of the stress function (membrane) on the rectangle section occurs at $x = \pm h, y = 0$, where $\sigma_{zx} = 0$ and σ_{zy} has the maximum value ($h \le b$):

$$\tau_{\max} = \sigma_{zy}\bigg|_{\substack{x=h\\y=0}} = 2G\vartheta_t h\underbrace{\left[1 - \frac{8}{\pi^2}\sum_{n=1,3,5,\ldots}\frac{1}{n^2\cosh\frac{n\pi b}{2h}}\right]}_{k} = 2G\vartheta_t hk$$

(9.132)

If we introduce $k_2 = \frac{k_1}{k}$, the maximum shear stress can be computed alternatively using

$$\tau_{\max} = \frac{T}{k_2(2b)(2h)^2}$$

(9.133)

Coefficient k_1 and k can be computed using the code given below. Tabulated k_1 and k_2 are given in Ref. [4].

9.5.7 Python code for computing k_1

```
# Define the function for computing k1:
def k1_f(b_h, n_end):
    sum1 = 0.0
    for n in range(1, n_end, 2):
        sum1 += np.tanh(n*np.pi*b_h/2)/n**5
    return 1.0/3.0 - 192.0/(3*b_h*np.pi**5)*sum1

# Define the function for computing k:
def k_f(b_h, n_end):
    sum1 = 0.0
    for n in range(1, n_end, 2):
        sum1 += 1./(np.cosh(n*np.pi*b_h/2))/n**2
    return 1.0 - 8.0/(np.pi**2)*sum1
```

9.5.7.1 An example

First, we compute coefficients k_1, k, and k_2 for two special values of $b/h = 1$ and $b/h = 10$. These values may be referenced in later analysis.

```
1  np.set_printoptions(formatter={'float': '{: 0.5e}'.format})
2  n_end = 21
3  b_h = [1, 10]                           # at special values, b=h, b=10h
4
5  k1 = np.array([k1_f(bhi, n_end) for bhi in b_h])
6  print(f"Coefficient k1 for  J    with b/h of {b_h} = {k1}")
7  k  = np.array([ k_f(bhi, n_end) for bhi in b_h])
8  print(f"Coefficient k  for τ_max with b/h of {b_h} = {k}")
9  k2 = k1/k
10 print(f"Coefficient k2 for τ_max with b/h of {b_h} = {k2}")
```

```
Coefficient k1 for   J    with b/h of [1, 10] = [ 1.40577e-01  3.12325e-01]
Coefficient k  for τ_max with b/h of [1, 10] = [ 6.75314e-01  1.00000e+00]
Coefficient k2 for τ_max with b/h of [1, 10] = [ 2.08165e-01  3.12325e-01]
```

9.5.7.2 Cut-off terms in the solution series

We write the following code to study the required terms needed for solution of k_1 with desired accuracy. Because we assume $h \leq b$, we study the worst (slowest convergence) case of $b = h$.

```
1  np.set_printoptions(precision=7,suppress=True)  # Digits in print-outs
2
3  b_h, n_end_max = 1.0, 10
4  x_data = list(range(1, n_end_max, 1))
5  y_data = []
6  for ni in x_data:
7      y_data.append(k1_f(b_h, ni))                # function values
8
9  print(f"Cut-off term n_max: {np.array(x_data)}")
10 print(f"The corresponding k1 values:\n {np.array(y_data)}")
```

```
Cut-off term n_max: [1 2 3 4 5 6 7 8 9]
The corresponding k1 values:
[0.3333333 0.141523  0.141523  0.1406625 0.1406625 0.1405955 0.1405955
 0.1405831 0.1405831]
```

As shown when $n = 9$ is used, the error is at the 6th digit. Let us use the following code to plot the convergence curve.

```
1  X_data = [x_data]
2  Y_data = [y_data]                       # put the function values to a list
3  labels = ['k1']                                         # define the label
4  gr.plot_fs(X_data, Y_data, labels, 'n',    'k1',    'k1_n')
5                                          # x-label, y-label, file-name
```

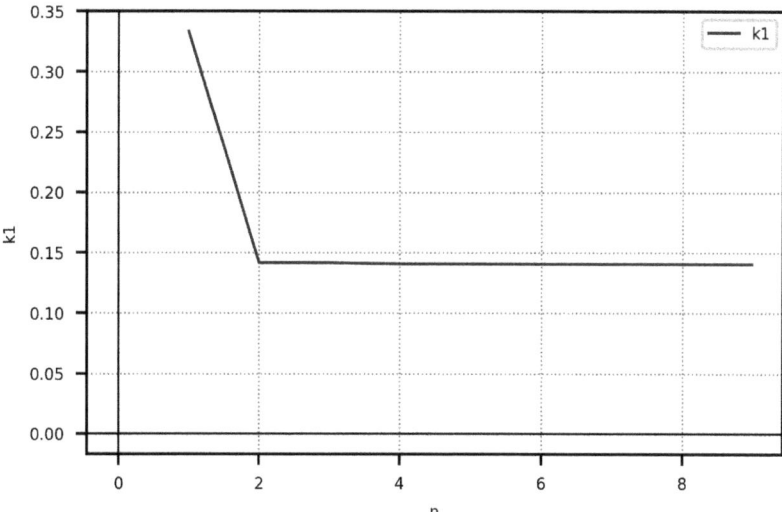

We observed a fast convergence from the plot above.

Note that these formulas work also for $h > b$, but the convergence is a little slower. Since we are using computer code to do it, it is really not a problem. Even if we take 101 terms, the computation takes less than one second.

In conclusion, with the formulas derived in this section, one can solve all torsion bars with a rectangular cross-section.

9.5.7.3 Values as function of b/h

We write the following code to compute the values of k_1, k and k_2 for given a list of b/h.

```
n_end = 25
dx = 0.01
b_h   = np.arange(0.1, 10.0, dx)             # b/h as x_data

k1_data = []
k_data  = []
for i in range(len(b_h)):
    k1_data.append(k1_f(b_h[i], n_end))      # function values
    k_data.append(  k_f(b_h[i], n_end))

k2_data = [k1_data[i]/k_data[i] for i in range(len(b_h))]
step = max(int(len(b_h)/5), 1)
print(f"Ratio of b/h: {np.array(b_h)[::step]}")
print(f"The corresponding k1 values:\n {np.array(k1_data)[::step]}")
print(f"The corresponding k values:\n {np.array(k_data)[::step]}")
print(f"The corresponding k2 values:\n {np.array(k2_data)[::step]}")
```

```
Ratio of b/h: [0.1  2.08 4.06 6.04 8.02]
The corresponding k1 values:
[0.003124  0.2326236 0.2815891 0.2985514 0.3071385]
The corresponding k values:
[0.074386  0.9382967 0.9972449 0.9998771 0.9999945]
The corresponding k2 values:
[0.0419976 0.2479212 0.282367  0.2985881 0.3071401]
```

```
1  k_3 = [k_data[i]/3. for i in range(len(b_h))]
2
3  X_data = [b_h, b_h, b_h]              # put the x (b/h) values to a list
4  Y_data = [k1_data, k_3, k2_data]      # put the function values to a list
5
6  labels=['k1', 'k/3', 'k2']                          # define the label
7  gr.plot_fs(X_data, Y_data, labels, 'b/h',    'k1, k/3, k2',   'ks_bh')
8                                        # x-label,    y-label, file-name
```

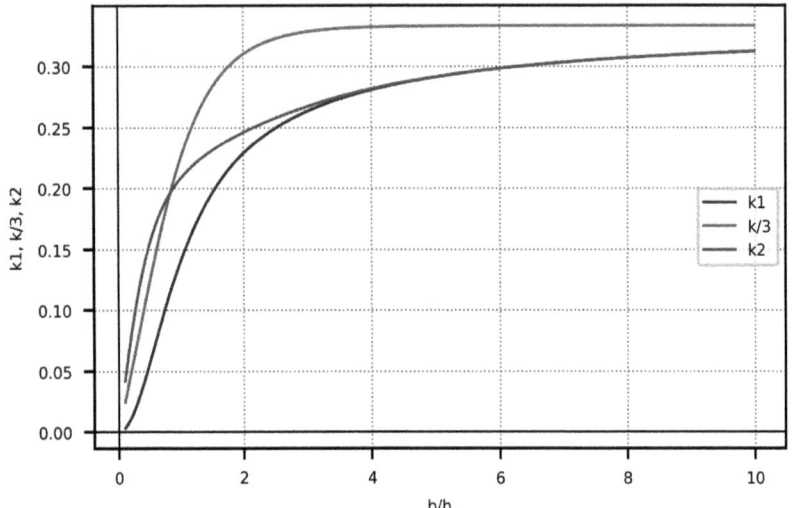

9.5.8 Comparison of equivalent J and polar moment of area J_z

Consider the bar has a square cross-section. Equation (9.131) becomes

$$J = (2h)^4 k_{1(b=h)} = 0.140577(2h)^4 \qquad (9.134)$$

Here, we used $k_{1(b=h)} = 0.140577$ obtained earlier. Compared to the polar moment of area of a square bar, using perpendicular axis theorem (see Chapter 10), $J_z = 2I_{xx} = \frac{(2h)^4}{6}$, the corresponding k_1 should be calculated using

$$k_1^{\text{square}} = \frac{1}{6} = 0.166667 \qquad (9.135)$$

This shows that the equivalent J value of the same square bar subjected to torsion is about 15.65% lower than J_z value of a square bar of $(2h) \times (2h)$.

For narrow channel of $b/h \to \infty$, $J_z = (I_{xx} + I_{yy}) \to \frac{(2b)(2h)^3}{12}$, the corresponding k_1 should be calculated using

$$k_1^{\text{narrow}} = \frac{1}{12} \qquad (9.136)$$

For the same infinitely long channel bar subjected to torsion, we have $k_1 = 1/3$, which is 4 times k_1^{narrow}.

9.5.9 Comparison of torsional stiffness between rectangular and square bars

Let us conduct an analysis on torsional stiffness of rectangular bar of $(2b) \times (2h)$ ($b > h$) and square bars of $(2a) \times (2a)$. They both are solid and made of the same material. We set $(2a)^2 = (2b)(2h)$ or $a^2 = bh$ to ensure their areas are the same. The stiffness ratio of these two bars becomes

$$\alpha_{\text{stiffness}} = \left.\frac{J_{\text{rectangle}}}{J_{\text{square}}}\right|_{a^2=bh} = \left.\frac{(2b)(2h)^3 k_1}{(2a)^4 k_{1(b=h)}}\right|_{a^2=bh}$$

$$= \frac{bh^3 k_1}{(bh)^2 k_{1(b=h)}} = \frac{h}{b}\frac{k_1}{k_{1(b=h)}} \qquad (9.137)$$

To find out how $\alpha_{\text{stiffness}}$ changes with b/h, we use the following code to plot the curve.

```python
1  n_end = 25; dx = 0.01
2  b_h  = np.arange(1.0, 10.0+dx, dx)              # b/h as x_data
3
4  X_data = [b_h]                                  # put the x (b/h) values to a list
5  k1_data = []
6  for i in range(len(b_h)):
7      k1_data.append(k1_f(b_h[i], n_end))         # function values
8
9  a_stiffness = [k1_data[i]/b_h[i]/k1_data[0] for i in range(len(b_h))]
10 Y_data = [a_stiffness]                          # put the function values to a list
11
12 print(f"b/h={b_h[-1]}, a_stiffness={a_stiffness[-1]}")
13 labels=['$a$_stiffness']                        # define the label
14 gr.plot_fs(X_data, Y_data, labels, 'b/h',  '$a$_stiffness', 'a_stiff')
15                                                 # x-label,   y-label,           file-name
```

b/h=10.000000000000007, a_stiffness=0.22217349780492124

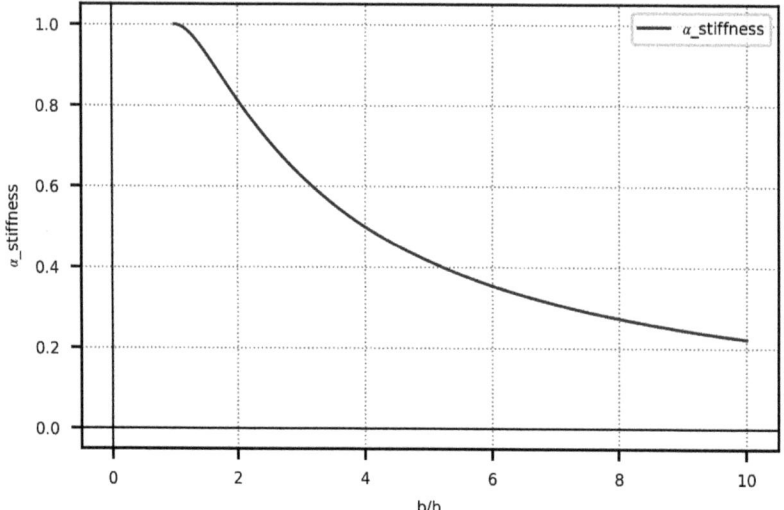

It is clear that $\alpha_{\text{stiffness}} < 1$ for $b > h$. This means that the torsional stiffness of a rectangular bar is smaller than that of the square bar with the same area. They are equal when $h = b$. When $b/h = 10$, the stiffness reduces to only about 22%. This significant reduction can be explained using Fig. 9.18. When the height is very small, the distance between these force couple (one near the top and another near the bottom) of the rectangular channel becomes small. In addition, for the same twist angle per bar length, the shear stress on both top and bottom surfaces is also small. The double reduction effect results in significantly reduction in torque, and hence the stiffness.

This finding is similar to that given in Section 9.3.11, when comparing the narrow elliptic and circular bars.

9.5.10 Comparison of torsional stiffness between circular and square bars

Let us also compare the torsional stiffness of a circular bar of radius R with a square bar of $(2a) \times (2a)$. Assume both bars are solid and made of the same material. We set $\pi R^2 = (2a)(2a)$ or $R^2 = \frac{4}{\pi}a^2$ to ensure their areas are the same. The stiffness ratio of these two bars becomes

$$\alpha_{\text{stiffness}} = \left.\frac{J_{\text{circular}}}{J_{\text{square}}}\right|_{R^2=\frac{4}{\pi}a^2} = \left.\frac{\pi R^4/2}{(2a)^4 k_{1(b=h)}}\right|_{R^2=\frac{4}{\pi}a^2} \quad (9.138)$$

$$= \frac{\pi(\frac{4}{\pi}a^2)^2}{2(2a)^4 k_{1(b=h)}} = \frac{1}{2\pi k_{1(b=h)}} = \frac{1}{2\pi \times 0.140577} = 1.132$$

This means that the torsional stiffness of a circular bar is about 13.2% higher than that of the square bar with the same area.

9.5.11 Code to compute the stress function on rectangular cross-section

We write the following code to compute the stress function for torsion bar with rectangular cross-section, using Eq. (9.126).

It computes effectively a series solution for any function governed by a 2D Laplace PDE with a constant right-hand side, and fixed boundary condition.

```
1  def sf_phi(b, h, Gt, x, y, n_end = 21):
2      '''Compute the stress function phi for torsion bar with rectangular
3      cross-section. It is effectively a series solution to 2D Laplace PDE
4      with a constant right-hand-side, and fixed boundary condition.
5      b:       half-height of the domain
6      h:       half-width  of the domain
7      Gt:      constant at the right-hand-side of the Laplace.
8      n_end:   terms used in the series of the solution
9      x and y: point at which the solution to be found.'''
10
11     sum1 = 0.0
12     for n in range(1, n_end, 2):
13         sum1 += (-1)**((n-1)/2)/(n**3*np.cosh(n*np.pi*b/(2*h)))*\
14                 np.cos(n*np.pi*x/(2*h))*np.cosh(n*np.pi*y/(2*h))
15
16     return Gt*(h**2-x**2) - (32*Gt*h**2/np.pi**3)*sum1
```

9.5.12 Plot of the stress function on rectangular cross-section

```
1  plt.rcParams['figure.dpi'] = 500
2  h, b, Gt, n_end = 2., 1., 1., 21        # arbitrarily set values
3
4  xl, xr, dx = -h, h, 0.02
5  yl, yr, dy = -b, b, 0.02
6
7  X = np.arange(xl, xr+dx, dx)
8  Y = np.arange(yl, yr+dy, dy)
9  X, Y = np.meshgrid(X, Y)
10 xn, yn = X.shape
11 W = X*0
12
13 for xk in range(xn):
14     for yk in range(yn):
15         W[xk, yk] = sf_phi(b, h, Gt, X[xk,yk], Y[xk,yk], n_end)
16
17 fig = plt.figure()       # Create a 3D plot
18 ax = fig.add_subplot(111, projection='3d')
19
20 surf = ax.plot_surface(X,Y,W,rstride=1,cstride=1,cmap='viridis')
21 ax.set_xlim(-4, 2)
22 ax.set_ylim(-3.5, 1.)
23 ax.set_zlim(-5, 2.5)
24 ax.grid(False);
25 #plt.show()
```

Figure 9.23 plots the stress function of a rectangular cross-section using the code given above. It is the deformed membrane under pressure.

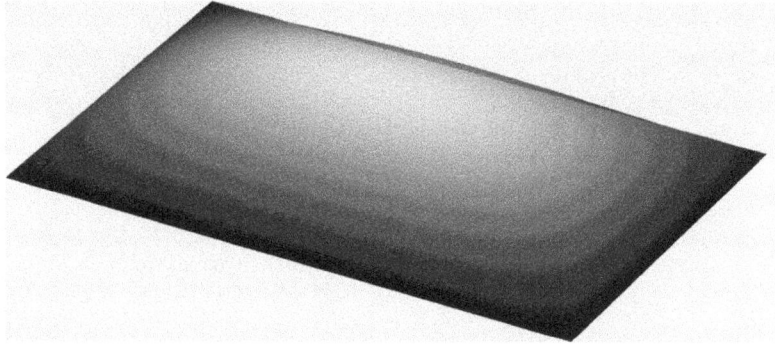

Figure 9.23. A schematic view of the stress function (deformed membrane) for a rectangular cross-section.

9.5.13 Code to compute the shear stresses on rectangular cross-section

We write the following code to compute the shear stresses on a rectangular cross-section bar subjected to torsion, using Eqs. (9.128) and (9.129).

It computes effectively a series solution for the derivatives of the function governed by any 2D Laplace PDE with a constant right-hand side, and fixed boundary condition.

```
def stress_rect(b, h, Gt, x, y, n_end = 21):
    '''compute the shear stresses on a rectangular cross-section
    bar subjected to torsion. It is effectively a series solution for
    the derivatives of the function governed by any 2D Laplace PDE
    with a constant right-hand-side, and fixed boundary condition.
    b:       half-height of the domain
    h:       half-width  of the domain
    Gt:      constant at the right-hand-side of the Laplace.
    n_end:   terms used in the series of the solution
    x and y: point at which the solution to be found.

    Return: sigam_zx, sigam_zy
    '''
    factor = 16*Gt*h/np.pi**2
    sum_x = 0.0;    sum_y = 0.0
    for n in range(1, n_end, 2):
        coef = (-1)**((n-1)/2)/(n**2*np.cosh(n*np.pi*b/(2*h)))
        sum_x += coef*np.cos(n*np.pi*x/(2*h))*np.sinh(n*np.pi*y/(2*h))
        sum_y += coef*np.sin(n*np.pi*x/(2*h))*np.cosh(n*np.pi*y/(2*h))

    sigam_zx =          - factor*sum_x
    sigam_zy = 2*Gt*x   - factor*sum_y

    return sigam_zx, sigam_zy
```

9.5.14 Plot of the shear stresses on rectangular cross-section

```
from matplotlib import cm
plt.rcParams['figure.dpi'] = 500
h, b, Gt, n_end = 2., 1., 1., 21        # arbitrarily set values

xl, xr, dx = -h, h, 0.2
yl, yr, dy = -b, b, 0.1

X = np.arange(xl, xr+dx, dx)
Y = np.arange(yl, yr+dy, dy)
X, Y = np.meshgrid(X, Y)
```

```
11  xn, yn = X.shape
12  Szx = X*0
13  Szy = Y*0
14
15  for xk in range(xn):
16      for yk in range(yn):
17          Szx[xk, yk], Szy[xk, yk] = stress_rect(b,h,Gt,X[xk,yk],Y[xk,yk])
18
19  magnitude = np.sqrt(Szx**2 + Szy**2)
20  plt.quiver(X, Y, Szx, Szy, magnitude, scale=20, cmap=cm.jet)#'viridis')
21  plt.axis('equal')   #plt.colorbar()
22  plt.savefig('images/shearOnRect.png', dpi=500)    # save plot to a file
23  plt.show()
```

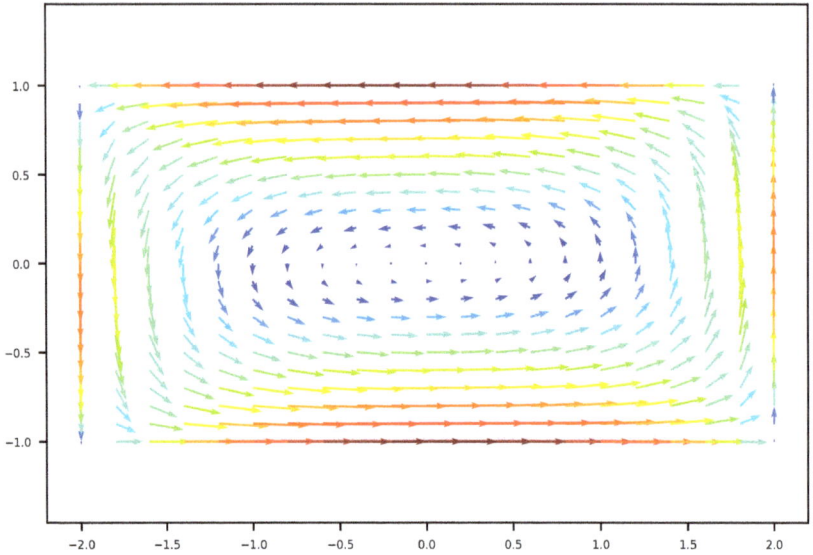

It is seen that the shear stress vector is circulating on the cross-section (the x–y plan) of the bar. The total circulation is the curl of the stress vector integrated over the cross-section, relating in the torque with respect to the z-axis. Because of the circulating nature, the total stress along both x- and y-direction will be zero. The equilibrium conditions are all satisfied. In addition, the shear stress is always along the edges of the cross-section, there will be no shear stress on the circumferential surface of the bar. The traction boundary conditions are also satisfied.

The stress distribution given in Fig. 9.20 is produced using the same codes given above.

9.5.15 Example: An I-beam

Consider an I-beam (W760 × 220) with basic shape shown in Fig. 9.21. Its dimensions (mm) are given as follows:

beam height: $h_b = 779$; flange width: $w_f = 266$; flange thickness: $t_f = 30$; web thickness: $t_w = 16.5$.

It is subjected to a torque $T = 5000$ Nm. Assume the shear modulus $G = 200$ GPa.

1. Determine the maximum shear stress and its location (Ignore the effects of fillets and stress concentrations). 2. Determine the twist angle of per unit length.

Solution to 1: There are of 3 rectangles. Their parameters of h, b, and J are computed using:

```
1  hb =.779; wf=.266; tf=.030; tw=.0165    # (m)
2  T, G = 5000, 200e9                      # (Nm), (Pa)
3  n_end = 11                              # terms to use
4
5  h2 = [tf, tw,          tf]              # 2h
6  b2 = [wf, (hb-2*tf), wf]                # 2b
7
8  b_k= [b2[i]/h2[i] for i in range(len(h2))]
9  print(f" Ratios of bi/hi = {b_k} ")
10
11 k1 = [k1_f(b2[i]/h2[i], n_end) for i in range(len(h2))]
12 k  = [k_f( b2[i]/h2[i], n_end) for i in range(len(h2))]
13 Ji = [k1[i]*b2[i]*h2[i]**3      for i in range(len(h2))]
14
15 J  = sum(Ji)
16 print(f" 2h values  = {h2} ")
17 print(f" k1 values  = {k1} ")
18 print(f" k values   = {k} ")
19 print(f" J values   = {np.array(Ji)} (m^4)")
20 print(f" Total J values = {J} (m^4)")
```

Ratios of bi/hi = [8.866666666666667, 43.57575757575758, 8.866666666666667]
2h values = [0.03, 0.0165, 0.03]
k1 values = [0.30964005201232053, 0.3285122943107051, 0.30964005201232053]
k values = [0.9999985509443367, 1.0, 0.9999985509443367]
J values = [0.0000022 0.0000011 0.0000022] (m^4)
Total J values = 5.508711157672834e-06 (m^4)

The maximum shear stress is computed using Eq. (9.132):

```
1  hi_max = np.argmax(h2)
2  τmax = k[hi_max]*T*max(h2)/J
3  print(f"The maximum shear stress = {τmax*1.0e-6} (MPa)")
```

The maximum shear stress = 27.229560299730473 (MPa)

Solution to 2: After the effective polar 2nd moment of area J is known, the twist angle of per unit length is computed using genetic Eq. (9.14):

```
1  ϑt = T/G/J
2  print(f"The twist angle of per unit length = {ϑt} (rad/m)")
```

The twist angle of per unit length = 0.004538266626156035 (rad/m)

9.6 Hollow thin-wall torsion members, use of membrane analogy*

9.6.1 *Solution strategy*

Consider a bar with a cross-section that is a hollow enclosed thin-wall, as shown in Fig. 9.24. The thickness of the wall need not be constant, and the shape of the cross-section can be arbitrary. Our formulation and notation follow largely the textbook [4]. Codes are developed for computations.

Before our analysis, we make the following assumptions:

1. The wall thickness is small compared to the other dimensions of the cross-section.
2. The shear stress is constant through the wall thickness (It may not be constant along the boundary, unless the thickness t is constant).

Based on the membrane analogy studied earlier, we know that the stress function ϕ relates to the deformed membrane w just by a constant factor. Because ϕ must satisfy Eq. (9.65) on the boundary of the cross-section, it must be either zero or a constant there. We set $\phi = 0$ on the outer boundary of the thin-wall. On the inner boundary it must be a constant, and it is noted as ϕ_1. Correspondingly, w must also be zero on the outer boundary, and a constant on the inner boundary and it is noted as w_1. This leads to the drawing on the bottom of Fig. 9.24.

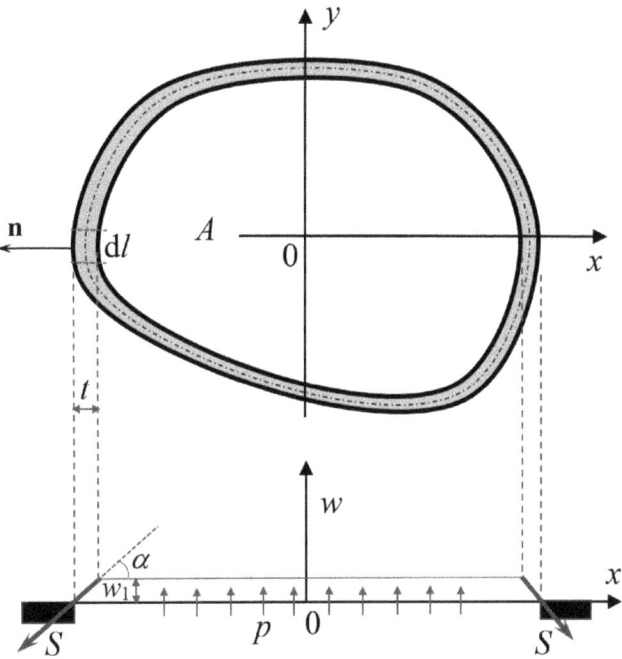

Figure 9.24. A hollow enclosed thin-wall cross-section for bars subjected to torsion (top), and the idealized deformed membrane w on the cross-section (bottom).

Using Eq. (9.69), we have $\tau = -\frac{\partial \phi}{\partial n}$, where n is normal to contour of w (or ϕ), we shall have

$$\tau = \frac{1}{c}\underbrace{\left(-\frac{\partial z}{\partial n}\right)}_{\tan \alpha} = \frac{1}{c}\tan\alpha = \frac{1}{c}\frac{w_1}{t} \qquad (9.139)$$

Introducing q as follows:

$$q = \tau t = \frac{w_1}{c} = \phi_1 \qquad (9.140)$$

Clearly, q has a unit of [Nm], and is called **shear flow**. It must be a constant because ϕ_1 is a constant. Equation (9.139) shows that the shear stress τ is inversely proportional to the thickness of the wall t, and hence the maximum shear stress is at the smallest t on the wall.

For a thin-wall member with perimeter segments $l_1, l_2, \ldots, l_n$, each of which with constant thickness $t_1, t_2, \ldots, t_n$, the corresponding shear stresses become: $\tau_1 = q/t_1, \tau_2 = q/t_2, \ldots, \tau_n = q/t_n$.

9.6.2 Formula for twist angle

We need to relate the twist angle with the shear stress. For this, we use the again the membrane analogy. Considering the equilibrium of force in the z-direction, we have

$$\sum F_z = 0 : pA - \oint S \tan \alpha \, dl = 0 \qquad (9.141)$$

where dl is the infinitely small length segment of the mean perimeter of the cross-section. Equation (9.141) is rewritten as

$$pA = S \oint \frac{w_1}{t} dl = 0$$

$$\frac{pA}{S} = w_1 \oint \frac{1}{t} dl$$

$$\frac{pA}{cS} = \frac{w_1}{c} \oint \frac{1}{t} dl \qquad (9.142)$$

$$\frac{pA}{cS} = \phi_1 \oint \frac{1}{t} dl$$

$$2GA\vartheta_t = q \oint \frac{1}{t} dl$$

We thus have

$$\boxed{\vartheta_t = \frac{q}{2GA} \oint \frac{1}{t} dl = \frac{1}{2GA} \oint \tau \, dl} \qquad (9.143)$$

This equation is the same as Eq. (9.51).

If t is piecewise constant, and there are n pieces, we have

$$\oint \frac{1}{t} dl = \frac{l_1}{t_1} + \frac{l_2}{t_2} + \cdots + \frac{l_n}{t_n} = \sum_i^n \frac{l_i}{t_i} \qquad (9.144)$$

The twist angle becomes

$$\vartheta_t = \frac{q}{2GA} \sum_i^n \frac{l_i}{t_i} \qquad (9.145)$$

Further, if the thickness of the wall is uniformly t, we have

$$\vartheta_t = \frac{ql}{2GAt} = \frac{\tau l}{2GA} \qquad (9.146)$$

where l is the length of mean perimeter of the cross-section.

9.6.3 Torque stress relation

Using now Eq. (9.73), the torque is twice the volume of the stress function. We obtain torque written in stress:

$$T = 2\int_A \phi\, dxdy = 2A\phi_1 = 2Aq = 2A\tau t \tag{9.147}$$

where A is the area enclosed by the mean perimeter of the cross-section. This is Eq. (9.42).

The twist angle can be written in torque:

$$\vartheta_t = \frac{\tau l}{2GA} = \frac{Tl}{4GtA^2} \tag{9.148}$$

9.6.4 Equivalent polar 2nd moment of area

This can be written in the conventional form $\vartheta_t = \frac{T}{GJ}$, if we write J as

$$J \underset{\text{general}}{=} \frac{4A^2}{\oint \frac{1}{t}dl} \underset{\text{uniform}}{=} \frac{4tA^2}{l} \tag{9.149}$$

This is Eq. (9.54).

Assuming the thin-wall cross-section is a thin hollow cylinder with uniform thickness t, Eq. (9.149) becomes

$$J = \frac{4tA^2}{l} \approx \frac{4t(\pi R)^2}{2\pi R} = 2\pi R^3 t \tag{9.150}$$

where R is the mean radius. This formula is Eq. (9.26) we obtained earlier. This means that Eq. (9.149) is the general form for computing the equivalent polar moment of area for closed thin-wall cross-sections.

If the thin-wall cross-section is a hollow rectangle with uniform thickness t, Eq. (9.149) becomes

$$J = \frac{4tA^2}{l} \underset{\text{rectangle}}{\approx} \frac{4t(hb)^2}{h+b} \underset{\text{square}}{=} 8b^3 t \tag{9.151}$$

where h is the mean half-height, and b is the mean half-width of the hollow rectangle.

9.6.5 Example: Bars with thin-wall circular cross-sections

Consider two bars of thin-wall circular cross-section. Bar-1's cross-section is closed, and bar-2's cross-section has a small cut, as shown in Fig. 9.25.

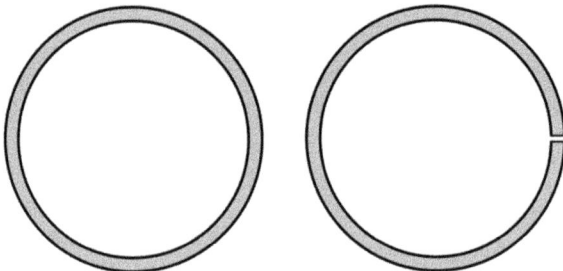

Figure 9.25. Two thin-wall circular cross-sections. One is closed (left), and one has a small cut (right).

Its dimensions are:

outer diameter is 22 mm; inner diameter is 18 mm.

1. Calculate the J values of bar-1 without assuming it is a thin-wall.
2. Calculate the J values of bar-1 assuming it is a thin-wall.
3. Calculate the J values of bar-2 assuming it is a narrow channel.
4. Assume the material shear stress is allowed at 150 MPa. Calculate the torque, based on the three assumptions in items 1, 2, and 3.
5. Assume the material shear stress is allowed at 150 MPa. Calculate the twist angle per unit length, based on the three assumptions in items 1, 2, and 3.

Solutions: We write the following code to solve and answer these questions.

```
1   d_outer = .022; d_inner = .018
2   t = (d_outer - d_inner)/2                              # thickness
3   r = d_outer/2                          # radius of the outer circle
4   R_mean = (d_outer + d_inner)/4
5   A_mean = np.pi*R_mean**2
6   l = 2*np.pi*R_mean
7
8   J_exact = np.pi/32*(d_outer**4-d_inner**4)       # Q1
9   J_thinW = 4*t*A_mean**2/l                        # Q2
10  k1 = k1_f(l/t, 21)
11  J_chnnl = k1*l*t**3                              # Q3
12
13  print(f"k1 = {k1} ")
14  print(f"J_exact = {J_exact} (mm^4)")
15  print(f"J_thinW = {J_thinW} (mm^4)")
16  print(f"J_chnnl = {J_chnnl} (mm^4)")
17  print(f"J_chnnl/J_thinW = {J_chnnl/J_thinW}")
```

```
k1 = 0.32664619022207575
J_exact = 1.2692034320502761e-08 (mm^4)
J_thinW = 1.2566370614359166e-08 (mm^4)
J_chnnl = 1.6419028344396277e-10 (mm^4)
J_chnnl/J_thinW = 0.013065847608883036
```

It is found that the torsional stiffness of the bar with a cut is only ∼1.3% of the closed one.

```
1  # Q4
2  τ_allow = 70.0e6              # Pa
3  T_exact = τ_allow * J_exact / r
4  T_thinW = τ_allow * J_thinW / R_mean      # or use: 2*A_mean*τ_allow*t
5  T_chnnl = τ_allow * J_chnnl / t
6
7  print(f"T_exact = {T_exact} (Nm)")
8  print(f"T_thinW = {T_thinW} (Nm)")
9  print(f"T_chnnl = {T_chnnl} (Nm)")
10 print(f"T_chnnl/T_thinW = {T_chnnl/T_thinW}")
```

```
T_exact = 80.76749113047212 (Nm)
T_thinW = 87.96459430051418 (Nm)
T_chnnl = 5.746659920538697 (Nm)
T_chnnl/T_thinW = 0.06532923804441518
```

It is found that the loading capacity of the bar with a cut is only ∼6.5% of the closed one.

```
1  # Q5
2  G = 77.5e9                    # Pa
3  ϑ_exact = T_exact / J_exact / G
4  ϑ_thinW = T_thinW / J_thinW / G
5  ϑ_chnnl = T_chnnl / J_chnnl / G
6
7  print(f"ϑ_exact = {ϑ_exact} (Nm)")
8  print(f"ϑ_thinW = {ϑ_thinW} (Nm)")
9  print(f"ϑ_chnnl = {ϑ_chnnl} (Nm)")
10 print(f"ϑ_chnnl/ϑ_thinW = {ϑ_chnnl/ϑ_thinW}")
```

```
ϑ_exact = 0.08211143695014664 (Nm)
ϑ_thinW = 0.0903225806451613 (Nm)
ϑ_chnnl = 0.45161290322580644 (Nm)
ϑ_chnnl/ϑ_thinW = 4.999999999999999
```

It is found that the twist angle of the bar with a cut is about 5 times the closed one, for the same allowable stress level.

In conclusion, a cut significantly reduces the loading capacity and the stiffness of a torsional bar with a thin-wall cross-section.

9.7 Remarks

This chapter presents theory, formulation, techniques, and codes for computing twist-angles, shear stress, and stiffness of bars with various cross-section shapes, subjected to torsion. It is clear that the shape of the cross-section makes differences.

The following chapter presents theory, formulation, techniques, and codes for computing various properties of different shapes of cross-sections, which are useful for stress analysis and design for bars subject to torsion and beams subject to bending.

References

[1] G.R. Liu and S.S. Quek, *The Finite Element Method: A Practical Course*. Butterworth-Heinemann, New York, 2013.
[2] G.R. Liu and T.T. Nguyen, *Smoothed Finite Element Methods*. Taylor and Francis Group, New York, 2010.
[3] G.R. Liu, *Mesh Free Methods: Moving Beyond the Finite Element Method*. Taylor and Francis Group, New York, 2010.
[4] A.P. Boresi and R.J. Schmidt, *Advanced Mechanics of Materials*. John Wiley & Sons, New York, 2002.
[5] J.M. Gere and S.P. Timoshenko, *Mechanics of Materials*. Van Nostrand Reinhold Company, New York, 1972.
[6] S. Timoshenko and J.N. Goodier, *Theory of Elasticity*. McGraw Hill, New York, 1970. http://books.google.com/books?id=yFISAAAAIAAJ&dq=theory+of+elasticity&ei=ICiKSsr3G4jwkQSbxMyPCg.
[7] Z.L. Xu, *Elasticity*, Vols. 1&2. People's Publisher, China, 1979.
[8] T.H.G. Megson, *Structural and Stress Analysis*, 1996. Available online https://books.google.com/books?id=qZykmEz4nzkC.
[9] G.R. Liu, *Vector Calculus: Formulation, Applications, and Python Codes*, World Scientific, New Jersey, 2025.

Chapter 10

Property of Areas

```
1  # Place cursor in this cell, and press Ctrl+Enter to import dependences.
2  import sys                          # for accessing the computer system
3  sys.path.append('../grbin/')   # Change to the directory in your system
4
5  from commonImports import *        # Import dependences from '../grbin/'
6  import grcodes as gr                # Import the module of the author
7  importlib.reload(gr)                # When grcodes is modified, reload it
8
9  from continuum_mechanics import vector
10 init_printing(use_unicode=True)     # For Latex-like quality printing
11 #np.set_printoptions(precision=6,suppress=True)  # Digits in print-outs
12 np.set_printoptions(precision=4,suppress=True,
13                     formatter={'float_kind': '{:.4e}'.format})
```

Area in general refers to any planar region that may have an arbitrary shape. In mechanics, many analyses involve areas of various shapes resulting from structural components. A typical case is the cross-section area of beams undergoing bending, bars under torsion, and bars undergoing tension. This chapter deals with issues on properties of areas, including area, the first moment of area, and the second moment of area, which are frequently used in mechanics.

10.1 Area of areas

10.1.1 *Integral formulas*

The most essential property of an area (planar region) is its area, often denoted as A. It measures the extent of the coverage of the area and carry a unit of [m²]. In our study, we discuss regions that cover a nonzero area, and hence A is always nonzero and positive.

Because an area can have different shapes, its measure needs to be done generally using integrals [1]:

$$A = \int_A dA \qquad (10.1)$$

The subscript A for the area integral stands for the planar region area. This means that we can cut the region into small pieces of area dA and then sum all these areas up leading to the area measure A. If a sheet of material is uniform in thickness and density, it measures the amount of material. Therefore, the area of an area is the physical property of the area. It does not depend on the coordinate system. Any coordinate system can be used to evaluate the area, and we often choose the one that is easy to work with.

If the shape of the area is regular, we would then have more explicit integrals, with the help of a coordinate system. For example, it can be a rectangle with base b and height h, as shown in Fig. 10.1.

We can use the 2D Cartesian coordinate system (x', y') with the x'-axis along its base b and the origin at its left-bottom corner. Equation (10.1) becomes

$$A \underset{\text{general}}{=} \int_A dA \underset{\text{rectangle}}{=} \int_0^h \left[\int_0^b dx' \right] dy' = b \int_0^h dy' = bh \qquad (10.2)$$

In the foregoing integral process, the area integral becomes a so-called double integral, which is integrated in two steps: one along the horizontal

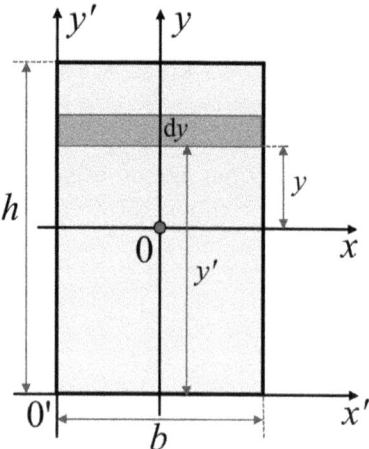

Figure 10.1. A rectangular area and the Cartesian coordinates set for evaluating its area. Two Cartesian coordinate systems are used.

axis and the other along the vertical axis. We can also use the 2D Cartesian coordinate system (x, y) with the origin passing through its center and the x-axis parallel to its base b. Equation (10.1) becomes

$$A = \int_A dA \Big|_{\text{general}} = \int_{-h/2}^{h/2} \left[\int_{-b/2}^{b/2} dx \right] dy \Big|_{\text{rectangle}} = b \int_{-h/2}^{h/2} dy = bh \quad (10.3)$$

The use of two different coordinates gives the same value. This coordinate independence feature of area evaluation is important to note. The code to get all these done is as follows:

```
1  x, y, xp, yp, b, h = sp.symbols("x, y, xp, yp, b, h")
2  Ap = sp.integrate(sp.integrate(1, (yp, 0, h)), (xp, 0, b))
3  A  = sp.integrate(sp.integrate(1, (xp, -b/2, b/2)), (yp, -h/2, h/2))
4  print(f"Area of rectangle, in (x',y') = {Ap};  in (x,y) = {A}")
```

Area of rectangle, in (x',y') = b*h; in (x,y) = b*h

The code can be modified for computing the area of more complicated shapes.

If the shape is **circular** with a radius of a, we can use the polar coordinate system (r, θ) with its origin at the center of the circle, and the integral becomes

$$A \Big|_{\text{circular}} = \int_0^a \left(\int_0^{2\pi} r d\theta \right) dr = 2\pi \int_0^a r dr = \pi a^2 \quad (10.4)$$

If the shape is totally irregular, we need to cut it into very small pieces without gaps and overlaps, called meshing, typically into rectangles and/or triangles, each of whose area can be computed with ease, and then sum all these areas up. This is essentially a numerical discrete evaluation of Eq. (10.1). Assume the area is then divided into n small areas ΔA_i. The total area becomes

$$A = \sum_{i=1}^{n} \Delta A_i \underset{\Delta A_i \to 0}{\overset{n \to \infty}{=}} \int_A dA \quad (10.5)$$

which is essentially Eq. (10.1) under the limits.

It is important to note that A is a global measure of an area. It does not give any information on the shape of the area. In other words, many areas with different shapes can have the same A value. It provides only 0th order of information on a given area.

10.1.2 Area of a composite area

Because of the summation form of the definition, the area calculation can be done for a complicated area consisting of multiple regular sub-areas for which the area can be found with ease. Assuming there are m regular sub-areas, the general formula is as follows:

$$A = \sum_{i=1}^{m} A_i \tag{10.6}$$

where A_i is the area of the ith sub-area. A list of formulas for calculating the areas of various relatively regular areas can be found at the Wikipedia page on area (https://en.wikipedia.org/wiki/Area).

10.2 First moment of area

10.2.1 *Definition*

The concept of area and its calculation are quite straightforward and taught in elementary schools. The concept, calculation, and property of the "first moment of area" are, however, much less trivial, and their use is quite confusing in the open literature. They are typically introduced in high schools.

The first moment of area is a measure of the spatial distribution of an area in relation to an axis. Therefore, it depends on the reference axis of a coordinate system used to evaluate it. It gives the first order of information about a given area. The shape of it matters. Its definition is given as follows.

Consider a given area A of arbitrary shape.

1. First, a Cartesian coordinate system (x, y) is introduced, and the coordinates for any point in the area can be found.
2. The area is then divided into n small areas ΔA_i. The coordinates of the center within ΔA_i are found and denoted as (x_i, y_i).

The first moments of area A are defined as

$$\begin{aligned} Q_x &= \sum_{i=1}^{n} \underbrace{y_i}_{\text{arm wrt } x} \Delta A_i \underset{\Delta A_i \to 0}{\overset{n \to \infty}{=}} \int_A y \, dA \\ Q_y &= \sum_{i=1}^{n} \underbrace{x_i}_{\text{arm wrt } y} \Delta A_i \underset{\Delta A_i \to 0}{\overset{n \to \infty}{=}} \int_A x \, dA \end{aligned} \tag{10.7}$$

The unit for the first moment of area is $[m^3]$.

We reiterate that the first moment of an area is with respect to (wrt) the reference axis in a coordinate system. **A coordinate transformation, translational or rotational, will affect its values.** The first moment of an area has two components: Q_x and Q_y, and it needs one index. **It is a**

10.2.2 First moment of a composite area

Because of the summation form of the definition, the first moment of area calculation can be done for a complicated area consisting of multiple regular sub-areas (for which the center and area can be found with ease). Assuming there are m regular sub-areas, the general formulas are as follows:

$$Q_x = \sum_{i=1}^{m} \bar{y}_i A_i$$

$$Q_y = \sum_{i=1}^{m} \bar{x}_i A_i$$
(10.8)

where A_i is the area of the ith sub-area and x_i, y_i are coordinates of the center of the ith sub-area with respect to the coordinate system (x, y).

10.2.3 Example: The first moment of area of a rectangle

Consider a rectangle with base b and height h, as shown in Fig. 10.2. We use two Cartesian coordinate systems: (x', y') that pass through its two edges, and (x, y) that pass through its two middle lines and the center of the rectangle.

1. Calculate the first moment of area of the rectangle with respect to x'- and y'-axis.
2. Calculate the same with respect to the x- and y-axis.

Solution to 1: Using the coordinate system (x', y') and Eq. (10.7), we obtain

$$Q_{x'} = \int_A y' \, dA = \int_0^b dx' \int_0^h y' \, dy' = b \int_0^h y' \, dy' = \frac{bh^2}{2}$$

$$Q_{y'} = \int_A x' \, dA = \int_0^h dy' \int_0^b x' \, dx' = h \int_0^b x' \, dx' = \frac{hb^2}{2}$$
(10.9)

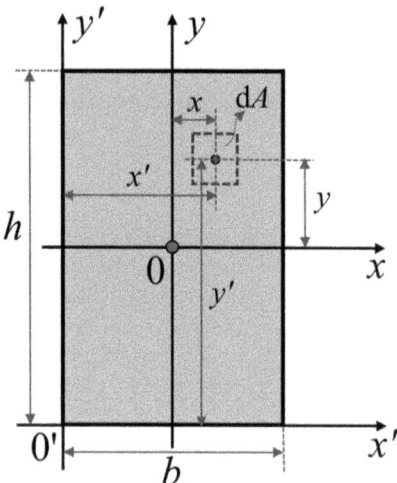

Figure 10.2. A rectangular area and the Cartesian coordinates set for evaluating its first moment of area. Two Cartesian coordinate systems are used.

We see clearly that the shape, the values of a and b, of the rectangle will affect the values of the first moments. The **shape does matter** for the first moments of area.

Solution to 2: Using the coordinate system (x, y) and Eq. (10.7), we obtain

$$Q_x = \int_A y\, dA = \int_{-b/2}^{b/2} dx \int_{-h/2}^{h/2} y\, dy = b \int_{-h/2}^{h/2} y\, dy = 0$$

$$Q_y = \int_A x\, dA = \int_{-h/2}^{h/2} dy \int_{-b/2}^{b/2} x\, dx = h \int_{-b/2}^{b/2} x\, dx = 0$$

(10.10)

The code to get all these done is as follows:

```
1  x, y, xp, yp, b, h = sp.symbols("x, y, xp, yp, b, h")
2  Q_xp = sp.integrate(sp.integrate(yp, (yp, 0, h)), (xp, 0, b))
3  Q_yp = sp.integrate(sp.integrate(xp, (xp, 0, b)), (yp, 0, h))
4  Qp = sp.Matrix([Q_xp, Q_yp])
5  gr.printM(Qp, "1st moments of area of rectangle: Q_x', Q_y':")
```

1st moments of area of rectangle: Q_x', Q_y':

$$\begin{bmatrix} \frac{bh^2}{2} \\ \frac{b^2 h}{2} \end{bmatrix}$$

```
1  Q_x = sp.integrate(sp.integrate(y, (y, -h/2, h/2)), (x, -b/2, b/2))
2  Q_y = sp.integrate(sp.integrate(x, (x, -b/2, b/2)), (y, -h/2, h/2))
3  Q = sp.Matrix([Q_x, Q_y])
4  gr.printM(Q, "1st moments of area of rectangle: Q_x, Q_y:")
```

1st moments of area of rectangle: Q_x, Q_y:

$$\begin{bmatrix} 0 \\ 0 \end{bmatrix}$$

We find the following:

1. The moments of an area are coordinate dependent.
2. When the reference axis passes through the center of the area, the first moment of the area becomes zero. This leads to the definition of the concept of a centroid.

10.2.4 Centroid of an area

The centroid of an area, denoted as $(\bar{x}, \bar{y})$, is defined based on the first moment of area. The formula is as follows:

$$\bar{x} = \frac{\int_A x \, dA}{\int_A dA}$$

$$\bar{y} = \frac{\int_A y \, dA}{\int_A dA}$$
(10.11)

From this definition and because the area A is not zero (we are not interested in discussing any area that is zero), it is clear that

the centroid is zero if and only if the first moments are zero.

This means that if the coordinates are set at the centroid of the area, the first moment vanishes.

For the rectangular area given in the previous section, in the (x', y') coordinate system, the centroid is found as

$$\bar{x}' = \frac{\int_A x' \, dA}{\int_A dA} = \frac{\frac{b^2 h}{2}}{bh} = \frac{b}{2}$$

$$\bar{y}' = \frac{\int_A y' \, dA}{\int_A dA} = \frac{\frac{bh^2}{2}}{bh} = \frac{h}{2}$$
(10.12)

Therefore, the centroid is the coordinates of the center of the area in the given reference coordinate system. Using the (x, y) coordinate system, the centroid is then found as

$$\bar{x} = \frac{\int_A x\, dA}{\int_A dA} = \frac{0}{bh} = 0$$

$$\bar{y} = \frac{\int_A y\, dA}{\int_A dA} = \frac{0}{bh} = 0$$
(10.13)

This confirms that the center of the area is right at the origin of the (x, y) coordinate system.

In many applications, it is often convenient to set our reference coordinate system with the origin at the center of the area of interest. In this case, the first moments become zero, and the formulation can be significantly simplified. However, we do not usually know the location of the center for a given arbitrary area. This is what we can do:

> First, set up any coordinate system as a reference, and use Eq. (10.11) to find the centroid. Then, move the coordinate system to have its origin at the centroid.

Equation (10.11) is thus useful in many applications involving areas.

10.2.5 Centroid for composite areas

Consider a composite area with m regular sub-areas. Its centroid can be computed using

$$\bar{x} = \frac{\sum_{i=1}^{m} \bar{x}_i A_i}{\sum_{i=1}^{m} A_i}$$

$$\bar{y} = \frac{\sum_{i=1}^{m} \bar{y}_i A_i}{\sum_{i=1}^{m} A_i}$$
(10.14)

where A_i is the area of the ith sub-area, and $\bar{x}_i$ and $\bar{y}_i$ denote the centroid of the ith sub-area with respect to the common coordinate system (x, y).

10.2.6 Example: Centroid of an I-beam

Consider an I-beam with a basic shape, as shown in Fig. 10.3.

Its dimensions (mm) are given as follows:

> beam height: $h_b = 200$; width of flange on top: $w_{ft} = 100$; width of flange at bottom: $w_{fb} = 150$; thickness of flange on top: $t_{ft} = 20$; thickness of flange at bottom: $t_{fb} = 18$; web thickness: $t_w = 10$.

Determine the centroid of the I-beam.

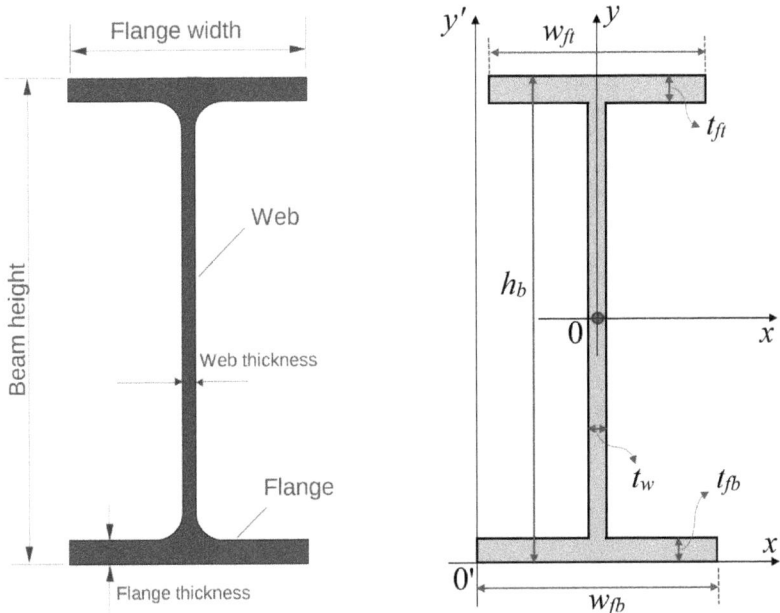

Figure 10.3. Schematic view of an I-beam. Left: generic cross-section with rounded concave corners to avoid stress concentration. Right: idealized cross-section for analysis. The image on the left is from Poutrelle metallique at Wikimedia Commons under the CC BY 3.0 license.

Solution: We write the following code to get it done using Eq. (10.14). First, we write a code function for computing the centroid of an I-beam, which will be used multiple times in this book.

```
def centroid_I(hb, wft, wfb, tft, tfb, tw):
    '''Compute the centroid of an I-beam made of rectangles.
    The reference coordinate is on the bottom edge with the left-bottom
    corner as the origin. The code can be used for any composite area
    with up to 3 rectangles. For a T beam, for example, just set tfb=0.
    The whole area must be left-right symmetric.
    Inputs:
    hb:  beam height (total)
    wft: width of flange on top
    wfb: width of flange at bottom
    tft: thickness of flange on top
    tfb: thickness of flange at bottom
    tw:  web thickness
    Return:
    x_bar, y_bar: the centroid of the composite area.
```

```
16          A: The total area
17          Ai: areas of each rectangles
18          '''
19          width_ai  = np.array([wft, tw,              wfb])
20          height_bi = np.array([tft, hb-tft-tfb, tfb])
21          xi_bar    = np.array([wfb/2,  wfb/2,   wfb/2])
22          yi_bar    = np.array([hb-tft/2, (hb-tft-tfb)/2+tfb, tfb/2])
23
24          Ai = width_ai*height_bi
25          A = np.sum(Ai)
26          print(f"Areas of sub-areas = {Ai} [L^2]")
27          print(f"Areas of I-beam = {A} [L^2]")
28
29          x_bar = np.sum(xi_bar*Ai)/A
30          y_bar = np.sum(yi_bar*Ai)/A
31
32          print(f"The centroid [L]):\n x = {x_bar} \n y = {y_bar}")
33          return x_bar, y_bar, A, Ai
```

Using the code function given above, the centroid can be found with ease.

```
1 # preparing data:
2 hb =.200; wft=.100; wfb=.150; tft=.020; tfb=.018; tw=.010    # (m)
3
4 # call the code function:
5 x_bar, y_bar, A, Ai = centroid_I(hb, wft, wfb, tft, tfb, tw)
```

Areas of sub-areas = [2.0000e-03 1.6200e-03 2.7000e-03] [L^2]
Areas of I-beam = 0.00632 [L^2]
The centroid [L]):
x = 0.075
y = 0.0893481012658228

10.2.7 Example: Compute the first moment of an area, using centroid

Consider again a rectangle with base b and height h shown in Fig. 10.2. Because of its simple geometry, we know its centroid based on a Cartesian coordinate system. For example, when we use (x', y') as the reference coordinate system, its centroid is at $\bar{x}' = b/2, \bar{y}' = h/2$. Calculate the first moment of the rectangle using the known centroid.

Solution: Using Eq. (10.11), we have

$$Q_{x'} = \bar{y}' \int_A dA = \bar{y}' A$$

$$Q_{y'} = \bar{x}' \int_A dA = \bar{x}' A$$

(10.15)

Substituting $\bar{x}' = a/2$, $\bar{y}' = b/2$, and $A = ab$ for a rectangle into Eq. (10.15), we obtain

$$Q_{x'} = \bar{y}'A = \frac{h}{2}bh = \frac{bh^2}{2}$$

$$Q_{y'} = \bar{x}'A = \frac{b}{2}bh = \frac{hb^2}{2}$$
(10.16)

which is the same as the results given in Eq. (10.9).

Equation (10.15) is useful for regular areas for which the centroid and area are readily available.

10.2.8 *First moment of area, coordinate rotation*

Since the first moment of an area has two components: Q_x and Q_y with one index, the coordinate rotational transformation rule follows that for vectors discussed in Chapter 2. Consider a new Cartesian coordinate system (x', y') that is rotated by an angle θ from the current coordinate system (x, y). Assume the first moments of area are known in (x, y). We would like to find the first moments of the same area with respect to the rotated (x', y'). Using the transformation matrix defined in Chapter 2, the new coordinates are given by

$$\begin{bmatrix} x' \\ y' \end{bmatrix} = \underbrace{\begin{bmatrix} \cos\theta & \sin\theta \\ -\sin\theta & \cos\theta \end{bmatrix}}_{\mathbf{T}} \begin{bmatrix} x \\ y \end{bmatrix}$$
(10.17)

The first moments of area in (x', y') becomes

$$Q_{x'} = \int_A y' \, dA = \int_A (-x\sin\theta + y\cos\theta) dA = -Q_y \sin\theta + Q_x \cos\theta$$

$$Q_{y'} = \int_A x' \, dA = \int_A (x\cos\theta + y\sin\theta) dA = Q_y \cos\theta + Q_x \sin\theta$$
(10.18)

In matrix form, they become

$$\begin{bmatrix} Q_{x'} \\ Q_{y'} \end{bmatrix} = \underbrace{\begin{bmatrix} \cos\theta & -\sin\theta \\ \sin\theta & \cos\theta \end{bmatrix}}_{\mathbf{T}^\top} \begin{bmatrix} Q_x \\ Q_y \end{bmatrix}$$
(10.19)

This means that the transformed first moments use the transpose of the transformation matrix $\mathbf{T}$. If both Q_x and Q_y are zero, both $Q_{x'}$ and $Q_{y'}$ will also be zero. This always remains true because if the origin of the original coordinate is at the centroid, any rotated coordinate will have the same origin.

10.3 Second moment of area

10.3.1 *Definition*

One can increase the order of the integrand by one when performing area integral. This constructs the second moments of an area. The formula is as follows:

$$I_{xx} = \int_A y^2 \, dA$$

$$I_{yy} = \int_A x^2 \, dA \qquad (10.20)$$

$$I_{xy} = \int_A xy \, dA$$

It has three components and carries two indexes. The first moment of an area extracts the first order of information from the area. One can naturally expect that a second moment of an area shall extract second-order information from it. We need to find a way to calculate it.

10.3.2 *Second moment of area for regular areas*

For areas with regular shape, we can derive formulas, and the calculation can be done analytically. Consider again the rectangular area shown in Fig. 10.2. We use the Cartesian coordinate system (x, y), and let the x- and y-axis be on the middle lines of the rectangle. We have

$$I_{xx} = \int_A y^2 \, dA = \int_{-b/2}^{b/2} dx \int_{-h/2}^{h/2} y^2 \, dy = b \int_{-h/2}^{h/2} y^2 \, dy = \frac{bh^3}{12}$$

$$I_{yy} = \int_A x^2 \, dA = \int_{-h/2}^{h/2} dy \int_{-b/2}^{b/2} x^2 \, dy = h \int_{-b/2}^{b/2} x^2 \, dy = \frac{hb^3}{12} \qquad (10.21)$$

$$I_{xy} = \int_A xy \, dA = \int_{-b/2}^{b/2} x \, dx \int_{-h/2}^{h/2} y \, dy = 0$$

Here, we see the benefit of having the coordinate going through the centroid of the area. Using a similar integration technique, the second moment of area for other regular shapes can be found. The code to get this done is as follows:

```
1  Ixx = sp.integrate(sp.integrate(y**2, (y, -h/2, h/2)), (x, -b/2, b/2))
2  Iyy = sp.integrate(sp.integrate(x**2, (x, -b/2, b/2)), (y, -h/2, h/2))
3  Ixy = sp.integrate(sp.integrate( x*y, (y, -h/2, h/2)), (x, -b/2, b/2))
4  I = sp.Matrix([[Ixx, -Ixy], [-Ixy, Iyy]])
5  gr.printM(I, "2nd moments of area of rectangle:")
```

2nd moments of area of rectangle:

$$\begin{bmatrix} \frac{bh^3}{12} & 0 \\ 0 & \frac{b^3h}{12} \end{bmatrix}$$

We found in the rectangle case, $I_{xy} = 0$. This is because the area is symmetric. For a general shape, it is usually not.

Consider next a box area that is symmetric in both directions, shown in Fig. 10.4. We use the Cartesian coordinate system (x, y) with its origin at the centroid of the box.

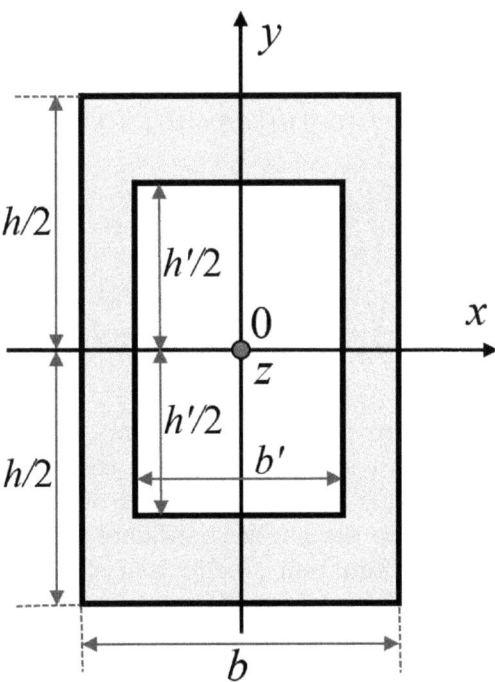

Figure 10.4. A rectangular box.

Because of the additive property of integral operations, the second moment of the box area can be simply obtained using

$$I_{xx} = \int_A y^2 \, dA - \int_{A'} y^2 \, dA = \frac{1}{12}(bh^3 - b'h'^3)$$

$$I_{yy} = \int_A x^2 \, dA - \int_{A'} x^2 \, dA = \frac{1}{12}(hb^3 - h'b'^3)$$

(10.22)

Similar techniques can be used to calculate the I values of areas of various shapes. A list of second moments of area for various regular shapes can be found at the wikipage (https://en.wikipedia.org/wiki/List_of_second_moments_of_area).

10.3.3 Second moment of area, coordinate translation

Consider a new Cartesian coordinate system (x', y') that is created by a translation of (x, y), and we found the centroid of the area as $\bar{x}$ and $\bar{y}$ in reference to (x', y'). The second moment of the area in reference to (x', y') can be found as

$$I_{x'x'} = \int_A y'^2 \, dA = \int_A (y + \bar{y})^2 \, dA = I_{xx} + \bar{y}'^2 A$$

$$I_{y'y'} = \int_A x'^2 \, dA = \int_A (x + \bar{x})^2 \, dA = I_{yy} + \bar{x}'^2 A$$

(10.23)

$$I_{x'y'} = \int_A y'^2 \, dA = \int_A (x + \bar{x})(y + \bar{y}) \, dA = I_{xy} + \bar{x}'\bar{y}' A$$

where I_{xx}, I_{yy}, and I_{xy} are the second moments of area with respect to coordinates (x, y) that are passing through the centroid. In the derivation above, we used the fact that the first moments of area with respect to x and y are all zero, leading to the simple coordinate translation rule for these second moments of area. Equation (10.23) is useful when calculating the second moment of a composite area consisting of multiple regular sub-areas for which their second moments of area and centroids are readily available.

10.3.4 Second moments of composite areas

Consider a composite area with m regular sub-areas, its second moments of area can be computed using

$$I_{x'x'} = \sum_{k=1}^{m}(I_{xx,k} + \bar{y'}_k^{\,2} A_k)$$

$$I_{y'y'} = \sum_{k=1}^{m}(I_{yy,k} + \bar{x'}_k^{\,2} A_k) \qquad (10.24)$$

$$I_{x'y'} = \sum_{k=1}^{m}(I_{xy,k} + \bar{x'}_k \bar{y'}_k A_k)$$

where $I_{ij,k}(i=x,y;j=x,y)$ is the second moments of the kth sub-area with respect to its own coordinates (x_k, y_k) that go through its own center. And $\bar{x'}_k$ and $\bar{y'}_k$ are the centroid of the kth sub-area with respect to the common coordinate system (x', y') for the whole area.

Note that in practice, we often set the common coordinate system (x', y') going through the centroid of the whole area, so that the obtained second moments are in reference to the center of the whole area. This can be easily done by shifting the coordinates, after the global centroid is found, using for example centroid_I(). We will demonstrate this in the example section.

10.3.5 Example: Second moments of an I-beam

Consider an I-beam with a basic shape shown in Fig. 10.3.

Its dimensions (mm) are given as follows:

beam height: $h_b = 200$; width of flange on top: $w_{ft} = 100$; width of flange at bottom: $w_{fb} = 150$; ; thickness of flange on top: $t_{ft} = 20$; thickness of flange at bottom: $t_{fb} = 18$; web thickness: $t_w = 10$.

Determine the second moments of the I-beam.

Solution: We write a code to get it done using Eq. (10.24). First, we write a Python function for computing the second moments of an I-beam using the centroids of sub-areas.

```python
def secondMoments_I(hb, wft, wfb, tft, tfb, tw, cxi, cyi, sy=True):
    '''Compute the second moments of I-beam cross-sections made of
    rectangles. The reference coordinate is on the bottom edge with
    the left-bottom corner as the origin and x-axis on the bottom edge.
    The code can be used for any composite area
    with up to 3 rectangles. For a T beam, for example, just set tfb=0.
    Inputs:
    hb:  beam height (total)
    wft: width of flange on top
    wfb: width of flange at bottom
    tft: thickness of flange on top
    tfb: thickness of flange at bottom
    tw:  web thickness
    cxi, cyi: centers of each sub-area wrt the common coordinates.
    sy:  if True (I-, T-beam), cxi, cyi are not used, computed here in.
         Otherwise, the provided cxi, cyi will be used here.
    Return:
    Ixx, Iyy, Ixy: 2nd moments
    x_bar, y_bar: the centroid
    A: The total area
    Ai: areas of each rectangles
    '''
    width_ai  = np.array([wft, tw,          wfb])
    height_bi = np.array([tft, hb-tft-tfb, tfb])

    # when whole area is symmetric with respect to y-axis:
    if sy:   #compute the centers (cxi, cyi) for each sub-area.
        cxi = np.array([wfb/2,  wfb/2,  wfb/2])
        cyi = np.array([hb-tft/2, (hb-tft-tfb)/2+tfb, tfb/2])

    print(f"Center of sub-areas [L]:\n x = {cxi} \n y = {cyi}")

    Ai = width_ai*height_bi
    A = np.sum(Ai)
    print(f"Areas of sub-areas = {Ai} [L^2]")
    print(f"Areas of I-beam = {A} [L^2]")

    x_bar = np.sum(cxi*Ai)/A
    y_bar = np.sum(cyi*Ai)/A

    print(f"The centroid = ({x_bar}, {y_bar}) [L]")

    # shifts the coordinates to the centroid of the whole area:
    xi_bar = cxi - x_bar
    yi_bar = cyi - y_bar

    print(f"New center of sub-areas [L]:\n x = {xi_bar}\n y = {yi_bar}")

    Ixx = np.sum(width_ai*height_bi**3/12 + yi_bar**2 * Ai)
    Iyy = np.sum(height_bi*width_ai**3/12 + xi_bar**2 * Ai)
    Ixy = np.sum(xi_bar*yi_bar * Ai)

    print(f"2nd moments [L^4]:\n Ixx = {Ixx}\n Iyy = {Iyy}\n Ixy = {Ixy}")

    return Ixx, Iyy, Ixy, x_bar, y_bar, xi_bar, yi_bar, A, Ai
```

```
1  # preparing data:
2  #hb =.060; wft=.050; wfb=.050; tft=.01; tfb=.0; tw=.01    # T-beam (m)
3  hb =.200; wft=.100; wfb=.150; tft=.02; tfb=.02; tw=.01    # I-beam (m)
4
5  # call the code function for 2nd moment calculation:
6  I_xx, I_yy, I_xy, x_bar, y_bar, xi_bar, yi_bar, A, Ai =\
7        secondMoments_I(hb, wft, wfb, tft, tfb, tw, [0]*2, [0]*2, sy=True)
```

```
Center of sub-areas [L]:
x = [7.5000e-02 7.5000e-02 7.5000e-02]
y = [1.9000e-01 1.0000e-01 1.0000e-02]
Areas of sub-areas = [2.0000e-03 1.6000e-03 3.0000e-03] [L^2]
Areas of I-beam = 0.0066 [L^2]
The centroid = (0.075, 0.08636363636363638) [L]
New center of sub-areas [L]:
x = [0.0000e+00 0.0000e+00 0.0000e+00]
y = [1.0364e-01 1.3636e-02 -7.6364e-02]
2nd moments [L^4]:
Ixx = 4.2852727272727287e-05
Iyy = 7.304999999999999e-06
Ixy = 0.0
```

Because the cross-section of the I-beam is symmetric with respect to the y-axis, the product moment of area I_{xy} found is zero. If the cross-section of the beam is asymmetric, I_{xy} will not be zero. The flowing code computes the second moments of an S-beam without left-right symmetry.

10.3.6 Example: Second moments of an S-beam

Consider an S-beam with a basic shape shown in Fig. 10.5.

Its dimensions (mm) are given as follows:

beam height: $h_b = 200$; width of flange on top: $w_{ft} = 80$; width of flange at bottom: $w_{fb} = 80$; thickness of flange on top: $t_{ft} = 15$; thickness of flange at bottom: $t_{fb} = 15$; web thickness: $t_w = 10$.

Determine the centroid and the second moments of the S-beam.

Solution: Because the cross-section is asymmetric, we need to compute the centroids of these three rectangles of the S-beam, and then feed these to secondMoments_I() to compute the second moments of areas. The code is as follows:

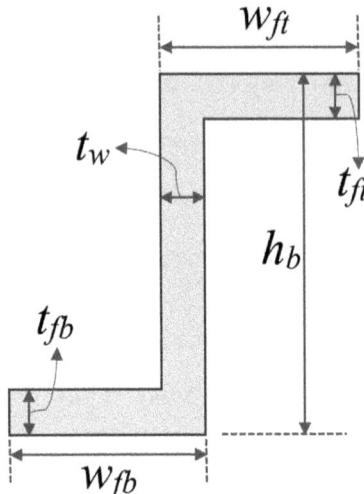

Figure 10.5. Schematic view of an S-beam.

```
1  # preparing data:
2  # hb =.2; wft=.1; wfb=.1; tft=.02; tfb=.02; tw=.02          # S-beam
3  hb =.2; wft=.08; wfb=.08; tft=.015; tfb=.015; tw=.01        # S-beam
4
5  # whole area is asymmetric with respect to y-axis, compute the centroid:
6  cxi = np.array([(wfb-tw)+wft/2, wfb-tw/2, wfb/2]) # from top to bottom
7  cyi = np.array([hb-tft/2, (hb-tft-tfb)/2+tfb, tfb/2])
8
9  # call the code function:
10 I_xx, I_yy, I_xy, x_bar, y_bar, xi_bar, yi_bar, A, Ai =\
11         secondMoments_I(hb, wft, wfb, tft, tfb, tw, cxi, cyi, sy=False)
```

Center of sub-areas [L]:
 x = [1.1000e-01 7.5000e-02 4.0000e-02]
 y = [1.9250e-01 1.0000e-01 7.5000e-03]
Areas of sub-areas = [1.2000e-03 1.7000e-03 1.2000e-03] [L^2]
Areas of I-beam = 0.0040999999999999995 [L^2]
The centroid = (0.075, 0.1) [L]
New center of sub-areas [L]:
 x = [3.5000e-02 0.0000e+00 -3.5000e-02]
 y = [9.2500e-02 -1.3878e-17 -9.2500e-02]
2nd moments [L^4]:
 Ixx = 2.467416666666666e-05
 Iyy = 4.2341666666666675e-06
 Ixy = 7.770000000000002e-06

Indeed, the product moment of area of this S-beam is nonzero.

10.3.7 Second moment of area, coordinate rotation

Consider a new Cartesian coordinate system $\mathbf{x}' = (x', y')$ that is rotated by an angle θ from the current coordinate $\mathbf{x} = (x, y)$ under which all the second moments of area are known. Suppose (x, y) pass through the centroid of the whole area. We would like to find the second moment of the same area with respect to the rotated (x', y'). The formula can be derived using the coordinate transformation rule for the coordinate vector discussed in Chapter 2. The new coordinates are given by Eq. (10.17). Substituting them into Eq. (10.20) gives

$$I_{x'x'} = \int_A y'^2 \, dA = \int_A (-x \sin\theta + y \cos\theta)^2 \, dA$$
$$= I_{xx} \cos^2\theta + I_{yy} \sin^2\theta - 2I_{xy} \sin\theta \cos\theta$$

$$I_{y'y'} = \int_A x'^2 \, dA = \int_A (x \cos\theta + y \sin\theta)^2 \, dA \qquad (10.25)$$
$$= I_{xx} \sin^2\theta + I_{yy} \cos^2\theta + 2I_{xy} \sin\theta \cos\theta$$

$$I_{x'y'} = \int_A x'y' \, dA = \int_A (-x \sin\theta + y \cos\theta)(x \cos\theta + y \sin\theta) \, dA$$
$$= (I_{xx} - I_{yy}) \sin\theta \cos\theta + I_{xy}(\cos^2\theta - \sin^2\theta)$$

where I_{xx}, I_{yy}, and I_{xy} are the second moments of area with respect to coordinates (x, y) that are passing through the centroid. Equation (10.25) is useful when calculating the second moment of an area when coordinates rotate.

Comparing Eq. (10.25) with the expression for 2D stresses transformation given in Chapter 3, we found that these are exactly the same if I_{xx} corresponds to σ_{xx}, I_{yy} corresponds to σ_{yy}, and $-I_{xy}$ corresponds to σ_{xy}. In other words, if we form a 2D tensor,

$$\begin{bmatrix} I_{xx} & -I_{xy} \\ -I_{xy} & I_{yy} \end{bmatrix} \qquad (10.26)$$

The same coordinate transformation (rotation) rule apply. Therefore, the double angle formulas for stress transformation given in Eq. (3.74) can also be used, with the sign flip for I_{xy} and $I_{x'y'}$, which gives

$$I_{x'x'} = \frac{1}{2}(I_{xx} + I_{yy}) + \frac{1}{2}(I_{xx} - I_{yy}) \cos 2\theta - I_{xy} \sin 2\theta$$

$$I_{y'y'} = \frac{1}{2}(I_{xx} + I_{yy}) - \frac{1}{2}(I_{xx} - I_{yy}) \cos 2\theta + I_{xy} \sin 2\theta \qquad (10.27)$$

$$I_{x'y'} = \frac{1}{2}(I_{xx} - I_{yy}) \sin 2\theta + I_{xy} \cos 2\theta$$

Because of the double angle formula, it is clear that if the rotation of the coordinates is 180°, there is no effect on all the second moments. For example, the x-right, y-up coordinate pair is the same as the x-left and y-down coordinate pair.

10.3.8 Principal second moments of area

As we did for stress tensor, we can rotate the coordinate to an angle at which $I_{x'y'}$ vanishes. Such an angle is found from the third equation in Eq. (10.27), which gives

$$\tan 2\theta = -\frac{2I_{xy}}{I_{xx} - I_{yy}} \tag{10.28}$$

The axis with $I_{x'y'} = 0$ is called the **principal axis**. The corresponding second moments of area are called the **principal moments of area**. This double angle formula is convenient for calculating the rotation angle needed to arrive at the principal axis. A list of principal moments of area for various regular shapes can be found at the wikipage (https://en.wikipedia.org/wiki/List_of_second_moments_of_area).

10.3.9 Example: Second moments of area after coordinate rotation

Using the second moments of area obtained in the previous example, compute the new second moments of area when the coordinates are rotated by 30°.

Solution: We write first the following code function to do this work, which can be used for future works.

```
def Ixx2IXX(Ixx, Iyy, Ixy, θ):
    '''Coordinate transformation for second moments of area.
    Ixx, Iyy, Ixy: for original (x,y)
    IXX, IYY, IXY: for (X,Y) that is θ-rotated from (x,y)
    '''
    c = np.cos(θ); s = np.sin(θ)
    c2= c**2;      s2= s**2;   cs = c*s
    IXX = Ixx*c2 + Iyy*s2 - 2*Ixy*cs
    IYY = Ixx*s2 + Iyy*c2 + 2*Ixy*cs
    IXY = (Ixx-Iyy)*cs + Ixy*(c2 - s2)

    return IXX, IYY, IXY
```

Find the new second moments of area using Ixx2IXX().

```
1  #Ixx = 3.936e-05; Iyy = 9.840e-06; Ixy = 1.440e-06      # (m^4)
2  Ixx = I_xx
3  Iyy = I_yy
4  Ixy = I_xy
5
6  θ = 30.0
7  IXX, IYY, IXY = Ixx2IXX(Ixx, Iyy, Ixy, np.deg2rad(θ))
8  print(f"2nd moment of area wrt the principal axis (m^4):\n"
9        f"IXX= {IXX:.4e}, IYY= {IYY:.4e},IXY= {IXY:.4e}")
```

2nd moment of area wrt the principal axis (m^4):
IXX= 1.2835e-05, IYY= 1.6073e-05,IXY= 1.2736e-05

10.3.10 Example: The principal second moments of area

Using the second moments of area obtained in the previous example, compute the principal second moments of area and their direction.

Solution:

```
1  # Compute the angle θ to principal axis for the 2nd moment:
2  #Ixx = 3.936e-05; Iyy = 9.840e-06; Ixy = 1.440e-06      # (m^4)
3
4  tan2θ = -2*Ixy/(Ixx-Iyy)
5  θ = np.arctan(tan2θ)/2
6  print(f"Angle to the principal axis θ = {θ} (rad)")
7  print(f"Angle to the principal axis θ = {θ*90/np.pi}°")
```

Angle to the principal axis θ = -0.3250220453336827 (rad)
Angle to the principal axis θ = -9.311195723164865

```
1  # Compute the principal 2nd moment of area:
2
3  IXX, IYY, IXY = Ixx2IXX(Ixx, Iyy, Ixy, θ)
4  print(f"2nd moment of area wrt the principal axis (m^4):\n"
5        f"IXX= {IXX:.4e}, IYY= {IYY:.4e},IXY= {IXY:.4e}")
```

2nd moment of area wrt the principal axis (m^4):
IXX= 2.7292e-05, IYY= 1.6159e-06,IXY= 0.0000e+00

10.3.11 Use of eigenvalue solver

Although not recommended, we can also use the eigenvalue solver to find the principal axis for the second moments. However, we need some extra work to determine the sign of the angle of the direction cosines. The details are as follows.

First, form the I tensor, and invoke an eigenvalue solver.

```
1  #Ixx = 3.936e-05; Iyy = 9.840e-06; Ixy = 1.440e-06           # (m^4)
2
3  I = np.array([[ Ixx, -Ixy],
4                [-Ixy,  Iyy]])
5  e, v = lg.eig(I)
6  print(f"Eigenvalues of I = {e}")       # This gives principal IXX and TYY
7
8  print(f"Eigenvectors of I: \n{v}")     # This gives the direction cosines
9  print(f"Principal axis (arccos(x,x') =±{np.arccos(v[0,0])} (rad)")
10 print(f"Principal axis (arccos(x,y') =±{np.arccos(v[1,0])} (rad)")
```

```
Eigenvalues of I = [2.7292e-05 1.6159e-06]
Eigenvectors of I:
[[ 9.4764e-01  3.1933e-01]
 [-3.1933e-01  9.4764e-01]]
Principal axis (arccos(x,x') =±0.3250220453336826 (rad)
Principal axis (arccos(x,y') =±1.8958183721285793 (rad)
```

We obtained the same values for the principal second moment of area. Because the cosine value can correspond to both positive and negative angles, we need to determine the sign of the angle between x and x'. This can be done by computing I_{XX}, I_{YY}, and I_{XY} (or just I_{XY}) from the positive angle and negative angle using our code function Ixx2IXX(). We should choose the one that gives $I_{XY} = 0$. The code is as follows:

```
1  n1 = np.arccos(v[0,0])    # arccos(x,x')
2
3  IXX, IYY, IXY = Ixx2IXX(Ixx, Iyy, Ixy, n1)           # take positive
4  print(f"2nd moment of area wrt the principal axis (m^4):\n"
5        f"IXX= {IXX:.4e}, IYY= {IYY:.4e},IXY= {IXY:.4e}")
6
7  IXX, IYY, IXY = Ixx2IXX(Ixx, Iyy, Ixy, -n1)          # take negative
8  print(f"\n2nd moment of area wrt the principal axis (m^4):\n"
9        f"IXX= {IXX:.4e}, IYY= {IYY:.4e},IXY= {IXY:.4e}")
```

```
2nd moment of area wrt the principal axis (m^4):
IXX= 1.7887e-05, IYY= 1.1021e-05,IXY= 1.2371e-05

2nd moment of area wrt the principal axis (m^4):
IXX= 2.7292e-05, IYY= 1.6159e-06,IXY= 8.4703e-22
```

We shall then choose the negative sign. We can compare the results with that obtained using Eq. (10.28).

```
1  print(f"Principal axis (arccos(x,x') ={np.arccos(v[0,0])} (rad)")
2  print(f"Principal axis (arccos(x,x') ={np.arccos(v[0,0])*180/np.pi}°")
3  print(f"θ1 to the principal axis={np.arctan(-2*Ixy/(Ixx-Iyy))/2}(rad)")
```

```
Principal axis (arccos(x,x') =-0.3250220453336826 (rad)
Principal axis (arccos(x,x') =-18.622391446329726°
θ1 to the principal axis=-0.3250220453336827(rad)
```

We obtained the same results.

The other principal axis should be orthogonal to θ_1, and we are done. If interested, readers can also do the same to determine the sign of the second cosine angle between x and y':

```
1  n2 = np.arccos(v[1,0])    # arccos(x,y')
2
3  IXX, IYY, IXY = Ixx2IXX(Ixx, Iyy, Ixy, n2)         # take positive
4  print(f"2nd moment of area wrt the principal axis (m^4):\n"
5        f"IXX= {IXX:.4e}, IYY= {IYY:.4e},IXY= {IXY:.4e}")
6
7  IXX, IYY, IXY = Ixx2IXX(Ixx, Iyy, Ixy, -n2)        # take negative
8  print(f"\n2nd moment of area wrt the principal axis (m^4):\n"
9        f"IXX= {IXX:.4e}, IYY= {IYY:.4e},IXY= {IXY:.4e}")
```

```
2nd moment of area wrt the principal axis (m^4):
IXX= 1.1021e-05, IYY= 1.7887e-05,IXY= -1.2371e-05

2nd moment of area wrt the principal axis (m^4):
IXX= 1.6159e-06, IYY= 2.7292e-05,IXY= 0.0000e+00
```

We shall take that negative sign also.

```
1  print(f"Principal axis (arccos(x,y') ={n2} (rad)")
2  print(f"Principal axis (arccos(x,y') ={n2*180/np.pi}°")
```

```
Principal axis (arccos(x,y') =-1.8958183721285793 (rad)
Principal axis (arccos(x,y') =-108.62239144632973°
```

This additional work is because the eigenvalue solver gives only the direction cosine values. However, the arccos function is multivalued in $[-\pi, \pi]$. Only one of which is the correct answer to our mechanics problem. All these computations are very fast and takes only milliseconds, but we do need to figure out how to find the correct one. In contrast, Eq. (10.28) is easier to use because $\arctan(x)$ is single-valued for x in $[-\infty, \infty]$.

10.4 Polar moment of area

We have seen the polar moment of area, J, in Chapter 9 when studying bars subjected to torsion. It is the stiffness of the bar from the shape and size of the cross-sectional area. We have derived the formula for compute its value for a circular cross-sectional area: $J_z = \frac{\pi R^4}{2}$, where R is the radius of the cross-section. We now present a general approach for computing J_z for other shapes of areas using the second moments of areas I_{xx} and I_{yy} and the perpendicular axis theorem.

10.4.1 Perpendicular axis theorem

The theorem is expressed as

$$J_z = I_{xx} + I_{yy} \tag{10.29}$$

Because we know how to compute I_{xx} and I_{yy} for a given area, this theorem gives a straightforward method to compute J_z for the same area. On the other hand, if we know J_z, it can also help to obtain I_{xx} and I_{yy}.

10.4.2 Proof of the perpendicular axis theorem

The proof is straightforward and needs only the use of the Pythagorean theorem:

$$\begin{aligned} J_z &= \int_A r^2 dA \\ &= \int_A (x^2 + y^2) dA \\ &= \int_A x^2 dA + \int_A y^2 dA \\ &= I_{xx} + I_{yy} \end{aligned} \tag{10.30}$$

This equation shows also that J_z is an **invariant** with respect to the coordinate rotation about the z-axis. This makes sense because the property of an area in the x–y plane should not be affected by the rotation of the x–y plane about the z-axis.

10.4.3 Other useful cases

If the area is **rotational symmetric**, such as circles and hollow circles, we have

$$J_z = 2I_{xx} \quad \text{or} \quad J_z = 2I_{yy} \tag{10.31}$$

For an **annulus** (hollow circles) area with inner radius R_I and outer radius R_O, the second polar moment of area J is calculated using

$$J_z = \int_A r^2 dA = 2\pi \int_a^b r^3 dr = \frac{\pi(R_O^4 - R_I^4)}{2} \tag{10.32}$$

Using Eq. (10.31), we have the second moment of area for an annulus area:

$$I_{xx} = I_{yy} = \frac{1}{2}J_z = \frac{\pi(R_O^4 - R_I^4)}{4} \tag{10.33}$$

For a **circular** (solid) area with radius R, we have

$$I_{xx} = I_{yy} = \frac{1}{2}J_z = \frac{\pi R^4}{4} \tag{10.34}$$

A list of second moments and polar moments of area of some other shapes can be found at the wikipage (https://en.wikipedia.org/wiki/List_of_second_moments_of_area).

10.4.4 Equivalent polar second moment of area

The polar second moment of area is defined as $J_z = \int_A r^2 dA$. It is directly used in torsional problems for bars with a circular (solid or hollow) cross-section and is denoted as J (see Chapter 9). This is because the shear stress on a circular cross-sectional area is in the polar angle θ direction when the bar is subjected to torsion. However, for any non-circular bars, J_z cannot be used, because the shear stress on a non-circular cross-sectional area is not necessarily in the polar angle θ direction. Instead, the shear stress must always flow along the edge of the cross-section. Therefore, we need to follow the principles of mechanics to derive a J specifically for a given cross-section, so that we can still use the same formulas derived for circular bars. Such a J for a non-circular cross-section is called the **equivalent polar second moment of area** in this book. It is thus important to distinguish them. Section 9.5.8 presents a comparison of the equivalent J with the polar moment of area J_z for the cases of rectangular cross-sections.

10.5 Remarks

This chapter studies the properties of areas related to mechanics of materials. Area is the most basic property and used in all mechanics problems. The first and second moments of area are particularly useful in beam members. For torsional problems, the equivalent polar second moment is used when the cross-sectional area is non-circular, which needs to be derived following the principles of mechanics, as presented in Chapter 9.

For irregular shapes of areas, we do not usually have closed-form formulas for their properties. Because all these properties can be computed via integration, one can always compute these via numerical integration without much difficulty.

For standard beams produced in the industry, the cross-sections may not be regular. Often these properties are provided by the manufactures or standardized database. Designers can directly make use of those.

Except area, all these properties change when the coordinates transform. The formulas are essentially the same as those we derived for vectors (first-order tensors) and stress tensors (second-order tensors). The rule for transformation is provided together with codes.

Reference

[1] G.R. Liu, *Calculus: A Practical Course with Python*, World Scientific, New Jersey, 2025.

Chapter 11

Normal Stress by Moments

```
1  # Place cursor in this cell, and press Ctrl+Enter to import dependences.
2  import sys                          # for accessing the computer system
3  sys.path.append('../grbin/')   # Change to the directory in your system
4
5  from commonImports import *    # Import dependences from '../grbin/'
6  import grcodes as gr              # Import the module of the author
7  importlib.reload(gr)              # When grcodes is modified, reload it
8
9  from continuum_mechanics import vector
10 init_printing(use_unicode=True)    # For Latex-like quality printing
11
12 # Digits in print-outs
13 np.set_printoptions(precision=4,suppress=True,
14                 formatter={'float_kind': '{:..4e}'.format})
```

We studied beams undergoing bending in Chapter 8 and developed methods to compute the moment at any location along the beam axis. This chapter presents techniques for computing the normal stress and its distribution on a cross-section of an arbitrary shape resulting from a moment with respect to an arbitrary direction. This chapter is written with reference to Refs. [1–4].

11.1 Loading plane

This chapter considers beams with arbitrary cross-sectional shapes. We thus need to set our coordinate system carefully. When a beam is subjected to transverse forces (concentrated or distributed), the direction of these forces

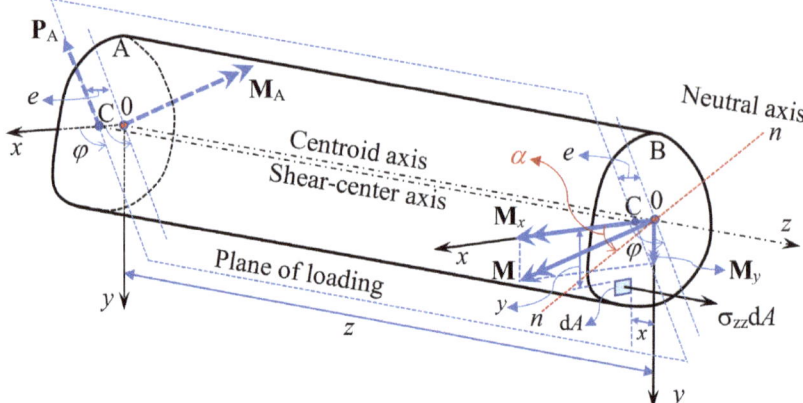

Figure 11.1. A beam subjected to forces and moments, resulting in a moment on its cross-section. When a force is applied in the loading plane, it results in a moment without torque. The moment results in normal stress σ_{zz} distributed on the cross-section.

are perpendicular to the beam's axial z-axis, which results in moments, as shown in Fig. 11.1. The z-axis is chosen, passing through the centroids of all the cross-sections along the beam. Using the technique and code given in Chapter 10, the centroid of a given cross-section area can be found with ease.

The location of the force in relation to the cross-section of the beam can, however, also result in a torque on the cross-section, and the torque vector is in the z-axis. Only when these forces pass through the so-called **shear center** on the cross-section no torque is introduced. The location of the shear center will be discussed in the following chapter. For now, we assume that it is known. We thus have an axis called shear-center axis that goes through the shear centers of all the cross-sections of the beam. The shear-center axis is also called the bending axis. It is parallel with the centroid axis. All these little complicated positional relations are shown in the busy drawing Fig. 11.1 [1].

This chapter assumes that the transverse forces pass through the shear center. The **loading plane** is a plane formed by the shear-center axis and the load force vector $\mathbf{P}_A$, as shown in Fig. 11.1. We only consider forces that are in this plane and perpendicular to the z-axis, such as P_A, so that the cross-section of the beam experiences moments without any torque. If a moment is applied, the loading plane is perpendicular to the moment vector, as shown also in Fig. 11.1. Essentially, the loading plane determines the direction of the moment applied on the cross-section resulting from a transverse force.

The loading plane is measured by φ counterclockwise from the x–z plane as viewed from the positive z axis.

11.2 Setting of the problem

Consider a cross-section B of the beam at distance z from the left end A. Assume there are forces and/or moments at the left end A. The equilibrium of the free-body diagram shall lead to a moment **M** on section z. The moment vector **M** can always be decomposed into two components along the x- and y-axis, using the direction cosines:

$$M_x = M \cos(\varphi - 90°) = M \sin \varphi$$
$$M_y = M \cos(180° - \varphi) = -M \cos \varphi \quad (11.1)$$

where φ defines the direction of the loading plane and hence the direction of the moment with respect to the x-axis. Our problem now becomes finding the normal stress and its distribution on the cross-section for given two orthogonal moment components M_x and M_y on the cross-section.

11.3 Symmetrical bending

11.3.1 Superposition of stresses

Consider first simpler cases where the x- and y-axis are the principal axes for the second moments of the area of the beam's cross-section, and the origin is at the centroid. In this case, the bending by each component of the moment gives a linearly distributed stress that is zero at the neutral axis that coincides with a principal axis. The final stress results and distributions are a superposition of these two stress distributions. A typical case is shown in Fig. 11.2.

For the problem given in Fig. 11.2, assume that the beam has a rectangular cross-section with base b and height h. The formulas for calculating the normal stress in the z-direction are as follows. For moment component M_x,

$$\sigma_{zz} = \frac{M_x y}{I_{xx}} \quad (11.2)$$

where I_{xx} is the second moment of area of the rectangular cross-section, which is $\frac{bh^3}{12}$. The normal stress varies linearly with respect to y and is constant along x. On the line of $y = 0$, the stress is zero, implying that the $y = 0$ line is the neutral axis.

Similarly, for moment component M_y, we have

$$\sigma_{zz} = -\frac{M_y x}{I_{yy}} \quad (11.3)$$

where I_{yy} is the second moment of area of the rectangular cross-section: $\frac{hb^3}{12}$. The normal stress varies linearly with respect to x and is constant along y,

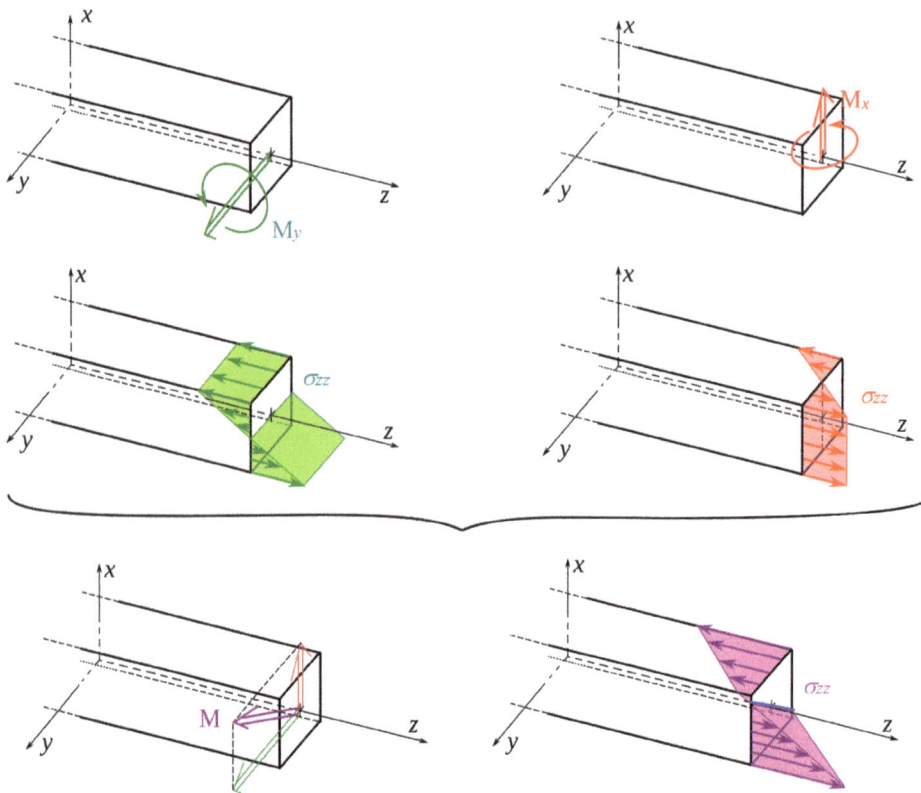

Figure 11.2. Stress distributions on the cross-section of a beam with a symmetric cross-section by two orthogonal moment components and the superposed stress distribution. Images adapted from the version by Cdang on Wikipedia under the CC BY-SA 3.0 license.

and the $x = 0$ line is the neutral axis, on which the stress are zero. The minus sign is because positive M_y results in negative normal stress (at a positive y in the positive cross-section), as shown in Fig. 11.1 or also in Fig. 11.2.

The superposed stress becomes

$$\sigma_{zz} = \frac{M_x y}{I_{xx}} - \frac{M_y x}{I_{yy}} \qquad (11.4)$$

which is a plane varying linearly with respect to both x and y, similar to the case shown in the bottom-right corner of Fig. 11.2.

The maximum stress occurs at $y = \pm h/2$ and $x = \pm b/2$. Depending on the sign of M_x and M_y, two of the four vertexes of the rectangular cross-section will register the maximum stresses, one tension and one compression, similar to the case shown in the bottom-right corner of Fig. 11.2.

If we integrate the stress over the entire cross-section, we obtain

$$\int_A \sigma_{zz} dA = \frac{M_y}{I_{yy}} \underbrace{\int_A x dA}_{Q_y=0} - \frac{M_x}{I_{xx}} \underbrace{\int_A y dA}_{Q_x=0} = 0 \qquad (11.5)$$

These first moments Q_x and Q_y are zero because the origin of the coordinates (x, y) is set at the centroid of the cross-section. In essence, the integral of the stress in the z-direction vanishes because all the forces applied on the beam are perpendicular to the z-axis.

11.3.2 Neutral axis of the combined normal stress

Because the distribution of σ_{zz} is a plane in the (x, y) coordinates on the cross-section and the integral of it is zero, there must be a line on the cross-section along which $\sigma_{zz} = 0$. Such a line is called the **neutral axis**. It can be easily found by setting $\sigma_{zz} = 0$ in Eq. (11.4), and the function for the neutral axis can be written as

$$y = \underbrace{\frac{M_y I_{xx}}{M_x I_{yy}}}_{\tan \alpha} x = x \tan \alpha \qquad (11.6)$$

where $\tan \alpha$ is the slope of the neutral axis, in which α is the angle measured from the x-axis, as shown in Fig. 11.1. It is computed using

$$\tan \alpha = \frac{M_y I_{xx}}{M_x I_{yy}} \qquad (11.7)$$

The blue line on the cross-section at the bottom-right corner of Fig. 11.2 is the neutral axis for the combined stress distribution.

11.3.3 Example: A cantilever beam subjected to a force and a moment

Consider a cantilever of length L shown in Fig. 11.3. Its cross-section is rectangular with base b and height h. It is clamped at its left end and subjected to a force P_A and a moment M_A at its right end marked with A. The force P_A passes through the shear center and hence no torque is generated. Assuming $M_A = \frac{P_A L}{3}$, derive the formula to compute the maximum stress on the beam.

Solution: First, we need to find out where along the beam axis the moment is the largest. Here, we use the usual method: free-body diagram and equilibrium conditions.

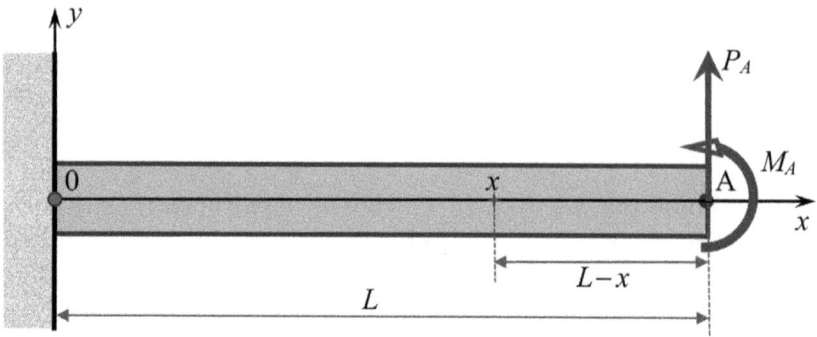

Figure 11.3. A cantilever beam subjected to a force and a moment at its free end A.

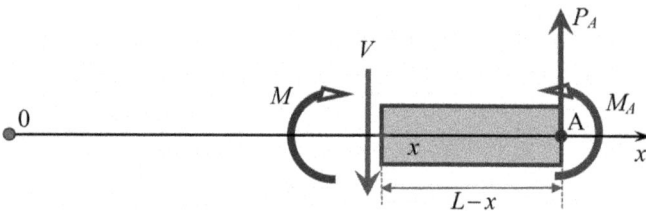

Figure 11.4. A free-body diagram of the cross-section at x for a cantilever beam subjected to a force and a moment.

Consider an arbitrary point x on the beam axis. We draw a free-body diagram in Fig. 11.4.

We put the moment M and shear force V on the cross-section at x, following the sign convention. Because the direction of the cross-section is leftward (the negative x-direction), the moment must be clockwise, and the shear force must point downward. The equilibrium equation of $\sum M_z = 0$ with respect to the point at x gives

$$-M + M_A + P_A(L - x) = 0 \tag{11.8}$$

Using the given condition $M_A = \frac{P_A L}{3}$, we have

$$M = M_A + P_A(L - x) = \frac{4P_A L}{3} - P_A x \tag{11.9}$$

This is a linear function of x. Its maximum can only be at its two boundaries: $x = 0$ or $x = L$. Clearly, the maximum moment is at $x = 0$ and is given as

$$M_{\max} = \frac{4P_A L}{3} \tag{11.10}$$

The stress distribution on the cross-section by the maximum moment becomes

$$\sigma_{xx} = -\frac{M_{\max} y}{I_{zz}} = -\frac{4P_A L y}{3 I_{zz}} \tag{11.11}$$

This formula works for any cross-sectional shape, as long as it is symmetric with respect to $x = 0$. For a rectangular cross-section, we have $I_{zz} = \frac{bh^3}{12}$, and the stress solution becomes

$$\sigma_{xx} = -\frac{4P_A L y}{3 I_{zz}} = -\frac{16 P_A L y}{bh^3} \tag{11.12}$$

It varies linearly along y. The maximum stresses are at both the top and bottom surfaces, $y = \pm h/2$, of the beam at $x = 0$:

$$\sigma_{xx} = \pm \frac{8 P_A L}{bh^2} \tag{11.13}$$

Note that the stress is inversely proportional to h^2. Therefore, it is much more effective to reduce the stress by increasing the height of the beam compared to increasing the base b for a given load. Ideally, we want to design beams with as large a height as possible.

11.3.4 Use gr.solver1D4() to find the moment

We can also simply use our general solver gr.solver1D4() for beams developed in Chapter 8 to find the moments.

```
1  E, Izz, A, b, h, L = symbols('E, I_zz, A, b, h, L', nonnegative=True)
2  x, y, PA, MA = symbols('x,y,  P_A, M_A')                # geometry & force
3
4  # for displacement boundary conditions (DBCs):
5  v0, θ0, v1, θ1 = symbols('v0, θ0, v_1, θ_1')            # DBCs
6
7  # for force boundary conditions (FBCs):
8  V0, M0, V1, M1 = symbols('V0, M0, V_1, M_A')            # FBCs
9
10 by = sp.Function('b_y')(x)                              # for force function
```

```
1  # Consider constant distributed force
2  by = 0                    # [F/L], this is zero for this example problem
3
4  v = gr.solver1D4(E, Izz, by, L, v0, θ0, v1, θ1, V0, M0, V1, M1,key='c-f')
5  print(f'Is solution correct? {by/E/Izz==v[4]}; The solution u(x) is:')
6  v[0]                      # solution of general displacement function
```

```
Outputs form solver1D4(): v, θ, M, V, qy
Is solution correct? True; The solution u(x) is
```

$$v_0 + x\theta_0 + \frac{LV_l x^2}{2EI_{zz}} + \frac{M_A x^2}{2EI_{zz}} - \frac{V_l x^3}{6EI_{zz}}$$

Use now all the prescribed displacements and moments given in this example problem:

```
1 # Set all the pre-scribed conditions:
2
3 vx=[v[i].subs({v0:0, θ0:0, Vl:PA, Ml:MA}) for i in range(len(v))]
4 vx                                    # Outputs are: v, θ, Mx, Vx, qy
```

$$\left[\frac{LP_A x^2}{2EI_{zz}} + \frac{M_A x^2}{2EI_{zz}} - \frac{P_A x^3}{6EI_{zz}}, \frac{LP_A x}{EI_{zz}} + \frac{M_A x}{EI_{zz}} - \frac{P_A x^2}{2EI_{zz}},\right.$$
$$\left. LP_A + M_A - P_A x, \; P_A, \; 0\right]$$

This gives the full set of solutions. For our example problem, we need only the moment.

```
1 # Get only the solution for the moment:
2 gr.printM(vx[2], 'Solution of moment distribution along the beam:')
```

Solution of moment distribution along the beam: $LP_A + M_A - P_A x$

```
1 # Substitute the M_A with known value in terms of P_A:
2 gr.printM(vx[2].subs({MA:PA*L/3}),'Solution of moment along the beam:')
```

Solution of moment along the beam: $\dfrac{4LP_A}{3} - P_A x$

```
1 # Set x=0, for the maximum moment:
2 Mmax = vx[2].subs({MA:PA*L/3, x:0})
3 gr.printM(Mmax,'Solution at the beam root, Mmax:')
```

Solution at the beam root, Mmax: $\dfrac{4LP_A}{3}$

We have obtained exactly the same solution as we did using the free-body diagram with equilibrium conditions. Our general solver gr.solver1D4() is powerful. It works for not only this problem but also many other types of problems for beams. If the problem is indeterminate, using a free-body diagram alone will not work. In such cases, our general solver gr.solver1D4() must be used. Note also that gr.solver1D4() gives a full set of solutions, including deflection, rotation, moment, and shear force! All are in closed-form formulas. The computation took only a few seconds on the author's laptop.

The formulas for the normal stress are finally found using:

```
1  σxx = - Mmax*y/Izz
2  print('σxx=', σxx)
3  print('σxx (in b and h) = ', σxx.subs(Izz, b*h**3/12))
4  print('σxx (at upper surface) = ', σxx.subs({Izz:b*h**3/12, y:h/2}))
```

σxx= -4*L*P_A*y/(3*I_zz)
σxx (in b and h) =   -16*L*P_A*y/(b*h**3)
σxx (at upper surface) =   -8*L*P_A/(b*h**2)

11.3.5 Example: A cantilever T-beam subjected to a force and a moment

Consider again the cantilever of length L shown in Fig. 11.3. All the settings are the same, except for its cross-section becoming T shape, as shown in Fig. 11.5.

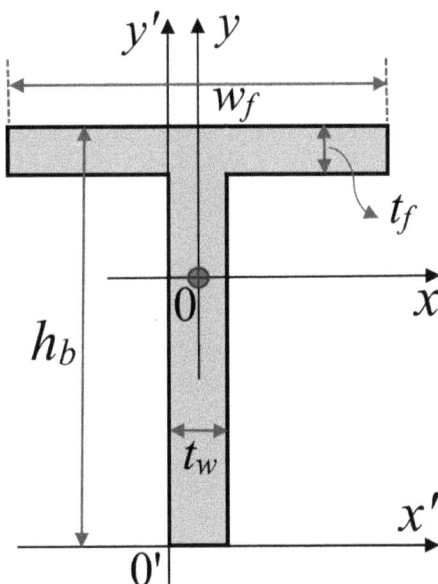

Figure 11.5. Schematic drawing of the cross-section of a typical T-beams.

Its dimensions (mm) are given as follows:

> beam height: $h_b = 200$; width of flange on top: $w_f = 100$; thickness of flange on top: $t_f = 20$; web thickness: $t_w = 10$.

1. Find the centroid of the cross-section.
2. Find the second moments of the cross-section.
3. Determine the maximum stress of the beam. Assume $P_A = 5000$ N and $L = 600$ mm.

Solution to 1: To find the centroid of the cross-section of the T-beam, we use the Python code function gr.centroid_I() developed in Chapter 10, which is also in the grcodes module. The gr.centroid_I() is made for I-beams consisting of three rectangles. The T-beam cross-section has only two rectangles and can be treated as a special case of the I-beam by setting the thickness of the flange at the bottom zero.

```
1  # preparing data: Data from Example 7.2 AMM by Boresi and Schmidt
2  #hb =.060; wft=.050; wfb=.010; tft=.010; tfb=.0; tw=.010    # (m) √
3  # This example:
4  hb =.200; wft=.100; wfb=.010; tft=.020; tfb=.0; tw=.010    # (m)
5
6  # call the code function:
7  x_bar, y_bar, A, Ai = gr.centroid_I(hb, wft, wfb, tft, tfb, tw)
```

```
Areas of sub-areas = [2.0000e-03 1.8000e-03 0.0000e+00][L^2]
Areas of I-beam = 0.0038000000000000004[L^2])
The centroid [L]):
x = 0.005
y = 0.14263157894736841
```

Solution to 2: Find the second moments of the cross-section.

This time we use the Python code function gr.secondMoments_I() developed in Chapter 10. It also computes the centroid while computing the second moments. A T-beam can be treated as a special case of the I-beam with zero thickness for the bottom flange.

```
1  # call the code function for 2nd moment calculation:
2  I_xx, I_yy, I_xy, x_bar, y_bar, xi_bar, yi_bar, A, Ai =\
3      gr.secondMoments_I(hb, wft, wfb, tft, tfb, tw, [0]*2, [0]*2, sy=True)
```

```
Center of sub-areas [L]:
x = [5.0000e-03 5.0000e-03 5.0000e-03]
y = [1.9000e-01 9.0000e-02 0.0000e+00]
Areas of sub-areas = [2.0000e-03 1.8000e-03 0.0000e+00] [L^2]
Areas of I-beam = 0.0038000000000000004 [L^2]
The centroid = (0.005, 0.14263157894736841) [L]
New center of sub-areas [L]:
x = [0.0000e+00 0.0000e+00 0.0000e+00]
y = [4.7368e-02 -5.2632e-02 -1.4263e-01]
2nd moments [L^4]:
Ixx = 1.4400350877192984e-05
Iyy = 1.6816666666666672e-06
Ixy = 0.0
```

One needs to ignore the third entry that corresponds to sub-areas with $A = 0$. The centroid of the T-beam is also found as x_bar, y_bar. In addition, the new centroids of the two sub-rectangles are also found.

Solution to 3: Determine the maximum stress of the beam. Assume $P_A = 5000$ N and $L = 600$ mm.

We can use Eq. (11.11) derived in the previous example, but we need to do a little more work to compute the distance from the centroid of the T-beam cross-section and find the y values at the top and bottom of the cross-section.

```
1  y_top = hb - y_bar                    # y_bar: the centroid of the T-beam
2  y_bot = - y_bar
3  print(f" y_top= {y_top}, y_bot ={y_bot}")
4
5  PA = 5000.  # (N)
6  L  = 0.600 # (m)
7  Mmax  = 4*PA*L/3                      # from previous example
8  print(f" Mmax = {Mmax} (Nm)")
9
10 σ_top = Mmax*y_top /I_xx
11 σ_bot = Mmax*y_bot /I_xx
12
13 print(f" σ_top= {σ_top*1e-6}, σ_bot ={σ_bot*1e-6} (MPa)")
```

```
y_top= 0.057368421052631596, y_bot =-0.14263157894736841
Mmax = 4000.0 (Nm)
σ_top= 15.935284227967157, σ_bot =-39.61891766769815 (MPa)
```

It is found that the largest normal stress is located at the bottom edge of the T-beam.

11.4 Beams under non-symmetrical bending*

In practical applications, beams with a **non-symmetrical cross-section** are used. In addition, the loading plane or the moment on the cross-section can be arbitrary. We need to find the normal stress on the cross-section and often the maximum one.

A general procedure is introduced in this section to compute the normal stresses on the non-symmetrical cross-section of a beam, subjected to moments in any given direction.

11.4.1 *General formula for computing normal stress distribution*

On the cross-section, the coordinates (x, y) are set with the centroid of the cross-section at the origin. Based on our discussion so far, we know that the stress σ_{zz} is linearly distributed on the cross-section defined in (x, y) coordinates, as shown in Fig. 11.1. This is true for all beams that obey the Euler–Bernoulli assumption. Thus, the normal stress can be written in the following general form:

$$\sigma_{zz} = a + bx + cy \qquad (11.14)$$

where a, b, and c are three constants to be determined.

Now, on the cross-section, we shall have the following three equilibrium conditions:

$$0 = \int_A \sigma_{zz} dA$$

$$M_x = \int_A y\sigma_{zz} dA \qquad (11.15)$$

$$M_y = -\int_A x\sigma_{zz} dA$$

This minus sign is because positive M_y results in negative σ_{zz} at positive y on the positive cross-section (see Fig. 11.1).

Using these three conditions in Eq. (11.15), we can determine the three constants a, b, and c:

$$0 = \int_A (a + bx + cy)dA = a\underbrace{\int_A dA}_{A} + b\underbrace{\int_A xdA}_{=0} + c\underbrace{\int_A ydA}_{=0}$$

$$M_x = \int_A (ay + bxy + cy^2)dA = a\underbrace{\int_A ydA}_{=0} + b\underbrace{\int_A xydA}_{I_{xy}} + c\underbrace{\int_A y^2 dA}_{I_{xx}}$$

$$M_y = -\int_A (ax + bx^2 + cxy)dA = -a\underbrace{\int_A xdA}_{=0} - b\underbrace{\int_A x^2 dA}_{I_{yy}} - c\underbrace{\int_A xydA}_{I_{xy}}$$
(11.16)

We used the fact that first moments of the cross-section area are zero because (x, y) is set at the centroid. The foregoing equation simply becomes

$$0 = aA$$
$$M_x = bI_{xy} + cI_{xx} \qquad (11.17)$$
$$M_y = -bI_{yy} - cI_{xy}$$

The first equation in Eq. (11.17) gives $a = 0$ because $A \neq 0$. The last two equations can be easily solved using the following code.

```
1  Mx, My, Ixx, Iyy, Ixy, b, c = symbols('M_x, M_y, I_xx, I_yy, I_xy, b,c')
2  sln = sp.solve([b*Ixy+c*Ixx-Mx, -b*Iyy-c*Ixy-My], [b, c])
3  sln
```

$$\left\{b : \frac{-I_{xx}M_y - I_{xy}M_x}{I_{xx}I_{yy} - I_{xy}^2}, \; c : \frac{I_{xy}M_y + I_{yy}M_x}{I_{xx}I_{yy} - I_{xy}^2}\right\}$$

We write down the results for later reference:

$$b = -\frac{M_y I_{xx} + M_x I_{xy}}{I_{xx}I_{yy} - I_{xy}^2}$$

$$c = \frac{M_x I_{yy} + M_y I_{xy}}{I_{xx}I_{yy} - I_{xy}^2}$$
(11.18)

Substituting Eq. (11.18) and $a = 0$ back into Eq. (11.14), we obtain

$$\sigma_{zz} = bx + cy = -\left(\frac{M_y I_{xx} + M_x I_{xy}}{I_{xx} I_{yy} - I_{xy}^2}\right) x + \left(\frac{M_x I_{yy} + M_y I_{xy}}{I_{xx} I_{yy} - I_{xy}^2}\right) y \quad (11.19)$$

This is the general formula to compute the normal stress distributed on the cross-section of a beam that undergoes non-symmetric bending. It is a linear function of x and y. If $I_{xy} = 0$, the bending becomes symmetric, and Eq. (11.19) becomes Eq. (11.4). This means that **symmetric bending** occurs only when $I_{xy} = 0$.

11.4.2 Neutral axis: general formula

The neutral axis is the line on the cross-section on which $\sigma_{zz} = 0$. Using Eq. (11.19), the function for the neutral axis can be written as

$$y = \underbrace{\frac{M_y I_{xx} + M_x I_{xy}}{M_x I_{yy} + M_y I_{xy}}}_{\tan \alpha} x = x \tan \alpha \quad (11.20)$$

where $\tan \alpha$ is the slope of the linear function and α is the angle that determines the neutral axis. It is measured from the x-axis, as shown in Fig. 11.1. It is computed using

$$\tan \alpha = \frac{M_y I_{xx} + M_x I_{xy}}{M_x I_{yy} + M_y I_{xy}} \quad (11.21)$$

It is clear that the slope of the neutral axis depends on both the property of the cross-section and the moment direction. Using Eq. (11.1), we have

$$\cot \varphi = -\frac{M_y}{M_x} \quad (11.22)$$

Substituting Eq. (11.22) into Eq. (11.21), we have, alternatively,

$$\tan \alpha = \frac{I_{xy} - I_{xx} \cot \varphi}{I_{yy} - I_{xy} \cot \varphi} \quad (11.23)$$

Once α is found, the stress can also be written alternatively as

$$\sigma_{zz} = \frac{M_x(y - x \tan \alpha)}{I_{xx} - I_{xy} \tan \alpha} \quad (11.24)$$

It is clear that the normal stress depends on both the moment M and its direction and the second moments of area of the cross-section of the beam, which is an area property depending also on orientation. Its distribution is a linear function of x and y, and hence the maximum stress should register on the extreme points that are the farthest from the neutral axis.

11.4.3 Example: A cantilever L-beam subjected to a force and a moment

Consider again the cantilever of length L shown in Fig. 11.3. All the settings are the same, except for its cross-section becoming an L shape that is not symmetric, as shown in Fig. 11.6.

Its dimensions (mm) are given as follows:

beam height: $h_b = 100$; width of foot flange: $w_f = 70$; thickness of foot flange: $t_f = 10$; web thickness: $t_w = 10$.

Assume $P_A = 5000$ N, $L = 600$ mm, and the angle of the loading plane is $\varphi = 2\pi/3$.

1. Find the centroid of the cross-section.
2. Find the second moments of the cross-section.
3. Determine the maximum stress of the beam using the formulas for non-symmetric bending.
4. Find the principal axis of the cross-section.
5. Determine the maximum stress of the beam using the formulas for symmetric bending.

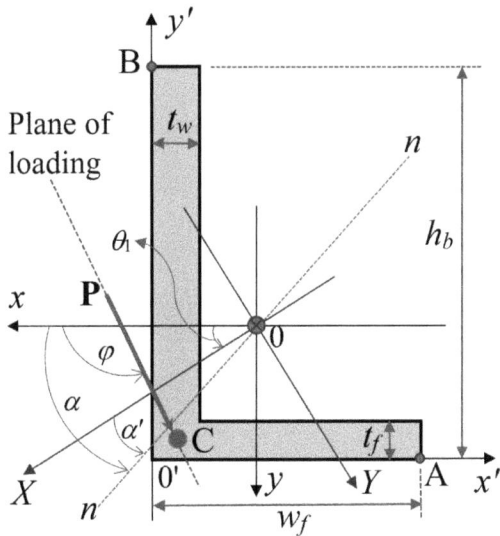

Figure 11.6. Schematic drawing of the cross-section of a typical L-beams.

11.4.3.1 Solution to 1 and 2

We use the Python code function gr.secondMoments_I() developed in Chapter 10 to compute the centroids and second moments.

```
1  # preparing data. Data from Example 7.4 AMM by Boresi and Schmidt
2  #hb =.120; wfb=.080;   tfb=.010; tw=.010; wft=.0; tft=.0    # (m) √
3
4  # This example:
5  hb =.100; wfb=.070; tfb=.010; tw=.010; wft=.0; tft=.0;    # (m)
6
7  # L-shape area is not symmetric with respect to y-axis, we need to
8  # provide the center of each sub-area of the L-beam
9  cxi = np.array([wft/2,  tw/2,  wfb/2])        # from top to bottom
10 cyi = np.array([hb, (hb-tft-tfb)/2+tfb, tfb/2])
11
12 # call the code function:
13 Ixx, Iyy, Ixy, xbar, ybar, xibar, yibar, A, Ai =\
14      gr.secondMoments_I(hb, wft, wfb, tft, tfb, tw, cxi, cyi, sy=False)
```

```
Center of sub-areas [L]:
x = [0.0000e+00 5.0000e-03 3.5000e-02]
y = [1.0000e-01 5.5000e-02 5.0000e-03]
Areas of sub-areas = [0.0000e+00 9.0000e-04 7.0000e-04] [L^2]
Areas of I-beam = 0.0016000000000000003 [L^2]
The centroid = (0.018125000000000002, 0.033125) [L]
New center of sub-areas [L]:
x = [-1.8125e-02 -1.3125e-02 1.6875e-02]
y = [6.6875e-02 2.1875e-02 -2.8125e-02]
2nd moments [L^4]:
 Ixx = 1.597708333333334e-06
 Iyy = 6.477083333333335e-07
 Ixy = -5.906250000000002e-07
```

One needs to ignore results corresponding to sub-areas with $A = 0$.

11.4.3.2 Solution to 3: Determine the maximum stress of the beam, using the formulas for non-symmetric bending

Because the cross-section is not symmetric, we must first find the neutral axis using Eq. (11.21).

```
1  # Mmax=4.8e3; Data from Example 7.4 AMM by Boresi and Schmidt for test √
2
3  PA   = 5000. # (N)
4  L    = 0.600 # (m)
5  Mmax = 4*PA*L/3
6
7  φ  = 2*np.pi/3
8  Mx = Mmax*np.sin(φ)
9  My = -Mmax*np.cos(φ)
10
11 tana = (My*Ixx + Mx*Ixy)/(Mx*Iyy + My*Ixy)
12 α = np.arctan(tana)
13
14 print(f" tana= {tana:.4f}, neutral axis α = {α:.4f} (rad),"\
15       f"{np.rad2deg(α):.4f}°")
```

tana= 1.0818, neutral axis α = 0.8247 (rad),47.2512°

Using the computed **tana**, the neutral axis is computed using Eq. (11.20). The stress is given in Eq. (11.24). Points A and B are the possible points having maximum stresses. To find the stress there, we need to find their coordinates in terms of (x, y).

```
1  #xA = wfb-xbar; yA = -ybar              # other possibilities
2  #xB = -xbar;    yB = hb-ybar
3
4  xA = -wfb+xbar; yA = ybar
5  xB = xbar;      yB = -hb+ybar
6
7  dm = Ixx - Ixy*tana
8  σzz_A= Mx*(yA - xA*tana)/dm
9  σzz_B= Mx*(yB - xB*tana)/dm
10
11 print(f" σzz_A= {σzz_A*1e-6:.4f}, σzz_B= {σzz_B*1e-6:.4f} (MPa)")
```

σzz_A= 138.2213, σzz_B= -133.9434 (MPa)

The maximum normal stress is found at point A, the lower-right vertex of the L-beam.

11.4.3.3 Solution to 4: Find the principal axis of the cross-section

To use the symmetric bending beam formulas for a non-symmetric beam, the principal axes $X-Y$ of the beam cross-section should be found first (see Fig. 11.6). For this, we use $\tan 2\theta_1 = -2I_{xy}/(I_{xx} - I_{yy})$ to find the angle of rotation θ_1 between X and x, and then compute the new second moments with respect to the principal axis. These can be done using the code function gr.Ixx2IXX() given in Chapter 10.

```
1  tan2θ = -2*Ixy/(Ixx-Iyy)
2  θ = np.arctan(tan2θ)/2
3  print(f"Angle to the principal axis θ1={θ} (rad); {np.rad2deg(θ)}°")
4
5  IXX, IYY, IXY = gr.Ixx2IXX(Ixx, Iyy, Ixy, θ)
6  print(f"2nd moment of area wrt the principal axis (m^4):\n"
7        f"IXX= {IXX:.4e}, IYY= {IYY:.4e},IXY= {IXY:.4e}")
```

Angle to the principal axis θ1=0.4467398651270534 (rad); 25.596308812023786°
2nd moment of area wrt the principal axis (m^4):
IXX= 1.8806e-06, IYY= 3.6478e-07,IXY= -5.2940e-23

We see that $I_{XY} = 0$, and hence the formulas for symmetric bending can be used. Note that after the coordinate rotation, the loading plane angle φ needs to be updated.

```
1  ψ = φ - θ
2  print(f" Angle of the loading plane wrt (X, Y) = {ψ:.4f}, "\
3        f"{np.rad2deg(ψ):.4f}°")
4
5  MX =  Mmax*np.sin(ψ)          # compute the moment components on (X, Y)
6  MY = -Mmax*np.cos(ψ)
7
8  tanaX = MY*IXX /(MX*IYY)
9  αX = np.arctan(tanaX)
10
11 print(f" tana= {tanaX:.4f}, neutral axis αX = {αX:.4f} (rad) or "\
12       f"{np.rad2deg(αX):.4f}°")
```

Angle of the loading plane wrt (X, Y) = 1.6477, 94.4037°
tana= 0.3970, neutral axis αX = 0.3779 (rad) or 21.6549°

Using tanaX, the neutral axis is given by Eq. (11.6). The stress is given in Eq. (11.4). We need to find the coordinates of points A and B with respect to the principal (X, Y). This is done through coordinate transformation studied in Chapter 2.

```
1  XA =  xA*np.cos(θ) + yA*np.sin(θ)
2  YA = -xA*np.sin(θ) + yA*np.cos(θ)
3  print(f" XA= {XA*1e3:.6f}, YA= {YA*1e3:.6f} (mm)")
4
5  XB =  xB*np.cos(θ) + yB*np.sin(θ)        # coordinate transformation
6  YB = -xB*np.sin(θ) + yB*np.cos(θ)        # for position vectors
7  print(f" XB= {XB*1e3:.6f}, YB= {YB*1e3:.6f} (mm)")
8
9  σzz_AX= MX*YA/IXX - MY*XA/IYY
10 σzz_BX= MX*YB/IXX - MY*XB/IYY
11
12 print(f" σzz_AX= {σzz_AX*1e-6:.4f}, σzz_BX= {σzz_BX*1e-6:.4f} (MPa)")
```

XA= -32.473090, YA= 52.285559 (mm)
XB= -12.545630, YB= -68.142413 (mm)
σzz_AX= 138.2213, σzz_BX= -133.9434 (MPa)

The maximum stress results are exactly the same as those found in item 3.

As an exercise, readers may compute the following:

1. The stresses and their distributions if the L-beam is upside down.
2. Find the neutral axis.

11.5 Effects of error in loading plane angle*

We know that for better performance in terms of higher stiffness and lower normal stress for the same moment, we would like to design the beam in such a way that the loading plane is normal to the principal axis that has the larger second moment of area. For example, if the loading plane is vertical (its normal is in the x-direction), we would want the beam as tall as possible, leading to a large I_{xx}. However, if it is too tall, the performance can be unstable, implying that it may be sensitive to small errors in the loading plane angle φ. Let us use the following example to demonstrate this effect.

11.5.1 Example: A tall beam with rectangular cross-section

Consider a beam with rectangular cross-section. It has a height h that is 10 times its base $b = 50$ mm. The cross-section is defined in (x, y), with x leftward and y downward. It is designed for transverse loads in the loading plane of $x = 0$ and perpendicular to the z-axis (out of the paper), implying that the angle of the loading plane is $\varphi = \pi/2$. Assume the moment created by a transverse force P is $M = 2000$ Nm at $\varphi = \pi/2$, as shown in Fig. 11.7.

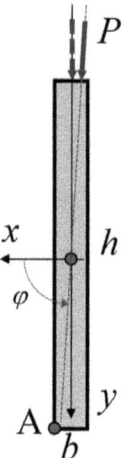

Figure 11.7. A tall rectangular beam designed for a moment with $\varphi = \pi/2$ and a small error in φ in the actual implementation.

Perform this following analysis:

1. Compute the stress σ_{zz} at point A $(b/2, h/2)$.
2. Assuming in a service situation, the value of angle φ of the loading plane is reduced by 1% due to error, determine the angle of the neutral axis resulting from the error.
3. Determine the percentage error in stress σ_{zz} at point A.

11.5.1.1 Python code for the solution

We write the follow code to complete the task, with proper comment lines.

```
# Compute the properties of the cross-section:
b = 0.05         # mm
h = 10*b

Ixx = b*h**3/12
Iyy = h*b**3/12

x_A = b/2; y_A = h/2                      # coordinates at point A
print(f" x_A= {x_A}, y_A ={y_A}")
```

x_A= 0.025, y_A =0.25

The following code computes and prints out the requested results for all these questions. The results for both the perfect and error cases are placed in an array and computed in one go. The perfect case is for reference.

```
1   M  = 2000.                       # (Nm)
2   φ0 = np.pi/2                     # angle of loading plane by design
3   coef = np.array([1.0, 0.99])     # reduction rates: perfect, 1% reduction
4
5   φ  = coef*φ0                     # angles (rad) perfect and 1% reduction
6   cotφ = np.cos(φ)/np.sin(φ)
7
8   Mx = M*np.sin(φ)                         # components of the moment
9   My = -M*np.cos(φ)
10  print(f"Components of moment:\n  Mx= {Mx} (Nm)\n  My= {My} (Nm)")
11
12  tana = My*Ixx/(Mx*Iyy)
13  α = np.arctan(tana)                      # angle of the neutral axis
14  print(f"Angles of the neutral axis:\n    α= {α} (rad) \n"\
15        f"    {np.rad2deg(α)}°")
16
17  σzz_A0=Mx[0]*y_A/Ixx                     # when no error in φ0
18  print(f"Normal stress at point A (no error), σzz_A0={σzz_A0*1e-6}(MPa)")
19
20  σzz_A =Mx*y_A/Ixx - My*x_A/Iyy
21  print(f"Normal stress at point A, σzz_A= {σzz_A*1e-6} (MPa)")
22
23  dσR = (σzz_A-σzz_A0)/σzz_A0*100
24  print(f"Percentage error in stress dσR = {dσR} (%)")
25
26  dφR = (φ0 - φ)/φ0*100
27  print(f"Percentage error in angle  dφR = {dφR} (%)")
```

Components of moment:
  Mx= [2.0000e+03 1.9998e+03] (Nm)
  My= [-1.2246e-13 -3.1415e+01] (Nm)
Angles of the neutral axis:
    α= [-6.1232e-15 -1.0039e+00] (rad)
        [-3.5084e-13 -5.7520e+01]°
Normal stress at point A (no error), σzz_A0=0.96(MPa)
Normal stress at point A, σzz_A= [9.6000e-01 1.1107e+00] (MPa)
Percentage error in stress dσR = [6.0633e-14 1.5695e+01] (%)
Percentage error in angle dφR = [0.0000e+00 1.0000e+00] (%)

11.5.1.2 An important finding

Although the angle error of the loading plane is only 1%, the resultant error in the neutral axis is significant, and the error in the normal stress is ~15.7% for this example. There is about 16 times magnification in error. If the h/b ratio becomes higher, the magnification will be even bigger. Readers may test this using the code given above.

11.5.2 Underlying reasons for error magnification

Let us examine in more detail the underlying reasons for the error magnification.

Assume (x, y) is the principal axis, and hence $I_{xy} = 0$. Equation (11.23) becomes

$$\tan \alpha = \frac{-I_{xx}}{I_{yy}} \cot \varphi \tag{11.25}$$

Substituting Eq. (11.25) into Eq. (11.24) and using $I_{xy} = 0$, we have

$$\sigma_{zz} = \frac{M_x y}{I_{xx}} + \frac{M_x y}{I_{xx}} \underbrace{\frac{x I_{xx}}{y I_{yy}}}_{\beta_s} \cot \varphi \tag{11.26}$$

The second term on the right-hand side of Eq. (11.26) is the increased part of the normal stress, in which β_s is the magnification factor (per $\cot \varphi$) for the increased part of the normal stresses:

$$\beta_s = \frac{x I_{xx}}{y I_{yy}} \tag{11.27}$$

Note that at φ close to but less than $\pi/2$, $\cot \varphi \approx \varphi - \pi/2$.

Equation (11.26) depends on the property of the beam cross-section and the interested point location on it. If the maximum stress is concerned, say at point A, x is replaced by x_A and y is replaced by y_A, and we obtain

$$\beta_s = \frac{x_A I_{xx}}{y_A I_{yy}} \tag{11.28}$$

For beams with rectangular cross-section of height h and base b, we have

$$\beta_s = \frac{b}{2h} \frac{2}{12} \frac{bh^3}{hb^3} \frac{12}{hb^3} = \frac{h}{b} \tag{11.29}$$

The stress magnification factor is proportional to the height–base ratio of a rectangular beam. If $h = b$, there is no magnification.

This study shows that when the beam is made too high, care is needed in actual implementations. Almost no error is allowed in the loading plane angle.

Using the code given above, readers may set different ratios of h/b to see the results. One may also set $h = b$ to see what the code produces.

For other shapes of cross-section, β_s will be different from that given in Eq. (11.29) and can be quite complicated to derive.

11.6 Remarks

1. This chapter focuses on computing the normal stresses on the cross-section of a beam for a given moment. The beam can have an irregular cross-section shape, and the loading can be in any direction as long as it is in the loading plane.
2. The moment can be easily computed using codes given in Chapter 8. If it is statically determinant, the moment can be obtained using a free-body diagram and equilibrium conditions.
3. The computation of the normal stress can be done in two ways: using either the asymmetric bending formula or the symmetric bending formulas. The latter requires finding the principal axis of the cross-section followed by a coordinate transformation.
4. An analysis is provided to demonstrate that although it is beneficial to have tall beams, but there is a limit because of stability and robustness issues.

References

[1] A.P. Boresi and R.J. Schmidt, *Advanced Mechanics of Materials*. John Wiley & Sons, New York, 2002.
[2] J.M. Gere and S.P. Timoshenko, *Mechanics of Materials*. Van Nostrand Reinhold Company, New York, 1972.
[3] S. Timoshenko and J.N. Goodier, *Theory of Elasticity*. McGraw-Hill, New York, 1970. http://books.google.com/books?id=yFISAAAAIAAJ&dq=theory+of+elasticity&ei=ICiKSsr3G4jwkQSbxMyPCg.
[4] Z.L. Xu, *Elasticity*, Vols. 1&2. People's Publisher, China, 1979.

Chapter 12

Shear Stress by Shear Force

Let us import the necessary external modules or dependencies for later use in this chapter.

```
1  # Place cursor in this cell, and press Ctrl+Enter to import dependences.
2  import sys                          # for accessing the computer system
3  sys.path.append('../grbin/')   # Change to the directory in your system
4
5  from commonImports import *         # Import dependences from '../grbin/'
6  import grcodes as gr                # Import the module of the author
7  importlib.reload(gr)                # When grcodes is modified, reload it
8
9  from functools import partial   # create function with single variable
10
11 from continuum_mechanics import vector
12 init_printing(use_unicode=True)     # For latex-like quality printing
13
14 # Digits in print-outs
15 np.set_printoptions(precision=4,suppress=True,
16                     formatter={'float_kind': '{:.4e}'.format})
```

When a beam is subjected to transverse loading, there will be shear force on its cross-section, in addition to the moments. Both moments and shear forces on the cross-section at any location along the beam axis can be computed using the techniques developed in Chapter 8, and gr.solver1D4() for various types of loaded thin beams. In the previous chapter, we found the normal stress on the cross-section caused by the moment.

This chapter zooms-in on a cross-section of a beam and examines the shear stress distribution on the cross-section resulting from the shear force. The concept of shear-center and techniques to find it will also be studied. It is written with reference to [1–4].

12.1 Setting of the problem

Figure 12.1 shows a cantilever beam subjected to a transverse point force P at its right-end, resulting in both moment and shear force on a cross-section at any location along the beam. The x-axis starts at the fixed end of the beam, and goes through the neutral axis of the cross-sections where the normal stress is zero.

For convenience in analysis, the transverse force is along the y-axis and passing thorough the shear-center of the cross-sections, so that the force does not generate any torque. We assume also that the bending is symmetric, implying that $I_{zy} = 0$ and $M_y = 0$. The z-axis can then be chosen on the neutral axis on the cross-section.

In this simple case with such a coordinate system, the shear force at $x = L$ is $V_y = P$, and moment at $x = L$ is $M = 0$. As usual, we use our standard sign convention. Using equilibrium conditions, we find at arbitrary location x, the shear force on the cross-section is $V_y = P$, and the moment becomes $M = (L-x)P$. It is easy to see that the relation between the shear force and moment satisfies the following equation:

$$\frac{dM_z}{dx} = -V_y \qquad (12.1)$$

This equation was derived in Chapter 8 for general cases. Equation (12.1) means that the change in moment is compensated by the shear force. If the shear force is zero, the change in moment will be zero, and hence the moment will be constant, which is the case of pure bending. In this chapter, we discuss cases that are not pure bending, and hence the shear force is nonzero. Otherwise, the shear stress will be simply zero, and we are done.

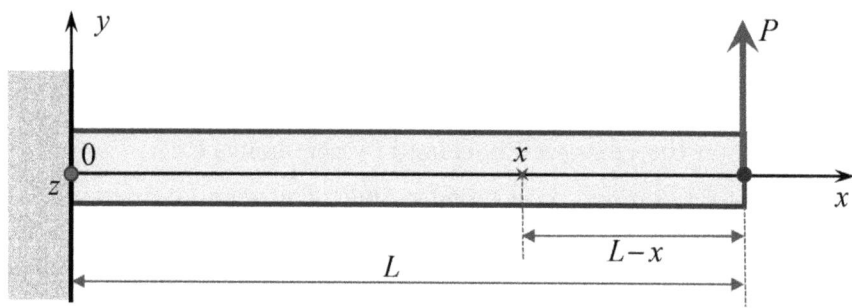

Figure 12.1. A cantilever beam subjected to a transverse point force at its right-end.

We use the coordinate system shown in Fig. 12.1, and consider only one shear force V_y and one moment M_z involved. Thus, there is only one normal stress σ_{xx} and one shear stress σ_{xy}. Thus in this chapter, we

drop the subscript z, y, and x for V, M, and σ.

The normal stress σ_{xx} is denoted as σ, and the shear stress σ_{xy} is denoted as τ. We put the subscript on when it is needed for avoiding confusion.

12.2 General formulation

12.2.1 Free-body diagram

To derive the formula for computing the shear stress on the beam cross-section, we must analyze the stress states in more detail, using again the usual trick: free-body diagram. It is shown in Fig. 12.2. We assume that the Euler–Bernoulli beam theory holds so that we can still use the moment shear-force relation and moment normal-stress relation. We require also the thickness t_y does not change too drastically with y, so that the shear stress boundary conditions on the free surfaces can be reasonably well observed.

The free-body is an infinitely small segment from the beam with length dx and throughout its entire depth (in the z-axis). It is taken from beam

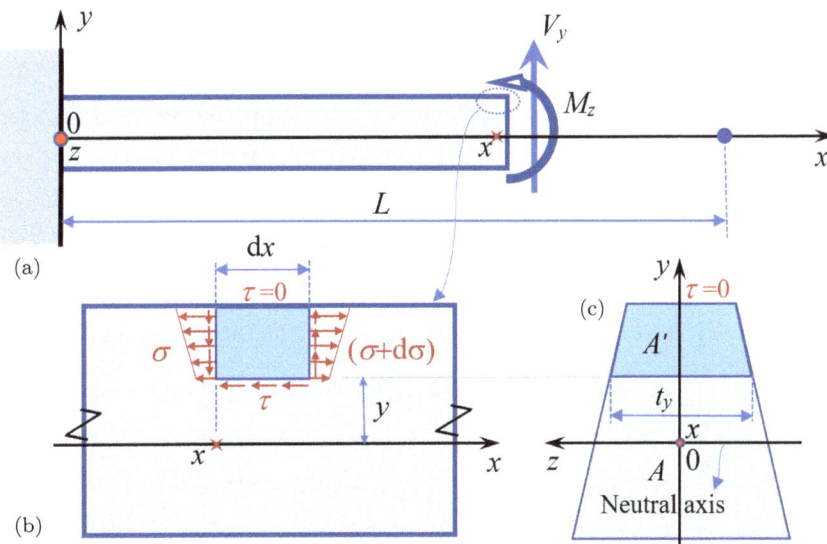

Figure 12.2. (a) Internal forces on a cross-section of the beam. (b) A free-body diagram, the shaded area, with an infinitely small length segment. (c) The free-body diagram viewed on the cross-section.

top surface (where the shear stress is zero) until y where there is unknown shear stress denoted as τ. The free-body is the shaded area in Fig. 12.2(b) and (c). We can then put all the nonzero stresses on the free-body, as shown on Fig. 12.2(b). All these variables are uniform in the z-direction.

12.2.2 Equilibrium conditions

Consider the equilibrium of free-body for all forces in the x-direction, we have

$$\sum F_x = 0: \quad -\int_{A'} \sigma dA + \int_{A'} (\sigma + d\sigma)dA - \tau t_y dx = 0 \quad (12.2)$$

where $d\sigma$ is the incremental normal stress with respect to dx, A' is the shaded area on the cross-section shown on Fig. 12.2(c), and t_y is the thickness (in the depth z-direction) of the beam at y. Equation (12.2) can be simplified to

$$\tau t_y dx = \int_{A'} d\sigma dA \quad (12.3)$$

12.2.3 Formula for shear stress

Because the normal stress on any given cross-section at x relates moment as

$$\sigma = -\frac{My}{I} \quad (12.4)$$

Substituting Eq. (12.4) into Eq. (12.3) gives

$$\tau t_y dx = -\int_{A'} \frac{dMy}{I} dA = -\frac{dM}{I} \underbrace{\int_{A'} y dA}_{Q_{A'}} \quad (12.5)$$

where $Q_{A'}$ is the first moment of area A' and I is the second moment of the entire cross-section. They all are with respect to the z-axis. In deriving Eq. (12.5), we used the fact that $\frac{dM}{I}$ is fixed at a given cross-section, does not relate to dA, and hence can be moved out of the integral. Equation (12.5) is re-arranged to

$$\tau = -\frac{dM}{dx}\frac{Q_{A'}}{It_y} \quad (12.6)$$

Using Eq. (12.1), we finally obtain the formula for computing the shear stress:

$$\boxed{\tau = \frac{VQ_{A'}}{It_y}} \quad (12.7)$$

This result shows that the shear stress at location y on any cross-section of the beam depends on, naturally, the shear force on the cross-section. It varies in the y-direction through the first moment of area A'. The calculation of the first moment of an area is discussed in Chapter 10. When y changes, the shaded area changes, and hence the value of $Q_{A'}$ changes. When y gets to the top surface (or bottom surface), the shaded area becomes zero (or the entire cross-section), and $Q_{A'} = 0$, leading to correct result: zero shear stress on the top (and bottom) surface of the beam.

12.2.4 Shear-flow resulted from shear force

Using the shear stress given in Eq. (12.5), we can define the **shear-flow** q as

$$q = \tau t_y = \frac{VQ_{A'}}{I} \tag{12.8}$$

Both V and I are fixed values at a cross-section, but $Q_{A'}$ changes with y, the shear-flow resulted from a shear force varies with the vertical location on the cross-section. Its integration over the cross-section becomes constant, which is the shear force at that cross-section.

The shear-flow q carries a unit of [N/m]. It does not change with z, but changes continuously with y because there is no concentrated force in the y-direction at any point in the cross-section. The shear-flow has no circulation, and its curl is zero. Due to these properties, we often work with q. Let us call it **q-value**, and its distribution on the cross-section a **q-map**, which is important to obtain when dealing with cross-sections with complicated geometry.

Note that the shear-flow caused by shear force and that caused by torsion behave differently. The shear stress resulting from torsion has a circulation (a curl) round the torsion axis, as discussed in Chapter 9.

12.3 Shear stress on rectangular cross-section

12.3.1 Parabolic distribution

Consider a beam with rectangular cross-section, as shown in Fig. 12.3.

To evaluate the shear stress at location y, we compute the first moment with respect to z-axis of the shaded area A' with bottom edge located at y. Using the last part of Eq. (12.5), it is found as

$$Q_{A'} = y_c A' = \underbrace{\frac{1}{2}\left(\frac{h}{2} + y\right)}_{y_c} \underbrace{\left(\frac{h}{2} - y\right) b}_{A'} = \frac{b}{2}\left[\left(\frac{h}{2}\right)^2 - y^2\right] \tag{12.9}$$

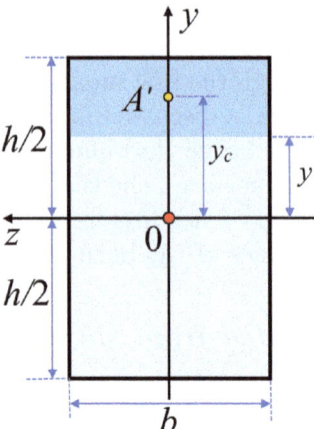

Figure 12.3. Rectangular cross-section for thin beams, on which the shear stress is evaluated. The centroid of the shaded area A' is denoted by y_c.

This equation will be frequently used for computing Q of an area A' for rectangular areas. It is a parabolic function in y.

Substitute Eq. (12.9) into Eq. (12.8), and note that for rectangular cross-section, $t_y = b$, we obtain the shear-flow as

$$q = \frac{bV}{2I}\left[\left(\frac{h}{2}\right)^2 - y^2\right] = \frac{6V}{h^3}\left[\left(\frac{h}{2}\right)^2 - y^2\right] \qquad (12.10)$$

The shear stress becomes

$$\tau = \frac{V}{2I}\left[\left(\frac{h}{2}\right)^2 - y^2\right] = \frac{6V}{bh^3}\left[\left(\frac{h}{2}\right)^2 - y^2\right] \qquad (12.11)$$

It is clear that shear stress is zero at $y = \pm\frac{h}{2}$. It has a maximum value at $y = 0$:

$$\tau_{max} = \frac{Vh^2}{8I} = \frac{3V}{2A} \qquad (12.12)$$

It is 1.5 times the average shear stress. In the foregoing equation, we used $I = \frac{bh^3}{12}$, and the area $A = bh$.

The shear stress distributed parabolically along the y-axis and uniform along the z-axis. It is drawn in Fig. 12.4.

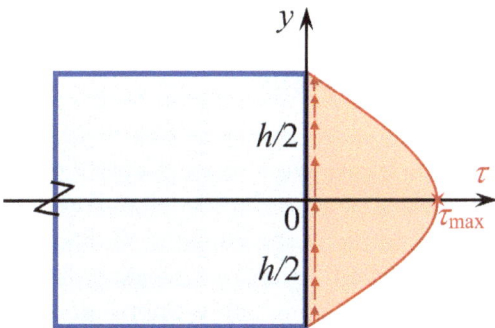

Figure 12.4. Parabolic distribution of shear stress in the vertical direction on a rectangular cross-section.

We have a useful finding:

> the shear stress has a parabolic profile in y, if the cross-section is rectangular.

This finding can be used to draw the profile if a cross-section consists of rectangles, such as in I-beams.

If we integrate it over the cross-section, it should give the shear stress V on the cross-section. Let us use the following Sympy code to confirm this.

```
1  V, y, h, b = symbols("V, y, h, b")
2  I = b*h**3/12
3  sp.integrate(V*b*(h**2/4 - y**2)/(2*I), (y, -h/2, h/2))
```

V

12.3.2 Alternative approach

It may be mentioned here that there is a simple approach to obtain Eq. (12.11) if the cross-section is rectangular shape. First, from the boundary condition, the shear stress must be zero at $y = \pm\frac{h}{2}$. In addition, it must be a continuous function of y. Thus, it can be written in the following form:

$$\tau = c\left[\left(\frac{h}{2}\right)^2 - y^2\right] \qquad (12.13)$$

where c is a constant. Next, from the equilibrium condition, the integral of it must be the shear force V. We can then integrate the foregoing expression over the cross-section, and force it to be V to determine c. The result is exactly the same as Eq. (12.11).

12.4 Shear stress on thin-wall cross-sections

12.4.1 *Setting of the problem*

The formulation derived in the previous section works well for beams with cross-sections on which the thickness t_y does not change significantly. In engineering applications, however, we often use beams with thickness changing significantly. A typical example is the I-beams. Its flange width is far larger than the web thickness. In these cases, the detailed distribution of the shear stress (in the flange for example) is not properly quantified by Eq. (12.7). For these type of beams, we need new approaches.

Many widely used beams (such as I-beams, T-beams and L-beams) has a common significant feature: the cross-section consists of thin walls, as shown in Fig. 12.5. Such designs are particularly good for efficient use of materials and the structure can be made light.

For this type of thin-wall beams, the cross-section are typically consisting of narrow channels. Each of the channel has a thickness t and length l with $t \ll l$. For most beams used in engineering, the channels are largely straight (narrow rectangles), but can also be curved (e.g. circular or elliptic). Since the thickness is small, we often use the middle line of the channel to represent it.

At any location in the channel, we can always set up a local orthogonal Cartesian coordinate system, say (s, r, x) with s along with the length direction of channel, and x along with the centroid axis of the beam perpendicular to the cross-section plane, as shown in Fig. 12.5.

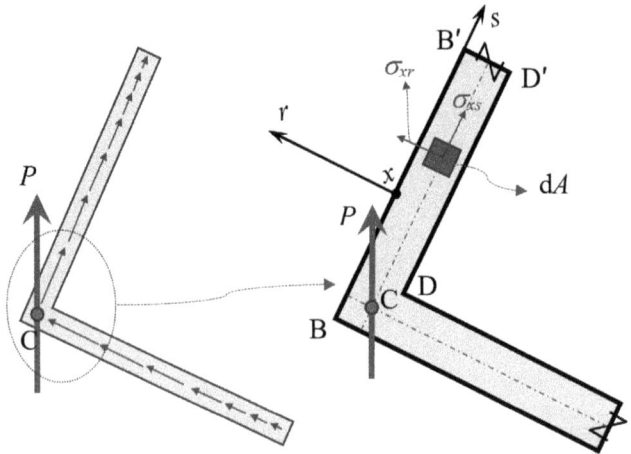

Figure 12.5. Shear stress distribution along thin walls of the cross-section of an L beam. The transverse force P goes through the shear-center.

12.4.2 Assumptions

We can make the following assumptions to each of the channels:

1. Because shear stress component σ_{xr} is zero on surfaces BB' and DD', and the channel thickness t is small with respect to the channel length, σ_{xr} should not have significant value. Hence, σ_{xr} is assumed as zero at any point in the channel.
2. The shear stress component σ_{xs} shall be approximately constant through the channel thickness, and equals to the average shear stress in the channel. Because there is only one stress component, we denote σ_{xs} as τ.
3. In addition, τ at any point in a channel is assumed to be along the tangential direction of the channel that may be curved.

12.4.3 Free-body diagram and formulation

With these assumptions, we can then have a free-body diagram shown in Fig. 12.6, which is essentially the same as that given in Fig. 12.2. Therefore, the derivation and the formulas obtained in Section 12.2 are all applicable to thin-wall cross-section beams.

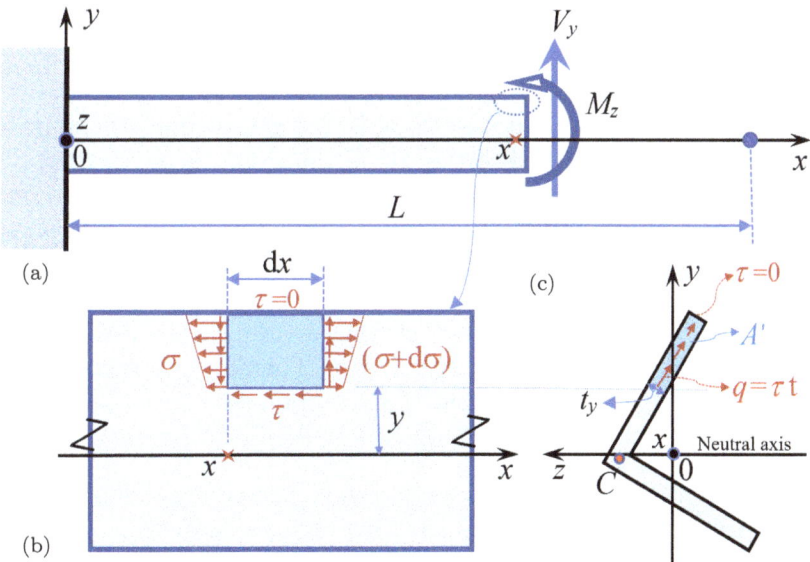

Figure 12.6. (a) Internal forces on a cross-section of a beam of thin-wall cross-section with two channels. (b) A free-body diagram with an infinitely small length segment dx for a thin-wall cross-section. It is the shaded area starting from beam top surface to y, at which the shear stress is evaluated. (c) The free-body diagram viewed on the cross-section.

The difference is that the shear stress in a thin-wall flows along the tangential direction of the wall curve, and not necessarily in the y-direction. In addition, because the wall is thin, we can use its middle line for measuring its length, when calculating the first moment of area $Q_{A'}$ in Eq. (12.7).

12.4.4 Shear-flow on thin wall cross-sections

Because of the geometric complexity of thin wall cross-sections, the shear stress distribution is expected to be complicated. Shear-flow q defined in Eq. (12.8) is thus particularly useful for thin wall cross-sections. This is because a flow map of q, called q-map, determines all the major characteristics of cross-section under a shear force. The major characteristics are listed as follows:

1. The thicknesses of the channels forming the cross-section are more or less the same. The shear stress will not change too drastically.
2. The distribution of q follows always the channel length direction, and varies continuously, even if the thickness of the channels are different.
3. Most importantly, q is conserved at any location in the thin-wall. When a channel branches to multiple channels at a juncture, the shear-flow out of a channel at the juncture must be the net-in to all remaining channels at the juncture. This is useful in computing the shear-flows in each of the channels.
4. The integral of q for any channel on a cross-section gives a force on the channel. The sum of the forces of all the channels is the given shear force V on the cross-section. Because V is in the y-direction, the sum of the forces projected on the z-direction of all the channels will be zero.
5. The moment about x-axis (in reference to a point on the cross-section) resulting from all the channel forces forms the torque with respect to the x-axis. This is useful to find out the location of the shear-center of the cross-section.
6. Calculation of the shear stress is simple: dividing the shear-flow by the local thickness of the channel t: $\tau = q/t$. Therefore, it is convenient to work out the shear-flow first and then compute the shear stress in these channels.

12.5 Example: Shear stress on the cross-section of a C-beam

Consider a C-beam with thin-wall cross-section of uniform thickness t subjected to a vertical shear force V that passes through the shear-center C, as shown in Fig. 12.7(a). The cross-section is symmetric with respect to the

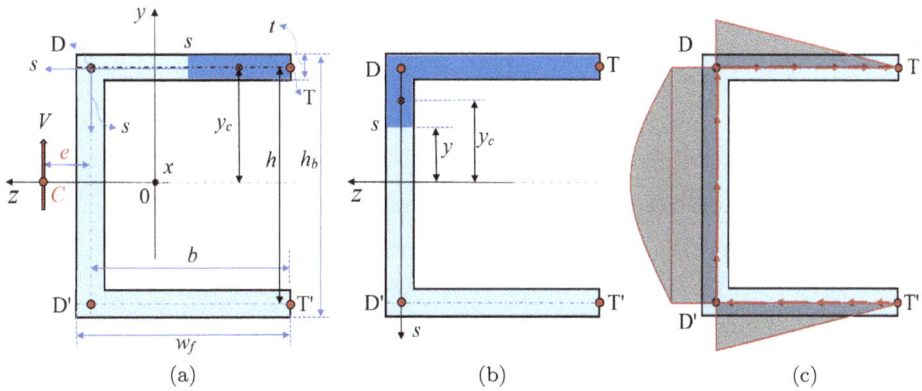

Figure 12.7. A thin-wall cross-section with three thin channels. (a) Dimensions and coordinates; local s coordinate located in the flange and web; (b) local s coordinate located in the web; (c) shear stress distribution (q-map).

z-axis. It has a base b and height h measured from middle lines. Derive the formulas for calculating the shear stress distributed on the cross-section.

12.5.1 Shear-flow and stress at any point in the flange

To compute the shear-flow, we need to compute the first moment of the area A' starting from a point where the shear stress is zero. Let us work from the flange on the top. We know at point T, the shear stress is zero because of free surface there. Set the local coordinate s starting from T, and move by distance s in the top flange shown in Fig. 12.7(a). The area A' is shaded darker. Because in the flange, the y coordinate of the middle line does not change, A' shall grow linearly with increase of s. Therefore, for any point on the flange, the Q value, $Q_{A'}$ (with respect to z-axis), must be linear in s. It can be calculated as

$$Q_{s-\text{flange}} = t y_c s = \frac{th}{2} s \qquad (12.14)$$

where y_c is the y value of the centroid of A', and t is the thickness, of the flange. The shear-flow is

$$q_{s-\text{flange}} = \frac{V}{I} Q_{s-\text{flange}} = \frac{thV}{2I} s \qquad (12.15)$$

where I is the second moment of area for the cross-section that can be computed using the techniques or code given in Chapter 10. Because the channels are thin, I can also be approximated using

$$I = \underbrace{\frac{1}{12}th^3}_{\text{for web}} + \underbrace{2\ tb\left(\frac{h}{2}\right)^2}_{\text{for flange}} = \frac{1}{12}th^3 + \frac{1}{2}tbh^2 = \frac{th^2}{12}(h+6b) \qquad (12.16)$$

The shear stress becomes

$$\tau_{s-\text{flange}} = \frac{q_{s-\text{flange}}}{t} = \frac{hV}{2I}s \qquad (12.17)$$

It is found that the shear stress (or shear-flow) is a linear function of s. At point D, $s = b$, we have

$$q_D = \frac{bhtV}{2I}$$

$$\tau_D = \frac{bhV}{2I} \qquad (12.18)$$

The distribution is a triangle shown in Fig. 12.7(c).

12.5.2 Shear stress at any point s in the web

Inside the web, area A' is the sum of that of the flange and the additional area in the web: ts, as shown in Fig. 12.7(b). To compute the the first moment of ts, we shall use the work done for the rectangular cross-section in Section 12.3 because the web is a rectangle. The Q value at y in the web becomes (see, Eq. (12.9)):

$$Q_{s-\text{web}} = \underbrace{tb\frac{h}{2}}_{Q_{\text{flange}}} + \frac{t}{2}\left[\left(\frac{h}{2}\right)^2 - y^2\right] \qquad (12.19)$$

The shear-flow at y in the web can be found as

$$q_{\text{web}} = \frac{V}{I}Q_{s-\text{web}} = \frac{tV}{2I}\left[bh + \left(\frac{h}{2}\right)^2 - y^2\right] \qquad (12.20)$$

The shear stress at y becomes

$$\tau_{\text{web}} = \frac{V}{2I}\left[bh + \left(\frac{h}{2}\right)^2 - y^2\right] \qquad (12.21)$$

The foregoing three formulas work for any point in the web, including point D' at the bottom of the web. At point D', we have exactly the same results given in Eq. (12.18), by substituting $y = -h/2$ in these three equations. This is due to the symmetry of the cross-section with respect to the

z-axis. In addition, the shear stress distribution along the bottom flange must also be mirror image of that on the top flange. The distribution of the shear-flow in the web is parabolic in y, as studied in Section 12.3. The complete results of shear stress and shear-flow are schematically drawn in Fig. 12.7(c).

12.6 Example: Shear stress on the cross-section of an L-beam

Consider an L-beam (inclined 45°) with thin-wall cross-section of uniform thickness t subjected to a vertical shear force V that passes through the shear-center C, as shown in Fig. 12.8(a). The cross-section is symmetric with respect to z-axis. The leg length is h_L measured from middle lines. Derive the formulas for calculating the shear stress distributed on the cross-section.

Because the shear force is in y-direction, the first and second moments of area of the cross-section is with respect to z-axis. Therefore we can project the area on the y-axis without affecting these area moments, as long as the area is unchanged. This gives a rectangular cross-section as shown in Fig. 12.8(b), for which equations for shear-flow given in Section 12.3 apply. To ensure the area unchanged, the base of the rectangle become $\sqrt{2}t$, and the height becomes $\sqrt{2}h_L$. The second moment of area of the L-beam can now be calculated using

$$I_L = \frac{\sqrt{2}t(\sqrt{2}h_L/)^3}{12} = \frac{th_L^3}{3} \qquad (12.22)$$

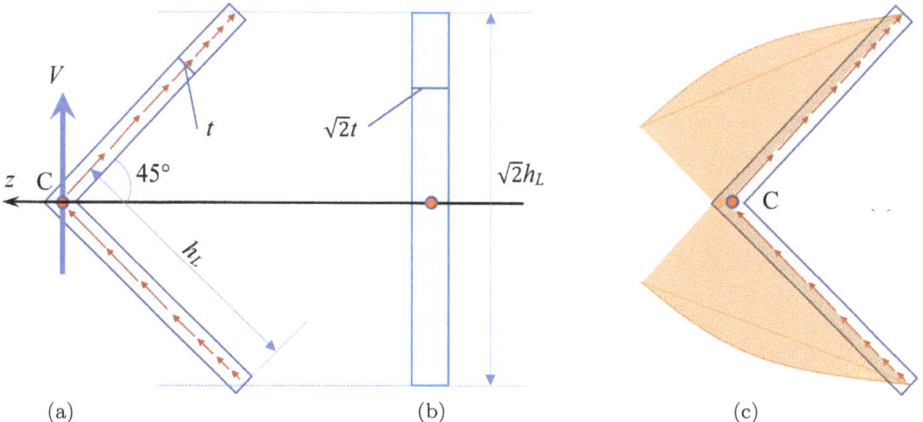

Figure 12.8. A thin-wall L-shaped cross-section with two thin channels of equal length. (a) Dimensions and coordinates; (b) area projected in y-axis; (c) shear stress shear-flow distribution (q-map).

The shear-flow can be calculated using Eq. (12.10):

$$q = \frac{\sqrt{2}tV}{2I}\left[\left(\frac{2h_L/\sqrt{2}}{2}\right)^2 - y^2\right] = \frac{3V}{\sqrt{2}h_L^3}\left[\left(\frac{h_L^2}{2}\right) - y^2\right] \quad (12.23)$$

These formula can be derived using the following code.

```
1  V, y, h, hL, b, t, I, IL = symbols("V, y, h, h_L, b, t, I, I_L")
2  A_Lbeam = 2*(t*hL)                                    # original area
3  A_Rctgl = sp.sqrt(2)*t*sp.sqrt(2)*hL                  # projected area
4  print(f'Is A_Lbeam = A_Rctgl? {A_Lbeam==A_Rctgl}')    # these must equal
5
6  I_R = b*h**3/12                          # formula for I value for rectangles
7  ILp = I_R.subs({b:sp.sqrt(2)*t,h:sp.sqrt(2)*hL}) # I for projected area
8  print(f'2nd moment of area of L-beam with 45 angle = {ILp}')
```

Is A_Lbeam = A_Rctgl? True
2nd moment of area of L-beam with 45 angle = h_L**3*t/3

```
1  q = V*b*(h**2/4 - y**2)/(2*I_R)    # formula for q value for rectangles
2  q_L = q.subs({h:sp.sqrt(2)*hL})
3  print(f'The maximum shear-flow = {q_L.subs(y,0)}')
4  gr.printM(q_L, "The formula for calculating shear-flow q:")
```

The maximum shear-flow = 3*sqrt(2)*V/(4*h_L)
The formula for calculating shear-flow q:

$$\frac{3\sqrt{2}V\left(\frac{h_L^2}{2} - y^2\right)}{2h_L^3}$$

The shear stress in the L-beam cross-section becomes

$$\tau = \frac{q}{t} = \frac{\sqrt{2}V}{2I}\left[\left(\frac{2h_L/\sqrt{2}}{2}\right)^2 - y^2\right] = \frac{3V}{\sqrt{2}h_L^3 t}\left[\left(\frac{h_L^2}{2}\right) - y^2\right] \quad (12.24)$$

Note here that we need to use the original thickness of the L-beam. The maximum value at $y = 0$:

$$\tau_{max} = \frac{3\sqrt{2}V}{4th_L} = \frac{3}{\sqrt{2}}\frac{V}{A_L} \quad (12.25)$$

It is about 2.12 times the average shear stress. The distribution of the shear-flow (or shear stress) is schematically plotted in Fig. 12.8(c).

12.7 Shear stress on an I-beam cross-section*

12.7.1 *Problem analysis*

Consider an I-beam made of three thin channels of possibly different lengths, as shown in Fig. 12.9. Because the beam is symmetric with respect to y-axis, the shear-center C is on the y-axis. We assume that the dimensions of the I-beam are all given.

Let the Cartesian coordinate have its origin at the cross-section centroid that can be found using gr.centroid_I() presented in Chapter 10. The cross-section has a given shear force V in the y-direction and hence passing through the shear-center. The 2nd moment of area I (with respect to z-axis) can be computed, using gr.secondMoments_I() developed in Chapter 10.

The cross-section consists of three rectangles. For convenience in discussion, we divide it into five rectangular channels marked with boxed number in Fig. 12.9(a). Each channel has a constant thickness. Channel-1 branches to channel-2 and -3 at juncture point D, and channel-4 and -5 at juncture point D'.

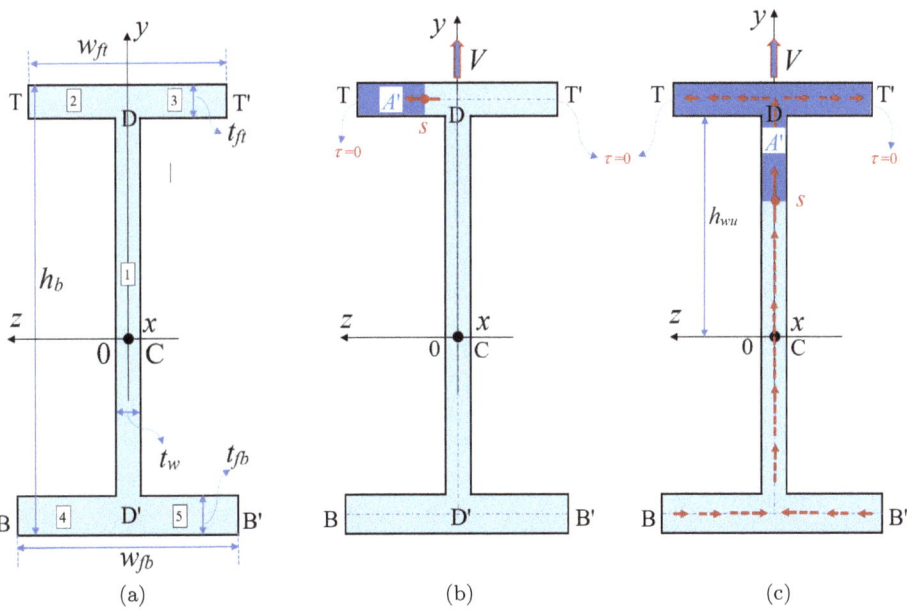

Figure 12.9. (a) A typical thin-wall cross-section of an I-beam with 5 channels. (b) Area A' for computing the shear flow in the flange. (c) Area A' for computing the shear flow in the web.

Because the geometry of the I-beam is discontinuous, we shall introduce local coordinate s for each of the channels. For channel-1, for example, s starts from its left vertical edge where shear stress is zero. For the web, it starts at point D on the top-most-point of the web. Because the rectangles are assumed thin, local coordinate s stays on the middle line of the rectangles.

12.7.2 Calculation procedure

12.7.2.1 Shear-flow and stress at any point in the flange

Let us work from the flange on the top. We know at points T and T', the shear stress is zero. Thus, we can form area A' starting from either points. Let us start from T, and move by distance s in channel-2 shown in Fig. 12.9(b). The area formed is shaded darker. Because on the flange, the y coordinate does not change, the area grows linearly with increase in s. Therefore, for any point on the flange, the Q value, $Q_{A'}$ (with respect to z-axis), must be linear in s. It can be calculated as

$$Q_{s-\text{flange}} = t_{ft} s y_{ftc} \tag{12.26}$$

where y_{ftc} is the y value of the centroid, and t_{ft} is the thickness of the flange on the top. The shear-flow is

$$q_{s-\text{flange}} = \frac{V}{I} Q_{s-\text{flange}} = \frac{V}{I} t_{ft} s y_{ftc} \tag{12.27}$$

and the shear stress becomes:

$$\tau_{s-\text{flange}} = \frac{q_{s-\text{flange}}}{t_y} = \frac{V}{I} y_{ftc} s \tag{12.28}$$

Here, we used $t_y = t_{ft}$ because it is in the flange.

12.7.2.2 Shear-flow and stress at point D in the flange

The largest stress value within the flange should be at point D where $s = w_{ft}/2$. The Q value there is computed using

$$Q_{D-\text{flange}} = t_{ft} \frac{w_{ft}}{2} y_{ftc} \tag{12.29}$$

The shear-flow is found as

$$q_{D-\text{flange}} = \frac{V}{I} Q_{D-\text{flange}} = \frac{V}{2I} w_{ft} t_{ft} y_{ftc} \tag{12.30}$$

The shear stress becomes

$$\tau_{D-\text{flange}} = \frac{V}{I t_{ft}} Q_{D-\text{flange}} = \frac{V}{2I} w_{ft} y_{ftc} \tag{12.31}$$

The same can be done starting from point T' and working leftwards to point D along channel-3. The shear-flow value will be the same if the I-beam is symmetric.

The shear stress, and hence the shear-flow is linear in s as shown in Eq. (12.28). The distribution profile is a right triangle over both channel-2 and -3. It is zero at points T and T'. This will be plotted in Fig. 12.10.

12.7.2.3 Shear stress at point D in the web

At point D in the web, area A' is suddenly doubled because channel-2 and -3 merge to channel-1. Therefore, $Q_{A'}$ value is doubled:

$$Q_{D-web} = \underbrace{2y_{ftc}t_{ft}(w_{ft}/2)}_{Q_{A'} \text{ at D at top of web}} = t_{ft}w_{ft}y_{ftc} \tag{12.32}$$

The shear-flow there should be twice of that at point D in the flange:

$$q_{D-web} = \frac{V}{I}Q_{A'} = \frac{V}{I}t_{ft}w_{ft}y_{ftc} \tag{12.33}$$

The stress at point D in the web becomes

$$\tau_{D-web} = \frac{q_{D-web}}{t_y} = \frac{V}{I}\frac{t_{ft}}{t_w}w_{ft}y_{ftc} \tag{12.34}$$

Here, we used $t_y = t_w$.

12.7.2.4 Shear stress at any point s in the web

Inside the web, channel-1, area A' is the sum of that of the flange and the additional area in the web: $t_w s$. Therefore, $Q_{A'}$ value is the sum of the Q value for the flange, both channel-2 and -3, and the Q value of the additional area in the web. To compute the Q value of the additional area in the web, we shall use the work done for the rectangular cross-section in Section 12.3 because the web is a rectangle. To do so, we need to find out the height of the upper part of the web above $y = 0$, which is denoted as h_{wu} shown in Fig. 12.9(c). The formula can be found using length relations:

$$h_{wu} = y_{wc} + (h_b - t_{ft} - t_{fb})/2 \tag{12.35}$$

where y_{wc} is the y value of the centroid of the web. Now, the Q value at s in the web becomes (see, Eq. (12.9)):

$$Q_{A'-web} = \frac{t_{ft}}{t_w}y_{ftc}w_{ft} + \underbrace{t_w(h_{wu}-y)}_{s}\frac{1}{2}(h_{wu}+y) \tag{12.36}$$

The shear-flow at s in the web can be found as

$$q_{\text{web}} = \frac{V}{I} Q_{A'-\text{web}} \tag{12.37}$$

The shear stress at the same location becomes

$$\tau_{\text{web}} = \frac{V}{It_y} Q_{A'-\text{web}} = \frac{V}{It_w} Q_{A'-\text{web}} \tag{12.38}$$

The foregoing three formulas work for any point in the web, including point D' at the bottom of the web. The distribution of the shear-flow in the web is quadratic in y, as studied in Section 12.3. It will be plotted in Fig. 12.10.

12.7.2.5 Shear stress at D' on the top of the bottom flange

Because the shear-flow branches into two flows in channel-4 and -5, when gets into the flange (or two shear-flows from the flange converge to one flow). The shear stress at D' on the top of the bottom flange can be computed using the shear-flow at D' obtained in Eq. (12.37), by dividing it by two:

$$q_{Dp-\text{flange}} = \frac{1}{2} q_{\text{web}} \tag{12.39}$$

where $q_{Dp-\text{flange}}$ is for one of the branches: channel-4 or -5. The stress in the flange becomes

$$\tau_{Dp-\text{flange}} = \frac{q_{Dp-\text{flange}}}{2} \frac{1}{t_{fb}} = \frac{1}{2} q_{\text{web}} / t_{fb} \tag{12.40}$$

Alternatively, we can do this in the same way as in Eqs. (12.29) and (12.31):

$$Q_{Dp-\text{flange}} = y_{fbc} t_{fb} \frac{w_{fb}}{2} \tag{12.41}$$

Note that y_{fbc} is the y coordinate of the centroid of the bottom flange, which is negative. And t_{ft} is the thickness of the bottom flange. The shear-flow is

$$\tau_{Dp-\text{flange}} = \frac{V}{It_{fb}} Q_{Dp-\text{flange}} = \frac{V}{2I} y_{fbc} w_{fb} \tag{12.42}$$

The shear-flow (and the stress) is linear, and the distribution profile is a right triangle over both channel-4 and -5. It is zero at points B and B'. This will be plotted in Fig. 12.10.

All the complicated tedious formulas given above can be directly coded in Python.

12.7.3 Example: Python code for computing the shear-flow and stress in an I-beam

Consider an I-beam with basic shape shown in Fig. 12.9(a). Its dimensions (mm) are given as follows:

beam height: $h_b = 200$; width of flange on top: $w_{ft} = 100$; width of flange at bottom: $w_{fb} = 150$; thickness of flange on top: $t_{ft} = 20$; thickness of flange at bottom: $t_{fb} = 18$; web thickness: $t_w = 10$.

Assume the shear force $V = 0.1MN$. Compute the distribution of the shear stress in the I-beam.

Solution:

12.7.3.1 Data preparation

First, prepare the input data, and compute the area properties.

```
1  # partial checked with Fig.6.8 in the textbook by Beer using data:
2  #hb =.140; wft=.100; wfb=.100; tft=.02; tfb=.02; tw=.02   # I-beam (m)
3
4  # preparing date:
5  # Symmetric case for test:
6  #hb =.200; wft=.100; wfb=.100; tft=.020; tfb=.020; tw=.010   # I-beam (m)
7
8  # data for this example:
9  hb =.200; wft=.100; wfb=.150; tft=.020; tfb=.018; tw=.010   # I-beam (m)
10
11 # call the code function to compute the sectional properties:
12 Ixx, Iyy, Ixy, x_bar, y_bar, xi_bar, yi_bar, A, Ai =\
13     gr.secondMoments_I(hb, wft, wfb, tft, tfb, tw, [0]*2, [0]*2, sy=True)
```

```
Center of sub-areas [L]:
x = [7.5000e-02 7.5000e-02 7.5000e-02]
y = [1.9000e-01 9.9000e-02 9.0000e-03]
Areas of sub-areas = [2.0000e-03 1.6200e-03 2.7000e-03] [L^2]
Areas of I-beam = 0.00632 [L^2]
The centroid = (0.075, 0.0893481012658228) [L]
New center of sub-areas [L]:
x = [0.0000e+00 0.0000e+00 0.0000e+00]
y = [1.0065e-01 9.6519e-03 -8.0348e-02]
2nd moments [L^4]:
  Ixx = 4.152574084388186e-05
  Iyy = 6.742666666666666e-06
  Ixy = 0.0
```

12.7.3.2 Compute shear-flow and stress at D in the top flange

```
1  # Compute shear-flow and stress at D in the top flange
2
3  V = 0.1e6      # (N)
4  Izz = Ixx      # in our coordinates, z is x in the gr.secondMoments_I()
5
6  Q_D_flange = yi_bar[0]*tft*wft/2
7
8  q_D_flange = V/Izz * Q_D_flange
9  print(f"shear-flow at D in top flange,  q = {q_D_flange*1e-6} [MPa/m]")
10
11 τ_D_flange = V/(Izz*tft) * Q_D_flange
12 print(f"Shear stress at D in top flange, τ = {τ_D_flange*1e-6} [MPa]")
```

shear-flow at D in top flange,   q = 0.24238435411082282 [MPa/m]
Shear stress at D in top flange, τ = 12.119217705541141 [MPa]

12.7.3.3 Compute the shear-flow and stress at D on the top of the web

```
1  # compute the shear-flow and stress at D on the top of the web
2  Q_Dweb = yi_bar[0]*wft*tft
3  q_D_web = V/Izz * Q_Dweb
4  print(f"shear-flow at D in web,  q = {q_D_web*1e-6} [MPa/m]")
5
6  τ_D_web = V/(Izz*tw) * Q_Dweb
7  print(f"Shear stress at D in web, τ = {τ_D_web*1e-6} [MPa]")
```

shear-flow at D in web,   q = 0.48476870822164575 [MPa/m]
Shear stress at D in web, τ = 48.47687082216458 [MPa]

12.7.3.4 Compute shear-flow and stress at any s in web

```
1  # Compute shear-flow and stress at key points in web
2  hw = hb - tft/2 - tfb/2                  # the height of the web
3  hwu = yi_bar[1] + hw/2                   # height above z=0
4
5  ys = 0            # at the neutral axis, τ_max
6  Q_sweb = Q_Dweb + tw*(hwu - ys)*(hwu + ys)/2
7
8  q_0_web = V/Izz * Q_sweb
9  print(f"shear-flow at the neural axis,  q = {q_0_web*1e-6} [MPa/m]")
10
11 τ_0_web = V/(Izz*tw) * Q_sweb
12 print(f"Shear stress at the neural axis, τ = {τ_0_web*1e-6} [MPa]")
```

shear-flow at the neural axis,   q = 0.6055420239557431 [MPa/m]
Shear stress at the neural axis, τ = 60.55420239557431 [MPa]

Shear Stress by Shear Force

```
1  # Compute shear-flow and stress at key points in web
2
3  hw1 = yi_bar[1] - hw/2              # height below z=0
4  ys = hw1                            # D' at the bottom of the web
5  Q_sweb = Q_Dweb + tw*(hwu-ys) * (hwu+ys)/2
6
7  q_Dp_web = V/Izz * Q_sweb
8  print(f"shear-flow at D' in web,  q = {q_Dp_web*1e-6} [MPa/m]")
9
10 τ_Dp_web = V/(Izz*tw) * Q_sweb
11 print(f"Shear stress at D' in web, τ = {τ_Dp_web*1e-6} [MPa]")
```

shear-flow at D' in web, q = 0.5268388467762832 [MPa/m]
Shear stress at D' in web, τ = 52.68388467762832 [MPa]

12.7.3.5 Compute shear stress at D' at the top of the bottom flange

```
1  # Compute shear stress at D' at the top of the bottom flange,
2  # using the shear-flow conservation property:
3
4  q_Dp_flange = q_Dp_web/2            # q branches to two flows
5  print(f"shear-flow at D' in flange,  q = {q_Dp_flange*1e-6} [MPa/m]")
6
7  τ_Dp_flange = q_Dp_flange/tfb
8  print(f"Shear stress at D' in flange, τ = {τ_Dp_flange*1e-6} [MPa]")
```

shear-flow at D' in flange, q = 0.2634194233881416 [MPa/m]
Shear stress at D' in flange, τ = 14.634412410452313 [MPa]

```
1  # Alternatively, using the standard VQ/I formula:
2
3  Q_Dp_flange = -yi_bar[2]*tfb*wfb/2
4  q_Dp_flange = V/Izz * Q_Dp_flange
5  print(f"shear-flow at D' in flange,  q = {q_Dp_flange*1e-6} [MPa/m]")
6
7  τ_Dp_flange = V/(Izz*tfb) * Q_Dp_flange
8  print(f"Shear stress at D' in flange, τ = {τ_Dp_flange*1e-6} [MPa]")
```

shear-flow at D' in flange, q = 0.2612113221932849 [MPa/m]
Shear stress at D' in flange, τ = 14.511740121849163 [MPa]

12.7.4 Plot of shear-flow in the web

We use the following code to plot the shear-flow distribution in the web of the I-beam.

```
1  dx = 0.005
2  y_web = np.arange(hwl, hwu+dx, dx)             # y_web values as x_data
3
4  Q_sweb = Q_Dweb + tw*(hwu - y_web)*(hwu + y_web)/2
5  q_0_web = V/Izz * Q_sweb
6
7  step = max(int(len(y_web)/5), 1)
8  print(f"y values in the web (m):\n {np.array(y_web)[::step]}")
9  print(f"The shear-flow in the web (MPa/m):\n {q_0_web[::step]*1e-6}")
10
11 x_data = [y_web]
12 y_data = [q_0_web*1e-6]                        # put the function values to a list
13 labels=['shear-flow (MPa/m)']                  # define the label
14 gr.plot_fs(x_data, y_data, labels, 'y', 'shear-flow q (MPa/m)', 'temp')
15                                                # x-label,       y-label,       file-name
```

y values in the web (m):
 [-8.0848e-02 -4.5848e-02 -1.0848e-02 2.4152e-02 5.9152e-02 9.4152e-02]
The shear-flow in the web (MPa/m):
 [5.2684e-01 5.8023e-01 6.0413e-01 5.9852e-01 5.6341e-01 4.9881e-01]

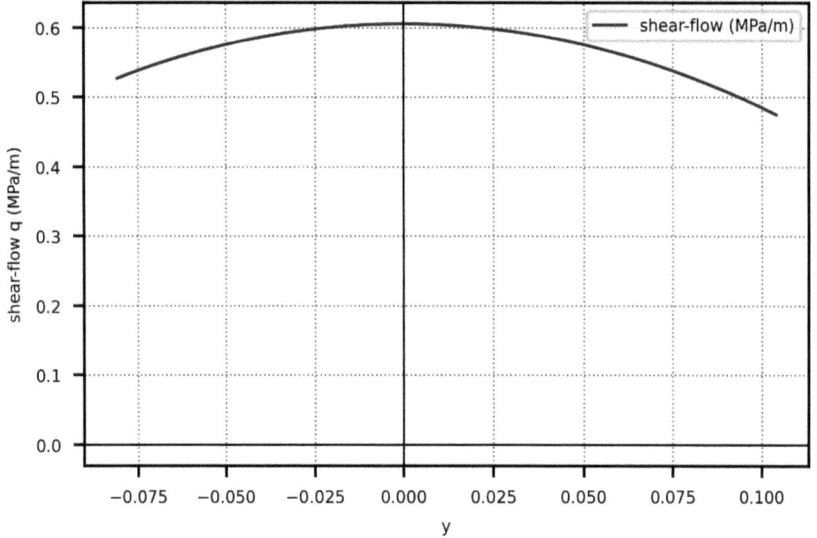

The final results on the shear-flow map are given in Fig. 12.10.

12.7.5 Compute forces in channels

After shear-flow is obtained, we can then compute the resultant force in each of the channels in the cross-section. This is done by integrating the shear-flow over the channel length. The general formula can be written as

$$F = \int_l q \, ds \qquad (12.43)$$

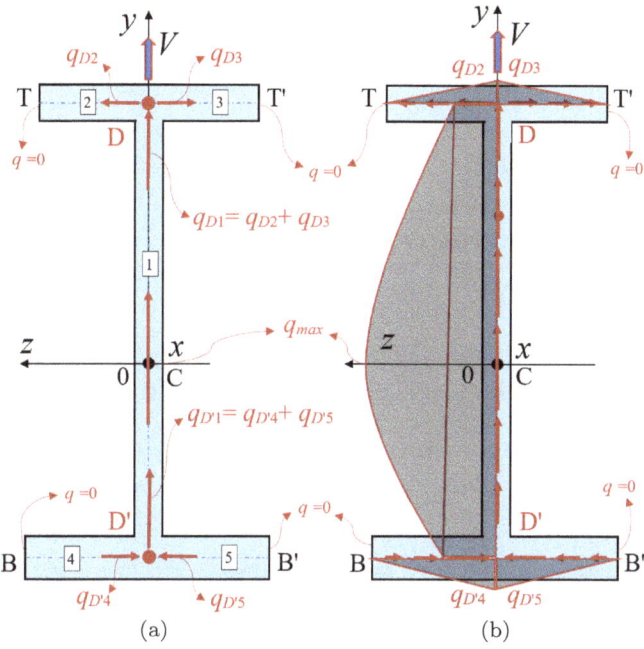

Figure 12.10. Typical shear-flow in an I-beam. (a) Shear-flow conservation at junctions of branches; (b) q-map on the channels of the I-beam.

where l is the length of the channel, and q is the shear-flow in the channel. The direction of the force follows the shear-flow direction. If the channel is not orientated orthogonal to the coordinates, one may need to project the resultant force to the coordinate directions to obtain the components of the force. For most of the beams used in engineering, the channels are rectangular and are arranged orthogonal to the coordinate. In such cases, the integration can be done by calculating the area under the shear-flow profile. For example, in the results given in Fig. 12.10(b), the force on the flanges is the area of the triangles. The formula for computing the area under a parabolic function can also be done easily. Following are some Python codes to do such computations. We use the same I-beam example studied before.

```
1  # Computing the force in the flange of I-beam.
2  F_flangeTop = -0.5*q_D_flange*wft
3  print(f"Force in the right top flange (kN): {F_flangeTop*1e-3}")
4
5  F_flangeBot = 0.5*q_Dp_flange*wfb
6  print(f"Force in the right bottom flange (kN): {F_flangeBot*1e-3}")
```

```
Force in the right top flange (kN): -12.119217705541143
Force in the right bottom flange (kN): 19.59084916449637
```

```
1  # Sometimes it is more convenient using numerical integration:
2
3  qt_s = lambda s: -q_D_flange/wft * s           # define a function for q
4  F_topFlange, er = si.quad(qt_s, 0, wft)        # use Gauss int.
5  print(f"Force in the right top flange={F_topFlange*1e-3}(kN),\
6     int_error= {er:.4e}")
7
8  qb_s = lambda s: q_Dp_flange/wfb * s
9  F_botFlange, er = si.quad(qb_s, 0, wfb)        # use Gauss int.
10 print(f"Force in the right bottom flange={F_botFlange*1e-3}(kN),\
11    int_,error={er:.4e}")
```

Force in the right top flange=-12.119217705541141(kN), int_error= 1.3455e-10
Force in the right bottom flange=19.59084916449637(kN), int_,error=2.1750e-10

Let us compute the force in the web of I-beam. We define first a function for computing the shear-flow at any location in the web.

```
1  # Define the distribution function:
2  qw_s = lambda y: V/Izz * (Q_Dweb + tw*(hwu - y)*(hwu + y)/2)
3  F_webG, error = si.quad(qw_s, hwl, hwu)        # use Gauss int.
4
5  print(f"Force in the web (kN): {F_webG*1e-3}, int_error= {error:.4e}")
6  print(f"The true shear force V (kN): {V*1e-3}")
```

Force in the web (kN): 103.45021322740982, int_error= 1.1485e-09
The true shear force V (kN): 100.0

```
1  # Computing the force in the web of I-beam. Use simpler trapezoidal rule
2  dx = 0.001
3  y_web = np.arange(hwl, hwu+dx, dx)             # y_web values as x_data
4
5  step = max(int(len(y_web)/10), 1)
6  print(f"y values in the web (m):\n {np.array(y_web)[::step]}")
7  print(f"The shear-flow in the web (MPa/m):\n {q_0_web[::step]*1e-6}")
8
9  F_web = np.trapz(qw_s(y_web), y_web)
10 print(f"Force in the web     : {F_web*1e-3}(kN)")
11 print(f"Actual shear force V: {V*1e-3}(kN)")
```

y values in the web (m):
 [-8.0848e-02 -6.2848e-02 -4.4848e-02 -2.6848e-02 -8.8481e-03 9.1519e-03
  2.7152e-02 4.5152e-02 6.3152e-02 8.1152e-02 9.9152e-02]
The shear-flow in the web (MPa/m):
 [5.2684e-01 6.0453e-01 4.8717e-01]
Force in the web   : 103.44984999889122(kN)
Actual shear force V: 100.0(kN)

We obtained the force in the web via numerical integration. Compared to the given true shear force V, we got about 3% error. It is a good approximation. The error is because of the thin wall approximation and use of the middle line to compute the Q values for these channels. Readers may confirm this by artificially setting the thickness of all the channels at very small value, and observing the reduction of error.

Readers may modify the codes given above to compute shear-flows and stresses of beam with other types of cross-sections.

12.8 Shear-center of a cross-section

12.8.1 *Setting of the problem*

We mentioned several times that transverse forces on beams should go through the so-called **shear-center**, so that no torque will be introduced to the beam. Figure 12.11 shows the location of shear-center for three different cross-sections, and twist deformations when the transverse force does not go through the shear-center. The major observations are:

1. When a transverse force passes through the shear-center C of a cross-section of the beam, it causes bending without torsion.
2. If the cross-section has either an axis of symmetry or an axis of antisymmetry, C is located on that axis. If the cross-section has two or more axes of symmetry or antisymmetry, the shear-center is located at the intersection of these axes, such as the cases in Figs. 12.11(a) and (c). In such cases, locating the shear-center is easy.
3. If the cross-section has only one or no axis of symmetry (or antisymmetry), as shown typically in Fig. 12.11(b), effort is needed to locate the shear-center.

We now discuss about the procedure and technique to locate the shear-center for a given cross-section.

Consider a C-beam shown in Fig. 12.12.

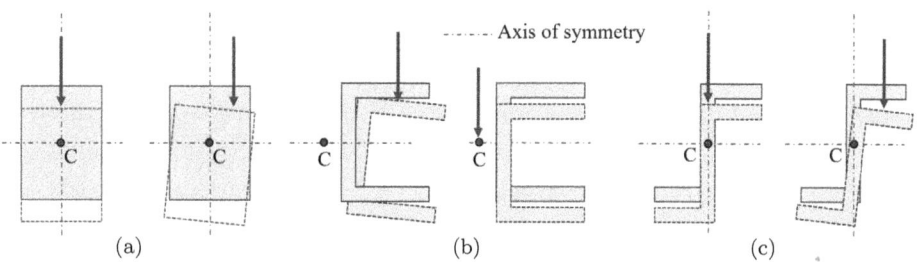

Figure 12.11. Schematic drawings of shear-center C on three typical shapes of cross-section of beams.

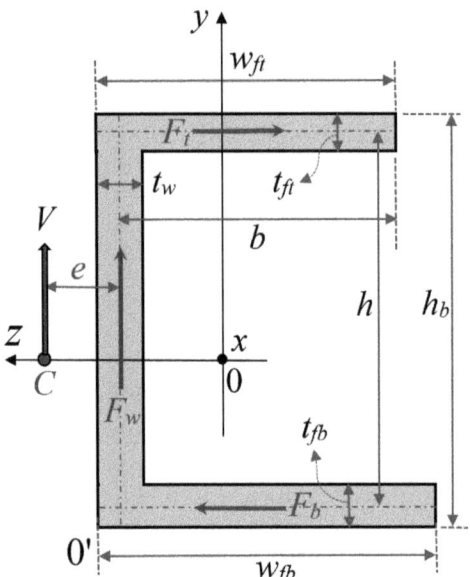

Figure 12.12. Cross-section of a C-beam. It does not have any symmetric axis. The shear-center is located on the left side of the web with an unknown distance e.

The C-beam has three rectangular channels. Its cross-section may not have any symmetric axis, and hence it is not obvious where the shear-center is located. By intuition, the shear-center should be on the left side of the web of the cross-section, but the distance e needs to be determined.

12.8.2 Strategy to find the shear-center

Based on our analysis in the previous section, we know how to calculate the shear-flow on each of the channels of the cross-section, and we demonstrated this for a general I-beam. The shear-flow can then be integrated to compute the shear force on these channels. This is done based on the assumption that the transverse shear force passing through the shear-center and hence no torque is generated. Therefore, the shear forces in these channels are the results of shear force only. We also demonstrated that the shear force in the vertical web channel equals the shear force V on the cross-section. In other words, the shear force V creates the shear force on the vertical channel.

For the C-beam cross-section, we can do exactly the same calculation and find shear-flow and then a force in the vertical web channel for any given shear force V on the cross-section. The channel force is created by V, under the assumption that it does not generate any net torque. At the same time, we will also find the two channel forces: each on a horizontal channel as shown in Fig. 12.12. Force F_t is on the top flange and F_b is on the bottom flange. These two channel forces must be equal in magnitude and opposite in direction because of the equilibrium condition for all forces in the horizontal z-direction. These two forces form a couple that gives a torque about the x-axis, $F_t h$. Because these two forces are also generated by the given vertical force V (just like the F_w is created by V), V must be placed at a distant e from the middle line of the vertical web, so that the torque of eV can create the torque of $F_t h$. Therefore, we have

$$eV = hF_t \tag{12.44}$$

which gives the formula for e.

$$e = \frac{hF_t}{V} \tag{12.45}$$

Note that the value of F_t dependents linearly on V. Therefore, e value will not depend on the value of V. It is a property of the geometry of the cross-section, and its orientation in relation to the direction of V. Therefore, one can use any nonzero amount of V to compute e, such as a unit force, $V = 1$.

12.8.3 Procedure to find the shear-center

The procedure to find e is quite straightforward:

1. Compute the shear-flow and the shear forces on each of the channels on the cross-section, for a unit (or any nonzero) shear force applied on the cross-section.
2. Compute the torque generated by the channel forces with respect to a point on a vertical channel.
3. Use Eq. (12.45) to compute e.

12.8.4 Example: Find the formula to compute e for a C-beam

Consider a C-beam shown in Fig. 12.12. Assume $t_{ft} = t_{fb} = t_f$, and $w_{ft} = w_{fb}$. The channel force F_t can be calculated using Eq. (12.43).

$$F_t = \int_0^b q_t ds = \frac{1}{2}b \, \underbrace{\frac{V}{I} \, \underbrace{t_f b}_{A'} \, \underbrace{\frac{h}{2}}_{\bar{y}'}}_{q_t} = \frac{V t_f b^2 h}{4I} \tag{12.46}$$

where b is the length of the top flange measured by middle lines of the channels, q_t is the shear-flow in the top flange, t_f is the thickness of the top flange, and h is the height of the beam measured between the two middle lines in the two flanges. Because the channels are thin, the second moment of area of the cross-section I can be approximated using

$$I = \frac{1}{12}t_w h^3 + 2t_f b \left(\frac{h}{2}\right)^2 = \frac{1}{12}t_w h^3 + \frac{1}{2}t_f b h^2 \qquad (12.47)$$

Substituting Eq. (12.47) into Eq. (12.46), and then into Eq. (12.45), we obtain:

$$e = \frac{3b}{6 + \alpha_A} \qquad (12.48)$$

where α_A is ratio of the web area and the area of one flange:

$$\alpha_A = \frac{t_w h}{t_f b} \qquad (12.49)$$

It is clear that shear force V is not involved in Eq. (12.48) at all. All we need is the geometric parameters for the cross-section, for the given direction of V.

The approximate locations of the shear-center for several other thin-wall sections are given in Table 8.1 in Ref. [1]. Those are all dependent on geometric parameters of the cross-section.

12.8.5 Example: Python code for computing the shear-center location for a box-beam*

Consider a box-beam with basic shape shown in Fig. 12.13(a). It is from the textbook [1]. Its dimensions (mm) measured based on the middle lines of the edges are given as follows:

---

beam height: $h = 500$; width of top- and bottom-edge: $b = 300$; thickness of top-edge: $t_T = 10$; thickness of bottom-edge: $t_B = 10$; thickness of left-edge: $t_L = 20$; thickness of right-edge: $t_R = 10$.

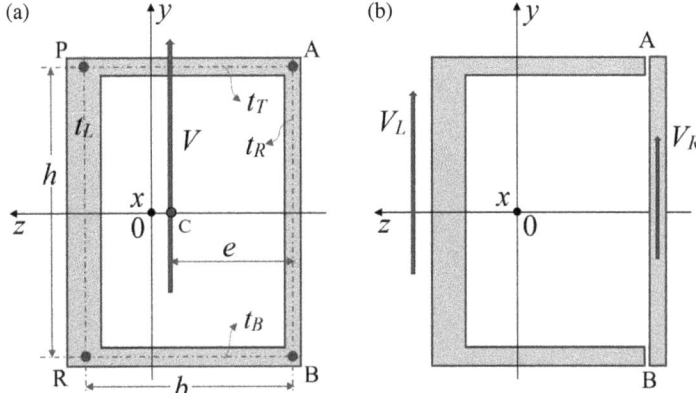

Figure 12.13. Cross-section of a rectangular box-beam with thin walls. It has four edges. The shear-center is located inside of the box with an unknown distance e measured from the middle line of the right-edge. (a) Geometry description with shear force direction given; (b) two divided sections.

Compute the shear-center of this box-beam, with respect to the middle line of the right-edge.

Solution: We shall use the following strategy to solve this problem.

12.8.5.1 Dividing the box to two open cross-sections

Because the cross-section of the beam is closed, we do not know where is the point on the edges that has zero shear-flow. The technique given earlier cannot be used directly to obtain the q-map. Therefore, we perform the analysis in following steps:

1. Cut the box into two sections: one on the left becomes a C-beam and another one on the right is a rectangular beam, as shown in Fig. 12.13(b). Therefore, the shear stress (hence shear-flow) at point A for both beams is now zero. Formulas given in the previous sections apply.
2. Find the q-map for the "open box" with divided two sections separately.
3. Because the shear-flow at point A is not zero in the original box section, superpose an unknown shear-flow, say q_A, uniformly on all edges, and then use the following condition to determine q_A:

$$\oint_{\text{box}} \frac{q_{\text{final}}}{t} dl = 0 \qquad (12.50)$$

where q_{final} is the final superposed q-map, and t is the thickness of the edges. This condition must be satisfied because the torque on the closed box is zero, based on our discussion in Chapter 9.

Because $q_{final} = q_{open\text{-}box} + q_A$, Eq. (12.50) can be written as

$$\oint_{box} \frac{q_{open\text{-}box}}{t} dl + q_A \oint_{box} \frac{1}{t} dl = 0 \qquad (12.51)$$

where $q_{open\text{-}box}$ is the q-map for the "open box" (with divided two sections). The foregoing equation gives,

$$q_A = \frac{\oint_{box} \frac{q_{open\text{-}box}}{t} dl}{\oint_{box} \frac{1}{t} dl} \qquad (12.52)$$

After q_A is found, the superposed q-map is the final q-map for the closed box section. We can then compute the forces on all edges, and then find the shear-center, by enforcing the moment equilibrium about the axial x-axis.

12.8.5.2 Splitting the shear force

Because the shear-center of any cross-section is a property of the geometry of the section, once the direction of the shear force is given (in our case, it is fixed in the y-direction). Therefore, the location of the shear-center on the box cross-section should not depend on the shear force V value, we can simply set it to the I value, just for conveniences in evaluation:

$$V = I \qquad (12.53)$$

In this special setting, the q value can be evaluated simply using

$$q = Q_{A'} \qquad (12.54)$$

Note that this equation is only used for convenience in locating the shear-center. The shear-flow has a special unit of $\frac{V}{I}$.

Using the relation of the shear force with the deflection (Eq. 8.17), we have

$$\frac{V}{I} = -E \frac{\partial^3 v}{\partial x^3} \qquad (12.55)$$

When the box is divided into two open sections, these two parts must have the same deflection to represent the original problem, implying that the right-hand-side of Eq. (12.55) must be equal for these two divided beams. Thus, the ratio of $\frac{V}{I}$ for both parts must be equal. Because the box is divided vertically, the second moment of area must satisfy:

$$I = I_L + I_R \tag{12.56}$$

where I_L is the I value for the left, and I_R is the I value for the right section, all with respect to the z-axis. Using Eq. (12.53) the shear force should also be divided as

$$V = V_L + V_R \tag{12.57}$$

where V_L is the shear force for the left, and V_R is the shear force for the right section, all in the y-direction.

Note that in computing the q values for the purpose of locating the shear-center, we can use Eq. (12.53) for both sections:

$$V_L = I_L; \quad V_R = I_R \tag{12.58}$$

Also because torsion is involved in the process for the closed section, we use the standard sign conversion for involved variables: All q vectors along the counterclockwise direction with respect to x is positive.

With this setting, we are now ready to write a Python code to solve the problem. We write a Python function for the general purpose of computing the shear-center of any given rectangular box section, together with the relative shear-flow values at key positions where values are the largest. We provide detailed comments in the code, and it should be self-explanatory. Figure 12.14 can be a good guide, when reading the following codes.

12.8.5.3 General purpose code for rectangular box sections

Based on the strategy outlined above, computing q values is to compute the first moment of area $Q_{A'}$ (see Eq. (12.54)). This can be easily done for these two cutoff sections, starting from point A (or B). This gives q-maps for all these four channels of the box-section. The obtained q-maps are given in Fig. 12.14(a). The integrals in Eq. (12.52) can be computed using the following equations. For channel i, we compute first:

$$\tau_{l,\text{open-box},i} = \int_0^{l_i} \frac{q_i}{t_i} dl \tag{12.59}$$

where q_i, t_i, l_i (i = top, left, bottom, right) are, respectively, the q-map, thickness, and length of these four channels of the open-box. The positive

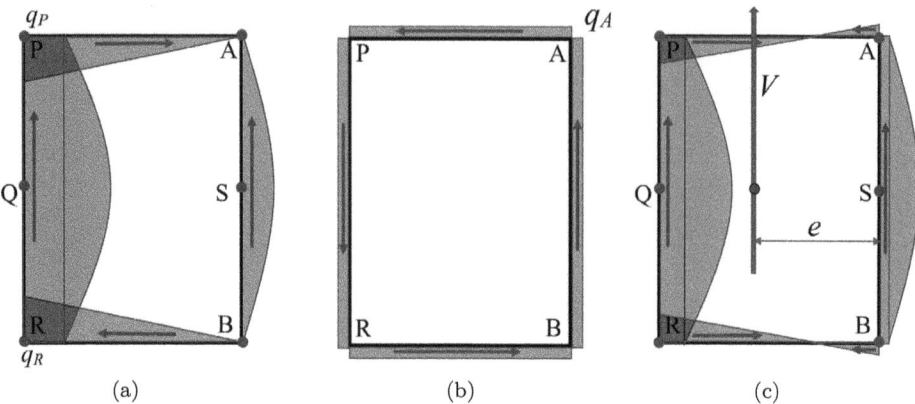

Figure 12.14. q-maps on the rectangular box section. (a) q-map for the two divided sections; (b) q-map for the uniformly distributed q_A on the box section, whose value is to be found in the process locating the shear-center; (c) the superposed final solution of q-map on the box section. The shear force in the vertical y-direction is set as $V = 1$. The positive direction is counterclockwise about x-axis.

direction follows the counterclockwise with respect to x-axis. Next, we compute their sum:

$$T_{l,\text{box}} = \sum_i T_{l,\text{open-box},i} \tag{12.60}$$

Similarly, for the uniform q-map given in Fig. 12.14(b), we set $q_i = 1$ (all positive) for all channels, and then compute for channel i:

$$T_{l,1,i} = \frac{q_1}{t_i} l_i \tag{12.61}$$

and

$$T_{l,1} = \sum_i T_{l,1,i} \tag{12.62}$$

Finally, q_A is found using

$$q_A = -\frac{T_{l,\text{box}}}{T_{l,1}} \tag{12.63}$$

After q_A is found, we can the produce the final q-map via superposition for the closed-box. The result is given in Fig. 12.14(c). Using this final q-map, we compute the shear forces in each of the channels, and finally the shear center, as we did in the previous examples.

The following Python function follows exactly the foregoing equations.

```python
def box_shearCenter(h, b, t_L, t_R, t_T):
    '''Compute the shear-center of a box-section with 4 thin edges wrt
    the middle line of the right-edge. Its dimensions are measured
    based on the middle lines of the edges.
    h:    beam height; b: width of top- and bottom-edge;
    t_T: thickness of top-edge; t_B thickness of bottom-edge: t_B=t_T
    t_L, t_R: thickness of left-edge and right-edges.
    Return:
    e: shear-center; qP, qQ, qS, qA: q values at key points:
    P(north-west), Q(west), S(east), A(north-east).
    '''
    t_B = t_T
    # compute Izz of whole box section:
    Izz = (1./12)*(t_L+t_R)*h**3 + b*t_T*(h/2)**2 + b*t_B*(h/2)**2
    print(f"2nd moment of area of whole box section Izz = {Izz} [L^4]")

    # Compute shear-flow at P and R (north-west, south-west corners)
    qP = b*t_T*(h/2)                                    # after cutting
    qR = b*t_B*(h/2)                                    # after cutting
    # qQ = qP + (h/2)*t_L*(h/4); qS = (h/2)*t_R*(h/4)
    #print(f"shear-flows at key points [Pa/m]:qP={qP:.4e};qR={qR:.4e}")

    # Define τ = q/t as functions of s for all four edges:
    τt_s = lambda s:-qP/b/t_T * s                       # top edge
    τb_s = lambda s:-qR/b/t_B * s                       # bottom edge
    τl_s = lambda s:-(qP + t_L*(h/2 - s)*(h/2 + s)/2)/t_L   # left edge
    τr_s = lambda s: t_R*(h/2 - s)*(h/2 + s)/2/t_R      # right edge

    # Compute τl_{box,i} via integration for all four edges:
    τl_t, _ = si.quad(τt_s, 0, b)                       # use Gauss integration
    τl_b, _ = si.quad(τb_s, 0, b)
    τl_l, _ = si.quad(τl_s, -h/2, h/2)
    τl_r, _ = si.quad(τr_s, -h/2, h/2)

    τl_box = τl_t + τl_b + τl_l + τl_r                  # the sum

    # Compute total area of q/t for the unit q=1 on all four edges:
    τl_q1 = b/t_T + b/t_B + h/t_L + h/t_R

    qA = -τl_box/τl_q1
    #print(f"q value at point A (north-east corner) qA = {qA:.4e}")

    # Compute the channel forces:
    # Define (q+qA) as functions of s for all four edges:
    qt_sA = lambda s:-qP/b*s + qA                       # for top edge
    qb_sA = lambda s:-qR/b*s + qA                       # for top edge
    ql_sA = lambda s:-(qP + t_L*(h/2-s)*(h/2+s)/2) + qA # left edge
    qr_sA = lambda s: t_R*(h/2 - s)*(h/2 + s)/2 + qA    # right edge
```

```
49
50      # Compute shear forces for the left and right channels:
51      F_l = si.quad(ql_sA, -h/2, h/2)[0]         # use Gauss integration
52      F_r = si.quad(qr_sA, -h/2, h/2)[0]
53      #print(f"Shear forces: F_l={F_l:.4e}; F_r={F_r:.4e}")
54      V = -F_l+F_r
55      #print(f"The total shear forces in box-beam (V=Izz): V={V:.4e}")
56
57      # Compute shear-center wrt the middle line of the right edge.
58      F_t, _ = si.quad(qt_sA, 0, b)              # use Gauss int.
59      F_b, _ = si.quad(qb_sA, 0, b)
60      #print(f"Shear forces: F_t={F_t:.4e}; F_b={F_b:.4e}")
61      e = -(F_t*h + F_l*b)/V
62      print(f"e wrt the middle line of the right edge: {e:.4e}")
63      return e, qt_sA(b), ql_sA(0), qr_sA(0), qr_sA(h/2)
```

12.8.5.4 Data preparation

Because the box-beam is top-down symmetric, its centroid is on z-axis. The input data for this example are given as follows:

```
1 # Given length of middle lines.  Assume the box is top-down symmetric.
2 h    =.500    # all in m                    # height of the beam
3 t_L =.020;  t_R =.010      # thickness of the left and Right edge
4 b    =.300                                  # base (width) of the beam
5 t_T =.010                  # thickness of the top (=bottom) edge
```

Using box_shearCenter() to obtain the solution.

```
1 e, qP, qQ, qS, qA = box_shearCenter(h, b, t_L, t_R, t_T)
2 print(f"q/t value at key points:\n qP = {qP:.4e}; qQ = {qQ:.4e};"
3        f"\n qS = {qS:.4e}; qA = {qA:.4e}")
```

```
2nd moment of area of whole box section Izz = 0.0006875 [L^4]
e wrt the middle line of the right edge: 2.0303e-01
q/t value at key points:
 qP = -4.4444e-04; qQ = -1.0694e-03;
 qS = 6.1806e-04; qA = 3.0556e-04
```

12.8.5.5 Distribution profile of the shear-flow

Finally, the distribution profile of the shear-flow on the box cross-section at different stages is given in Fig. 12.14.

12.8.6 Example: The shear-center location for a C-beam

Consider a C-beam shown in Fig. 12.12. Its dimensions (mm) measured based on the middle lines of the edges are given as follows:

> beam height: $h_b = 500$; width of top- and bottom-edge: $b = 300$ with $t_{wfb} = t_{wft}$; thickness of top-edge: $t_{tft} = 10$; thickness of bottom-edge: $t_{tfb} = t_{tft}$; thickness of web on the left: $t_w = 20$.

Compute the shear-center of this C-beam, with respect to the middle line of the right-edge.

Solution: One may modify a little the Python function box_shearCenter() to make it for C-beams. The easiest is to use it as it is, by simply setting the thickness of the right-edge t_R to a very small value. The code is given in the following.

```
1  # Given length of middle lines.     The C-section is top-down symmetric.
2  h   =.500       # all in m                        # height of the beam h_b
3  t_L =.020                                         # this is same as $t_w$
4  t_R =1.0e-12    # Set to a small value but not 0 to avoid zero division
5  b   =.300                                         # base (width) of the beam
6  t_T =.010                            # thickness of the top (=bottom) edge
```

Using box_shearCenter() to obtain the solution.

```
1  e, qP, qQ, qS, qA = box_shearCenter(h, b, t_L, t_R, t_T)
2              # or: gr.box_shearCenter(h, b, t_L, t_R, t_T)
3
4  # change the reference to the middle line of the web on the left:
5  eL = e - b
6  print(f"e wrt the middle line of the left edge: {eL:.4e}")
7  print(f"q/t value at key points:\n qP = {qP:.4e}; qQ = {qQ:.4e};"
8        f"\n qS = {qS:.4e}; qA = {qA:.4e}")
```

```
2nd moment of area of whole box section Izz = 0.0005833333333437501 [L^4]
e wrt the middle line of the right edge: 3.9643e-01
e wrt the middle line of the left edge: 9.6429e-02
q/t value at key points:
 qP = -7.5000e-04; qQ = -1.3750e-03;
 qS = 1.1375e-13; qA = 8.2500e-14
```

We can confirm the solution using Eq. (12.48) derived earlier for C-beams.

```
1  aA = t_L*h/(t_T*b)
2  e = 3*b/(6+aA)
3  print(f"e wrt the middle line of the right edge: {e:.4e}")
```

e wrt the middle line of the right edge: 9.6429e-02

As shown, we obtained the same solution for the shear-center. The code box_shearCenter() provides additional results on relative shear-flows at key points on the C-beam cross-section. It can be useful for other similar problems.

12.9 Remarks

This chapter focuses on computing the shear stress or shear-flows on the cross-section of a beam subjected to transverse loading. The value of the shear stress depends on the first moment of the area formed from the surface free from shear stress to the surface where the shear stress is in question. Techniques and codes are developed for computing the q-map, the distribution of the shear stresses, as well as the shear forces. These techniques and codes work for beams with irregular cross-section, including the section with multiple narrow channels.

With the techniques and codes for computing the shear-flows and shear forces, the shear-center can then be determined with ease. This enables proper application of transverse loading without generating torsion to the beam.

References

[1] A.P. Boresi and R.J. Schmidt, *Advanced Mechanics of Materials*. John Wiley & Sons, New York, 2002.
[2] J.M. Gere and S.P. Timoshenko, *Mechanics of Materials*, Van Nostrand Reinhold Company, New York, 1972.
[3] S. Timoshenko and J.N. Goodier, *Theory of Elasticity*, McGraw-Hill, New York, 1970. http://books.google.com/books?id=yFISAAAAIAAJ&dq=theory+of+elasticity&ei=ICiKSsr3G4jwkQSbxMyPCg.
[4] Z.L. Xu, *Elasticity*, Vols. 1&2. People's Publisher, China, 1979.

Chapter 13

Energy Methods

```
1  # Place cursor in this cell, and press Ctrl+Enter to import dependences.
2  import sys                          # for accessing the computer system
3  sys.path.append('../grbin/')  # Change to the directory in your system
4
5  from commonImports import *    # Import dependences from '../grbin/'
6  import grcodes as gr            # Import the module of the author
7  importlib.reload(gr)            # When grcodes is modified, reload it
8
9  from continuum_mechanics import vector
10 init_printing(use_unicode=True)    # For Latex-like quality printing
11 np.set_printoptions(precision=4,suppress=True)  # Digits in print-outs
12 np.set_printoptions(formatter={'float_kind': '{:.4e}'.format})
```

A stressed solid stores energy. The level of the energy stored shall naturally depend on the externally applied loads, as well as the states of the loaded solid, including the displacement and/or stress field in it. Therefore, there should be possible methods that make use of the energy to find the states of the solid in response to the external loads. This type of technique is called **energy methods**, which are the focus of this chapter.

Because such methods are useful in dealing with large-scale structural systems, especially in powerful numerical methods for approximated solutions, we thus place emphasis on understanding the principles and concepts of energy methods. This chapter is written with reference to [1–8].

13.1 Energy principles and fundamental concepts

Energy methods are used widely to obtain solutions to elasticity problems. When a loaded solid is in a stable or equilibrium state, the energy should be at stationary. Otherwise, the solid shall still be in a deformation process. The solution or important equations can thus be obtained via essentially a minimization process. Energy methods discussed in this chapter focuses on linearly elastic solids with small displacements, unless specified explicitly. We will study two energy principles that are most fundamentally essential:

1. Potential energy principle (PEP); and
2. Complementary energy principle (CEP).

Both can be derived from the principle of virtual work, which uses virtual displacements or virtual forces. The former leads to PEP and the latter to CEP. This chapter will elaborate both in detail.

Using energy method requires a conceptual shift in dealing with mechanics problems. Thus, we will first use a one degree of freedom system (1-DOF) to elaborate the fundamental concepts in great detail. Once these concepts are established, we will then naturally extend the formulation to complex solid structure systems.

13.2 Principle of virtual work by virtual displacement

Consider a mechanically loaded solid constrained without rigid body movement. Assume there are no other effects involved, such as electromagnetic, thermal, etc. The principle of virtual work states as follows.

> The work done by the external forces to a solid that undergoes an arbitrary infinitely small virtual displacement equals the variation in strain energy in the solid caused by the virtual displacement.

Mathematically, it is expressed as

$$\delta W_d = \delta U \tag{13.1}$$

where W denotes the work done by external forces applied on the solid, U stands for the strain (also called potential) energy stored inside the solid, and δ stands for small variation.

Equation (13.1) can be expressed as

$$\delta \underbrace{(W_d - U)}_{\text{total potential energy}} = 0 \tag{13.2}$$

where $(W_d - U)$ is often called the **total potential energy**. Its stationary point corresponds to an equilibrium state of the solid.

Energy Methods

The principle of virtual work is an energy conservation law for solids that are loaded only mechanically. This principle is the basis for deriving PEP.

The best way to discuss the concepts and formulations is to use a simple example that is already familiar to us.

13.2.1 Example: A loaded linear elastic spring

Let us consider the simplest possible 1-DOF system: a linear elastic spring fixed at point A and loaded at point B. It has a stiffness K known as spring constant. The force f is (externally) applied at point B in the longitudinal direction of the spring, resulting in displacement d at the same point B. On the other hand, the spring experiences an internal force denoted as N, and shall have an elongation denoted as e.

Because the principle of virtual work relates two energies: δW_d that is the work-done by the external force, and δU potential energy change inside the solid/structure. The best way to clearly discuss about this relation is to draw two graphs: $f \sim d$ giving the work-done by the external force shown in Fig. 13.1(a), and $N \sim e$ which gives the internal potential energy shown in Fig. 13.1(b). The former is the external action, and the latter is the internal reaction.

Let us use this simple system to study in detail these energy principles.

13.2.2 Evaluation of potential energy

At such a loaded state, inside the spring, there must be an internal force N and an elongation e. The **constitutive equation** for the linear elastic spring can be written as

$$N = Ke \qquad (13.3)$$

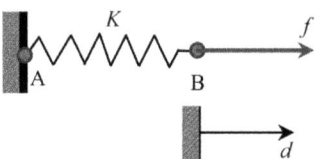

Figure 13.1. A spring loaded by a force f at point B resulting in a displacement d also at point B.

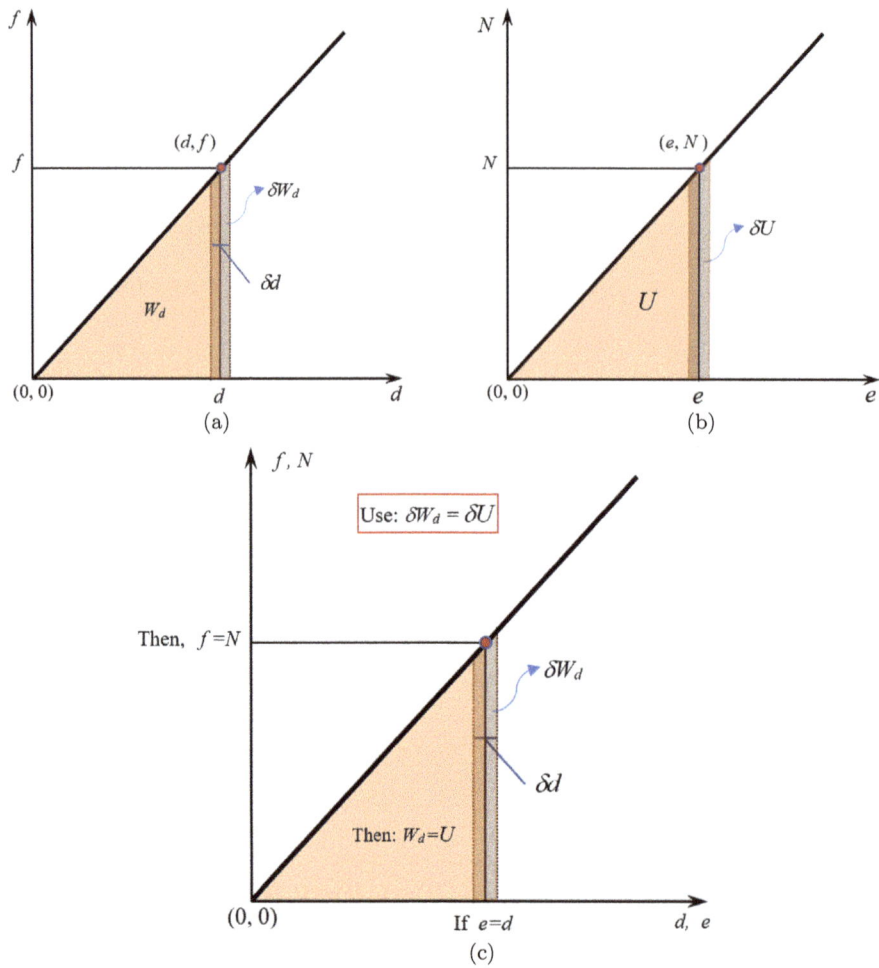

Figure 13.2. (a) The force displacement relation at point B of the spring system that is subjected to small virtual displacement δd; **The external action**: work done by the force f under δd is evaluated. (b) The relationship of internal-force N and elongation e of the spring body that stores strain energy; **The internal reaction**: variation in strain energy δU resulting from δd is evaluated. (c) The outcome form the PEP.

This implies that the internal force in the spring is proportional to its elongation with a constant K. This $N \sim e$ relationship can be plotted in Fig. 13.2(b). Note that we need to carefully distinguish the internal and external forces because of the need for proper energy evaluation. The external work done is by the external force f, but the potential energy should be evaluated using the internal force N.

In the loaded spring, some energy is stored within the spring. One measure is the potential energy denoted as U. It is the area under the black line in

Fig. 13.2(b), which is the triangular area below the straight line. Thus, it can easily be calculated using

$$U = \frac{1}{2}Ne = \frac{1}{2}Ke^2 \tag{13.4}$$

Equation (13.4) is the formula for computing the potential energy in a spring in terms of its elongation e. Also, we need to carefully distinguish the displacement d at point B, and the elongation of e of the spring. The displacement d goes with the external force f, resulting in the work done.

13.2.3 Imposition of infinitely small virtual displacement

Next, at the actually loaded state, we impose (imaginarily) an infinitely small virtual displacement δd at point B on the spring, at which f is on. The work done by f should be

$$\delta W_d = f\, \delta d \tag{13.5}$$

where δW_d is the virtual work done by f as a result of the imposed virtual displacement δd. It is the area of the shaded strip shown in Fig. 13.2(a).

13.2.4 Compatibility condition

In the process of imposing δd, if the virtual displacement (at point B) is compatible with the elongation e (of the spring body), implying that the spring is not broken, we must have

$$e = d \tag{13.6}$$

This is known as the **compatibility condition**. As will see in later examples, the compatibility condition can get complicated for general solids and structures. It is simple for this spring system because there is only one displacement (1-DOF).

Equation (13.4) can now be re-written as

$$U = \frac{1}{2}Kd^2 \tag{13.7}$$

Note here that U is a function of the elongation d (in fact, quadratic in d), and hence U is differentiable with respect to d. We are essentially treating d as an independent variable of function U. This is a shift of concept of functions, and d is called a generalized coordinate.

13.2.5 Potential energy principle for 1-DOF systems

Inside the spring, in response to the imposed virtual displacement δd, the strain energy must have a **variation**, which is denoted as δU. Because U is differentiable with respect to d, δU can be expressed using its gradient $\frac{\partial U}{\partial d}$ as

$$\delta U = \frac{\partial U}{\partial d} \delta d \tag{13.8}$$

We see now how δU relates δd: through the gradient of U with respect to d.

Substituting Eqs. (13.5) and (13.8) into the virtual work principle Eq. (13.2), we obtain

$$f \delta d = \frac{\partial U}{\partial d} \delta d \quad \text{or} \quad \left[f - \frac{\partial U}{\partial d} \right] (\delta d) = 0 \tag{13.9}$$

Now, because δd is arbitrary, for the foregoing equation to hold, the term in the square bracket must vanish, which gives

$$\boxed{f = \frac{\partial U}{\partial d}} \tag{13.10}$$

This is the **PEP** for the spring, which is derived based on the principle of virtual work, using virtual displacement. Equation (13.10) means that if we can write the potential energy U in terms of displacement d, we can obtain the force f that causes the displacement d.

13.2.6 Results of potential energy principle: Equilibrium equation

Substituting U given in Eq. (13.7) into Eq. (13.10), we obtain:

$$f = Kd \tag{13.11}$$

This is the relation between the external force f and the displacement d, obtained from the potential energy principle. Using Eqs. (13.6) and (13.3), Eq. (13.11) can be rewritten as

$$f = Kd = Ke = N \tag{13.12}$$

This is the **equilibrium equation** for the spring system. This means that *the potential energy principle is equivalent to the the equilibrium equation, as long as the compatibility equation is satisfied.* The results from the PEP is shown in Fig. 13.2(c).

Note that Eq. (13.12) is the same as that we would obtain via the usual means of using a free-body diagram.

> **Remark:** The above derivation process can be confusing to beginners. However, one should be convinced that the entire process never used the equilibrium equation. We used only the PEP, constitutive equation for the spring, and the compatibility Eq. (13.6). The outcome is indeed the equilibrium Eq. (13.12) or its equivalent Eq. (13.11).

Using Eqs. (13.4) and (13.3), U can also be written as

$$U = \frac{1}{2K}N^2 \tag{13.13}$$

This is the formula for computing the potential energy in a spring in terms of internal force N.

13.2.7 On compatibility equation

We see also that the compatibility equation is a **prerequisite** to apply the PEP (or the virtual work principle using virtual displacements). An intuitive way to understand this is as follows. Because the virtual work needs to impose an arbitrary virtual displacement δd, if the internal elongation e is not responding to δd, such an imposition will have no effect on the spring, and hence the virtual work principle cannot work!

13.2.8 Clapeyron's theorem

Equation (13.13) can be re-written as

$$U = \frac{1}{2}N\underbrace{\frac{1}{K}N}_{e} = \frac{1}{2}Ne = \frac{1}{2}\underbrace{fd}_{\text{work done by } f \text{ over constant } d} \tag{13.14}$$

Here, we used again Eqs. (13.6) and (13.12). Equation (13.14) is known as Clapeyron's theorem, which states that "the potential energy of deformation of a body, which is in equilibrium under a given load, is equal to half the work done by the external forces computed assuming these forces had remained constant from the initial state to the final state" [9]. This is practically saying that the orange triangular area is half the square area shown in Fig. 13.2(c).

We can now extend our findings to all 1-DOF systems.

13.2.9 Formulas of potential energy for 1-DOF systems

As seen in the previous chapters, any one-dimensional (1D) structural member can have its relationship of the internal-force with the displacement written in the form of Eq. (13.3). For all these linear elastic structure members, the strain energy has the following generalized form:

$$U = \begin{cases} \frac{K}{2}e^2 & \text{in terms of elongation} \\ \frac{1}{2K}N^2 & \text{in terms of internal force} \end{cases} \quad (13.15)$$

where e stands for generalized deformation (including strain, elongation, twist angle, cross-section rotation, etc.), and N stands for generalized internal force (including stress, axial force, shear force, moment, torque, etc.). K is the stiffness and $N = Ke$. At the material level, K is Young's modulus E.

The relationships like $N = Ke$ are summarized as follows:

$N = Ke$, $K = $ string const. Spring with force N, e: elongation

$N = Ke$, $K = \dfrac{EA}{L}$ Bar (truss) with axial force N, e: elongation

$T = K\Theta$, $K = \dfrac{GJ}{L}$ Bar with torque T, Θ: twist angle

$M = K\theta$, $K = \dfrac{EI}{L}$ Beam with moment M, θ: cross-section rotation

$V = Kd_y$, $K = \dfrac{GA}{kL}$ Beam with shear force V, d_y: deflection in y-direction

(13.16)

Note that for beams under shear forces, the stiffness K is with a correction coefficient k. This is because when a beam is with a shear force on its cross-section, the resulting shear stress is not uniformly distributed, as discussed in Chapter 12. It depends on the cross-section shape of the beam. Exact values of k are not generally available, and hence estimation is needed. Some often used empirical values for thin beam with different cross-sections are given as follows [1]:

Rectangular cross-section: $k = 1.2$

Solid circular cross-section: $k = 1.33$

Thin-wall circular cross-section: $k = 2.0$

I-beam, box-beam, channel: $k = 1.0$

(13.17)

The shear correction coefficient k will be used in this book.

13.2.10 Formulas of potential energy for various members

In general, the internal forces in a 1D structure member can be a function of axial coordinate. In addition, the area properties (such as area, moments of area, etc.) can also change with the coordinate. Evaluation of the potential energy needs to be in integral forms. This section presents these formulas for various 1D structure members, including tension bar, torsion bar, and beams.

13.2.11 Strain energies U_N for bars with varying cross-section

Consider a bar with varying cross-section subjected to axial loads as shown in Fig. 13.3. It has length L, and a varying cross-section A. Young's modulus of the material is E. The loads generate an internal force N in the bar that can be a function of x. The potential energy can be evaluated as follows.

Consider a small cell dx taken from the bar. Using Eqs. (13.15) and (13.16), the potential energy for this dx cell can be given as

$$dU_N = \frac{1}{2 \times \text{Stiffness}} \times \text{Internal-force}^2 \underset{\text{for bars}}{=} \frac{dx}{2EA} N^2 \qquad (13.18)$$

The potential energy for the entire bar can then be obtained by integrating over the length of the bar:

$$U = \int_0^L dU_N = \int_0^L \frac{N^2}{2EA} dx \\ \underset{\text{if uniform}}{=} \frac{N^2 L}{2EA} = \frac{N^2}{2K} \text{ or } \frac{Ke^2}{2} \qquad (13.19)$$

Note that in this equation, E can be a function of x too, such as functionally graded materials. Also, if the tapered bar has multiple segments with different A, the integral should just be piecewise.

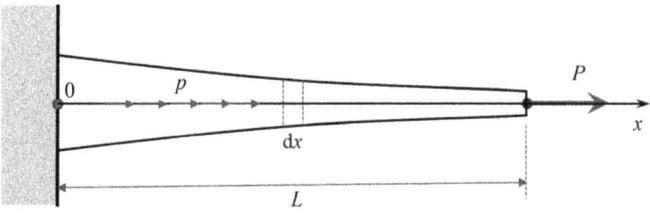

Figure 13.3. A tapered bar subject to axial forces in its axis x-direction.

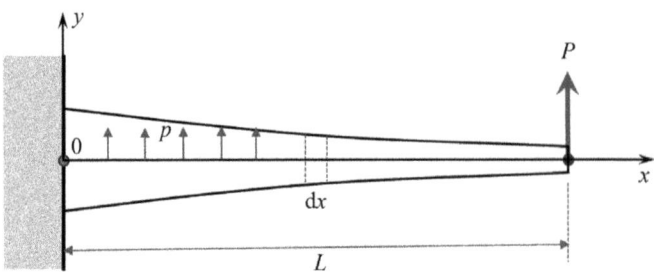

Figure 13.4. A tapered beam subject to transverse forces in y-direction.

13.2.12 Strain energies for beams with varying cross-section

Consider a beam with possibly varying cross-section subjected to transverse loads as shown in Fig. 13.4. It has length L, varying cross-section A, and 2nd moment of area I. The material has Young's modulus E and shear modulus G. The loads result in internal moments M and shear force V in the beam. The potential energy can be evaluated, by considering a small cell dx taken from the beam.

13.2.12.1 From moment, U_M

Using Eqs. (13.15) and (13.16), the potential energy for this dx cell resulting from the moment M on the cross-section of the beam is given as

$$dU_M = \frac{1}{2 \times \text{Stiffness}} \times \text{Internal-force}^2 \underset{\text{for beams}}{=} \frac{dx}{2EI} M^2 \qquad (13.20)$$

The potential energy for the entire beam is obtained by integrating over the length of the beam:

$$U_M = \int_0^L dU_M = \int_0^L \frac{M^2}{2EI} dx$$
$$\underset{\text{if uniform}}{=} \frac{M^2 L}{2EI} = \frac{M^2}{2K} \underset{\text{or}}{=} \frac{K\theta^2}{2} \qquad (13.21)$$

13.2.12.2 From shear force, U_V

Using also Eqs. (13.15) and (13.16), the potential energy for this dx cell resulting from the shear force V on the cross-section of the beam can be given as

$$dU_V = \frac{1}{2 \times \text{Stiffness}} \times \text{Internal-force}^2 \underset{\text{for beams}}{=} \frac{kdx}{2GA} V^2 \qquad (13.22)$$

The potential energy for the entire beam can then be obtained by integrating over the beam length:

$$U_V = \int_0^L dU_V = \int_0^L \frac{kV^2}{2GA} dx \qquad (13.23)$$

$$\underset{\text{if uniform}}{=} \frac{kV^2 L}{2GA} = \frac{V^2}{2K} \text{ or } \frac{Ke_y^2}{2}$$

Note that in these equations, E and G can be a function of x. Also, if the tapered beam has multiple segments with different A, the integral should just be piecewise.

13.2.13 Strain energies U_T for bars under torsion

Consider a bar with possibly varying cross-section subjected to torques in the x-direction as shown in Fig. 13.5. It has length L, possibly varying cross-section and a 2nd polar moment of area J. The shear modulus of the material is G. The loads generate an internal force T in the bar that could be a function of x. The potential energy can be evaluated as follows.

Consider a small disk cell dx taken from the bar. Using Eqs. (13.15) and (13.16), the potential energy for this cell can be given as

$$dU_T = \frac{dx}{2GJ} T^2 \qquad (13.24)$$

The potential energy for the entire bar is obtained by integrating over the length of the bar:

$$U_T = \int_0^L dU_T = \int_0^L \frac{T^2}{2GJ} dx \qquad (13.25)$$

$$\underset{\text{if uniform}}{=} \frac{T^2 L}{2GJ} = \frac{T^2}{2K} \text{ or } \frac{K\Theta^2}{2}$$

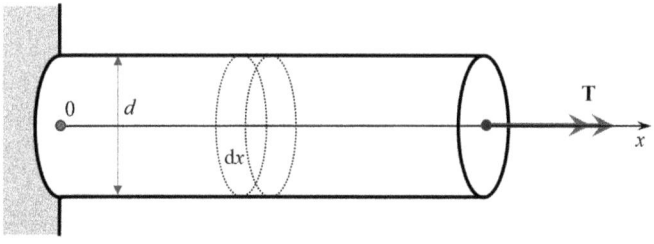

Figure 13.5. A bar subject to torsional forces in x-direction.

Note that in this equation, G can be a function of x. If the tapered bar has multiple segments with different GJ, the integral should be piecewise.

13.2.14 Summary of formulas for U for various structural members

We can now summarize the formulas as follows:

Struc. member	Stiffness	$U(N)$: Int.,	uniform	$U(e)$: Int.,	uniform
spring, $N-e$	K	$U_N = \frac{N^2}{2K}$		$\frac{Ke^2}{2}$	
truss, $N-e$	$\frac{EA}{L}$	$U_N = \int_0^L \frac{N^2}{2EA}dx,$	$\frac{LN^2}{2EA},$	$\int_0^L \frac{EAe^2}{2L^2}dx,$	$\frac{EAe^2}{2L}$
Tor. bar, $T-\Theta$	$\frac{GJ}{L}$	$U_T = \int_0^L \frac{T^2}{2GJ}dx,$	$\frac{LT^2}{2GJ},$	$\int_0^L \frac{GJ\Theta^2}{2L^2}dx,$	$\frac{GJ\Theta^2}{2L}$
Beam, $M-\theta$	$\frac{EI}{L}$	$U_M = \int_0^L \frac{M^2}{2EI}dx,$	$\frac{LM^2}{2EI},$	$\int_0^L \frac{EI\theta^2}{2L^2}dx,$	$\frac{EI\theta^2}{2L}$
Beam, $V-e_y$	$\frac{GA}{kL}$	$U_V = \int_0^L \frac{kV^2}{2GA}dx,$	$\frac{kLV^2}{2GA},$	$\int_0^L \frac{GAe_y^2}{2kL^2}dx,$	$\frac{GAe_y^2}{2kL}$

(13.26)

Note that if the structure member is uniform (the stiffness and the internal force do not change with the coordinate), it behaves simply like a spring. One can get its formulas for U simply by replacing the K in the formulas for spring. If the stiffness or internal force change with the coordinate, we need to do an integration, and one of the length L in the uniform formula should be changed to dx. Once this is understood, the foregoing tabulated equations are easy to write out.

13.3 Principle of virtual work by virtual force

Consider a mechanically loaded solid constrained without rigid body movement. Assume there are no other effects involved. The **principle of virtual work** by virtual force states as follows:

> The work done by arbitrary infinitely small virtual forces over the displacement field in a loaded solid in an equilibrium state equals the variation in complementary energy in the solid caused by the virtual forces.

Energy Methods

Mathematically, it is expressed as

$$\delta W_f = \delta C \tag{13.27}$$

where W_f denotes the work done by the virtual forces δf imposed on the solid over the displacement field, C stands for the complementary energy stored in the solid, and δ stands for small variation.

Equation (13.27) can be expressed as

$$\delta \underbrace{(W_f - C)}_{\text{total complementary energy}} = 0 \tag{13.28}$$

As will be seen in later examples, the stationary point of the total complementary energy gives the compatibility equations of the solid.

Consider the same spring system shown in Fig. 13.1. Assume the spring is loaded by f at point B resulting in a displacement d also at point B. The spring system is at state (d, f). We shall now use the similar trick and process that led to the PEP in the previous section.

The complementary energy stored in the spring is denoted as C, which is the shaded area above the black line shown in Fig. 13.6(b).

Let us examine all these energies.

13.3.1 Application of an infinitely small virtual force

Next, at the actually loaded state, we apply (imaginarily) an infinitely small virtual force δf at point B of the spring system, at which d is measured. The work done by δf should be

$$\delta W_f = d\, \delta f \tag{13.29}$$

where δW_f is the virtual work done as a result of the virtual force δf. It is the area of the shaded strip shown in Fig. 13.6(a).

13.3.2 Evaluation of complementary energy

At such a loaded state, inside the spring, there must be an internal force N and an elongation e. This $N \sim e$ relationship can be plotted in Fig. 13.6(b). Thus, some energy is stored within the spring. This time we use the complementary energy denoted as C. It is the shaded triangular area above the black line in Fig. 13.6(b), and computed using:

$$C = \frac{1}{2} Ne = \frac{1}{2K} N^2 \tag{13.30}$$

Here, we used the **constitutive equation** for the spring, Eq. (13.3).

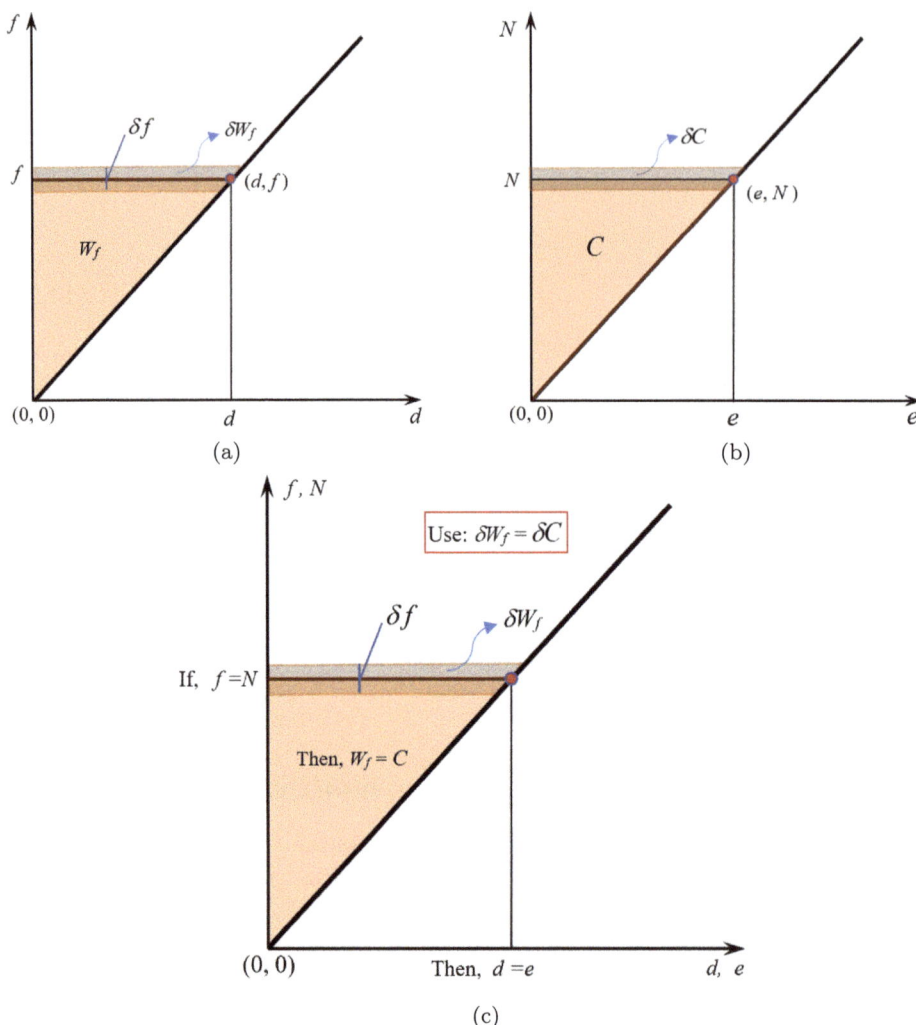

Figure 13.6. (a) The force displacement relation at point B of the spring system that is subjected to small virtual force δf; **The external action**: work done by the force δf under d is evaluated. (b) The relationship of internal-force N and elongation e of the spring body that stores complementary energy; **The internal reaction**: variation in strain energy δC resulting from δf is evaluated. (c) The result from the complementary energy principle.

Equation (13.30) is the formula for computing the complementary energy in a spring in terms of internal force N. The expression is the same as Eq. (13.4) because this spring system is linear elastic and the $N \sim e$ relation is a straight line. The area of the two triangles below and above the straight line in Fig. 13.6(b) is the same.

If the system is not linear, Eqs. (13.30) and (13.4) can differ.

13.3.3 Equilibrium condition

When the virtual force δf is applied, we require the external force f and the internal force N in the spring satisfying:

$$f = N \tag{13.31}$$

This is known as the **equilibrium condition**. For this spring system, it is simple. As we will see in later examples, the equilibrium condition can be complicated for general solids and structures.

Using Eqs. (13.31) and (13.30) can now be re-written as

$$C = \frac{1}{2K} f^2 \tag{13.32}$$

Note here that C is a function of the external force f (in fact, quadratic in f), and hence C is differentiable with respect to f. We treat f as an independent variable of function C. Hence, f is also called a generalized coordinate.

13.3.4 Complementary energy principle for 1-DOF systems

Inside the spring, in response to the applied virtual force δf, the complementary energy must have a **variation**, which is denoted as δC. Because C is differentiable with respect to f, δC can be expressed using its gradient $\frac{\partial C}{\partial f}$ as

$$\delta C = \frac{\partial C}{\partial f} \delta f \tag{13.33}$$

We see how δC relates δf: through the gradient of C with respect to f.

Substituting Eqs. (13.29) and (13.33) into the virtual work principle Eq. (13.28), we obtain

$$d\,\delta f = \frac{\partial C}{\partial f} \delta f \quad \text{or} \quad \left[d - \frac{\partial C}{\partial f} \right] \delta f = 0 \tag{13.34}$$

Now because δf is arbitrary, for the foregoing equation to hold, the term in the square bracket must vanish, which gives

$$\boxed{d = \frac{\partial C}{\partial f}} \tag{13.35}$$

This is the **CEP** for the spring, which is derived based on the principle of virtual work using virtual force. Equation (13.35) means that if we can write the potential energy C in terms of force f, we can obtain the displacement d that are caused by force f.

Substituting C given in Eq. (13.32) into Eq. (13.35), we obtain:

$$d = \frac{f}{K} \qquad (13.36)$$

This is the formula for computing the displacement at point B, when the spring is loaded there with f. We obtained this using the principle of complementary energy.

13.3.5 Compatibility equation

Using the equilibrium condition Eq. (13.31), the foregoing equation becomes

$$d = \frac{N}{K} = e \qquad (13.37)$$

This is the **compatibility equation** for the spring system: the displacement at point B equals the elongation of the spring. This means that *the CEP is equivalent to the compatibility equation, as long as the equilibrium equation is satisfied.* The outcome of the CEP is shown in Fig. 13.6(c).

> **Remark:** Note that the process leading to Eq. (3.37) never used the compatibility equation. We used only the complementary energy principle, constitutive equation for the spring, and the equilibrium Eq. (13.31). The outcome is the compatibility Eq. (13.37).

13.3.6 On equilibrium equation

We see also that the equilibrium equation is a **prerequisite** to apply the CEP (or the virtual work principle using virtual forces). An intuitive way to understand this is as follows. Since the virtual work needs to apply an arbitrary virtual force δf, if the internal force N is not responding to δf, such an application will have no effect on the spring, and hence the virtual work principle cannot work!

13.3.7 General formulas for displacements of 1-DOF members

Note that if the problem is linear elastic, $C = U$, Eq. (13.35) can also be written as

$$d = \frac{\partial U}{\partial f} \qquad (13.38)$$

This means that if we can write the potential energy in terms of the external force on a structure member of 1-DOF, we can obtain the displacement using Eq. (13.38).

Equation (13.38) is generalized as follows:

$$d_{\text{generalized}} = \frac{\partial U}{\partial f_{\text{generalized}}} \quad (13.39)$$

where $d_{\text{generalized}}$ is the generalized displacement including translational displacements, rotation, twist angle, etc., and $f_{\text{generalized}}$ is the generalized force including forces, moment, torque, etc. The condition is that $d_{\text{generalized}}$ and $f_{\text{generalized}}$ must be an energy conjugate pair, implying that $d_{\text{generalized}} \delta f_{\text{generalized}}$ produces work done δW_f, so that the corresponding potential energy U can be used. The often used formulas are listed as follows:

$$\begin{aligned}
d &= \frac{\partial U_N}{\partial f} \quad \text{for translational displacements} \\
\Theta &= \frac{\partial U_T}{\partial T} \quad \text{for twist angle of torsional bars} \\
\theta &= \frac{\partial U_M}{\partial M} \quad \text{for cross-sectional rotation of thin beams} \\
d_y &= \frac{\partial U_V}{\partial f} \quad \text{for vertical displacement of thin beams by shear}
\end{aligned} \quad (13.40)$$

In Eq. (13.40), the potential energy U is given in Eq. (13.26).

Equation (13.40) is more frequent used because writing potential energy in terms of external forces can often be more convenient. We will demonstrate this with examples.

13.3.8 Example: Cantilever beam subjected to a concentrated force at its tip

Consider a 1D bar/beam with uniform cross-section and subjected to point vertical load P and point horizontal load F at its tip, as shown in Fig. 13.7. The bar/beam has a cross-section area A, and the length is L that is much larger than its height, and hence can be treated as a thin beam. The 2nd moment of area of the cross-section is I. Young's modulus of the material is E, Poisson's ratio is ν, and all are given. Use the energy method to complete the following tasks:

1. Determine the tip displacement in the horizontal direction of the bar.
2. Determine the deflection of the beam at its tip, considering only the moment in the beam.
3. Determine the rotation of the beam cross-section at its tip, considering only the moment in the beam.

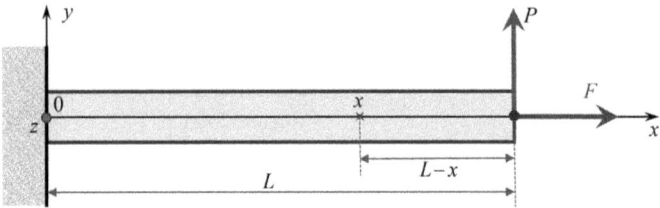

Figure 13.7. A 1D bar/beam subjected to a concentrated forces at its tip.

4. Determine the deflection of the beam at its tip, considering only the shear force in the beam. Assume the shear force correction coefficient k is given.
5. Determine the ratio of the D_P from the internal shear force and that from the internal moment on the cross-section of the beam. Assume the beam has a rectangular cross-section with height h known. Briefly discuss the results obtained.
6. Determine the rotation of the beam cross-section at its tip, considering only the shear force in the beam.

Solution to 1: Determine the tip displacement in the horizontal direction of the beam. Since only the horizontal force F is applied at the tip of the beam. We need only consider F. Using the 1st equation in Eq. (13.40) and the 2nd equation in Eq. (13.26). The formula for computing the tip displacement D_F in the x-direction resulting from the force F becomes

$$D_F = \frac{\partial U}{\partial F} = \frac{\partial}{\partial F}\left(\frac{LN^2}{2EA}\right) = \frac{\partial}{\partial N}\left(\frac{LN^2}{2EA}\right)\frac{\partial N}{\partial F} = \frac{LN}{EA}\frac{\partial N}{\partial F} \qquad (13.41)$$

Here, we used the chain rule of differentiation [11]. Using now the equilibrium equation,

$$N = F \qquad (13.42)$$

which gives $\frac{\partial N}{\partial F} = 1$. Equation (13.41) becomes

$$D_F = \frac{LF}{EA} \qquad (13.43)$$

This result is exactly that obtained via solving differential equation for bars (see Chapter 7). Note that if EA is a function of x, we shall use the integral form of the equation in Eq. (13.26) and carried out with the integration.

Solution to 2: Since only moment is considered, we shall use the 1st equation in Eq. (13.40) and the 4th integral equation in Eq. (13.26). The formula for computing the deflection at tip D_P resulting from the force P becomes

$$D_P = \frac{\partial U}{\partial P} = \frac{\partial}{\partial P}\left(\int_0^L \frac{M^2}{2EI}dx\right) = \int_0^L \frac{M}{EI}\frac{\partial M}{\partial P}dx \qquad (13.44)$$

Here, we used the chain rule of differentiation. The internal moment M in the beam is a function of x, and can be computed using the equilibrium equation:

$$M = P(L - x) \qquad (13.45)$$

Its derivative with respect to P becomes

$$\frac{\partial M}{\partial P} = (L - x) \qquad (13.46)$$

Substituting Eqs. (13.45) and (13.46) into (13.44), we obtain the solution:

$$D_P = \int_0^L \frac{P}{EI}(L-x)^2 dx = \frac{P}{EI}\frac{-1}{3}(L-x)^3\Big|_0^L = \frac{PL^3}{3EI} \qquad (13.47)$$

This result is exactly that obtained via solving differential equation for thin beams (see Chapter 8). Note that if EI are function of x, we shall keep it in the integrand and carried out with the integration.

Solution to 3: This time we shall also use the 3rd equation in Eq. (13.40) and the 4th integral equation in Eq. (13.26), but the differentiation should be with respect to M because we need to compute θ. The process is as follows:

$$\theta_P = \frac{\partial U}{\partial M} = \frac{\partial}{\partial M}\left(\int_0^L \frac{M^2}{2EI}dx\right) = \int_0^L \frac{M}{EI}\frac{\partial M}{\partial M}dx \qquad (13.48)$$

The moment M within the beam is already given in Eq. (13.45). Its derivative with respect to M itself is simply 1.

Equation (13.48) gives

$$\theta_P = \int_0^L \frac{P(L-x)}{EI}dx = \frac{PL^2}{2EI} \qquad (13.49)$$

This solution is also exact, which can be obtained using the codes given in Chapter 8.

Solution to 4: Since only shear force is considered, we shall use the 4th equation in Eq. (13.40) and the 5th integral equation in Eq. (13.26). The formula for computing the deflection at tip D_P becomes

$$D_P = \frac{\partial U}{\partial P} = \frac{\partial}{\partial P}\int_0^L \frac{kV^2}{2GA}dx = \int_0^L \frac{kV}{GA}\frac{\partial V}{\partial P}dx \qquad (13.50)$$

Due to equilibrium, the shear force $V = P$ within the beam, and is uniform. Its derivative with respect to P is simply 1. Using Eq. (13.50), we obtain the solution:

$$D_P = \int_0^L \frac{kP}{GA} dx = \frac{kPL}{GA} \tag{13.51}$$

where k is the shear force correction coefficient given in Eq. (13.17). Note for thin beams, L is much larger compared to its cross-sectional dimension (that affects I and A), the deflection resulting from the shear force is much smaller than that from the moment given in Eq. (13.47).

Solution to 5: The ratio of D_P from the internal shear force and that from the internal moment on the cross-section of the beam can be computed using:

$$R_{VM} = \frac{kPL}{GA} \div \frac{PL^3}{3EI} = 3\frac{kEI}{GAL^2} = 6(1+\nu)\frac{kI}{AL^2} \tag{13.52}$$

where ν is Poisson's ratio, and $2G = E/(1+\nu)$. Assuming the beam cross-section is rectangular, we have $I = bh^3/12$, $A = bh$, we obtain

$$R_{VM} = \frac{(1+\nu)k}{2}\left(\frac{h}{L}\right)^2 \tag{13.53}$$

Further, let $k = 1.2$ and $\nu = 0.3$, we obtain

$$R_{VM} = 0.78 \left(\frac{h}{L}\right)^2 \tag{13.54}$$

The result shows clearly that when $L \gg h$, the ratio is very small, implying that the deflection resulting from the shear force is very small because of the quadratic ratio. If $L = 5h$ (which is considered very thick), the ratio is about 3%. Therefore, for thin beams, we can often ignore the shear forces and consider only the moment, when computing the deflection. For thick beams, the shear force effect becomes more important. Also, the thin beam theory may not work properly, and alternative theory and method are needed. Or the problem should be treated as a 2D one, which becomes an advanced topic in solid mechanics.

Solution to 6: Since the shear force does not result in any change in the rotation of the beam cross-section, based on the thin beam theory. The answer to this question is 0.

13.3.9 Comparison of PEP and CEP, linear elastic systems

For linear systems, the $f \sim d$ relation is a straight line, and the $N \sim e$ relation is also a straight line, we have $U = C$: the potential energy and the complementary energy are the same. We state:

1. If the solid structure system satisfies the compatibility conditions, implying that the internal deformation and external displacement have no conflict (all parts and members staying together, no gap and overlap) when it is loaded, we use the PEP. The outcome will be automatically the equilibrium equations. This is the base of numerical methods [6, 7], and very powerful in dealing with large-scale solid structural systems for approximated solutions.
2. If the solid structure system satisfies the equilibrium conditions, implying that all the externally applied forces are balanced by the internal forces/stresses at a stable state, we use the CEP, and the outcome will be automatically the compatibility equations. This is particularly useful in numerical methods for obtaining upper-bound solutions [7].
3. If the solid structure system satisfies both the compatibility and the equilibrium conditions. We have already the solution, and it is exact.

The comparison is shown schematically in Fig. 13.8.

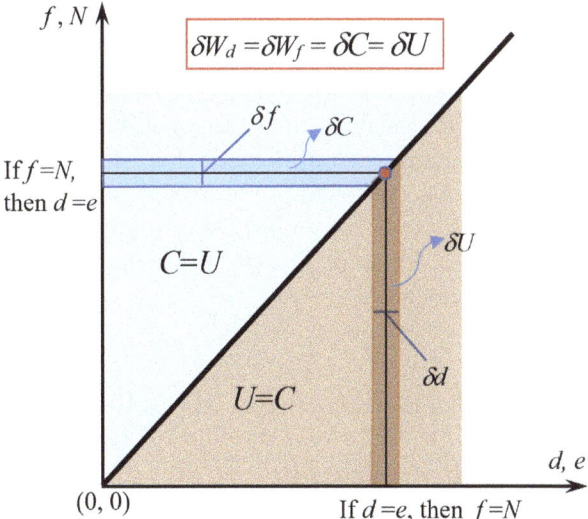

Figure 13.8. Comparison of the potential energy and the complementary energy principles. Linear elastic systems.

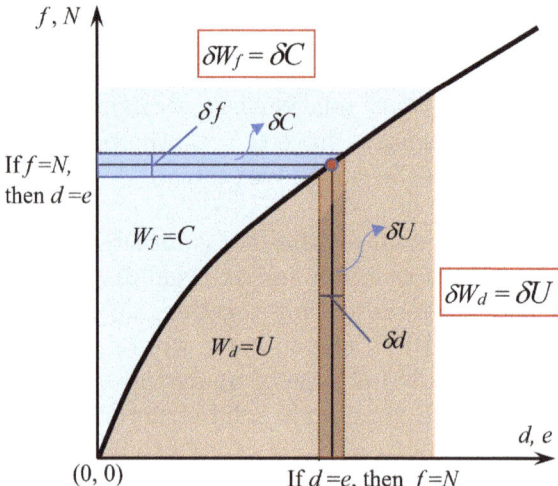

Figure 13.9. Comparison of the potential energy and the complementary energy principles. Nonlinear elastic systems.

13.3.10 Comparison of PEP and CEP, nonlinear systems

For nonlinear systems, the $f \sim d$ relation or $N \sim e$ relation or both is curved, we have in general $U \neq C$: the potential energy and the complementary energy are not the same. In general, these three points given above still holds, but only locally. Solutions can only be found via iterations [7]. The comparison is shown schematically in Fig. 13.9.

In the discussions for PEP, it is clear that how to find the potential energy is the key in applying PEP. This can be a challenge when the solid structure is higher dimensional and more complicated. The next section derives the general expression for strain (potential) energy and its density at any point in a stressed 3D solids. As one can expect, formulation will be a lot more complicated for 3D solids. Einstein summation will be used, to avoid over 3-page long derivation otherwise.

13.4 Strain energy formulation for 3D solids*

13.4.1 Mathematical derivation

Consider a solid with volume V and an enclosed surfaces S. A Cartesian coordinate system $(x_i, i = 1, 2, 3)$ is used. On the surface, there are external traction applied on it. The solid is also subjected to body forces over its volume, and is in an equilibrium state. We impose a small variation of displacement field ($\delta u_i, i = 1, 2, 3$) that is differentiable over the volume of the solid. The displacement variation is infinitely small so that the equilibrium state of the solid is not affected.

13.4.1.1 Work done by external forces

The work done by the external forces under (δu_i) shall consist of two parts: the work done by the surface stresses, and the work done by the body forces. The former, δW_{ds}, is evaluated using

$$\delta W_{ds} = \int_S (t_{ni}\delta u_i)dS \qquad (13.55)$$

where t_{ni} is the traction on surface S with normal $\mathbf{n}$ defined by the direction cosine n_i. Using the traction–stress relation Eq. (3.38), Eq. (13.55) can be rewritten as

$$\delta W_{ds} = \int_S (n_j\sigma_{ji}\delta u_i)dS \qquad (13.56)$$

The above integral is on the surface of volume. It can be converted to the volume integration using the well-known Green's **divergence theorem** [10], which gives

$$\begin{aligned}\delta W_{ds} &= \int_S (n_j\sigma_{ji}\delta u_i)dS = \int_V \frac{\partial}{\partial x_j}(\sigma_{ji}\delta u_i)dV \\ &= \int_V \frac{\partial \sigma_{ji}}{\partial x_j}(\delta u_i)dV + \int_V \sigma_{ji}\underbrace{\frac{\partial \delta u_i}{\partial x_j}}_{\sigma_{ji}\delta\varepsilon_{ij}}dV\end{aligned} \qquad (13.57)$$

Here, we used the the symmetry of the stresses, and the strain–displacement relation Eq. (4.14) because δu_i is assumed differentiable.

The work done by body forces, δW_{db}, is evaluated using

$$\delta W_{db} = \int_V (b_i\delta u_i)dV \qquad (13.58)$$

where b_i is the body forces applied over volume V. Adding up Eqs. (13.57) and (13.58), we obtain total work done:

$$\delta W_d = \delta W_{ds} + \delta W_{db} = \int_V \left(\underbrace{(\sigma_{ji,j} + b_i)}_{\text{equilibrium}=0} \delta u_i + \sigma_{ji}\delta\varepsilon_{ij} \right) dV = \int_V (\sigma_{ji}\delta\varepsilon_{ij})dV$$

$$(13.59)$$

Here, we used the equilibrium equation at any point inside the volume. This is because of our assumption that the solid is in equilibrium when δu_i is imposed.

13.4.1.2 *Strain energy expression*

Using the virtual work principle given in Eq. (13.1), we found the expression for the variation of the **strain energy** as

$$\delta U = \underbrace{\int_V (\sigma_{ij}\delta\varepsilon_{ij})dV}_{\delta W_d}$$

$$= \int_V (\sigma_{xx}\delta\varepsilon_{xx} + \sigma_{yy}\delta\varepsilon_{yy} + \sigma_{zz}\delta\varepsilon_{zz} + 2\sigma_{yz}\delta\varepsilon_{yz} + 2\sigma_{xz}\delta\varepsilon_{xz} + 2\sigma_{xy}\delta\varepsilon_{xy})\,dV$$

(13.60)

Here, we used the symmetry of the stress and strain components.

Finally, we obtained the widely used important expression for computing the **variation of the strain energy density** (the strain energy per unit volume):

$$\delta U_0 = \sigma_{ij}\delta\varepsilon_{ij}$$
$$= \sigma_{xx}\delta\varepsilon_{xx} + \sigma_{yy}\delta\varepsilon_{yy} + \sigma_{zz}\delta\varepsilon_{zz} + 2\sigma_{yz}\delta\varepsilon_{yz} + 2\sigma_{xz}\delta\varepsilon_{xz} + 2\sigma_{xy}\delta\varepsilon_{xy}$$
$$= \vec{\sigma}^\top \delta\vec{\varepsilon}$$

(13.61)

where $\vec{\sigma}$ and $\vec{\varepsilon}$ are the stress and strain vectors defined in Eq. (5.60). The variation of the strain energy density is now written as the dot product of the stress vector and the variation of the strain vector.

It is clear that δU_0 is a sum of the energy of each of the 9 stress and strain components. Each contribution is evaluated with individual term $\sigma_{ij}\delta\varepsilon_{ij}$ ($i, j = 1, 2, 3$, no summation). This proves that σ_{ij} and ε_{ij} are energy conjugate: σ_{ij} and ε_{ij} are in the same direction, and hence their product gives the energy.

Note that if the stress-strain pair is not conjugate, implying that they are not in the same direction. One needs to project one of them onto the other and use the projected component to compute the energy. If the stress and strain are orthogonal, the energy by them is zero, regardless of the values of the stress and strain components.

Also because our stresses and strains are all expressed in orthogonal coordinates, the stress component in one axis direction is orthogonal to those in other axis-directions. Therefore, the energy can be written in individual $\sigma_{ij}\delta\varepsilon_{ij}$ terms, and then summed up. One stress component is not coupled with any other ones in energy evaluation, as shown in Eq. (13.60).

13.4.1.3 *Compatibility conditions*

Note that the strains are subjected to variation. This implies that these strains must be compatible with the displacements, implying that these compatibility equations defined in the last section of Chapter 4 must be satisfied by these strains. This is similar to the discussion given in Section 13.2.7 for 1-DOF systems.

It turns out that in numerical methods based on domain discretization, we can find ways to construct displacement fields for 2D and 3D solids and various structures, while ensuring the compatibility (known as the admissible displacements). PEP can then be applied to establish equilibrium equations and then find approximated solutions in terms of displacements as well as in stresses for large-scale solid structure systems. Detailed discussion is out of the scope of this book. Interest readers may refer to [6–8].

13.4.2 *Pictorial derivation*

An alternative and easier derivation of Eq. (13.61) is a pictorial approach to show that σ_{ij} and ε_{ij} are in the same direction. This is done as follows.

First, for all the normal stresses and strains, it is obvious that σ_{ii} and ε_{ii} are in the same direction because ε_{ii} gives displacement in the i-direction which is the same as the direction of σ_{ii}.

For the shear stress and strain, let us consider $\sigma_{xy} \sim \varepsilon_{xy}$ pair and $\sigma_{yx} \sim \varepsilon_{yx}$ pair. We use a picture shown in Fig. 13.10, where a small square 2D solid abdc has a small deformation becoming ab'd'c'. The displacement by ε_{xy} is $\varepsilon_{xy}dx$ and is in the y-direction, which is the same as that of stress component σ_{xy}. Therefore, σ_{xy} and ε_{xy} are energy conjugate, and the energy can be evaluated by their product.

The displacement by ε_{yx} is $\varepsilon_{yx}dy$, and is in the x-direction, which is the same as that of σ_{yx}. Therefore, σ_{yx} and ε_{yx} are also energy conjugate, and the energy can be evaluated by its product.

The same arguments can be made for other shear components. This proves that the strain energy density can be directly written out in the form of Eq. (13.61).

13.4.3 *Strain energy density for 1D problem*

With the expression for 3D solid, we can re-visit the 1D problems. This leads to some familiar formulas given earlier, and also some general formulas.

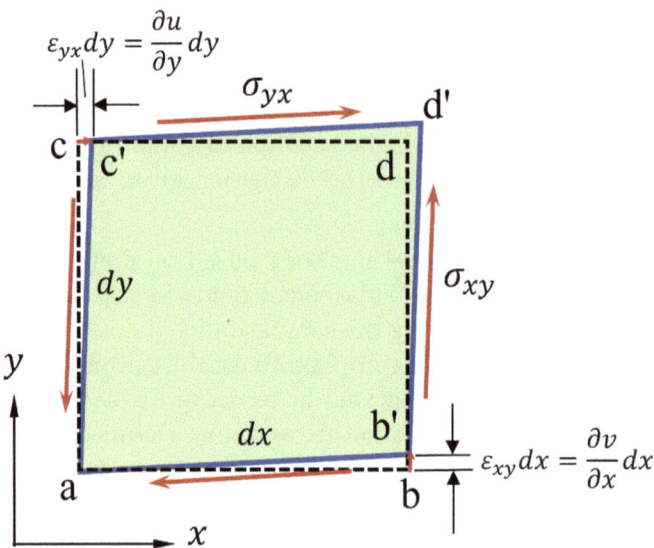

Figure 13.10. A small square solid abdc is deformed to ab'd'c', by pure shear stresses.

13.4.3.1 General formulas

For 1D problems, there is only one stress or strain. Note that variation and differentiation carry the same meaning if both treat ε as the independent variable. Equation (13.61) can be expressed as

$$dU_0 = \sigma d\varepsilon \tag{13.62}$$

or

$$\sigma = \frac{dU_0}{d\varepsilon} \tag{13.63}$$

This means that if the strain energy density is given as a function of the strain, one can easily obtain the stress via differentiation with respect to the strain. This is in fact the potential energy principle at material level.

Also, Eq. (13.62) gives

$$U_0 = \int_0^\varepsilon \sigma(\varepsilon')d\varepsilon' \tag{13.64}$$

where the prime denotes an integral variable, which disappears after the integration is done [11]. The strain energy density is the shaded orange area in Fig. 13.11(a).

Consider a 1D stress problem for bars with uniform cross-section of area A subject to axial loads. The strain (potential) energy of the bar at the

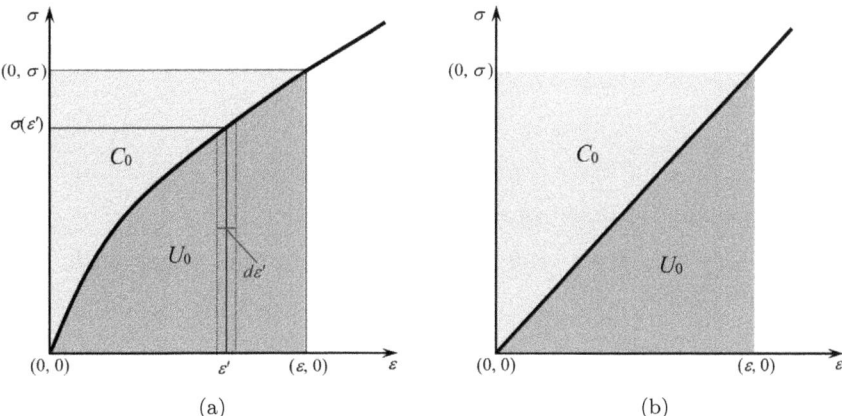

Figure 13.11. Strain energy density U_0 and complementary energy density C_0 at a point in a stressed solid. (a) Nonlinear stress-strain relation; (b) linear stress-strain relation.

structural member level can be evaluated using Eq. (13.64):

$$U = \int_V U_0 dV = \int_0^L \left(\int_0^\varepsilon \sigma d\varepsilon' \right) \underbrace{A dx}_{dV}$$

$$= \int_0^L \left(\int_0^\varepsilon \underbrace{A\sigma}_{N} d\varepsilon' \right) dx = \int_0^L \left(\int_0^\varepsilon N d\varepsilon' \right) dx \qquad (13.65)$$

where N is the internal force in the bar, L is the length, V is the volume. If N does not change along the bar, implying that there is no body force on the bar, the integral over the bar length can move inside, which gives

$$U = \int_V U_0 dV = \int_0^\varepsilon N \underbrace{\left(\int_0^L dx \right)}_{L} d\varepsilon' = \int_0^\varepsilon N \underbrace{dL\varepsilon'}_{de'} \qquad (13.66)$$

or

$$\boxed{U = \int_0^e N de'} \qquad (13.67)$$

where e is the elongation of the bar.

Note that the condition for Eq. (13.67) is that the internal force N does not change along the bar. If the internal force N changes along the bar, Eq. (13.65) should be used.

When the material is **linear elastic**, U_0 is a linear stress-strain curve. Equation (13.64) gives

$$U_0 = \frac{1}{2}\sigma\varepsilon \qquad (13.68)$$

which is the shaded orange area in Fig. 13.11(b).

In this case, the strain energy for a 1D linear elastic bar with Young's modulus E becomes

$$U = \int_V U_0 dV = \int_0^L \frac{1}{2}\sigma\varepsilon \underbrace{Adx}_{dV} = \int_0^L \frac{1}{2}\underbrace{\sigma A}_{N}\varepsilon dx$$

$$= \int_0^L \frac{1}{2}N\varepsilon dx = \int_0^L \frac{N^2}{2EA}dx \qquad (13.69)$$

In this equation, E, A, and E can all be a function of x. This is Eq. (13.19) obtained earlier.

13.4.3.2 Formulas for linear elastic materials

Consider first 1D stress problem for uniform bars with Young's modulus E, we have $\sigma = E\varepsilon$, and thus Eq. (13.64) becomes

$$U_0 = \int_0^\varepsilon E\varepsilon' d\varepsilon' = \frac{1}{2}E\varepsilon^2 \qquad (13.70)$$

where E is the stiffness of the material. Using $\sigma = E\varepsilon$, Eq. (13.70) can also be written in stress.

$$U_0 = \frac{1}{2}\sigma\varepsilon = \frac{1}{2E}\sigma^2 \qquad (13.71)$$

Using the first part of the foregoing equation, it is trivial to confirm Eq. (13.63).

For 1D uniform bars made of linear elastic materials with uniform internal force N, we have

$$N = \underbrace{\frac{EA}{L}}_{K} e = Ke \qquad (13.72)$$

where K is the stiffness of the bar. The bar practically behaves like a spring. Equation (13.67) becomes

$$U = \int_0^e \frac{EA}{L} e\, de = \frac{EA}{2L} e^2 = \frac{K}{2} e^2 \tag{13.73}$$

or in terms of force:

$$U = \frac{1}{2K} N^2 \tag{13.74}$$

This is also the expression of strain energy stored in an elastic spring, when it is subjected to force N. This is the same as Eq. (13.15).

13.4.4 Various formulas for strain energy density for 3D solids

For 3D linear elastic solids, the strain energy density can be written in various forms. We write all these different formulas together for easy comparison and future reference:

$$U_0 = \int_0^\varepsilon \sigma(\varepsilon')d\varepsilon' \quad \text{General definition. Prime stands for integration variable.}$$

$$= \frac{1}{2}\sigma_{ij}\varepsilon_{ij} = \frac{1}{2}\left(\sigma_{xx}\varepsilon_{xx} + \sigma_{yy}\varepsilon_{yy} + \sigma_{zz}\varepsilon_{zz} + \sigma_{yz}\gamma_{yz} + \sigma_{xz}\gamma_{xz} + \sigma_{xy}\gamma_{xy}\right)$$

$$= \frac{1}{2E}\left[\sigma_{xx}^2 + \sigma_{yy}^2 + \sigma_{zz}^2 - 2\nu(\sigma_{yy}\sigma_{zz} + \sigma_{xx}\sigma_{zz} + \sigma_{xx}\sigma_{yy})\right.$$
$$\left. - 2(1+\nu)(\sigma_{yz}^2 + \sigma_{xz}^2 + \sigma_{xy}^2)\right]$$

$$= \frac{1}{2E}\left[\sigma_1^2 + \sigma_2^2 + \sigma_3^2 - 2\nu(\sigma_2\sigma_3 + \sigma_1\sigma_3 + \sigma_1\sigma_2)\right]$$

$$= \frac{(\sigma_1 + \sigma_2 + \sigma_3)^2}{18K} + \frac{(\sigma_2-\sigma_3)^2 + (\sigma_1-\sigma_3)^2 + (\sigma_1-\sigma_2)^2}{12G} \tag{13.75}$$

$$= \frac{1}{2E}\left[I_1^2 - 2(1+\nu)I_2\right]$$

$$= \frac{1}{2}\lambda\left(\varepsilon_{xx} + \varepsilon_{yy} + \varepsilon_{zz}\right)^2 + G\left(\varepsilon_{xx}^2 + \varepsilon_{yy}^2 + \varepsilon_{zz}^2 + 2\varepsilon_{yz}^2 + 2\varepsilon_{xz}^2 + 2\varepsilon_{xy}^2\right)$$

$$= \frac{1}{2}\left(c_{11}\varepsilon_1^2 + c_{22}\varepsilon_2^2 + c_{33}\varepsilon_3^2\right) + \left(c_{23}\varepsilon_2\varepsilon_3 + c_{13}\varepsilon_1\varepsilon_3 + c_{12}\varepsilon_1\varepsilon_2\right)$$

$$= \frac{1}{2}\lambda\left(\varepsilon_1 + \varepsilon_2 + \varepsilon_3\right)^2 + G\left(\varepsilon_1^2 + \varepsilon_2^2 + \varepsilon_3^2\right)$$

$$= \left(\frac{1}{2}\lambda + G\right)\bar{I}_1^2 - 2G\bar{I}_2$$

13.5 Potential energy principle for structures

13.5.1 *Virtual work via virtual displacement for structures*

Consider a structure member that is properly constrained and loaded, and is in equilibrium. The generalized displacements are denoted as $(D_1, D_2, \ldots, D_n)$. These can be translational and rotational. For beams, for example, it is the deflection of the beam axis or the rotation of the cross-section of the beam. The number of the generalized displacements is the number of degrees of freedom (DoFs). All these displacements must be compatible, ensuring the integrity of the structure. In addition, the loads on the structure are also treated as generalized loads denoted as $(P_1, P_2, \ldots, P_n)$ that result in displacements D_i in an equilibrium state. For beams, the generalized loads include transverse load and moments.

Principle of virtual work is also a most general form of energy principle for solids and structures. It can be used to derive many formulations using energies in the system, including the principle of potential energy and the principle of complementary energy, as we did for the 1D systems. When virtual displacements are used, it leads to the principle of total potential energy. The process is as follows.

Consider a set of (infinitesimal) *arbitrary independent virtual* displacements $(\delta D_1, \delta D_2, \ldots, \delta D_n)$, subjected to the boundary and compatibility conditions. The virtual displacements are then imposed on the structure. Since the structure is loaded with P_i, these forces will do work under the virtual displacements, which is denoted as δW_P and can be evaluated as

$$\delta W_p = P_1 \delta D_1 + P_2 \delta D_2 + \cdots + P_n \delta D_n = P_i \delta D_i \qquad (13.76)$$

The virtual displacements δD_i will result in a virtual strain field. Also, within the loaded structure, there exists a stress field. Such a stress field will do virtual work under the virtual strain field resulting from δD_i, which should be the strain energy change caused by δD_i. Therefore, we have

$$\delta U = \frac{\partial U}{\partial D_1} \delta D_1 + \frac{\partial U}{\partial D_2} \delta D_2 + \cdots + \frac{\partial U}{\partial D_n} \delta D_n = \frac{\partial U}{\partial D_i} \delta D_i \qquad (13.77)$$

13.5.2 *Formulas of potential energy principle for structures*

It is also called PEP.

Substituting Eqs. (13.76) and (13.77) into (13.2) gives,

$$P_1 \delta D_1 + P_2 \delta D_2 + \cdots + P_n \delta D_n = \frac{\partial U}{\partial D_1} \delta D_1 + \frac{\partial U}{\partial D_2} \delta D_2 + \cdots + \frac{\partial U}{\partial D_n} \delta D_n \qquad (13.78)$$

which is re-grouped to give

$$\left(P_1 - \frac{\partial U}{\partial D_1}\right)\delta D_1 + \left(P_2 - \frac{\partial U}{\partial D_2}\right)\delta D_2 + \cdots + \left(P_n - \frac{\partial U}{\partial D_n}\right)\delta D_n = 0 \tag{13.79}$$

Now because each of the virtual displacements δD_i is **arbitrary** and δD_i are **independent** of each other. The foregoing equation holds, only if

$$\boxed{P_i = \frac{\partial U}{\partial D_i} \quad \text{for all } i = 1, 2, \ldots, n} \tag{13.80}$$

This states that if the potential energy U can be written as a function of generalized displacements $(D_1, D_2, \ldots, D_n)$ that are compatible in the structure, the load P_i at the same location and direction of D_i can be found using Eq. (13.80).

Equation (13.80) is known also as Castigliano's theorem. It is in fact the potential energy principle at the structural level, in comparison with Eq. (13.63) that is for material level. For 1DoF problems, Eq. (13.80) is Eq. (13.10).

All these loads P_i found using Eq. (13.80) will be in equilibrium, as will be demonstrated in the example sections.

13.5.3 Example: A two-bar truss subjected to two loads

13.5.3.1 Setting of the problem

Consider a truss structure with two bars: bar-1 (AB) and bar-2 (AC) of lengths l_1 and l_2, respectively. The truss is fixed to a rigid wall at points A and C, and is loaded with horizontal (x-direction) force P_1 and vertical (y-direction) force P_2 at point B, as shown in Fig. 13.12. Thus, point B undergoes finite displacements with components u_1 and u_2, respectively, in x- and y-directions. The lengths of these two bars become L_1 and L_2.

The cross-section area of bar AB is A_1 and that of bar AC is A_2. Their corresponding Young's moduli are E_1 and E_2. Assume these two bars remaining linearly elastic during the deformation.

(a) Derive formulas for forces P_1 and P_2 in terms of displacements u_1 and u_2, based on the potential energy principle.
(b) Using the formulas derived in part (a), determine the values of P_1 and P_2. Assume $\frac{E_1 A_1}{l_1} = k_1 = 3000$ N/m and $\frac{E_2 A_2}{l_2} = k_2 = 4300$ N/m, $b = 0.4$ m, $h = 0.4$ m, and set $u_1 = 0.05$ m and $u_2 = 0.03$ m.

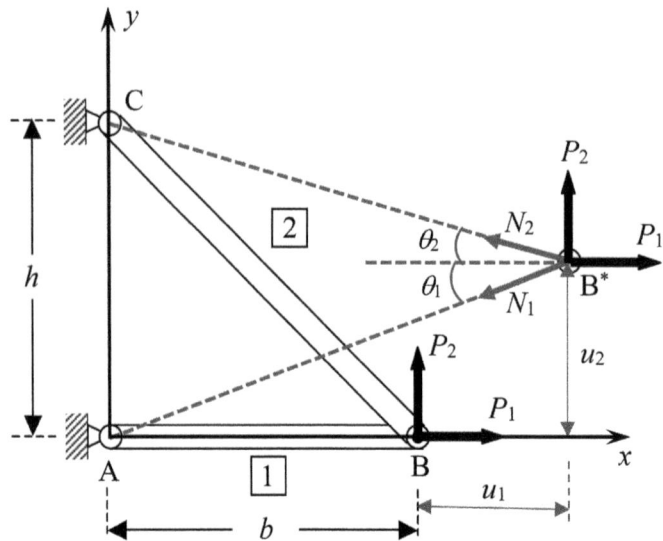

Figure 13.12. A truss structure with two bars AB and AC of initial lengths l_1 and l_2, and deformed lengths L_1 and L_2, which can all be computed using the geometric relations, ensuring compatibility (two bars remain supported and connected).

(c) Consider the equilibrium at the displaced position B* and verify the results of part (b).

(d) Considering small displacements ($u, v \ll b, h$), linearize the formulas for P_1 and P_2 derived in part (a).

13.5.3.2 Solutions to the problem

The initial lengths of these two bars l_1 and l_2, is calculated as

$$l_1 = b; \quad l_2 = \sqrt{b^2 + h^2} \tag{13.81}$$

Solution to question (a): Denote the elongations of these two bars, respectively, as e_1 and e_2. Using the geometry of the truss shown in Fig. 13.12, the lengths of these bars after deformation become

$$L_1 = l_1 + e_1 = \sqrt{(b + u_1)^2 + u_2^2}$$
$$L_2 = l_2 + e_2 = \sqrt{(b + u_1)^2 + (h - u_2)^2} \tag{13.82}$$

Solving the foregoing equation for e_1 and e_2 gives,

$$e_1 = L_1 - l_1 = \sqrt{(b + u_1)^2 + u_2^2} - l_1$$
$$e_2 = L_2 - l_2 = \sqrt{(b + u_1)^2 + (h - u_2)^2} - l_2 \tag{13.83}$$

These relations between the displacements and elongations ensure that the compatibility of the displacements so that these bar members of the structures are always connected during the deformation. We can thus apply the potential energy principle.

Since the bars remain linearly elastic, the elongations of these two bars can be calculated using

$$e_i = \frac{N_i l_i}{E_i A_i} \quad \text{for } i = 1, 2 \tag{13.84}$$

where N_1 and N_2 are the internal forces in these two bars. The strain energy can then be evaluated using

$$U_i = \frac{1}{2} N_i e_i = \frac{E_i A_i}{2 l_i} e_i^2 \quad \text{for } i = 1, 2 \tag{13.85}$$

where U_1 and U_2 are the strain energies for bar-1 and bar-2. The total strain energy U for the whole truss structure becomes

$$U = U_1 + U_2 = \frac{E_1 A_1}{2 l_1} e_1^2 + \frac{E_2 A_2}{2 l_2} e_2^2 \tag{13.86}$$

We have written the potential energy in terms of the elongations. Using the PEP Eq. (13.80), we obtain:

$$P_1 = \frac{\partial U}{\partial u_1} = \frac{E_1 A_1}{l_1} e_1 \frac{\partial e_1}{\partial u_1} + \frac{E_2 A_2}{l_2} e_2 \frac{\partial e_2}{\partial u_1}$$

$$P_2 = \frac{\partial U}{\partial u_2} = \frac{E_1 A_1}{l_1} e_1 \frac{\partial e_1}{\partial u_2} + \frac{E_2 A_2}{l_2} e_2 \frac{\partial e_2}{\partial u_2} \tag{13.87}$$

in which e_i is given in Eq. (13.83). Using this, the partial derivatives of e_i with respect to u_i can be obtained as

$$\frac{\partial e_1}{\partial u_1} = \frac{(b + u_1)}{L_1} = \frac{(b + u_1)}{\sqrt{(b + u_1)^2 + u_2^2}}$$

$$\frac{\partial e_2}{\partial u_1} = \frac{(b + u_1)}{L_2} = \frac{(b + u_1)}{\sqrt{(b + u_1)^2 + (h - u_2)^2}}$$

$$\frac{\partial e_1}{\partial u_2} = \frac{u_2}{L_1} = \frac{u_2}{\sqrt{(b + u_1)^2 + u_2^2}} \tag{13.88}$$

$$\frac{\partial e_2}{\partial u_2} = -\frac{h - u_2}{L_2} = -\frac{(h - u_2)}{\sqrt{(b + u_1)^2 + (h - u_2)^2}}$$

Solution to question (b): Using the formulas derived in part (a), determine the values of P_1 and P_2. Assume $\frac{E_1 A_1}{l_1} = K_1 = 3000$ N/m and $\frac{E_2 A_2}{l_2} = K_2 = 4300$ N/m, $b = 0.4$ m, $h = 0.4$ m, and set $u_1 = 0.05$ m and $u_2 = 0.03$ m. We write the following code to do this.

```
1  # External forces
2  K1 = 2.0e3; K2 = 3.0e3
3  b  = 0.4; h = 0.4
4  u1 = 0.05e-3; u2 = 0.03e-3
5
6  l1 = b; l2 = np.sqrt(b**2 + h**2)                       # initial lengths
7
8  L1 = np.sqrt((b+u1)**2 + u2**2)                         # deformed lengths
9  L2 = np.sqrt((b+u1)**2 + (h-u2)**2)
10 print(f'Initial  lengths of two bars: {l1:0.6e},{l2:0.6e}(m)')
11 print(f'Deformed lengths of two bars: {L1:0.6e},{L2:0.6e}(m)')
12
13 e1 = L1 - l1                                            # elongations
14 e2 = L2 - l2
15 print(f'Elongations of two bars: {e1:0.6e},{e2:0.6e}(m)')
16
17 pe1u1 = (b+u1)/L1              # partial derivatives of elongations
18 pe1u2 =    u2 /L1
19 pe2u1 = (b+u1)/L2
20 pe2u2 =-(h-u2)/L2
21
22 P1 = K1*e1*pe1u1 + K2*e2*pe2u1
23 P2 = K1*e1*pe1u2 + K2*e2*pe2u2
24
25 print(f'Two forces at point B: {P1:0.6e},{P2:0.6e}(N)')
```

```
Initial  lengths of two bars: 4.000000e-01,5.656854e-01(m)
Deformed lengths of two bars: 4.000500e-01,5.656996e-01(m)
Elongations of two bars: 5.000112e-05,1.414496e-05(m)
Two forces at point B: 1.300112e-01,-2.999550e-02(N)
```

Note that when using PEP, the displacements can be large as shown.

Solution to question (c): Use the equilibrium of at the displaced position B* and verify the results of part (b).

Using the values of P_1 and P_2 obtained in (b), the values of the internal normal forces N_1 and N_2 in these two bars, as shown in Fig. 13.12, can be found using the following code.

```
1  N1 = e1*K1
2  N2 = e2*K2
3  print(f'The internal forces in two bars: {N1:0.4e}, {N2:0.4e}(N)')
```

The internal forces in two bars: 1.0000e-01, 4.2435e-02(N)

The sine and cosine values of θ_1 and θ_2 (see, Fig. 13.12) can be found using:

```
1  sinθ1 =    u2 /L1; cosθ1 = (b+u1)/L1
2  sinθ2 = (h-u2)/L2; cosθ2 = (b+u1)/L2
```

Using the free-body diagram for point B*, shown in Fig. 13.12, we can now establish the equilibrium equations.

$$P_1 - N_1 \cos(\theta_1) - N_2 \cos(\theta_2) = 0 \quad \text{all forces in } x\text{-direction}$$
$$P_2 - N_1 \sin(\theta_1) + N_2 \sin(\theta_2) = 0 \quad \text{all forces in } y\text{-direction} \quad (13.89)$$

or

$$P_1 - \underbrace{\left(N_1 \tfrac{b+u_1}{L_1} + N_2 \tfrac{b+u_1}{L_2}\right)}_{N_x} = 0 \quad \text{all forces in } x\text{-direction}$$

$$P_2 - \underbrace{\left(N_1 \tfrac{u_2}{L_1} - N_2 \tfrac{h-u_2}{L_2}\right)}_{N_y} = 0 \quad \text{all forces in } y\text{-direction} \quad (13.90)$$

The total force from N_1 and N_2 in the x-direction becomes

```
1  N_x = N1*cosθ1 + N2*cosθ2
2  print(f'Total forces of N1 & N2 in x-direction: {N_x:0.6e}(N)')
```

Total forces of N1 & N2 in x-direction: 1.300112e-01(N)

The total force from N_1 and N_2 in the y-direction becomes

```
1  N_y = N1*sinθ1 - N2*sinθ2
2  print(sinθ1, sinθ2)
3  print(f'Total force of N1 & N2 in y-direction: {N_y:0.6e}(N)')
```

7.499062596087012e-05 0.707036068741148
Total force of N1 & N2 in y-direction: -2.999550e-02(N)

We found that $N_x = P_1$ and $N_y = P_2$. This confirms that the PEP results satisfy automatically the equilibrium equations. *The total potential energy principle is in place of the equilibrium equations.* This finding is extremely useful for complicated structural systems, for which equilibrium equations are difficult to establish. One can then use the total PEP to obtain (approximate) equilibrium equations. The widely used finite element method [6] is, in fact, established based on the total potential energy principle.

Solution to question (d): Considering small displacements ($u, v \ll b, h$), linearize the formulas for P_1 and P_2 derived in part (a).

Using the binomial expansion, for an arbitrary small variable α, we have

$$(1+\alpha)^{\frac{1}{2}} \approx 1 + \frac{1}{2}\alpha - \frac{1}{8}\alpha^2 + \cdots \quad (13.91)$$

Equations in (13.82) can be linearized as

$$L_1 = \sqrt{(b+u_1)^2 + u_2^2} = l_1\sqrt{1 + 2bu_1 + \frac{u_1^2 + u_2^2}{l_1}} \approx l_1 + \frac{bu_1}{l_1}$$

$$L_2 = \sqrt{(b+u_1)^2 + (h-u_2)^2} \approx l_2\sqrt{1 + 2\frac{bu_1}{l_2} - 2\frac{hu_2}{l_2}} = l_2 + \frac{bu_1}{l_2} - \frac{hu_2}{l_2}$$
(13.92)

In the foregoing derivation, we first remove the second-order terms in the square root, and then apply the binomial formula keeping only up to the first order.

Using Eqs. (13.92) and (13.83), the elongations are linearized:

$$e_1 = L_1 - l_1 = \sqrt{(b+u_1)^2 + u_2^2} - l_1 = \frac{bu_1}{l_1} \qquad (13.93)$$

$$e_2 = L_2 - l_2 = \sqrt{(b+u_1)^2 + (h-u_2)^2} - l_2 = \frac{bu_1 - hu_2}{l_2}$$

Further, the partial derivatives of e_i with respect to u_i, given in Eq. (13.88) are linearized as

$$\frac{\partial e_1}{\partial u_1} = \frac{b}{L_1}, \quad \frac{\partial e_2}{\partial u_1} = \frac{b}{L_2}$$

$$\frac{\partial e_1}{\partial u_2} = 0, \quad \frac{\partial e_2}{\partial u_2} = -\frac{h}{L_2}$$
(13.94)

Substituting Eqs. (13.93) and (13.94) into Eq. (13.87), we obtain the linearized expression for forces as follows:

$$P_1 = \frac{E_1 A_1 b^2}{l_1^3} u_1 + \frac{E_2 A_2 b}{l_2^3}(bu_1 - hu_2)$$

$$P_2 = -\frac{E_2 A_2 h}{l_2^3}(bu_1 - hu_2)$$
(13.95)

This is the set of two **equilibrium equations** for this truss structure. It works for small displacements compared to the length of the bars. The following code computes linearized forces for given small displacements.

```
1  P1_1 = K1*b**2*u1/l1**2 + K2*b*(b*u1-h*u2)/l2**2
2  P2_1 =-K2*h*(b*u1-h*u2)/l2**2
3  print(f'Two forces at point B (linearized): {P1_1:0.6e},{P2_1:0.6e}(N)')
```

Two forces at point B (linearized): 1.300000e-01,-3.000000e-02(N)

It is found that the forces are very close to those using the non-linearized formulas, which were (1.300112e-01, −2.999550e-02).

Readers may set larger displacements u_1 and u_2, and re-compute the forces using the foregoing codes. One should find that when the displacements get larger, the results from the linearized and non-linearized formulas depart.

13.6 Complementary energy formulation for 3D solids*

13.6.1 General formulations

Using the proven fact that the stress and strain components in orthogonal coordinates are energy conjugate, the variation of complementary energy C can be written out directly as

$$\delta C = \int_V (\varepsilon_{ij} \delta \sigma_{ij}) \, dV$$

$$= \int_V (\varepsilon_{xx} \delta \sigma_{xx} + \varepsilon_{yy} \delta \sigma_{yy} + \varepsilon_{zz} \delta \sigma_{zz} + 2\varepsilon_{yz} \delta \sigma_{yz} + 2\varepsilon_{xz} \delta \sigma_{xz} + 2\varepsilon_{xy} \delta \sigma_{xy}) \, dV$$
(13.96)

The variation of the complementary energy density (the complementary energy per unit volume) is given as

$$\delta C_0 = \varepsilon_{xx} \delta \sigma_{xx} + \varepsilon_{yy} \delta \sigma_{yy} + \varepsilon_{zz} \delta \sigma_{zz} + 2\varepsilon_{yz} \delta \sigma_{yz} + 2\varepsilon_{xz} \delta \sigma_{xz} + 2\varepsilon_{xy} \delta \sigma_{xy}$$
$$= \vec{\varepsilon}^{\top} \delta \vec{\sigma}$$
(13.97)

where $\vec{\sigma}$ and $\vec{\varepsilon}$ are the stress and strain vectors defined in Eq. (5.60). The variation of the strain energy density is now written as the dot product of the strain vector and the variation of the stress vector.

13.6.2 Equilibrium conditions

Note that the stresses are subjected to variation. This implies that these stresses must be in equilibrium with the externally applied forces over the solid body and over the boundary surface. These equilibrium equations and the stress boundary conditions derived in Chapter 6 for solids must be satisfied by these stresses, when the CEP is applied. This is similar to the discussion given in Section 13.3.6 for 1-DOF systems.

It turns out finding stresses that satisfy equilibrium equations is usually quite difficult for higher-dimensional solids and complex structures. Therefore, using CEP for such systems is much more challenging. On the other hand, methods based on CEP are particularly useful for obtaining upper-bound solutions. Therefore, methods based on PEP with so-called strain smoothing operations can have equilibrium equations satisfied locally, so as to soften the model and produce upper bound solutions. Detailed discussion is out of the scope of this book. Interested readers may refer to refs. [7, 8].

13.6.3 Complementary energy 1D problems

For 1D problems, there is only one stress or strain. The variation and differentiation carry the same meaning if both treat the stress σ as the independent variable. Equation (13.97) can be expressed as

$$dC_0 = \varepsilon d\sigma \tag{13.98}$$

or

$$\varepsilon = \frac{dC_0}{d\sigma} \tag{13.99}$$

This means that if the complementary energy density is given as a function of the stress, one can easily obtain the strain via differentiation. This is in fact the complementary energy principle at material level.

Also, Eq. (13.98) gives

$$C_0 = \int_0^\sigma \varepsilon(\sigma') d\sigma' \tag{13.100}$$

The complementary energy density is the shaded blue area in Fig. 13.11(a).

Consider a 1D stress problem for bars with uniform cross-section subject to an axial load. The complementary energy of the bar should be its density times its volume. Thus, at the structural member level, Eq. (13.100) becomes

$$C = \int_V C_0 dV = \int_0^L \left(\int_0^\sigma \varepsilon d\sigma' \right) \underbrace{A dx}_{dV}$$

$$= \int_0^L \left(\int_0^N \varepsilon \underbrace{dA\sigma'}_{dN'} \right) dx = \int_0^N \left(\underbrace{\int_0^L \varepsilon dx}_{e} \right) dN' \tag{13.101}$$

or

$$\boxed{C = \int_0^N e \, dN'} \tag{13.102}$$

Note that in deriving this equation, we did not impose any condition to the material. This means that Eq. (13.102) is applicable to nonlinear materials. We will use it in the example section for bars with nonlinear stress-strain relations.

If the material is linear elastic, we shall have

$$C_0 = \int_0^\sigma \varepsilon(\sigma')d\sigma' = \frac{1}{2}\sigma\varepsilon \qquad (13.103)$$

which is the shaded blue area in Fig. 13.11(b). Also, in this case we have $C_0 = U_0$. The shaded orange and blue areas are the same, as shown in Fig. 13.11(b).

13.7 Complementary energy principle for structures*

13.7.1 *Virtual work via virtual forces for structures*

When virtual forces are used in the principle of virtual work, it leads to the principle of complementary energy, similar to the 1D problems discussed earlier. The process for a structure is as follows.

Consider a set of (infinitesimal) *arbitrary independent virtual* forces $(\delta F_1, \delta F_2, \ldots, \delta F_n)$, subjected to the equilibrium conditions. These virtual forces are then applied to the loaded structure, which has a displacement field D_i. These virtual forces δF_i will do work under D_i, which is denoted as δW_F and can be expressed as

$$\delta W_F = D_1\delta F_1 + D_2\delta F_2 + \cdots + D_n\delta F_n = D_i\delta F_i \qquad (13.104)$$

The virtual forces δF_i will result in a virtual stress field. Also, within the loaded structure there exists a strain field. The virtual stresses will do virtual work under the strain field, which is the change in the complementary energy caused by δF_i. Therefore, we have

$$\delta C = \frac{\partial C}{\partial F_1}\delta F_1 + \frac{\partial C}{\partial F_2}\delta F_2 + \cdots + \frac{\partial C}{\partial F_n}\delta F_n = \frac{\partial C}{\partial F_i}\delta F_i \qquad (13.105)$$

13.7.2 *Formulas of complementary energy principle for structures*

Substituting Eqs. (13.104) and (13.105) into the foregoing equation gives,

$$D_1\delta F_1 + D_2\delta F_2 + \cdots + D_n\delta F_n = \frac{\partial C}{\partial F_1}\delta F_1 + \frac{\partial C}{\partial F_2}\delta F_2 + \cdots + \frac{\partial C}{\partial F_n}\delta F_n$$

$$(13.106)$$

which is re-grouped to give

$$\left(D_1 - \frac{\partial C}{\partial F_1}\right)\delta F_1 + \left(D_2 - \frac{\partial C}{\partial F_2}\right)\delta F_2 + \cdots + \left(D_n - \frac{\partial C}{\partial F_n}\right)\delta F_n = 0 \tag{13.107}$$

Now because each of the virtual force δF_i is **arbitrary** and δF_i are **independent** of each other. The foregoing equation holds only if

$$\boxed{D_i = \frac{\partial C}{\partial F_i} \quad \text{for all } i = 1, 2, \ldots, n} \tag{13.108}$$

This states that if the complementary energy C can be written as a function of generalized forces $(F_1, F_2, \ldots, F_n)$ acting on the structure, the displacement D_i at the location and direction of F_i can be found using Eq. (13.108).

Equation (13.108) is in fact the **CEP** at the structural level, in comparison with Eq. (13.99) that is for material level. For 1DoF structural members, Eq. (13.108) is Eq. (13.35).

We reiterate that all these forces F_i must be in equilibrium when using Eq. (13.108).

As the results of CEP, all these displacement D_i obtained using Eq. (13.108) will be compatible, as will be demonstrated in example section.

Note also that when the load is a moment M, the corresponding deformation will be rotation, and we shall have

$$\boxed{\theta_i = \frac{\partial C}{\partial M_i} \quad \text{for all } i = 1, 2, \ldots, p} \tag{13.109}$$

where p is the number of moments on the structure.

Let us look at some examples.

13.7.3 Example: Linear spring–mass system

Consider a spring–mass system with two linear elastic springs connecting sequentially two rigid masses. The left end of the first spring is fixed to a wall. The spring constants are k_1 and k_2, respectively, for the two springs. Two forces F_1 and F_2 are applied, respectively, on these two masses, as shown in Fig. 13.13. These two masses will have displacements, respectively, denoted as D_1 and D_2.

1. Using the PEP, determine the forces F_1 and F_2, assuming k_1, k_2, D_1, and D_2 are given.

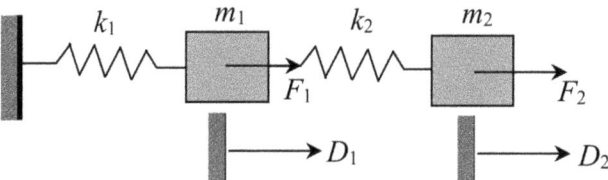

Figure 13.13. A spring mass system with two springs and two rigid masses.

2. Using the CEP, determine the displacements D_1 and D_2, assuming k_1, k_2, F_1, and F_2 are given.

Solution to 1: For the PEP, Eq. (13.80), strain (potential) energy must be written in terms of displacements. The expression for the strain energy can be found using the equation for spring given in Eq. (13.26), for which we need to find the elongations for these two springs: e_1 and e_2. From the geometry given in Fig. 13.13, we have the following compatibility equations:

$$e_1 = D_1; \quad e_2 = D_2 - D_1 \qquad (13.110)$$

The strain energy stored in the system of two springs can now be written in terms of displacements, as follows:

$$U = U_1 + U_2 = \frac{k_1 e_1^2}{2} + \frac{k_2 e_2^2}{2}$$
$$= \frac{k_1}{2} D_1^2 + \frac{k_2}{2} (D_2 - D_1)^2 \qquad (13.111)$$

Next, using Eq. (13.80), we obtain:

$$F_1 = \frac{\partial U}{\partial D_1} \Rightarrow F_1 = k_1 D_1 - k_2 (D_2 - D_1)$$
$$F_2 = \frac{\partial U}{\partial D_2} \Rightarrow F_2 = k_2 (D_2 - D_1) \qquad (13.112)$$

Substituting Eq. (13.110) into (13.112), we obtain:

$$F_1 = k_1 D_1 = k_1 e_1 - k_2 e_2 = N_1 - N_2$$
$$F_2 = k_2 (D_2 - D_1) = k_2 e_2 = N_2 \qquad (13.113)$$

where N_i is the internal force in the ith spring. This is in fact the equilibrium equations that one can obtain using free-body diagram for each of the masses. It is seen again that the potential energy principle produces the equilibrium equations. In the process of obtaining the results Eq. (13.112), we must use the compatibility equations Eq. (13.110).

Solution to 2: To use the CEP, Eq. (13.108), we need to write the complementary energy principle in terms of forces in the springs. Since the strings are linear elastic, the expression for the complementary energy is the same as the potential energy. However, we need to write it in terms of the internal forces. This can be done using the equation for spring given in Eq. (13.26), which gives,

$$C = U = U_1 + U_2 = \frac{N_1^2}{2k_1} + \frac{N_2^2}{2k_2} \qquad (13.114)$$

Note here that N_1 and N_2 are the internal forces in these two springs. These forces should be in equilibrium with the externally applied forces F_1 and F_2. These equilibrium equations should be:

$$N_1 = F_1 + F_2; \quad N_2 = F_2 \qquad (13.115)$$

Substituting Eq. (13.115) into (13.114), we obtain:

$$C = \frac{(F_1 + F_2)^2}{2k_1} + \frac{F_2^2}{2k_2} \qquad (13.116)$$

With the help of the equilibrium equations, the complementary energy is now in terms of external forces. We are ready to apply the CEP. Using Eq. (13.108), we arrive at:

$$D_1 = \frac{\partial C}{\partial F_1} \Rightarrow D_1 = \frac{F_1 + F_2}{k_1} \qquad (13.117)$$

$$D_2 = \frac{\partial C}{\partial F_2} \Rightarrow D_2 = \frac{F_1 + F_2}{k_1} + \frac{F_2}{k_2}$$

Careful examination should find that the foregoing equation is identical to Eq. (13.112).

Substituting Eq. (13.115) into (13.117), we obtain:

$$D_1 = \frac{N_1}{k_1} = e_1$$
$$D_2 = \frac{N_1}{k_1} + \frac{N_2}{k_2} = e_1 + e_2 \qquad (13.118)$$

These are compatibility equations: the displacement at mass-1 is the elongation of spring-1, and the displacement at mass-2 is the sum of the elongations of spring-1 and spring-2. The deformations of the springs agree with the displacement solutions. It is seen again that the complementary energy principle produces the compatibility equations. In the process of obtaining the results Eq. (13.117), we must use the equilibrium equations Eq. (13.115).

13.7.4 Example: Nonlinear spring–mass system

Consider the same spring–mass system studied in the previous example, but assume the two spring constants becoming nonlinear. The force-elongation relation or the constitutive equation is given by [1]:

$$N = k_i e^2 \quad \text{for } i = 1, 2 \tag{13.119}$$

where k_1 and k_2 are given constants.

Determine the displacements D_1 and D_2, assuming k_1, k_2, F_1, and F_2 are given.

Solution: Since the force-elongation relationship is nonlinear, the complementary energy principle should be used.

Using Eq. (13.119), we found e in terms of the internal force N for each of the string:

$$e = \left(\frac{N}{k_i}\right)^{1/2} \quad \text{for } i = 1, 2 \tag{13.120}$$

Using integral Eq. (13.102), the complementary energy becomes

$$\begin{aligned} C = C_1 + C_2 &= \int_0^{N_1} e\, dN + \int_0^{N_2} e\, dN \\ &= \int_0^{N_1} \left(\frac{N}{k_1}\right)^{1/2} dN + \int_0^{N_2} \left(\frac{N}{k_2}\right)^{1/2} dN = \frac{2}{3}\left(\frac{N_1^{3/2}}{k_1^{1/2}} + \frac{N_2^{3/2}}{k_2^{1/2}}\right) \end{aligned} \tag{13.121}$$

Next, substituting Eq. (13.115) into (13.121), we obtain:

$$C = \frac{2}{3}\left(\frac{(F_1 + F_2)^{3/2}}{k_1^{1/2}} + \frac{F_2^{3/2}}{k_2^{1/2}}\right) \tag{13.122}$$

The complementary energy is now in terms of external forces. Using Eq. (13.108), we arrive at:

$$\begin{aligned} D_1 &= \frac{\partial C}{\partial F_1} \Rightarrow D_1 = \left(\frac{F_1 + F_2}{k_1}\right)^{1/2} \\ D_2 &= \frac{\partial C}{\partial F_2} \Rightarrow D_2 = \left(\frac{F_1 + F_2}{k_1}\right)^{1/2} + \left(\frac{F_2}{k_2}\right)^{1/2} \end{aligned} \tag{13.123}$$

We see from this example that the CEP works well for nonlinear constitutive equations. Using the equilibrium equations Eq. (13.115), it is easy to see that Eq. (13.123) is the compatibility equation.

13.7.5 Example: A two-bar truss subjected to two loads

13.7.5.1 Setting of the problem

Consider the same setting of the previous example shown in Fig. 13.12. In this example, we assume the displacement is small. We would like to complete the following tasks, using the complementary energy principle.

(a) Derive formulas for displacements u_1 and u_2 in terms of forces P_1 and P_2.
(b) Given the same values for b, h, K_1, K_2, P_1, and P_2 as the previous example, compute the two displacement components u_1 and u_2 using the formulas obtained in part (a).

13.7.5.2 Solution to the problem

Solution to question (a): To use the CEP, we need derive the formulas for computing C in terms of the forces for the structure shown in Fig. 13.12. Using Eq. (13.101), the complementary energy C_i of the ith bar should be it volume V_i times C_0. We thus have (see, Eq. (13.102)):

$$C_i = \int_0^{N_i} e_i dN_i \tag{13.124}$$

Since the bars remain linearly elastic, the elongations of these two bars can be calculated using Eq. (13.84). Thus, Eq. (13.124) becomes

$$C_i = \int_0^{N_i} \frac{N_i l_i}{E_i A_i} dN_i = \frac{l_i}{2E_i A_i} N_i^2 = \frac{1}{2K_i} N_i^2 \tag{13.125}$$

where K_i is the stiffness of the ith bar.

The complementary energy for the structure becomes

$$C = C_1 + C_2 = \frac{1}{2K_1} N_1^2 + \frac{1}{2K_2} N_2^2 \tag{13.126}$$

Next, we need to find formulas for N_i in terms of P_i, so that we can use Eq. (13.108). This requires the use of equilibrium conditions. Since the displacement is small, the two equations in Eq. (13.90) become

$$\begin{aligned} P_1 - N_1 \tfrac{b}{L_1} - N_2 \tfrac{b}{L_2} = 0 \\ P_2 + N_2 \tfrac{h}{L_2} = 0 \end{aligned} \tag{13.127}$$

Let us check how well the equilibrium is satisfied.

```
1  Nx = -N1*b/l1 - N2*b/l2          # sum of the internal forces in bars
2  Ny = N2*h/l2                     # sum of the internal forces in bars
3  Fx = P1_1 + Nx                   # this should be close to zero
4  Fy = P2_1 + Ny                   # this should be close to zero
5
6  print(f'The internal forces in two bars: {N1:0.4e}, {N2:0.4e}(N)')
7  print(f'Two forces at point B (linearized): {P1_1:0.4e},{P2_1:0.4e}(N)')
8  print(f'Sum of all forces at point B (linear): {Fx:0.4e},{Fy:0.4e}(m)')
9  #1.3001e-01,-2.9995e-02(N)   u1 = 0.05e-3; u2 = 0.03e-3
```

The internal forces in two bars: 1.0000e-01, 4.2435e-02(N)
Two forces at point B (linearized): 1.3000e-01,-3.0000e-02(N)
Sum of all forces at point B (linear): -8.2496e-06,5.9998e-06(m)

It is seen that the equilibrium is reasonably well satisfied.

Equations (13.127) can be solved for N_1 and N_2, which gives

$$N_1 = \frac{P_1 L_1}{b} + \frac{P_2 L_1}{h}$$

$$N_2 = -\frac{P_2 L_2}{h}$$

(13.128)

Now use Eq. (13.108), we obtain the two displacement components at point B:

$$u_1 = \frac{\partial C}{\partial P_1} = \frac{\partial C}{\partial N_1}\frac{\partial N_1}{\partial P_1} + \frac{\partial C}{\partial N_2}\frac{\partial N_2}{\partial P_1} = \frac{N_1}{k_1}\frac{L_1}{b} + \frac{N_2}{k_2} \times 0$$

$$= \frac{L_1^2}{bk_1}\left(\frac{P_1}{b} + \frac{P_2}{h}\right)$$

(13.129)

$$u_2 = \frac{\partial C}{\partial P_2} = \frac{\partial C}{\partial N_1}\frac{\partial N_1}{\partial P_2} + \frac{\partial C}{\partial N_2}\frac{\partial N_2}{\partial P_2} = \frac{N_1 L_1}{k_1 h} - \frac{N_2 L_2}{k_2 h}$$

$$= \frac{L_1^2}{k_1 h}\left(\frac{P_1}{b} + \frac{P_2}{h}\right) + \frac{L_2^2}{k_2 h^2}P_2$$

(13.130)

Solution to question (b): Given the same values for b, h, K_1, K_2, P_1, and P_2, compute the two displacement components u_1 and u_2 using the formulas obtained in part (a).

This can be done by running the following code, with the results from the previous example. Note that the displacements must be small.

```
1  # print out the Displacements given in the previous example:
2  print(f'Displacements at point B (linear): {u1:0.4e},{u2:0.4e}(m)')
3
4  u1_ = l1**2/(b*K1)*(P1_1/b + P2_1/h)
5  u2_ = l1**2/(h*K1)*(P1_1/b + P2_1/h) + l2**2/(h**2*K2)*P2_1
6  print(f'Displacements at point B (by CEP): {u1_:0.4e},{u2_:0.4e}(m)')
```

```
Displacements at point B (linear): 5.0000e-05,3.0000e-05(m)
Displacements at point B (by CEP): 5.0000e-05,3.0000e-05(m)
```

As shown, the results obtained using the CEP are the same as those by PEP that are compatible. This confirms that the CEP results satisfy automatically the compatibility condition. This means that *the complimentary energy principle is in place of the compatibility equations.*

13.8 Displacements of statically determinate structures*

13.8.1 *Principles for linear elastic structure members*

Consider structures undergoes linear elastic deformation:

1. the material is linear elastic; and
2. the deformation is small.

For such a structure, the stress-strain curve for its material is shown in Fig. 13.11(b), which is a **straight line**. Hence, the strain energy density U_0 is equal to the complementary energy density C_0. When the structure undergoes small deformation, the displacement-load relation for such a structure is also linear. The strain energy U is equal to the complementary energy C. Therefore, U and C are exchangeable. This can be made of a good use.

Since the energy evaluation is additive, the energy in a structure is the sum of the energies in each of the members in the structure. In addition, the principle of **superposition** applies for linear problems. Therefore, formulas can be derived for computing the displacements for various structures with multiple different members subject to multiple loads.

This section presents some of the widely used formulas.

Since $C = U$, Eqs. (13.108) and (13.109) can be rewritten as

$$\boxed{D_i = \frac{\partial U}{\partial F_i} \quad \text{for } i = 1, 2, \ldots, n} \qquad (13.131)$$

$$\theta_i = \frac{\partial U}{\partial M_i} \quad \text{for } i = 1, 2, \ldots, p \tag{13.132}$$

These two equations allow the use of U for computing the displacements, which are of interest. In many cases, formulating U in terms of forces is more convenient than formulating C, and hence the foregoing equation is more often used.

13.8.2 Principles for structures with multiple multiple-type members

Many structures can be solved for internal forces using only the static equilibrium equations, and are said to be statically determinate structures. Consider a structure with m members of multiple types that may be coupled with each other. Assume it behaves linear elastic, so that $C = U$.

The strain energy U should be the sum of these of all members. For the jth member, its strain energy is denoted as U_j. In general, it can be expressed as

$$U_j = U_{Nj} + U_{Mj} + U_{Sj} + U_{Tj} \tag{13.133}$$

Using Eqs. (13.26) and (13.108), $C = U$, and the additive property of the potential energy, we shall have

$$\begin{aligned} D_i = \frac{\partial U}{\partial F_i} &= \sum_{j=1}^m \frac{\partial U_j}{\partial F_i} \\ &= \sum_{j=1}^m \left(\frac{\partial U_{Nj}}{\partial N_j} \frac{\partial N_j}{\partial F_i} + \frac{\partial U_{Tj}}{\partial T_j} \frac{\partial T_j}{\partial F_i} + \frac{\partial U_{Mj}}{\partial M_j} \frac{\partial M_j}{\partial F_i} + \frac{\partial U_{Sj}}{\partial V_j} \frac{\partial V_j}{\partial F_i} \right) \\ &= \sum_{j=1}^m \left(\int_0^{L_j} \frac{N_j}{E_j A_j} \frac{\partial N_j}{\partial F_i} dx + \int_0^{L_j} \frac{T_j}{G_j J_j} \frac{\partial T_j}{\partial F_i} dx \right. \\ &\quad \left. + \int_0^{L_j} \frac{M_j}{E_j I_j} \frac{\partial M_j}{\partial F_i} dx + \int_0^{L_j} \frac{kV_j}{G_j A_j} \frac{\partial V_j}{\partial F_i} dx \right) \end{aligned} \tag{13.134}$$

This computes the displacement at the same degree, location and direction where F_i is applied.

Similarly, to compute the rotation at the location where M_i is applied, we use:

$$\theta_i = \frac{\partial U}{\partial M_i} = \sum_{j=1}^{m} \frac{\partial U_j}{\partial M_i}$$

$$= \sum_{j=1}^{m} \left(\frac{\partial U_{Nj}}{\partial N_j} \frac{\partial N_j}{\partial M_i} + \frac{\partial U_{Tj}}{\partial T_j} \frac{\partial T_j}{\partial M_i} + \frac{\partial U_{Mj}}{\partial M_j} \frac{\partial M_j}{\partial M_i} + \frac{\partial U_{Sj}}{\partial V_j} \frac{\partial V_j}{\partial M_i} \right) \quad (13.135)$$

$$= \sum_{j=1}^{m} \left(\int_0^{L_j} \frac{N_j}{E_j A_j} \frac{\partial N_j}{\partial M_i} dx + \int_0^{L_j} \frac{T_j}{G_j J_j} \frac{\partial T_j}{\partial M_i} dx \right.$$

$$\left. + \int_0^{L_j} \frac{M_j}{E_j I_j} \frac{\partial M_j}{\partial M_i} dx + \int_0^{L_j} \frac{kV_j}{G_j A_j} \frac{\partial V_j}{\partial M_i} dx \right)$$

A similar derivation can also be done to compute the shear defection d_y at the location when F_i is applied, and the twist angle Θ at the location and direction where T_i is applied. We omit these formulas because they are very similar.

If the structure member does not have a type of internal force, the corresponding term in Eqs. (13.134) and (13.135) should be ignored. For example, for a pure truss structure, any member can only have internal normal force N_j. We only need to use the first integral term in these equations.

13.8.3 Examples: Cantilever beam subjected to two concentrated forces

Consider a cantilever beam with uniform cross-section and subjected to two point vertical forces, F_1 and F_2, as shown in Fig. 13.14. The length of the beam is given by L and is assumed as a thin beam. The 2nd moment of area of the cross-section is denoted as I, Young's modulus of the material is E, shear modulus is G, and all are given.

Determine the deflection of the beam at point C, consider both effects from the moment and shear forces in the beam.

Solution: We shall use the 3rd and 4th integral in Eq. (13.134). Due to the point force F_1 at the middle point, both the moment and shear force can only be expressed in two pieces of functions.

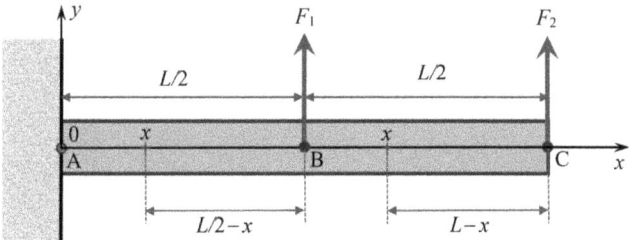

Figure 13.14. Cantilever beam subjected to two concentrated forces: one at its free tip, and one at the middle span.

In the A-B segment, we have

$$V = F_1 + F_2, \quad \frac{\partial V}{\partial F_2} = 1$$

$$M = F_1 \left(\frac{L}{2} - x\right) + F_2(L - x), \quad \frac{\partial M}{\partial F_2} = L - x$$

(13.136)

In the B-C segment, we get

$$V = F_2, \quad \frac{\partial V}{\partial F_2} = 1$$

$$M = F_2(L - x), \quad \frac{\partial M}{\partial F_2} = L - x$$

(13.137)

Considering both effects from the moment and shear force, the deflection at point C is computed using

$$D_2 = \frac{\partial U}{\partial F_2} = \frac{\partial U}{\partial M}\frac{\partial M}{\partial F_2} + \frac{\partial U}{\partial V}\frac{\partial V}{\partial F_2}$$

$$= \int_0^{L/2} \frac{F_1(\frac{L}{2}-x)+F_2(L-x)}{EI}(L-x)dx + \int_{L/2}^L \frac{F_2(L-x)}{EI}(L-x)dx$$

$$+ \int_0^{L/2} \frac{k(F_1+F_2)}{GA}dx + \int_{L/2}^L \frac{kF_2}{GA}dx$$

(13.138)

Note the integrals are over two segments. These integrations can be done easily by hand, but it is even easier using the following Sympy codes.

```
1  EI, L, x, F1, F2 = symbols('EI, L, x, F1, F2')    # define variables
2  f1 = Function('f1')(x)                             # define functions
3  f2 = Function('f2')(x)
4  f1 = (F1*(L/2-x)+ F2*(L-x))*(L-x)/EI
5
6  int_f1 = sp.integrate(f1,(x, 0,L/2))    # integration over 1st segment
7
8  f2 = F2*(L-x)*(L-x)/EI
9  int_f2 = sp.integrate(f2,(x, L/2, L))   # integration over 2nd segment
10
11 D_M =(int_f1.simplify()+int_f2.subs(F1,F2)).simplify()    # print out
12 gr.printM(D_M, "Deflection resulted from internal moments:")
```

Deflection resulted from internal moments: $\dfrac{L^3 \cdot (5F_1 + 16F_2)}{48EI}$

```
1  D_M = D_M.subs(F2, F1)
2  gr.printM(D_M, "Deflection resulted from internal moments, if F1=F2:")
```

Deflection resulted from internal moments, if F1=F2: $\dfrac{7F_1 L^3}{16EI}$

Codes for computing the deflection resulting from internal shear force are computed using:

```
1  GA, k = symbols('GA, k')                           # define variables
2  sf1 = Function('f1')(x)                            # define functions
3  sf2 = Function('f2')(x)
4
5  sf1 = k*(F1 + F2)/GA
6  int_sf1 = sp.integrate(sf1,(x, 0,L/2))   # integration over 1st segment
7
8  sf2 = k*F2/GA
9  int_sf2 = sp.integrate(sf2,(x, L/2,L))   # integration over 2nd segment
10
11 D_S =(int_sf1.simplify()+int_sf2.subs(F1,F2)).simplify()   # print out
12 gr.printM(D_S, "Deflection resulted from internal shear force:")
```

Deflection resulted from internal shear force: $\dfrac{Lk(F_1 + 2F_2)}{2GA}$

```
1  D_S = D_S.subs(F2, F1)
2  gr.printM(D_S, "Deflection resulted from internal shear, if F1=F2:")
```

Deflection resulted from internal shear, if F1=F2: $\dfrac{3F_1 Lk}{2GA}$

The final result adds up these two deflections from moment and shear force:

$$D_2 = \frac{L^3(5F_1 + 16F_2)}{48EI} + \frac{kL(F_1 + 2F_2)}{2GA} \qquad (13.139)$$

If $F_1 = F_2$, we have

$$D_2 = \frac{7L^3 F_1}{16EI} + \frac{3kLF_1}{2GA} \qquad (13.140)$$

The first part is from the internal moment and the second from the internal shear force on the cross-section resulting from the externally applied two forces.

13.8.4 Examples: Cantilever beam subjected to distributed force

Consider a cantilever beam with uniform cross-section and subjected to uniformly distributed force, p, as shown in Fig. 13.15. The length of the beam is given by L and is assumed as a thin beam. The 2nd moment of area of the cross-section is denoted as I, Young's modulus of the material is E, and all are given. Consider only the moment in the beam.

1. Determine the deflection at the beam's free end.
2. Determine the rotation of the cross-section at the beam's free end.

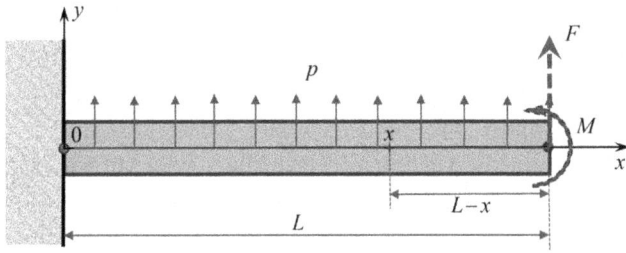

Figure 13.15. Cantilever beam subjected to uniformly distributed force. A fictitious force or moment is applied at the beam's free end when calculating the displacement there.

Solution to 1: Since only moment is considered, we shall use the 3rd integral in Eq. (13.134). However, at the free end, there is no concentrated force. We can apply a fictitious force F there to obtain the formula for the deflection. After that, we can set $F = 0$ to obtain the final solution.

The formula for computing the rotation at tip D_F resulting from the force F becomes

$$D_F = \frac{\partial U}{\partial F} = \int_0^L \frac{M_x}{EI} \frac{\partial M_x}{\partial F} dx \qquad (13.141)$$

The moment M_x is the internal moment within the beam, and is a function of x, which has contributions from both the originally applied distributed force and the fictitious one. The subscript x is not for rotation direction. The moment M_x can be computed by equilibrium condition using,

$$M_x = F(L-x) + \frac{1}{2}p(L-x)^2 \qquad (13.142)$$

Its derivative with respect to F becomes

$$\frac{\partial M_x}{\partial F} = (L-x) \qquad (13.143)$$

Substituting Eqs. (13.142) and (13.143) into (13.141), we obtain the solution:

$$D_F = \int_0^L \frac{F(L-x) + \frac{1}{2}p(L-x)^2}{EI} \bigg|_{F=0} (L-x) dx$$

$$= \int_0^L \frac{\frac{1}{2}p(L-x)^3}{EI} dx = \frac{p}{2EI} \frac{-1}{4}(L-x)^4 \bigg|_0^L = \frac{pL^4}{8EI} \qquad (13.144)$$

Readers may use gr.solver1D4() developed in Chapter 8 to compute the same.

Solution to 2: Since the rotation from the internal moment is requested, we shall use the 3rd integral in Eq. (13.135). However, at the free end, there is no concentrated moment. We need to apply a fictitious moment M there to obtain the formula for the rotation. After that, we can set $M = 0$ to obtain the final solution.

The formula for computing the rotation at tip θ_M resulting from the internal moment M_x becomes

$$\theta_M = \frac{\partial U}{\partial M_x} = \int_0^L \frac{M_x}{EI} \frac{\partial M_x}{\partial M} dx \qquad (13.145)$$

The moment M_x within the beam is a function of x, which has contributions from both the originally applied distributed force and the fictitious one. It can be computed by equilibrium condition using

$$M_x = M + \frac{1}{2}p(L-x)^2 \quad (13.146)$$

Its derivative with respect to M is simply 1.

Substituting Eq. (13.146) into (13.145), we obtain the solution:

$$\theta_M = \int_0^L \frac{M + \frac{1}{2}p(L-x)^2}{EI}\bigg|_{M=0} \times 1 dx$$

$$= \int_0^L \frac{\frac{1}{2}p(L-x)^2}{EI} dx = \frac{p}{2EI}\frac{-1}{3}(L-x)^3\bigg|_0^L = \frac{pL^3}{6EI} \quad (13.147)$$

Readers may also use gr.solver1D4() developed in Chapter 8 to compute the same. Such a comparison can enhance the understanding, and it is easy to do.

13.8.5 *Fictitious unit load method*

In the previous example we used fictitious load to obtain displacement at a location where there is no load. We then need to set back the fictitious load to zero. This means in computing U, we do not really need to include the fictitious load, but we need it for differentials for the internal forces with respect to the corresponding external forces. The improved approach is the so-called fictitious unit load method [1]. The detailed procedure as follows.

1. **Equilibrium:** Write an expression for each of the internal actions (axial force, shear, moment, and torque) in each member of the structure in terms of only the actually applied external forces (and moments, etc.).
2. Apply a unit fictitious force $F_i = 1$ (or moment $M_i = 1$. etc.) at the location and in the same direction as the displacement D_i (or rotation θ_i) to be determined.
3. Evaluate coefficients n_{ji}^F, m_{ji}^F, v_{ji}^F, t_{ji}^F, which are defined as

$$n_{ji}^F = \frac{\partial N_j}{\partial F_i}, \quad t_{ji}^F = \frac{\partial T_j}{\partial F_i}, \quad m_{ji}^F = \frac{\partial M_j}{\partial F_i}, \quad v_{ji}^F = \frac{\partial V_j}{\partial F_i}$$

$$n_{ji}^M = \frac{\partial N_j}{\partial M_i}, \quad t_{ji}^M = \frac{\partial T_j}{\partial M_i}, \quad m_{ji}^M = \frac{\partial M_j}{\partial M_i}, \quad v_{ji}^M = \frac{\partial V_j}{\partial M_i} \quad (13.148)$$

A coefficient given above is essentially the internal force in member j resulting from a unit fictitious force at location i, obtained via equilibrium.

4. Substitute these expressions into Eqs. (13.134) and (13.135), which gives

$$D_i = \sum_{j=1}^{m}\left(\int_0^{L_j} \frac{N_j n_{ji}^F}{E_j A_j}dx + \int_0^{L_j} \frac{T_j t_{ji}^F}{G_j J_j}dx + \int_0^{L_j} \frac{M_j m_{ji}^F}{E_j I_j}dx + \int_0^{L_j} \frac{kV_j v_{ji}^F}{G_j A_j}dx\right) \quad (13.149)$$

and

$$\theta_i = \sum_{j=1}^{m}\left(\int_0^{L_j} \frac{N_j n_{ji}^M}{E_j A_j}dx + \int_0^{L_j} \frac{T_j t_{ji}^M}{G_j J_j}dx + \int_0^{L_j} \frac{M_j m_{ji}^M}{E_j I_j}dx + \int_0^{L_j} \frac{kV_j v_{ji}^M}{G_j A_j}dx\right) \quad (13.150)$$

5. Perform the integrations, which gives D_i (or θ_i).

13.8.6 Example: Cantilever beam subjected to distributed force

We repeat the previous example, but using the fictitious unit load method.

Solution to 1: The internal moment M_x is computed by equilibrium condition using only the actual force.

$$M_x = \frac{1}{2}p(L-x)^2 \quad (13.151)$$

We set $F = 1$, and calculate the coefficient m_{ji}^F that is the moment at $j = x$ resulting from the unit force, and is obtained via equilibrium:

$$m_{xi}^F = 1 \times (L-x) \quad (13.152)$$

Substituting Eqs. (13.151) and (13.152) into (13.149), and using only the integral for moment, we obtain the solution:

$$D_F = \int_0^{L_j} \frac{M_x m_{xi}^F}{EI}dx = \int_0^L \frac{\frac{1}{2}p(L-x)^2}{EI}(L-x)dx$$

$$= \int_0^L \frac{\frac{1}{2}p(L-x)^3}{EI}dx = \frac{p}{2EI}\frac{-1}{4}(L-x)^4\Big|_0^L = \frac{pL^4}{8EI} \quad (13.153)$$

Solution to 2: The internal moment M_x is computed by equilibrium condition using only the actual force. Thus, it is the same given in Eq. (13.151).

Set now $M = 1$, and calculate the coefficient m_{ji}^M that is the moment at $j = x$ resulting from the unit moment, and is obtained via equilibrium:

$$m_{xi}^M = 1 \tag{13.154}$$

Substituting Eq. (13.151) and (13.151) into (13.150), we obtain the solution:

$$\begin{aligned}\theta_M &= \int_0^{L_j} \frac{M_x m_{xi}^M}{EI} dx = \int_0^L \frac{\frac{1}{2}p(L-x)^2}{EI} \times 1 dx \\ &= \int_0^L \frac{\frac{1}{2}p(L-x)^2}{EI} dx = \frac{p}{2EI}\frac{-1}{3}(L-x)^3 \bigg|_0^L = \frac{pL^3}{6EI}\end{aligned} \tag{13.155}$$

As shown, we obtained the same solution. The fictitious unit load method is more convenient to use because 1. it eliminates the differentiations, and 2. it simplifies the evaluation of the equilibrium equations. Essentially, one needs only to get the internal force via equilibrium considering only the actual external forces, and the internal force via equilibrium considering only the fictitious unit force.

13.8.7 Example: A planar truss structure with three bars

Consider a simple planar truss structure with three bars (numbered in small box), as shown in Fig. 13.16. These bars all have uniform cross-section with area A, and all made of the same material with E. The structure has three nodes (numbered in small circles). The structure is subjected to a vertical downward force of F_{2y} at node 2. Assume the structure behaves linear elastically. Using the energy method, determine:

1. the vertical displacement at node 2, D_{2y};
2. the horizontal displacement at node 2, D_{2x};
3. the values of D_{2x} and D_{2y}, letting $F_{2y} = 1000$ N, $E = 70.0$ GPa, and $A = 0.01$ m^2.

Solution to 1: We shall use the 1st integral in Eq. (13.134). Therefore, we need to find out the internal axial forces for each of the bars. Since this structure is statically determinate, these internal forces can be found using equilibrium conditions.

First, at node 2, we can easily find:

$$\begin{aligned}N_2 &= -\frac{F_{2y}}{\sin 45°} = -F_{2y}\sqrt{2} \\ N_1 &= -N_2 \cos 45° = F_{2y}\end{aligned} \tag{13.156}$$

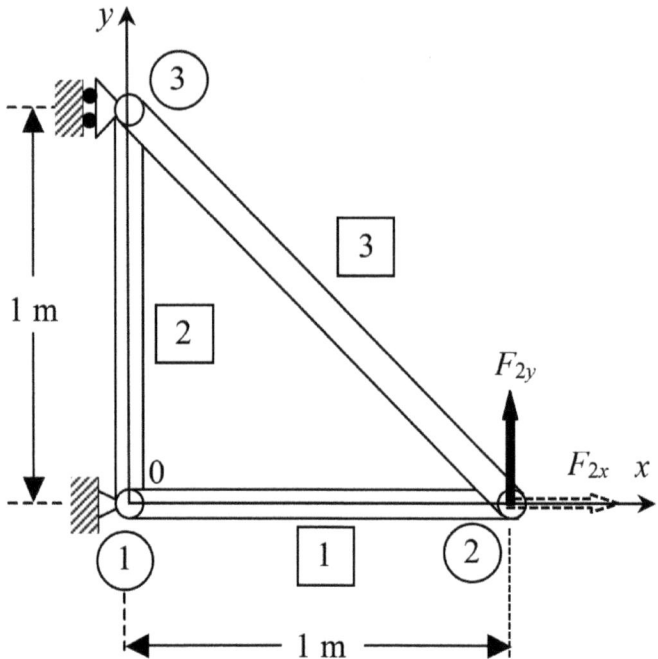

Figure 13.16. A planar truss structure with three bars hinged together. It is subjected to a vertical force.

Next, at node 3, we found

$$N_3 = -N_2 \cos 45° = F_{2y} \qquad (13.157)$$

The derivatives of these internal forces with respect to F_{2y} become

$$\frac{\partial N_2}{\partial F_{2y}} = -\sqrt{2}; \quad \frac{\partial N_1}{\partial F_{2y}} = 1; \quad \frac{\partial N_3}{\partial F_{2y}} = 1 \qquad (13.158)$$

Substitute Eqs. (13.156), (13.157), and (13.158) to the 1st integral in Eq. (13.134), we obtain

$$\begin{aligned} D_{2y} &= \int_0^{L_1} \frac{N_1}{EA} \times 1 dx + \int_0^{L_2} \frac{N_2}{EA}(-\sqrt{2}) dx + \int_0^{L_3} \frac{N_3}{EA} \times 1 dx \\ &= \frac{F_{2y} L_1}{EA} + \frac{-F_{2y} L_2 \sqrt{2}}{EA}(-\sqrt{2}) + \frac{F_{2y} L_3}{EA} \\ &= \frac{F_{2y}}{EA}(L_1 + 2L_2 + L_3) = \frac{2F_{2y}}{EA}(1 + \sqrt{2}) \end{aligned} \qquad (13.159)$$

Solution to 2: Since there is no force in the x-direction at node 2, we shall use the fictitious unit load method. We apply $F_{2x} = 1$. The internal forces have already been found and given in Eqs. (13.156) and (13.157). We need only to compute the coefficients, each of which is the internal force in a bar resulting from $F_{2x} = 1$. This can again be obtained via equilibrium.

First, at node 2, we shall find:

$$n_{2x}^{F_{2x}} = 0; \quad n_{1x}^{F_{2x}} = 1; \quad n_{3x}^{F_{2x}} = 0 \qquad (13.160)$$

Substitute Eqs. (13.160) and (13.157) to the 1st integral in Eq. (13.149), we obtain

$$D_{2x} = \int_0^{L_1} \frac{N_1 \times 1}{EA} dx + \int_0^{L_2} \frac{N_2 \times 0}{EA} dx + \int_0^{L_3} \frac{N_3 \times 0}{EA} dx$$

$$= \frac{F_{2y} L_1}{EA} \qquad (13.161)$$

Solution to 3: Substitute the given values for all these variables, we compute the numerical values for F_{2x} and F_{2y} using the following codes.

```
1  E = 70.0e9; A = 0.01; L1 = 1.0
2  F2y = 1000.
3
4  D2y = 2*F2y*L1/E/A*(1.+np.sqrt(2))
5  D2x = F2y*L1/E/A
6  print(f"Displacements at node 2: ({D2x}, {D2y})")
```

Displacements at node 2: (1.4285714285714286e-06, 6.8977530353517e-06)

13.9 Statically indeterminate structures*

13.9.1 *Issues in dealing with indeterminate structures*

In using the CEP, we need to find internal forces in a structure using equilibrium equations. This means that the structure needs to be statically determinate. However, many structures that are statically indeterminate. In fact, most of the structures used in practical engineering are indeterminate because of the important requirement on the structural stiffness. For an indeterminate structure, we need to find additional equations or constrain conditions to find the solution, when using CEP.

In this section, we use an example to demonstrate how an indeterminate structure can be solved.

13.9.2 Example: Cantilever beam with a simple support at the other end

Consider a cantilever thin beam with a support at its right-end, as shown in Fig. 13.17(a). It is thus an indeterminate structure, implying that the internal forces cannot be found via only the use of equilibrium conditions. The beam is subject to an uniformly distributed force p. Assume Young's modulus of the material is E, and the 2nd moment of area is I. Use energy method.

1. Determine the reaction R at the support at the right-end.
2. Determine the rotation of the cross-section of the beam at the right-end.

Solution to 1: Due to the redundant support to the beam, we cannot find the internal forces directly. Thus, we remove the support at the right-end, and replace it with an unknown force R, called reaction force. The new setting is given in Fig. 13.17(b). We now have a cantilever beam subjected to two forces: one distributed p, and one concentrated force R. This problem is solved in previous examples and the solutions are given in Eqs. (13.47) and (13.144). The deflection at the reaction force location D_R can be written as

$$D_R = \frac{RL^3}{3EI} + \frac{pL^4}{8EI} \qquad (13.162)$$

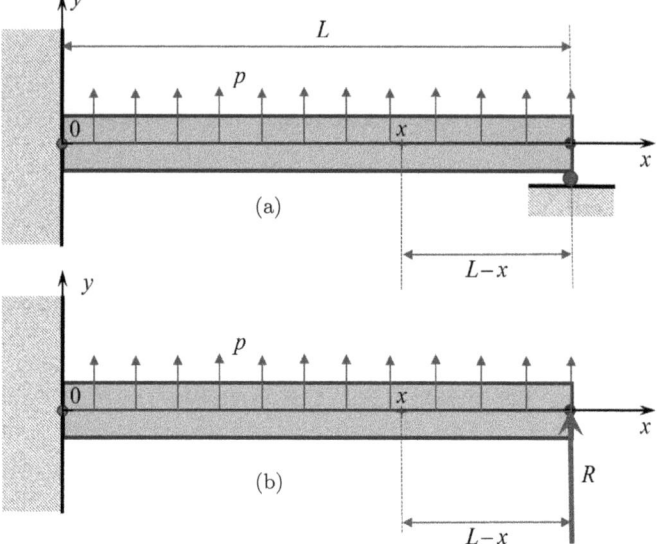

Figure 13.17. A cantilever beam with an additional support at its right-end. It is subject to a distributed force. (a) The beam with redundant support; (b) the support at the right-end is replaced by a reaction force.

Since the right-end of the beam should be zero, $D_R = 0$, we thus obtain a constraint equation:
$$\frac{RL^3}{3EI} + \frac{pL^4}{8EI} = 0 \qquad (13.163)$$
This gives the reaction force:
$$R = -\frac{3pL}{8} \qquad (13.164)$$

Solution to 2: Since the reaction force has now been found, computing the rotation of the cross-section of the beam at the right-end is straightforward. The solution is, in fact, found in our earlier examples, given in Eq. (13.49) for a concentrated load at the right-end, and in Eq. (13.147) for a distributed loading. The solution is the sum of those:
$$\theta_R = \frac{RL^2}{2EI} + \frac{pL^3}{6EI} = \frac{-3pL^3}{16EI} + \frac{pL^3}{6EI} = -\frac{pL^3}{48EI} \qquad (13.165)$$
Readers may compare the results with that obtained in Section 8.13.

13.10 Remarks

This chapter studies two major energy principles:

1. PEP, which is derived from the principle of virtual work using virtual displacement. It expresses the external force in terms of the partial differential of the potential (strain) energy with respect to displacement. Compatibility conditions for the displacement and elongation (or strains) are required to use PEP. The result is the equilibrium equation.
2. CEP, which is derived from the principle of virtual work using virtual force. It expresses the displacement in terms of the partial differential of the complementary energy with respect to the external force. Equilibrium conditions for the external and internal forces are required to use CEP. The result is the compatibility equation.

These energy methods are demonstrated using a number of examples with Python codes. The energy principles, especially the PEP, form a theoretical foundation for extremely powerful numerical methods, such as FEM [6], S-FEM [7] and meshfree methods [8] for large-scale solid mechanics problems.

References

[1] A.P. Boresi, R.J. Schmidt and O.M. Sidebottom, *Advanced Mechanics of Materials*. John Wiley & Sons, New York, 1985.
[2] A.P. Boresi, K. Chong and J.D. Lee, *Elasticity in Engineering Mechanics*. John Wiley & Sons, New York, 2010.

[3] J.M. Gere and S.P. Timoshenko, *Mechanics of Materials*, Van Nostrand Reinhold Company, New York, 1972.
[4] S. Timoshenko and J.N. Goodier, *Theory of Elasticity*. McGraw-Hill, New York, 1970. http://books.google.com/books?id=yFISAAAAIAAJ&dq=theory+of+elasticity&ei=ICiKSsr3G4jwkQSbxMyPCg.
[5] Z.L. Xu, *Elasticity*, Vols. 1&2. People's Publisher, China, 1979.
[6] G.R. Liu and S.S. Quek, *The Finite Element Method: A Practical Course*. Butterworth-Heinemann, New York, 2013.
[7] G.R. Liu and T.T. Nguyen, *Smoothed Finite Element Methods*. Taylor and Francis Group, New York, 2010.
[8] G.R. Liu, *Mesh Free Methods: Moving Beyond the Finite Element Method*. Taylor and Francis Group, New York, 2010.
[9] A.E.H. Love, *A Treatise on the Mathematical Theory of Elasticity*, 1944, http://www.worldcat.org/isbn/0486601749.
[10] G.R. Liu, *Vector Calculus: Formulation, Applications, and Python Codes*, World Scientific, New Jersey, 2025.
[11] G.R. Liu, *Calculus: A Practical Course with Python*, World Scientific, New Jersey, 2025.

Chapter 14

Failure Criteria and Failure Analysis

Let us import the necessary external modules or dependencies for later use in this chapter:

```
1  # Place cursor in this cell, and press Ctrl+Enter to import dependences.
2  import sys                          # for accessing the computer system
3  sys.path.append('../grbin/')   # Change to the directory in your system
4
5  from commonImports import *    # Import dependences from '../grbin/'
6  import grcodes as gr           # Import the module of the author
7  importlib.reload(gr)           # When grcodes is modified, reload it
8
9  from continuum_mechanics import vector
10 init_printing(use_unicode=True)     # For latex-like quality printing
11 np.set_printoptions(precision=4,suppress=True)  # Digits in print-outs
```

The major purpose of stress analysis is to assess whether the stressed material will fail. Such an assessment is called failure analysis. The most important stage in material failure process is yielding, which can often be clearly observed during experiments of many types of materials. When the stressed material gets to the yielding point, it starts to deform drastically without much increase in loading, and the load level is often very close to the ultimate failure load.

In addition, in many conservative designs of structure, material yielding is regarded as failure. Such yielding-based designs apply to all structures and devices that are subjected to multiple repetitive uses and often with large safety factors, implying that the service stress should be far below the yield stress. Therefore, the yield criterion of a material is particularly important. This chapter focuses on various types of yield criteria used for different types of materials. The chapter is written with reference to Refs. [1–3].

14.1 Basic concept

As discussed in previous chapters, a structural component can have multiple (as much as six) stress components at a point in the material. When a yielding criterion is defined, it is thus convenient to define a single equivalent or effective stress σ_e that is a combination of all the stress components. Since material yielding/failing is a physical property of the material and it should not depend on the coordinates used to compute the stresses, the effective stress should be a proper combination of the principal stresses or other stress invariants.

Clearly, different forms of yield criteria are needed for different types of materials. We define a yield function $f(\sigma_{ij}, \sigma_Y)$, where σ_{ij} represent the stress states and σ_Y is the yield strength of the material in uniaxial tension (or compression) obtained from an experiment. A stress-based yield criterion can be expressed as [4]

$$f(\sigma_{ij}, \sigma_Y) = 0 \tag{14.1}$$

at which the material is said to be on the yield surface. When $f(\sigma_{ij}, \sigma_Y) < 0$, the stress state is elastic, and $f(\sigma_{ij}, \sigma_Y) > 0$ is not allowed.

14.2 A case for this study: Yield analysis of a steel shaft

We use a typical case for studying various yielding criteria. The case chosen has a combined stress state resulting from bending, pressure, and torsion at the same time. The strategy is to introduce different yielding criteria one by one, and use it to assess whether or not the material will yield for the same given stress state at a point. We compute also the safety factor based on the yielding criterion used.

14.2.1 *Setting of the problem*

Consider a solid cylindrical shaft with a diameter of $d = 100$ mm. It is loaded by a torque $\mathbf{T} = 30$ kNm, a bending moment $\mathbf{M} = 15$ kNm, and a pressure $p = 50$ MPa on its outer surface. The directions of these loads are shown in Fig. 14.1. The shaft is made of steel treated as a linear elastic material before yielding. Its Young's modulus is $E = 200$ GPa, and Poisson's ratio is $\nu = 0.3$. Determine whether or not the shaft will yield at point A, assuming the safety factor as $S_f = 2.5$, and the material yield stress $\sigma_Y = 750$ MPa.

14.2.2 *Formulas for computing the stresses*

From the setting given in Fig. 14.1, the contributions to the stresses on the lateral surface of the shaft are from the torque and the uniform pressure.

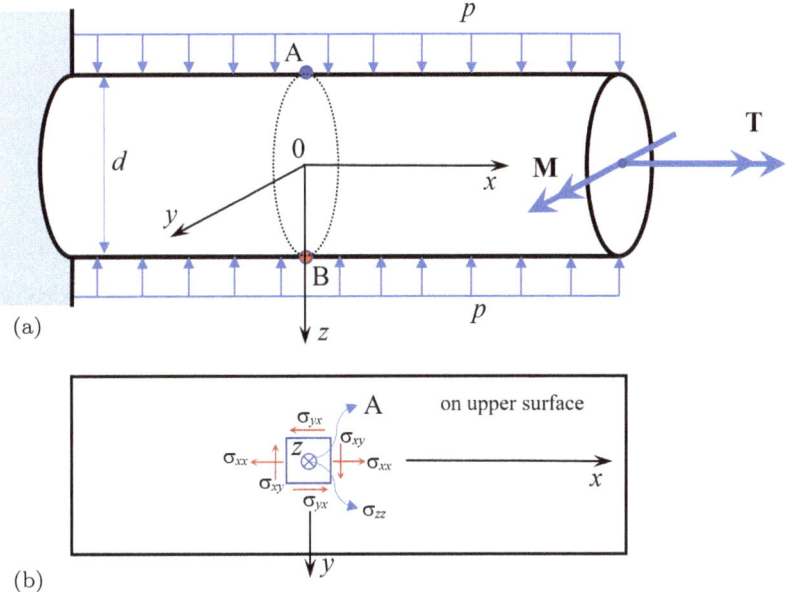

Figure 14.1. (a) A solid cylindrical shaft subjected to a torque **T**, moment **M**, and a pressure p on the outer surface. (b) Top-down view on the upper surface of the bar. All stress components shown are positive following the standard sign convention.

The normal stress in the x-direction caused by the moment will be largest at points A or B. Therefore, the critical points on this shaft should be points A or B. Let us examine the stress states at point A.

The normal stress in the x-direction at point A ($z = d/2$) caused by the moment is computed using

$$\sigma = \frac{zS_F M}{I} = -\frac{S_F M d}{2I} \qquad (14.2)$$

Here, we use the same safety factor S_F to scale up the moment applied to the bar, and I is the second moment of area of the cross-section of the shaft, computed using

$$I = \frac{\pi d^4}{64} \qquad (14.3)$$

The shear stress caused by the torque is computed by

$$\tau = \frac{S_F T d}{2J} \qquad (14.4)$$

Similarly, we use the safety factor S_F to scale up the torque applied to the bar, and J is the second polar moment of the cross-section of the shaft, computed using

$$J = \frac{\pi d^4}{32} \qquad (14.5)$$

The stress components at point A are then given by

$$\sigma_{xx} = \sigma; \quad \sigma_{xy} = \tau; \quad \sigma_{zx} = 0; \quad \sigma_{yy} = 0; \quad \sigma_{zy} = 0; \quad \sigma_{zz} = -S_F p \qquad (14.6)$$

14.2.3 Python code to compute the stresses

We write the following code to compute the stress tensor in matrix form, and then compute the principal stresses.

```python
# External forces
Torque = 30.0e3   # Nm
Moment = 15.0e3   # Nm
p = 50.0e6        # Pa, pressure on the surface
SF = 2.5          # safety factor

# Material property
E = 200.0e9       # Pa, Young's modulus, not used in this example
v = 0.3           # Poisson's ratio
sY= 750.0e6       # Pa, Yield stress

# Dimensions of the cylindrical shaft
d = 0.100         # m, diameter
#A = np.pi*((0.1/2)**2-(0.08/2)**2)

# Compute structure/member properties
I = (np.pi * d**4)/64.0    # 2nd moment of circular area
J = (np.pi * d**4)/32.0    # 2nd polar moment of circular area

# Compute the stresses
σ =-SF*Moment*(d/2.)/I     # stress by the moment at point A
τ = SF*Torque*(d/2.)/J     # stress by the Torque

σxx=σ; σxy=τ; σxz=0; σyy=0; σyz=0; σzz=-SF*p

Sij = np.array([[σxx, σxy, σxz],
                [σxy, σyy, σyz],
                [σxz, σyz, σzz]])
print(f'Stress state:\n{Sij}\n')

eigens, eigenVectors = gr.principalS(Sij)    # principal stresses
```

```
Stress state:
[[-3.8197e+08  3.8197e+08  0.0000e+00]
 [ 3.8197e+08  0.0000e+00  0.0000e+00]
 [ 0.0000e+00  0.0000e+00 -1.2500e+08]]

Principal stress/strain (Eigenvalues):
[ 2.3607e+08 -1.2500e+08 -6.1804e+08]

Principal stress/strain directions:
[[-0.5257  0.     -0.8507]
 [-0.8507  0.      0.5257]
 [ 0.      1.      0.    ]]

Possible angles (n1,x)=121.71747441146103○ or 58.282525588538974○
```

We got two angles because of the non-uniqueness of the eigenvectors and the multivalued feature of cosine functions. We use the following code to find out which one is correct.

```
1  # Form the transformation matrix using the first angle:
2  T0 = gr.transferM(58.282525588538974, about = 'z')[0]
3
4  # Perform the transformation to the stress tensor
5  T0@Sij@T0.T
```

```
array([[ 2.3607e+08,  4.1106e-07,  0.0000e+00],
       [ 4.1692e-07, -6.1804e+08,  0.0000e+00],
       [ 0.0000e+00,  0.0000e+00, -1.2500e+08]])
```

This angle of 58.28 degree is not correct. The angle of 121.72 or -58.28 degree is correct.

We now discuss various yield criteria to assess whether or not the material will yield at point A and the actual safety factor.

14.3 Maximum principal stress criterion

The maximum principal stress criterion states that yielding begins at a point where the maximum absolute value of the principal stress reaches the absolute yield stress σ_Y that may be obtained from a tensile or compressive test.

For given a stress state with σ_{ij} at a point in solid, we compute the principal stresses and rank them in the order of $|\sigma_1| \geq |\sigma_2| \geq |\sigma_3|$. The effective stress σ_e is chosen as

$$\sigma_e = |\sigma_1| \tag{14.7}$$

and the yield criterion is

$$\sigma_e - \sigma_Y = 0 \tag{14.8}$$

This equation is convenient to use for assessing whether or not the material is yielding at a point, and hence useful in design of structures.

14.3.1 Yield surface

We can also compute a surface, called the **yield surface**, in the $(\sigma_1, \sigma_2, \sigma_3)$ coordinate system. It is an envelope. If the principal stresses of any given stress state fall on the envelope, the material will yield. If it is within the envelope, it is still elastic and safe. Such a yield surface is useful to determine the bounds on which the material will yield. The function for the yield surface is called the **yield function**. For the maximum principal stress criterion, the yield function can be written as

$$f_{1 \text{ or } 2} = |\sigma_1 - \sigma_Y| = 0 \quad \text{or} \quad f_{1 \text{ or } 2} = \sigma_1 - \pm\sigma_Y = 0$$
$$f_{3 \text{ or } 4} = |\sigma_2 - \sigma_Y| = 0 \quad \text{or} \quad f_{3 \text{ or } 4} = \sigma_2 - \pm\sigma_Y = 0 \quad (14.9)$$
$$f_{5 \text{ or } 6} = |\sigma_3 - \sigma_Y| = 0 \quad \text{or} \quad f_{5 \text{ or } 6} = \sigma_3 - \pm\sigma_Y = 0$$

It has a total of six surfaces, and the yield function is the enclosed piecewise function formed by these six surfaces, denoted as

$$f = f_1 \cup f_2 \cup f_3 \cup f_4 \cup f_5 \cup f_6 \quad (14.10)$$

It is the six surfaces of a cube in the $(\sigma_1, \sigma_2, \sigma_3)$ coordinate system. For plane stress problems, it is a square in the (σ_1, σ_2) coordinate system.

Figure 14.2 shows a yield surface with four straight lines in the (σ_1, σ_2) coordinates for plane stress problems.

This yield criterion may not be good for ductile materials but can be good for brittle materials.

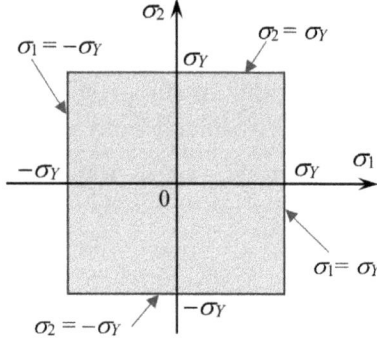

Figure 14.2. Schematic view of the yield surface of the maximum principal stress for plane stress problems. It consists of four straight lines.

14.3.2 Application to the case study problem

The following codes compute the maximum principal stress using the principal stresses obtained in Section 14.2.3. This gives the material yield stress value needed to have point A stay below yielding, which determines the yield state.

```
1  # maximum principal stress criterion
2  ps_max = max(abs(eigens))
3  print(f'The maximum principal stress = {ps_max:0.5e}')
4
5  print(f'The material yield stress needs to be > {ps_max:0.5e}')
6  print(f'The yield stress of the material = {sY:0.5e}')
7  print(f'Is yield? {ps_max>sY}')
8  print(f'Actual safety factor = {SF*sY/ps_max}')
```

```
The maximum principal stress = 6.18043e+08
The material yield stress needs to be > 6.18043e+08
The yield stress of the material = 7.50000e+08
Is yield? False
Actual safety factor = 3.0337672480085427
```

14.4 Maximum principal strain criterion

It states that yielding begins when the maximum absolute value of the principal strains at a point reaches the absolute value of the yield strain ε_Y that may be obtained from a tensile or compressive test.

For given a strain state with ε_{ij}, we compute the principal strains and rank them in the order of $|\varepsilon_1| \geq |\varepsilon_2| \geq |\varepsilon_3|$. The effective stress ε_e is chosen as

$$\varepsilon_e = |\varepsilon_1| \tag{14.11}$$

and the yield criterion is

$$\varepsilon_e - \varepsilon_Y = 0 \tag{14.12}$$

This yield function is strain-based. It may not be good for ductile materials but can fit better for brittle materials.

Using the constitutive equations for materials (see Chapter 5), one can express the principal strains in terms of principal stresses, and then the maximum principal strain criterion can be written in terms of stresses.

For example, using Eq. (5.73) for isotropic materials, we have

$$\varepsilon_1 = \frac{1}{E}[\sigma_1 - \nu\sigma_2 - \nu\sigma_3]$$
$$\varepsilon_2 = \frac{1}{E}[\sigma_2 - \nu\sigma_1 - \nu\sigma_3] \quad (14.13)$$
$$\varepsilon_3 = \frac{1}{E}[\sigma_3 - \nu\sigma_1 - \nu\sigma_2]$$

From the uniaxial load at the yield point ($\sigma_1 = \sigma_Y, \sigma_2 = \sigma_3 = 0$), we have

$$\varepsilon_Y = \frac{\sigma_Y}{E} \quad (14.14)$$

14.4.1 Yield Surface

Using Eqs. (14.12), (14.13), and (14.14), we shall have a yield surface with six flat planes in the ($\sigma_1, \sigma_2, \sigma_3$) coordinates:

$$\begin{aligned} f_{1 \text{ or } 2} &= \sigma_1 - \nu\sigma_2 - \nu\sigma_3 \pm \sigma_Y = 0 \\ f_{3 \text{ or } 4} &= \sigma_2 - \nu\sigma_1 - \nu\sigma_3 \pm \sigma_Y = 0 \\ f_{5 \text{ or } 6} &= \sigma_3 - \nu\sigma_1 - \nu\sigma_2 \pm \sigma_Y = 0 \end{aligned} \quad (14.15)$$

The yield surface is the union of the six surfaces of a parallelepiped. The effective yield stress can be written as a single expression as

$$\sigma_e = \max_{i \neq j \neq k}(\sigma_i - \nu\sigma_j - \nu\sigma_k) \quad (14.16)$$

The yield criterion can also be written as

$$\sigma_e - \sigma_Y \quad (14.17)$$

Figure 14.3 shows a yield surface with four straight lines in the (σ_1, σ_2) coordinates for plane stress problems.

14.4.2 Application to the case study problem

The following codes compute the maximum principal strain using the principal stresses obtained in Section 14.2.3. This gives the material yield stress

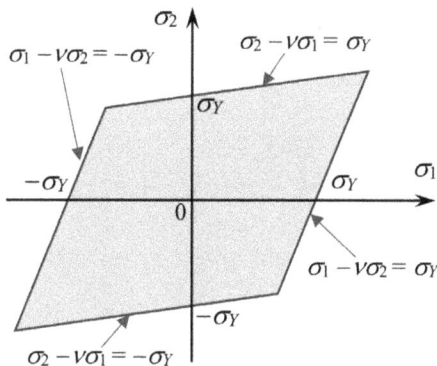

Figure 14.3. Schematic view of the yield surface of the maximum principal strain for plane stress problems. It is a parallelogram of four straight lines.

value needed to have point A stay below yielding and determines the yield state.

```
1  # maximum principal strain criterion
2  se1 = eigens[0]-v*eigens[1]-v*eigens[2]   # principal strains E ε1
3  se2 = eigens[1]-v*eigens[0]-v*eigens[2]   # principal strains E ε2
4  se3 = eigens[2]-v*eigens[0]-v*eigens[1]   # principal strains E ε3
5  e_max =max(abs(np.array([se1, se2, se3])))
6
7  print(f'The maximum principal strain is {e_max:0.5e}')
8  print(f'The material yield stress needs to be > {e_max:0.5e}')
9  print(f'The yield stress of the material = {sY:0.5e}')
10 print(f'Is yield? {e_max>sY}')
11 print(f'Actual safety factor = {SF*sY/e_max}')
```

The maximum principal strain is 6.51365e+08
The material yield stress needs to be > 6.51365e+08
The yield stress of the material = 7.50000e+08
Is yield? False
Actual safety factor = 2.878570669361039

14.5 Strain-energy density criterion

The strain-energy density is one of the coordinate-independent quantities and hence is often used for establishing failure criteria. This criterion states that yielding occurs at a point where the strain-energy density equals the strain energy density at yield in uniaxial tension (or compression). It is also known as the total strain energy criterion.

14.5.1 Strain-energy density

The strain-energy density can be defined in various forms, which are given in Eq. (13.75). The strain energy in principal stresses is given by

$$U_0 = \frac{1}{2E}\left[\sigma_1^2 + \sigma_2^2 + \sigma_3^2 - 2\nu(\sigma_2\sigma_3 + \sigma_1\sigma_3 + \sigma_1\sigma_2)\right] \quad (14.18)$$

For a uniaxial load at the yield point ($\sigma_1 = \sigma_Y, \sigma_2 = \sigma_3 = 0$), U_0 at failure is

$$U_{0Y} = \frac{\sigma_Y^2}{2E} \quad (14.19)$$

The failure criterion becomes

$$\sigma_1^2 + \sigma_2^2 + \sigma_3^2 - 2\nu(\sigma_2\sigma_3 + \sigma_1\sigma_3 + \sigma_1\sigma_2) - \sigma_Y^2 = 0 \quad (14.20)$$

The effective stress σ_e becomes

$$\sigma_e = \sqrt{\sigma_1^2 + \sigma_2^2 + \sigma_3^2 - 2\nu(\sigma_2\sigma_3 + \sigma_1\sigma_3 + \sigma_1\sigma_2)} \quad (14.21)$$

and the yield function is

$$f = \sigma_e^2 - \sigma_Y^2 = 0 \quad (14.22)$$

This yield criterion is an ellipsoid in the principal stress space. It is a circle when $\nu = 0$. The yield function is one piece.

14.5.2 Application to the case study problem

The following codes compute the strain-energy density using the principal stresses obtained in Section 14.2.3. This gives the material yield stress value needed to have point A stay below yielding, which determines the yield state.

```
1  # Strain-energy density criterion
2  U_stress = gr.energy_stress(eigens, v)
3  print(f'The strain-energy stress is {U_stress:0.5e}')
4  print(f'The material yield stress needs to be > {U_stress:0.5e}')
5  print(f'The yield stress of the material = {sY:0.5e}')
6  print(f'Is yield? {U_stress>sY}')
7  print(f'Actual safety factor = {SF*sY/U_stress}')
```

```
The strain-energy stress is 7.15700e+08
The material yield stress needs to be > 7.15700e+08
The yield stress of the material = 7.50000e+08
Is yield? False
Actual safety factor = 2.619813700936882
```

14.6 Yielding criteria for ductile materials

Many materials used in engineering applications are purposely made ductile. Yielding criteria for ductile materials are thus well studied. This section introduces these type of criteria.

14.6.1 *Maximum shear-stress (Tresca) criterion*

Many crystals have slip planes with relatively weak against shearing. Yield criteria based on shear stress make sense. The Tresca criterion states that the maximum shear stress equals the yield shear stress under the uniaxial load σ_Y.

With ranked principal stresses obtained, the maximum shear stress $\tau_{\max}$ at a point in a solid can be computed using

$$\tau_{\max} = (\sigma_1 - \sigma_3)/2 \tag{14.23}$$

Note that because the principal stresses are invariant, the maximum shear stress is also invariant. The effective stress σ_e is chosen as

$$\sigma_e = \tau_{\max} \tag{14.24}$$

For uniaxial stress states, we have

$$\sigma_1 = \sigma_Y, \sigma_3 = 0 \Longrightarrow \tau_Y = \sigma_1/2 = \sigma_Y/2 \tag{14.25}$$

Thus, the Tresca yield criterion becomes

$$\sigma_e - \sigma_Y/2 = 0 \tag{14.26}$$

14.6.2 *Tresca yield surface*

For an arbitrary state of stresses, we can compute all these principal stresses, and then draw a Tresca yield surface. This is done by computing all three possible maximum shear stresses in terms of these principal stresses. Considering each of the maximum shear stresses can be possibly either positive or negative, the entire yield surface shall consist of six planes. The detailed formulation is given as follows:

$$\tau_{\max 1} = \pm \frac{\sigma_2 - \sigma_3}{2}; \quad \tau_{\max 2} = \pm \frac{\sigma_1 - \sigma_3}{2}; \quad \tau_{\max 3} = \pm \frac{\sigma_1 - \sigma_2}{2} \tag{14.27}$$

The Tresca yield surface on which the material can yield becomes

$$\sigma_2 - \sigma_3 \pm \sigma_Y = 0; \quad \sigma_1 - \sigma_3 \pm \sigma_Y = 0; \quad \sigma_1 - \sigma_2 \pm \sigma_Y = 0 \tag{14.28}$$

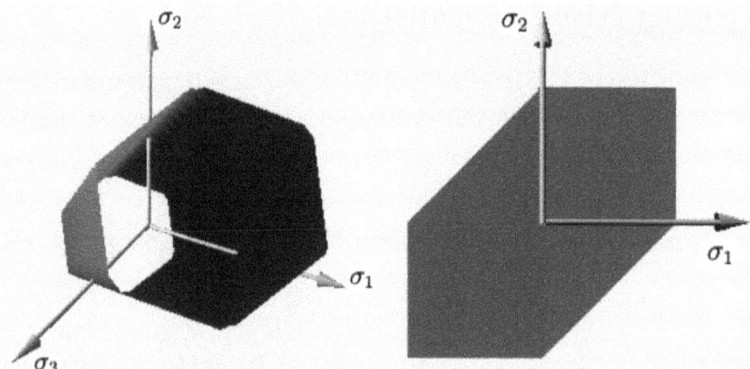

Figure 14.4. Schematic view of the Tresca yield surface. Left: cylinder with uniform hexagonal cross-section in 3D space of the principal stresses $(\sigma_1, \sigma_2, \sigma_3)$; right: elongated hexagon in 2D space (σ_1, σ_2). Transferred from en.wikipedia by User: Jaqen, original uploader was Janek Kozicki at en.wikipedia under the CC BY-SA 3.0 license.

These six planes in 3D space of the principal stresses are shown on the left image of Fig. 14.4. It can be better viewed in 2D space by setting $\sigma_3 = 0$, as shown on the right. In such a 2D case, the surface consists of six line segments. The expressions for the corresponding six lines are given by

$$\sigma_2 = \pm\sigma_Y; \quad \sigma_1 = \pm\sigma_Y; \quad \sigma_1 - \sigma_2 = \pm\sigma_Y \tag{14.29}$$

These are four vertical lines and two 45-degree inclined lines.

This maximum shear stress yield criterion is one of the most widely used criteria for ductile materials.

14.6.3 *Application to the case study problem*

The following codes compute the maximum shear stress using the principal stresses obtained in Section 14.2.3. This gives the material yield stress value needed to have point A stay below yielding and determines the yield state.

```
1  # Maximum shear stress (Tresca) criterion
2  Tau_max =0.5*(eigens[0]-eigens[2])
3  print(f'The maximum shear stress is {Tau_max:0.5e}')
4  print(f'The material yield stress needs to be > {(2*Tau_max):0.5e}')
5  print(f'The yield stress of the material = {sY:0.5e}')
6  print(f'Is yield? {(2*Tau_max)>sY}')
7  print(f'Actual safety factor = {SF*sY/(2*Tau_max)}')
```

The maximum shear stress is 4.27058e+08
The material yield stress needs to be > 8.54115e+08
The yield stress of the material = 7.50000e+08
Is yield? True
Actual safety factor = 2.195254603450228

14.6.4 Distortional energy density criteria

Yielding begins when the distortional strain-energy density is at that of the uniaxial limit. The distortional strain-energy density is associated with the shape distortion in a solid. The total strain-energy density U_0 can be written in two parts:

$$U_0 = \underbrace{\frac{(\sigma_1 + \sigma_2 + \sigma_3)^2}{18K}}_{U_V} + \underbrace{\frac{(\sigma_2 - \sigma_3)^2 + (\sigma_1 - \sigma_3)^2 + (\sigma_1 - \sigma_2)^2}{12G}}_{U_D} \quad (14.30)$$

where U_V and U_D are energies related to the volumetric and deviatoric deformation. In the distortional strain-energy density criterion, we discard U_V and assume that only the deviatoric energy density U_D controls the yielding.

Using the **von Mises stress** defined in Chapter 3, we have

$$U_D = \frac{\sigma_{vm}^2}{6G} \quad (14.31)$$

14.6.4.1 von Mises criterion

The effective stress σ_e is chosen as

$$\sigma_e = \sigma_{vm} = \sqrt{\frac{1}{2}[(\sigma_2 - \sigma_3)^2 + (\sigma_1 - \sigma_3)^2 + (\sigma_1 - \sigma_2)^2]} \quad (14.32)$$

For 2D stress cases, say $\sigma_3 = 0$, we have

$$\sigma_e = \sigma_{vm} = \sqrt{\sigma_1^2 + \sigma_2^2 - \sigma_1 \sigma_2} \quad (14.33)$$

Consider uniaxial loading ($\sigma_1 = \sigma_Y, \sigma_2 = \sigma_3 = 0$), the corresponding deviatoric energy density becomes $U_{DY} = \frac{\sigma_Y^2}{6G}$. Equaling U_D and U_{DY} gives the yield function:

$$f = \sigma_e^2 - \sigma_Y^2 = 0 \quad (14.34)$$

This is the well-known **von Mises yield criterion**.

It is clear that the von Mises yield surface is an ellipsoid in the principal stress space, as shown in Fig. 14.5.

The von Mises yield criterion is the most widely used criterion for ductile materials.

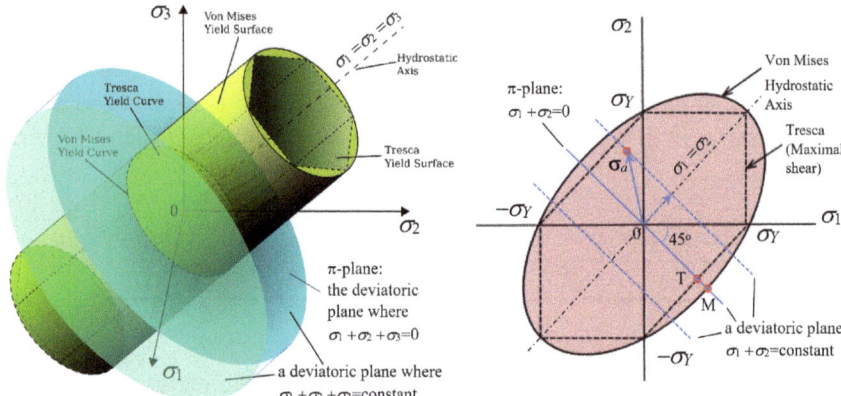

Figure 14.5. Schematic view of the von Mises yield surface together with the Tresca yield surface. Left: the von Mises yield surface is the cylindrical surface in 3D space of the principal stresses $(\sigma_1, \sigma_2, \sigma_3)$; The Tresca yield surface is a cylinder with hexagonal cross-section. Right: in 2D space (σ_1, σ_2). Images modified based on the one from en.wikipedia by Rswarbrick under the CC BY-SA 3.0 license (left); and the one by Melchoir to public domain (right).

14.6.4.2 Application to the case study problem

The following codes compute the von Mises stress using the principal stresses obtained in Section 14.2.3. This gives the material yield stress value needed to have point A stay below yielding and determines the yield state.

```
# von Mises criterion
svm = gr.von_Mises(eigens)
print(f'The von Mises stress is {svm:0.5e}')
print(f'The material yield stress needs to be > {svm:0.5e}')
print(f'The yield stress of the material = {sY:0.5e}')
print(f'Is yield? {svm>sY}')
print(f'Actual safety factor = {SF*sY/svm}')
```

```
The von Mises stress is 7.42623e+08
The material yield stress needs to be > 7.42623e+08
The yield stress of the material = 7.50000e+08
Is yield? False
Actual safety factor = 2.524835151818577
```

14.6.4.3 Octahedral shear-stress criterion

The von Mises stress relates to the octahedral shear-stress as discussed in Chapter 3, which is

$$\sigma_{vm} = \frac{3}{\sqrt{2}} \tau_{oct} \tag{14.35}$$

The yield criterion can also be written as

$$\tau_{oct} - \frac{\sqrt{2}}{3}\sigma_Y = 0 \tag{14.36}$$

This is known as the **octahedral shear-stress criterion**. It has a slightly different form, but it is essentially the same as the von Mises yield criterion.

Since octahedral shear-stress relates to the second invariant of the deviatoric stress tensor, J_2, in the form of

$$J_2 = -\frac{3}{2}\tau_{oct}^2 \tag{14.37}$$

the octahedral shear-stress function (or the von Mises function) can also be written as

$$f = \sqrt{|J_2|} - \frac{\sigma_Y}{\sqrt{3}} \tag{14.38}$$

14.6.4.4 Application to the case study problem

The following codes compute the octahedral shear stress using the principal stresses obtained in Section 14.2.3. This gives the material yield stress value needed to have point A stay below yielding and determines the yield state.

```
1  # Octahedral shear stress criterion
2  ⌄, tau_oct=gr.Oct_stress(eigens)
3  print(f'The Octahedral shear stress is {tau_oct:0.5e}')
4  Y = tau_oct*3/np.sqrt(2)
5  print(f'The material yield stress needs to be > {Y:0.5e}')
6
7  print(f'The yield stress of the material = {sY:0.5e}')
8  print(f'Is yield? {Y>sY}')
9  print(f'Actual safety factor = {SF*sY/Y}')
```

The Octahedral shear stress is 3.50076e+08
The material yield stress needs to be > 7.42623e+08
The yield stress of the material = 7.50000e+08
Is yield? False
Actual safety factor = 2.524835151818577

14.6.4.5 Difference between the von Mises and Tresca criteria

As shown in Fig. 14.5, the von Mises criterion is the upper bound of the Tresca criterion. In the 2D case, the von Mises criterion is a smooth ellipse, but the Tresca is piecewise straight lines. Their difference at points T and M shown in the right image of Fig. 14.5 can be quantified as follows.

Considering a state of pure shear ($\sigma_1 = -\sigma_2 = \sigma, \sigma_3 = 0$), which is the blue line shown in Fig. 14.5, we have

$$\tau_{max} = \frac{\sigma_1 - \sigma_2}{2} = \sigma \tag{14.39}$$

The failure τ_Y predicted by the Tresca criterion is

$$\tau_Y = \tau_{max} = \frac{\sigma_Y}{2} = 0.5\sigma_Y \tag{14.40}$$

The von Mises criterion Eq. (14.32) gives

$$\sigma = \tau_Y = \frac{\sigma_Y}{\sqrt{3}} = 0.5774\sigma_Y \tag{14.41}$$

In this case, the von Mises criterion prediction is approximately 15% greater than that by the Tresca criterion.

The von Mises criterion is often found to be more accurate for some materials than the Tresca criterion in predicting yield under pure shear. The Tresca criterion is on the conservative side.

14.6.5 π-plane

The π-plane is a plane on which the hydrostatic stress (or the means stress), $\sigma_m = (\sigma_1 + \sigma_2 + \sigma_3)/3 = 0$. It is the blue plane shown in the left image of Fig. 14.5. It passes through the origin of the coordinate system of $(\sigma_1, \sigma_2, \sigma_3)$. The direction cosines of the π-plane is $(1/\sqrt{3}, 1/\sqrt{3}, 1/\sqrt{3})$, which is the same direction as the octahedral plane. For 2D cases, the π-plane is a line on which $\sigma_m = (\sigma_1 + \sigma_2)/2 = 0$. It is the blue line shown in the right image of Fig. 14.5. It passes through the origin of the coordinate system of (σ_1, σ_2). It is a special case of the so-called deviatoric plane.

14.6.6 *Deviatoric planes*

A deviatoric plane is a plane on which the hydrostatic stress or the means stress, $\sigma_m = (\sigma_1 + \sigma_2 + \sigma_3)/3 = constant$. The direction cosines of the deviatoric plane is $(1/\sqrt{3}, 1/\sqrt{3}, 1/\sqrt{3})$, same as the octahedral plane. It is a plane parallel to the π-plane, and thus there are infinite number of them. One of them is shown as a light blue ellipse in the left image of Fig. 14.5. For 2D cases, is a line on a deviatoric plane $\sigma_m = (\sigma_1 + \sigma_2)/2 = constant$. In other words, the projection on the hydrostatic axis of an arbitrary stress state σ_a in a deviatoric plane is a constant.

Such a deviatoric plane is important because a distortional energy density criterion (such as the nov Mises) applies to stress states on any of these planes because the level of the hydrostatic stress does not affect the failure.

Therefore, the yielding surface is a cylinder as shown in the left image of Fig. 14.5. Due to this, we often draw yield surfaces on the π-plane, as often seen in the literature.

Note that because the direction of a deviatoric plane is the same as the octahedral plane, the von Mises criterion is the same as the octahedral shear stress criterion.

14.7 Yield criterion for cohesive materials*

In many applications in engineering and nature, we frequently encounter cohesive materials, including concrete and rock. It is found that:

1. Yielding of many such cohesive materials depends on hydrostatic stress.
2. An increase in hydrostatic compressive stress can increase the resistance to yield.
3. Such materials may exhibit different behaviors in tension and compression.

We discuss two widely used yield criteria for cohesive materials.

14.7.1 Mohr–Coulomb yield function

The Mohr–Coulomb yield criterion can be used for cohesive materials. It is a generalization of the Tresca criterion. It accounts for the effects of hydrostatic stress and uses two material parameters: the cohesion c and the angle of internal friction ϕ. For ordered principal stresses of $\sigma_1 > \sigma_2 > \sigma_3$, the Mohr–Coulomb yield function is [4]

$$f = \sigma_1 - \sigma_3 + (\sigma_1 + \sigma_3)\sin\phi - 2c\cos\phi \qquad (14.42)$$

14.7.2 Determination of material parameters

The material parameters c and ϕ can be determined via experiments.

First, we can set a uniaxial tension failure test, which gives $\sigma_1 = \sigma_{YT}, \sigma_2 = \sigma_3 = 0$, and the yield stress σ_{YT} can be found using Eq. (14.42):

$$\sigma_{YT} = \frac{2c\cos\phi}{1 + \sin\phi} \qquad (14.43)$$

Next, we set a uniaxial compression failure test, for which we have this time $\sigma_1 = \sigma_2 = 0, \sigma_3 = -\sigma_{YC}$, and the yield stress σ_{YC} can be found using Eq. (14.42):

$$\sigma_{YC} = \frac{2c\cos\phi}{1 - \sin\phi} \qquad (14.44)$$

Solving now Eqs. (14.43) and (14.44) simultaneously gives

$$c = \frac{1}{2}\sqrt{\sigma_{YT}\sigma_{YC}} \qquad (14.45)$$

and

$$\phi = \sin^{-1}\left(\frac{\sigma_{YC} - \sigma_{YT}}{\sigma_{YC} + \sigma_{YT}}\right) \qquad (14.46)$$

For a frictionless material ($\phi = 0$), the Mohr–Coulomb yield criterion reduces to the Tresca criterion.

14.7.3 Mohr–Coulomb yield surface

The Mohr–Coulomb yield surface is an irregular hexagonal pyramid. The axis of the pyramid is still the hydrostatic axis, as shown in Fig. 14.6.

14.7.4 Drucker–Prager yield criterion

The Drucker–Prager yield criterion is a generalization of the von Mises criterion, taking the hydrostatic stress into consideration [5]. In reference to Eq. (14.38), the Drucker–Prager yield function can be written as

$$f = \alpha I_1 + \sqrt{|J_2|} - K \qquad (14.47)$$

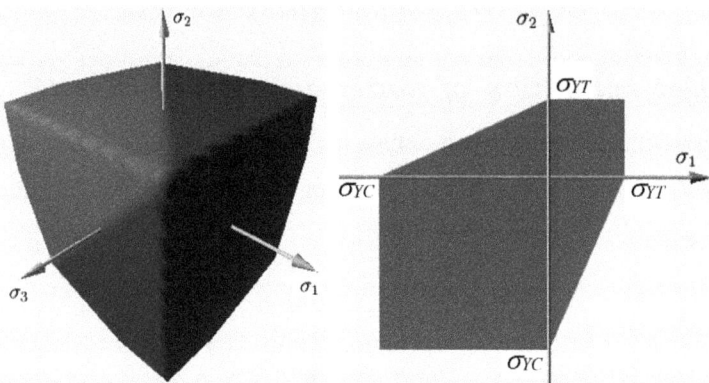

Figure 14.6. Schematic view of the Mohr–Coulomb yield surfaces. Left: an irregular hexagonal pyramid in 3D space of the principal stresses ($\sigma_1, \sigma_2, \sigma_3$); right: irregular hexagon in 2D space (σ_1, σ_2). Image modified base on the one from Wikimedia Commons by Janek Kozicki under the CC BY-SA 3.0 license.

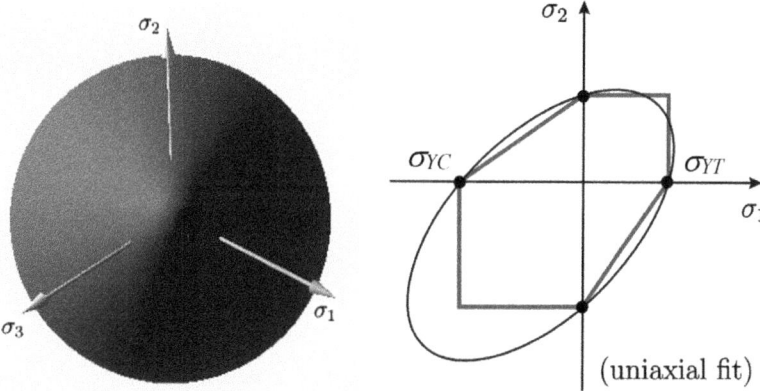

Figure 14.7. Schematic view of the Drucker–Prager yield surfaces. Left: the Drucker–Prager yield surfaces is a circular cone in 3D space of the principal stresses $(\sigma_1, \sigma_2, \sigma_3)$; right: the Drucker–Prager yield surfaces is an ellipse (black) in 2D space (σ_1, σ_2); the irregular hexagon is for the Mohr–Coulomb yield surface. Image on the left is from Wikimedia Commons by Janek Kozicki under the CC BY-SA 3.0 license. On the right is by Vtzslav tembera under the CC BY-SA 4.0 license.

where α and K are coefficients that depend on the cohesion c and the angle of internal friction ϕ:

$$\alpha = \frac{2\sin\phi}{\sqrt{3}(3-\sin\phi)} \quad (14.48)$$

and

$$K = \frac{6c\cos\phi}{\sqrt{3}(3-\sin\phi)} \quad (14.49)$$

For a frictionless material ($\phi = 0$), the Drucker–Prager criterion reduces to the von Mises criterion.

The Drucker–Prager yield surface is a circular cone, as shown in Fig. 14.7.

14.8 Comparison of different criteria

We finally use the case study problem defined in Section 14.2, and the principal stresses obtained in Section 14.2.3, to compute the material yield stress value needed to have point A stay below yielding and determines the yield state. This time we put all the results from the different criteria together for easy comparison.

```python
1   print(f'\nThe yield stress of the material: {sY:0.5e}\n')
2
3   # maximum principal stress criterion
4   ps_max = max(abs(eigens))
5   print(f'The maximum principal stress is {ps_max:0.5e}')
6   print(f'The material yield stress needs to be > {ps_max:0.5e}')
7   print(f'Is yield? {ps_max>sY}\n')
8
9   # maximum principal strain criterion
10  se1 = eigens[0]-v*eigens[1]-v*eigens[2]   # principal strains E ε1
11  se2 = eigens[1]-v*eigens[0]-v*eigens[2]   # principal strains E ε2
12  se3 = eigens[2]-v*eigens[0]-v*eigens[1]   # principal strains E ε3
13  e_max =max(abs(np.array([se1, se2, se3])))
14  print(f'The maximum principal strain is {e_max:0.5e}')
15  print(f'The material yield stress needs to be > {e_max:0.5e}')
16  print(f'Is yield? {e_max>sY}\n')
17
18  # Maximum shear stress (Tresca) criterion
19  Tau_max =0.5*(eigens[0]-eigens[2])
20  print(f'The maximum shear stress is {Tau_max:0.5e}')
21  print(f'The material yield stress needs to be > {(2*Tau_max):0.5e}')
22  print(f'Is yield? {(2*Tau_max)>sY}\n')
23
24  # von Mises criterion
25  svm = gr.von_Mises(eigens)
26  print(f'The von Mises stress is {svm:0.5e}')
27  print(f'The material yield stress needs to be > {svm:0.5e}')
28  print(f'Is yield? {svm>sY}\n')
29
30  # Octahedral shear stress criterion
31  _, tau_oct=gr.Oct_stress(eigens)
32  print(f'The Octahedral shear stress is {tau_oct:0.5e}')
33  Y = tau_oct*3/np.sqrt(2)
34  print(f'The material yield stress needs to be > {Y:0.5e}')
35  print(f'Is yield? {Y>sY}\n')
36
37  # Strain-energy density criterion
38  U_stress = gr.energy_stress(eigens, v)
39  print(f'The strain-energy stress is {U_stress:0.5e}')
40  print(f'The material yield stress needs to be > {U_stress:0.5e}')
41  print(f'Is yield? {U_stress>sY}')
```

```
The yield stress of the material: 7.50000e+08

The maximum principal stress is 6.18043e+08
The material yield stress needs to be > 6.18043e+08
Is yield? False

The maximum principal strain is 6.51365e+08
The material yield stress needs to be > 6.51365e+08
Is yield? False

The maximum shear stress is 4.27058e+08
The material yield stress needs to be > 8.54115e+08
Is yield? True

The von Mises stress is 7.42623e+08
The material yield stress needs to be > 7.42623e+08
Is yield? False

The Octahedral shear stress is 3.50076e+08
The material yield stress needs to be > 7.42623e+08
Is yield? False

The strain-energy stress is 7.15700e+08
The material yield stress needs to be > 7.15700e+08
Is yield? False
```

Readers my do the same analysis for point B on the shaft using the codes given and examine the differences.

14.9 Remarks

A number of failure criteria have been introduced in this chapter with detailed formulations and Python codes. The choice of the criteria shall depend on the material. Therefore, experiences on the material to be used and the databases of material properties for different types of materials are of importance.

The author was involved in one project that used a newly designed composite material. We did not have sufficient data on the material, so we decided to write a code to assess the safety using a number of criteria together to ensure the safety for possible scenarios we can think of.

There is a large class of materials called hyper-elastic materials, such as rubbers widely used in industry and soft issues involving biological systems. For these types of materials, the analyses are more on the large geometrical deformation, and yielding is not usually a concern. This is an important and currently active area of study. Examples of study on these types of materials from the author's group can be found in Refs. [6–11].

When the structure is designed for one-time use, we may stretch to use the full capacity of the material and allow the material to go beyond the yield point, into plastic deformation, until fracture. This involves plasticity analyses. It requires yield criterion together with: 1. a proper flow rule governing the plastic strain and stress evolution after yielding, and 2. a hardening rule that predicts changes in the yield surface. Examples for such analysis can be found in textbooks. Some studies on these types of materials in the authors; group can be found in Refs. [12–14]. Interested readers may pursue this topic, starting from the wikipage on flow_plasticity_theory (https://en.wikipedia.org/ wiki/Flow_plasticity_theory), and the references therein.

References

[1] A.P. Boresi and R.J. Schmidt, *Advanced Mechanics of Materials*. John Wiley & Sons, New York, 2002.

[2] J.M. Gere and S.P. Timoshenko, *Mechanics of Materials*. Van Nostrand Reinhold Company, New York, 1972.

[3] S. Timoshenko and J.N. Goodier. *Theory of Elasticity*. McGraw Hill, New York, 1970. http://books.google.com/books?id=yFISAAAAIAAJ&dq=theory+of+elasticity&ei=ICiKSsr3G4jwkQSbxMyPCg.

[4] J. Lubliner, *Plasticity Theory*. Macmillan, 1990. https://books.google.com/books?id=LUqYPwAACAAJ.

[5] W.F. Chen and D.J. Han, *Plasticity for Structural Engineers*. Gau Lih Company, Limited, 1995. https://books.google.com/books?id=EEITngEACAAJ.

[6] G.R. Liu and T.T. Nguyen, *Smoothed Finite Element Methods*. Taylor and Francis Group, New York, 2010.

[7] G.R. Liu, *Mesh Free Methods: Moving Beyond the Finite Element Method*. Taylor and Francis Group, New York, 2010.

[8] Wu Shao-Wei, Jiang Chen, Jiang Chao et al., A selective smoothed finite element method with visco-hyperelastic constitutive model for analysis of biomechanical responses of brain tissues. *International Journal for Numerical Methods in Engineering*, **121**(22): 5123–5149, 2020.

[9] S. Wu, C. Jiang, C. Jiang et al., A selective S-FEM with visco-hyperelastic model for analysis of biomechanical responses of brain tissues. In *The 11th International Conference on Computational Methods (ICCM2020)*, 2020.

[10] Wu Shao-Wei, Jiang Chao, Liu GR et al., An n-sided polygonal selective smoothed finite element method for nearly incompressible visco-hyperelastic soft materials. *Applied Mathematical Modelling*, **107**: 398–428, 2022.

[11] Li Yang, Sang Jianbing, Wei Xinyu et al., A novel constitutive parameters identification procedure for hyperelastic skeletal muscles using two-way neural networks. *International Journal of Computational Methods*, **19**(02): 2150060, 2022.

[12] Zeng W., Larsen J.M. and Liu G.R., Smoothing technique based crystal plasticity finite element modeling of crystalline materials. *International Journal of Plasticity*, **65**: 250–268, 2015.

[13] Cui X.Y., Liu G.R., Li G.Y. *et al.*, Elasto-plasticity analysis using edge-based smoothed finite element method (ES-FEM). *International Journal of Pressure Vessels and Piping*, **86**: 711–718, 2009.

[14] Zeng W., Liu G.R., Li D. *et al.*, A smoothing technique based beta finite element method (βFEM) for crystal plasticity modeling. *Computers & Structures*, **162**: 48–67, 2016.

Index

A

actual safety factor, 565
admissible BCs for beams, 292
algebraically closed, 11
angle of the principal axis, 118
angular change, 164
anti-commutativity, 31
anti-symmetric, 32
antisymmetric matrix, 152
apply_symmetry, 192
area of a composite area, 417
area projection, 88
axial flexibility of a bar, 251
axial stiffness of a bar, 251

B

bar with non-circular cross-section, 376
bars with varying cross-section, 509
basic property preservation, 72
beam equation, 290
beam solver, 294–295
beam with two spans, 321
bended beams, 283
bending deformation, 284
bending stiffness, 314
bi-linear function, 212
body force, 74, 77
boundary conditions for beams, 291
box cross-section, 372
box cross-section of thin-walls, 373
box-beam, 492
box_shearCenter, 497
bulk modulus, 195–196

C

C-beam, 474, 491, 499
C2toC4, 191
C4toC2, 192
cantilever beam, 292
carbon/epoxy composite, 228
Cauchy–Green deformation tensor, 158
centroid for composite areas, 422
centroid of an area, 421
centroid of an I-beam, 422
characteristic equation, 107
characteristic value, 107
clamped-clamped beam, 297, 317
Clapeyron's theorem, 507
closure, 10
code for computing various stresses, 112
cohesive materials, 577
collinearity, 31
comparison of different criteria, 579
compatibility condition, 505
compatibility equation, 177–178, 507, 516, 542, 546
complementary energy, 513, 537
complementary energy 1D problems, 538
complementary energy density, 527, 537

complementary energy in 3D, 537
complementary energy principle (CEP), 502, 542
complementary energy principle for structures, 539
complete and unique, 141
compliance equations for 2D problems, 208
compliance matrix, 200
compliance matrix for isotropic materials, 207
complimentary, 295
complimentary energy principle, 546
components of a force, 78
composite bar, 275
computation of large strains, 155
computation of various strains, 175
computing cross products, 33
concentrated force, 73
conditions for the stress function, 382
conservation law, 85
constitution equations for stress and strain invariants, 197
constitutive equation, 184, 513
constructing the stress function, 384
continuity and equilibrium conditions, 332
continuity equation, 276
continuity in rotation, 324
contraction of index, 67
contraction operations, 96
coordinate rotation, 425, 433
coordinate systems, 42
coordinate transformation, 41, 46, 171
coordinate transformation in 3D, 50
coordinate transformation rule of strain tensor, 172
coordinate translation, 428
cross product, 27–29, 31, 70, 107
cross-section, 75
cross-section of beams, 287
cross-section rotation assumption, 357
cross-section surface, 74
cubic materials, 194
curl of a vector function, 69

curl of the stress vector, 355
curl_vf, 70

D

1D bar problems, 244
1D beam member, 284
2D plane strain, 198
2D plane stress, 197
2D stress states, 116
3D free-body diagram, 234–235
deflection, 284
deflection in its third derivative, 290
deformation gradient, 155, 157
derivative of deflection, 286
derivative of the moment, 290
derivatives of displacement functions, 148
description of vectors, 42
deviatoric plane, 576
deviatoric stress, 200
deviatoric stress tensor, 575
diagonalization of a stress tensor in 3D, 133
differential equation solver, 299
dimensionality and stresses, 85
direction cosine, 19–20, 24, 48, 87, 163
direction cosines of the principal axes, 107
direction of the force, 75
direction of the surface, 75
direction-independent, 197
discontinuity of the solutions, 321, 328
discontinuous body forces, 264
displacement boundary conditions, 246
displacement components on the cross-section, 353
displacement functions, 142
displacement gradient, 148–150
displacement gradient caused by temperature, 270
displacement gradient matrix, 150
displacement method, 247
displacement vector, 140
displacement-driven, 261
displacements derivatives, 154
distortional energy density, 111

distortional energy density criteria, 573
distortional energy density criterion, 576
distortional strain-energy density, 573
distributed force, 74, 264
distribution of solutions, 342, 345
div_tensorF, 239
divergence-free tensor, 355
divergence of a vector function, 67
divergence of stress tensor function, 238
divergence of the stress, 77
divergence of the stress states, 355
divergence of the stress tensor, 240
divergence of the stress vector, 237
divergence of the tensor function, 239
divergence theorem, 523
division by zero, 11
dot product, 30, 37, 67
dot product of vectors, 18
double angle formula, 116
Drucker–Prager yield criterion, 578
Drucker–Prager yield surfaces, 579
dummy index, 52
dynamic equilibrium equation, 290

E

E_SnC2D_strain, 206
E_SnC2D_stress, 203
E_SnC3Dep, 202
effective stress, 562
eigenvalue problem, 106
eigenvalue solver, 435
elasticity tensor, 181
elasticity tensor transformation, 221
element-wise division, 18
element-wise multiplication, 18, 37
elliptical cross-section, 384–385
enclosed thin-wall, 369
enclosed thin-wall cross-sections, 368
energy conjugate, 524–525
energy methods, 501
energy principles, 501
energy_stress, 112
engineering constants, 195
engineering strain, 161

equilibrium, 68
equilibrium condition, 515, 537
equilibrium equation, 76–77, 233, 236, 506
equilibrium equation for beams, 289
equilibrium equation in terms of stresses, 234
equilibrium equations, 378, 535–536, 541
equilibrium equations for 2D plane stress, 244
equilibrium equations in displacements, 240
equilibrium in mechanics, 72
equilibrium in moment, 324
equivalence of force vector, 78
equivalent polar 2nd moment of area, 396
error magnification, 461
Euler–Bernoulli assumptions, 285
Euler–Bernoulli beam theory, 284
external action, 504, 514
external force, 237
external modules, 5
external packages, 4
externally applied force, 76
extreme values, 302

F

features of a vector, 13
fictitious unit load method, 553–555
first axis of the stress tensor, 89
first moment of a composite area, 419
first moment of area, 418, 425
first-order tensor, 43
force vector in transformed coordinates, 79
force vectors in 3D space, 78
force-driven problem, 255
forces in channels, 486
formula for shear stress, 468
forward transformation, 141
fourth-order differential equation, 284
fourth-order tensor, 184, 241
free-body diagram, 76, 276, 324, 467
free-clamp supported beams, 312

free-indexes, 52
frictionless material, 578
from stress tensor symmetry, 186
function concept, 46
function of coordinates, 60

G

general definition of stress, 80
general solution, 251, 293
generalization of the Tresca criterion, 577
generalized deformation, 508
generalized displacement, 517, 531
generalized Hooke's Law, 184
generalized internal force, 508
generating 2D Mohr circles, 121
geometric representation, 115
geometric representation of tensors, 115
governing equation, 233
gr.div_vf(), 68
gr.grad_vf(), 66
grad_vf, 149
gradient of a scalar function, 61, 64
gradient of the displacement vector, 149
gradient of vector functions, 66
gradient operator, 61
Green–Lagrange Strain, 153
Green–Lagrange strain tensor, 158

H

Hadamard product, 18
hollow cylindrical cross-section, 361, 363
hollow enclosed thin-wall, 408
hydrostatic axis, 576
hydrostatic stress, 199, 576
hyper-elastic materials, 581

I

I-beam, 392, 407, 479, 483
idealized deformed membrane, 409
in-plane strains, 242

indeterminate beams in temperature gradient, 340
indeterminate structures, 557
inner product, 18, 22
inner product associated norm, 20
internal, 295
internal force, 76
internal reaction, 504, 514
inverse transformation, 97, 105, 142, 185
isotropic materials, 194
Ixx2IXX, 434

K

k1_f, 397
k_f, 397
kinematic relations, 145
Kronecker delta, 52
Kronecker delta function, 21

L

L-beam, 455, 477
Laplace equation, 389
large deformation, 160, 164
large strains, 153
law of cosines, 21
linear body force, 260
linear combination, 15–16
linearized expression, 536
linearly distributed normal stress, 286
linearly independent, 15, 28
loading plane, 442
loading plane angle, 459

M

M_eign, 112
M_invariant, 112
magnification factor, 160–161
magnitude of the shear stress, 91
matrix-vector multiplication rule, 89
maximum moment, 446
maximum principal strain, 569
maximum principal strain criterion, 567
maximum principal stress, 566

maximum principal stress criterion, 565
maximum shear stress, 397, 571
maximum shear stress yield criterion, 572
maxminS, 301
mean stress, 199
meanDeviator, 112
means stress, 576
membrane analogy, 387
membrane deflection, 389
modern steam turbine, 350
Mohr circle for 2D, 119–120
Mohr circle for 3D, 126, 128
Mohr circle of a uniaxial stress state, 122
Mohr circles, 115
Mohr–Coulomb yield criterion, 577
Mohr–Coulomb yield function, 577
Mohr–Coulomb yield surface, 578–579
Mohr_circle2D, 121
Mohr_circle3D, 128
moment equilibrium, 84–85
multi-valued, 56
multiple multiple-type members, 547

N

narrow rectangular cross-section, 389
neutral axis, 285–286, 445, 454
non-circular cross-sections, 375
non-symmetrical bending, 452
non-uniqueness of the eigenvectors, 109
nonlinear spring–mass system, 543
normal direction of the surface, 90
normal strain, 147, 286
normal strain along a fiber, 160
normal stress, 75, 286, 294
normal stress by moments, 441
normal stress distribution, 452
norms of a vector, 24
notation convention for a stress component, 83
Numpy code for transformation matrices, 54

O

Oct_stress, 112
octahedral shear-stress, 574
octahedral shear-stress criterion, 575
octahedral stress, 110
operations for vectors, 17
orthogonal coordinate, 46
orthogonal coordinate system, 43, 79
orthogonal fibers, 164
orthogonal moment components, 443
orthogonal projection, 20, 43–44, 72, 88, 90–91, 141
orthogonal projection rule, 42, 45, 48, 78, 140
orthogonal vectors, 39
orthogonality of vectors, 40
orthotropic materials, 193
outer product, 25–26, 65, 149

P

π-plane, 576
P-wave modulus, 195
parabolic distribution, 469
parabolic distribution of shear stress, 471
partial differential equations (PDEs), 233
perpendicular axis theorem, 438
piecewise solutions, 319, 326, 335
planar truss structure, 555–556
plane strain, 204
plane strain problems, 242
plane stress, 203
plane stress problems, 243
plastic deformation, 582
plot2curveS, 256
point force, 74
Poisson's ratio, 183, 246
polar moment of area, 438
polar second moment of area, 358–359
possible BCs for beams, 291
possible boundary conditions, 291
potential energy, 503
potential energy for 1-DOF systems, 507

potential energy for various members, 509
potential energy principle (PEP), 502, 506, 541
potential energy principle for structures, 530
Prandtl stress function, 378
principal axis, 106, 434
principal axis angle, 131
principal directions, 108
principal moments of area, 434
principal plane, 106
principal second moments of area, 434–435
principal strains, 173
principal stresses, 105, 107, 564
principal stresses for 2D stress states, 117
principal stresses of 2D stress states, 118
principal2angle, 118
principalS, 108
principals_ypr, 134
principle of equilibrium, 44–45
principle of superposition, 546
principle of virtual work, 512, 530, 539
problem well defined, 253
property of an area, 415
pure bending, 316

Q

q-maps, 496

R

randomly generated matrix, 57
randomly generated vectors, 34
rank-one matrix, 26
rate of change in length, 146
rate of circulation, 70
rate of rotation, 70
reaction force, 266, 269
reactions with the rigid block, 276
rectangular box-beam, 493
rectangular cross-section, 393
redundant support, 558

representation of a vector, 141
resultant moment, 287
right-hand rule, 28, 46, 91, 107
rotation angle, 285
rotation matrix, 58–60
rotation-displacement relations, 152
roundS, 301

S

same contraction effects, 97
scalar product, 19
scalar triplet product, 29
scaling the vector, 16
second derivative of deflection, 286
second moment of area, 287, 426, 428, 433
second moment of area for regular areas, 426
second moments of an I-beam, 429
second moments of an S-beam, 431
second moments of composite areas, 429
secondMoments_I, 430
separation of variables, 394
sf_phi, 403
shaft undergoes torsional deformation, 350
shaft with two segments, 365
shear center, 442
shear correction coefficient, 508
shear flow, 369, 409
shear force, 294
shear force and moment, 466
shear force correction, 520
shear strains, 147
shear strains in non-circular bars, 377
shear stress by shear force, 465
shear stress distribution (q-map), 475
shear stress equivalence, 84–85
shear stresses on a surface, 81, 91
shear stresses on rectangular cross-section, 405, 469
shear-center, 489–490
shear-center location, 499

shear-flow, 469, 474
shear-flow in the web, 485
sign convention, 76
sign convention for 2D Mohr circles, 120
simple-clamped beams, 310, 340
simple-simple supported beams, 304
slender geometry, 350
slope of the deflection, 285
smoothness of the strain energy density, 187
solid cylindrical shaft, 562–563
solution exists, 11
solution for the deflection, 293
solution for the rotation, 294
solutions at specific locations, 303
solutions for bars, 252
solver1D2, 252
solver1D4, 296
spatial rate of loss of the stresses, 239
spring–mass system, 540
square cross-section, 350
standard arithmetic operations, 10
standard basis vectors, 15, 35
standard convention, 76
standard sign convention, 235, 295
static problem, 237
statically determinate structures, 546
statically indeterminate, 266
stiffness matrix of materials, 200
strain energy, 524
strain energy density, 188, 527
strain energy density for 1D, 525
strain energy density for 3D, 529
strain energy in 3D, 522
strain energy stress, 112
strain invariants, 174
strain tensor symmetry, 186
strain tensor transformation, 172
strain transformation, 185
strain–displacement relation, 148, 151, 159, 177, 353, 377
strain-energy density, 570
strain-energy density criterion, 569
strain-stress-temperature relation, 221
stress and strain symmetry, 186

stress boundary conditions, 89, 246
stress components, 82
stress components on surfaces, 83
stress divergence, 77
stress function, 379
stress function on rectangular cross-section, 403
stress invariants, 110
stress method, 247
stress tensor transformation rule, 96
stress transformation, 185
stress transformation via Yaw, Pitch, and Roll, 132
stress vector projection, 87, 90
stress-strain-temperature relation, 221
stress_rect, 405
stresses on an arbitrary surface, 88
stresses on the cross-section, 353
stressOnN, 91
superposed stress distribution, 444
superposition of stresses, 443
swap the order of the two indexes, 80
swapping the indexes, 54
symmetric bending, 454, 457
symmetric frame, 328–329
symmetric relations, 201
symmetric tensor, 85
system of equations, 247

T

T-beam, 449
tall beam, 459
tall rectangular beam, 460
temperature gradient, 338
tensile spacemen, 182
tensor concept, 44
tensor operation for transformation, 144
tensor vs. matrix, 190
Tensor1_transfer, 144
Tensor2_transfer, 99
Tensor4_transfer, 186
tensorT2angle, 117
thermal deflection, 340
thermal displacement, 267
thermal expansion coefficients, 279

thermal expansion effects, 220
thermal forces, 268
thermal strains, 220
thermal stress, 220, 267–268, 273, 337
thermal stress in beams, 337
thermal stresses, 269
thick beams, 520
thin beam theory, 284
thin beam under bending, 285
thin-wall assumptions, 375
thin-wall cross-sections, 362, 472
torque and shear-flow relation, 370
torque on the cross-section, 383
torque stress relation, 411
torque twist-angle solution, 396
torsional bar members, 349
torsional flexibility of the bar, 360
torsional stiffness, 386
torsional stiffness of the bar, 360
total angular change, 147
total derivative, 381
total potential energy, 502
total potential energy principle, 535
tractions on an arbitrary surface, 86
tractions on the surface, 87
transformation matrices, 131
transformation matrix, 47–48
transformation matrix for 3D, 49
transformation of fourth-order tensor, 185
transformation of strain tensor, 171
transformation of stress tensor in 2D, 99
transformation rules, 45
transversely isotropic materials, 193
Tresca criterion, 571, 576
Tresca yield surface, 571–572, 574
triangle cross-section, 386
twist angle, 349, 410
twist deformation, 352
twist-angle and torque relation, 360, 371

twist-torque relation, 358
two indexes for a stress component, 82
two-bar truss, 531, 544
two-side application, 58
twobars_bf, 270
types of forces, 73

U

U for various structural members, 512
under- or overflow, 11
uniaxial compression failure test, 577
uniaxial force, 101
uniaxial stress state, 102
uniaxial stress tensor, 101
uniaxial tension failure test, 577
unit vectors, 14
unitary matrix, 52–53
unitary property, 55
universal testing machine, 182
use of membrane analogy, 408

V

variation, 506
vector operations, 34
vector property preservation, 53
vector space, 17
vectors, 12
vectors in mechanics, 42
version mismatch, 4
virtual displacement, 502, 505
virtual displacement for structures, 530
virtual force, 512–513, 539
Voigt notation, 190
volume of a parallelepiped, 29
von_Mises, 112
von Mises criterion, 573, 577
von Mises stress, 111, 573
von Mises yield surface, 574

W

well posed, 11

work done by external forces, 523
work-done, 22

Y

Yaw, Pitch, and Roll, 131
yield function, 562, 566

yield surface, 566
yielding criterion, 562
Young's modulus, 182

Z

zero vector, 16

www.ingramcontent.com/pod-product-compliance
Ingram Content Group UK Ltd.
Pitfield, Milton Keynes, MK11 3LW, UK
UKHW052219130325
456149UK00004B/8